Scalar Wave Driven Energy Applications

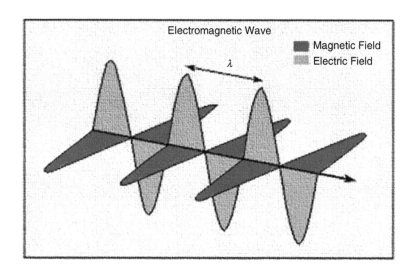

$$\vec{\nabla} \times \vec{E} = \frac{\partial \vec{B}}{\partial t}$$

$$\nabla^2 \phi - \frac{1}{c^2} \frac{\partial^2 \phi}{\partial t^2} = 0$$

$$\nabla^2 \vec{A} - \frac{1}{c^2} \frac{\partial^2 \vec{A}}{\partial t^2} = 0$$

Bahman Zohuri

Scalar Wave Driven Energy Applications

Bahman Zohuri
Department of Electrical and Computer Engineering
University of New Mexico,
Galaxy Advanced Engineering, Inc.
Albuquerque, NM, USA

ISBN 978-3-319-91022-2   ISBN 978-3-319-91023-9 (eBook)
https://doi.org/10.1007/978-3-319-91023-9

Library of Congress Control Number: 2018951776

© Springer International Publishing AG, part of Springer Nature 2019
This work is subject to copyright. All rights are reserved by the Publisher, whether the whole or part of the material is concerned, specifically the rights of translation, reprinting, reuse of illustrations, recitation, broadcasting, reproduction on microfilms or in any other physical way, and transmission or information storage and retrieval, electronic adaptation, computer software, or by similar or dissimilar methodology now known or hereafter developed.
The use of general descriptive names, registered names, trademarks, service marks, etc. in this publication does not imply, even in the absence of a specific statement, that such names are exempt from the relevant protective laws and regulations and therefore free for general use.
The publisher, the authors, and the editors are safe to assume that the advice and information in this book are believed to be true and accurate at the date of publication. Neither the publisher nor the authors or the editors give a warranty, express or implied, with respect to the material contained herein or for any errors or omissions that may have been made. The publisher remains neutral with regard to jurisdictional claims in published maps and institutional affiliations.

This Springer imprint is published by the registered company Springer Nature Switzerland AG
The registered company address is: Gewerbestrasse 11, 6330 Cham, Switzerland

*To my son Sasha and grandson Darius,
as well as my daughter Natalie and
Dr. Natasha Zohuri*

Preface

What is a "scalar wave" exactly? A scalar wave (hereafter SW) is just another name for a "longitudinal" wave. The term *scalar* is sometimes used instead because the hypothetical source of these waves is thought to be a "scalar field" of some kind, similar to the Higgs field for example.

There is nothing particularly controversial about longitudinal waves (hereafter LWs) in general. They are a ubiquitous and well-acknowledged phenomenon in nature. Sound waves traveling through the atmosphere (or underwater) are longitudinal, as are plasma waves propagating through space (i.e., Birkeland currents). LWs moving through the Earth's interior are known as "telluric currents." They can all be thought of as pressure waves of sorts.

SWs and LWs are quite different from a "transverse" wave (TW). You can observe TWs by plucking a guitar string or watching ripples on the surface of a pond. They oscillate (i.e., vibrate, move up and down or side-to-side) perpendicular to their arrow of propagation (i.e., directional movement). As a comparison, SWs/LWs oscillate in the same direction as their arrow of propagation.

Only the well-known (transverse) Hertzian waves can be derived from Maxwell's field equations, whereas the calculation of longitudinal SWs gives zero as a result. This is a flaw of the field theory because SWs exist for all particle waves (e.g., as plasma wave, as photon- or neutrino radiation). Starting from Faraday's discovery, instead of the formulation of the law of induction according to Maxwell, an extended field theory is derived. It goes beyond the Maxwell theory with the description of potential vortices (i.e., noise vortices) and their propagation as an SW but contains the Maxwell theory as a special case. With that the extension is allowed and does not contradict textbook physics.

William Thomson, who called himself Lord Kelvin after he had been knighted, already in his lifetime was a recognized and famous theoretical physicist. To him the airship seemed too unsafe and so he went aboard a steamliner for a journey from England to America in the summer of 1897. He was on the way for a delicate mission.

Eight years before his German colleague Heinrich Hertz had detected the electromagnetic wave (EW) in experiments in Karlsruhe and scientists all over the world had rebuilt his antenna arrangements. They all not only confirmed the wave as such, but also, they could show its characteristic properties. It was a TW, for which the electric and the magnetic field pointers oscillate perpendicular to the direction of propagation. This can be seen as the reason that the velocity of propagation is displays itself as field-independent and constant. It is the speed of light c.

Because Hertz had experimentally proved the properties of this wave, previously calculated in a theoretical way by Maxwell, and at the same time proved the correctness of the Maxwellian field theory. The scientists in Europe were just saying to each other: "Well Done!" While completely other words came across from a private research laboratory in New York: "Heinrich Hertz is mistaken, it by no means is a transverse wave but a longitudinal wave!"

Scalar waves also are called "electromagnetic longitudinal waves," "Maxwellian waves," or "Teslawellen" (i.e., Tesla waves). Variants of the theory claim that scalar electromagnetics, also known as scalar energy, is background quantum mechanical fluctuations and associated zero-point energies.

In modern-day electrodynamics (both classical and quantum), electromagnetic waves (EMW) traveling in "free space" (e.g., photons in the "vacuum") are generally considered to be TW. But then again, this was not always the case. When the preeminent mathematician James Clerk Maxwell first modeled and formalized his unified theory of electromagnetism in the late nineteenth-century, neither the EM SW/LW nor the EM TW had been experimentally proved, but he had postulated and calculated the existence of both.

After Hertz demonstrated experimentally the existence of transverse radio waves in 1887, theoreticians (e.g., Heaviside, Gibbs, and others) went about revising Maxwell's original equations; at this time, he was deceased and could not object. They wrote out the SW/LW component from the original equations because they felt that the mathematical framework and theory should be made to agree only with experiments. Obviously, the simplified equations worked—they helped make the AC/DC electrical age engineerable.

Then in the 1889 Nikola Tesla—a prolific experimental physicist and inventor of alternating current (AC)—threw a proverbial wrench into the works when he discovered experimental proof for the elusive electric SW. This seemed to suggest that SW/LW, as opposed to TW, could propagate as pure electric waves or as pure magnetic waves. Tesla also believed these waves carried a hitherto unknown form of excess energy he referred to as "radiant." This intriguing and unexpected result was said to have been verified by Lord Kelvin and others soon after.

Instead of merging their experimental results into a unified proof for Maxwell's original equations, however, Tesla, Hertz, and others decided to bicker and squabble over who was more correct because they all derived correct results. Nonetheless, because humans (even "rational" scientists) are fallible and prone to fits of vanity and self-aggrandizement, each side insisted dogmatically that they were right, and the other side was mistaken. The issue was allegedly settled after the dawn of the twentieth century when (1) the concept of the mechanical (i.e., passive/viscous)

Ether was purportedly disproved by Michelson-Morley and replaced by Einstein's Relativistic Space-Time Manifold, and (2) detection of SW/LWs proved much more difficult than initially thought; this was mostly because of the wave's subtle densities, fluctuating frequencies, and orthogonal directional flow. As a result, the truncation of Maxwell's equations was upheld. Nevertheless, SW/LW in free space are quite real.

Besides Tesla, empirical work carried out by electrical engineers (e.g., Eric Dollard, Konstantin Meyl, Thomas Imlauer, and Jean-Louis Naudin, to name only some) has clearly demonstrated SW/LWs' existence experimentally. These waves seem able to exceed the speed of light, pass through EM shielding (i.e., Faraday cages), and produce overunity—more energy out than in—effects. They seem to propagate in a yet unacknowledged counterspatial dimension (i.e., hyper-space, pre-space, false-vacuum, Aether, implicit order, etc.).

In addition to the mathematical calculation of SWs, this book contains a voluminous collection of material concerning the information's technical use of SWs; for example, if the useful signal and the usually interfering noise signal change their places, if a separate modulation of frequency and wavelength makes a parallel image transmission possible, if it concerns questions of the environmental compatibility for the sake of humanity (e.g., bioresonance, among others) or to harm humanity (e.g., electro-smog) or to be used as high-energy directed weapons—also known as Star Wars or the Strategic Defense Initiative (SDI)—as tomorrow's battlefield weapons.

Albuquerque, NM, USA B. Zohuri
2018

Acknowledgments

I am indebted to the many people who aided me, encouraged me, and supported me beyond my expectations. Some are not around to see the results of their encouragement in the production of this book, yet I hope they know of my deepest appreciation. I especially want to thank my friends, to whom I am deeply indebted, and have continuously given support without hesitation. They have always kept me going in the right direction.

My most gratitude goes to Dr. Edl Schamiloglu, the Associate Dean of Engineering and Distinguished Professor in Department of Electrical Engineering and Computer Science at the University of New Mexico, who first of all gave me the opportunity and funding for this research and guided me in right direction as well.

Above all, I offer very special thanks to my late mother and father and to my children, particularly, my son Sasha and grandson Darius. They have provided constant interest and encouragement without which this book would not have been written. Their patience with my many absences from home and long hours in front of the computer to prepare the manuscript are especially appreciated.

I would like to extend my gratitude to Dr. Horst Eckardt of A.I.A.S. and his valuable write up on the subject of scalar waves, which I found very helpful and useful for me to write the chapter in this book on the subject. My many thanks also go to the pioneer of this new subject area, Dr. Konstantin Meyl, professor of Computer and Electrical Engineering at Furtwangen University in Germany; he has written a few excellent books on the subject of scalar waves.

I also would like to take this opportunity to express my great appreciation and gratitude to Ms Cheyenne Stradinger, the Senior Librarian, and Ms Anne D. Schultz, the manager of Library Operation at the Engineering Library of the University of New Mexico–Albuquerque,. They constantly supported my research throughout by obtaining all the resource books and journals for me. Without their help this book could not have come to its final form as presented here.

Contents

1	**Foundation of Electromagnetic Theory**		1
	1.1	Introduction	1
	1.2	Vector Analysis	2
		1.2.1 Vector Algebra	2
		1.2.2 Vector Gradient	7
		1.2.3 Vector Integration	10
		1.2.4 Vector Divergence	12
		1.2.5 Vector Curl	13
		1.2.6 Vector Differential Operator	14
	1.3	Further Developments	15
	1.4	Electrostatics	18
		1.4.1 Coulomb's Law	18
		1.4.2 The Electric Field	21
		1.4.3 Gauss's Law	22
	1.5	Solution of Electrostatic Problems	25
		1.5.1 Poisson's Equation	25
		1.5.2 Laplace's Equation	26
	1.6	Electrostatic Energy	26
		1.6.1 Potential Energy of a Group of Point Charges	27
		1.6.2 Electrostatic Energy of a Charge Distribution	28
		1.6.3 Forces and Torques	29
	1.7	Description of Maxwell's Equations	33
	1.8	Time-Independent Maxwell's Equations	39
		1.8.1 Coulomb's Law	39
		1.8.2 The Electric Scalar Potential	44
		1.8.3 Gauss's Law	47
		1.8.4 Poisson's Equation	53
		1.8.5 Ampère's Experiments	55
		1.8.6 The Lorentz Force	57
		1.8.7 Ampère's Law	60

		1.8.8	Magnetic Monopoles	61
		1.8.9	Ampère's Circuital Law	63
		1.8.10	Helmholtz's Theorem	68
		1.8.11	The Magnetic Vector Potential	74
		1.8.12	The Biot-Savart Law	77
		1.8.13	Electrostatics and Magnetostatics	79
	1.9	Time-Dependent Maxwell's Equations		82
		1.9.1	Faraday's Law	83
		1.9.2	Electric Scalar Potential	86
		1.9.3	Gauge Transformations	87
		1.9.4	The Displacement Current	90
		1.9.5	Potential Formulation	95
		1.9.6	Electromagnetic Waves	96
		1.9.7	Green's Function	103
		1.9.8	Retarded Potentials	107
		1.9.9	Advanced Potentials	113
		1.9.10	Retarded Fields	116
		1.9.11	Summary	119
	References			121
2	**Maxwell's Equations—Generalization of Ampère-Maxwell's Law**			123
	2.1	Introduction		123
	2.2	Permeability of Free Space μ_0		127
	2.3	Generalization of Ampère's Law with Displacement Current		130
	2.4	Electromagnetic Induction		134
	2.5	Electromagnetic Energy and the Poynting Vector		137
	2.6	Simple Classical Mechanics Systems and Fields		147
	2.7	Lagrangian and Hamiltonian of Relativistic Mechanics		149
		2.7.1	Four-Dimensional Velocity	152
		2.7.2	Energy and Momentum in Relativistic Mechanics	153
	2.8	Lorentz versus Galilean Transformation		158
	2.9	The Structure of Spacetime, Interval, and Diagram		160
		2.9.1	Spacetime or the Minkowski Diagram	168
		2.9.2	Time Dilation	171
		2.9.3	Time Interval	172
		2.9.4	The Invariant Interval	172
		2.9.5	Lorentz Contraction Length	174
	References			176
3	**All About Wave Equations**			177
	3.1	Introduction		177
	3.2	The Classical Wave Equation and Separation of Variables		180
	3.3	Standing Waves		186
	3.4	Seiches		187
		3.4.1	Lake Seiches	189
		3.4.2	Sea and Bay Seiches	191

About the Author

Bahman Zohuri currently works for Galaxy Advanced Engineering, Inc., a consulting firm that he started in 1991 when he left both the semiconductor and defense industries after many years working as a chief scientist. After graduating from the University of Illinois in the field of physics and applied mathematics, he then went to the University of New Mexico, where he studied nuclear and mechanical engineering. He joined Westinghouse Electric Corporation after graduating; there he performed thermal hydraulic analysis and studied natural circulation in an inherent shutdown heat removal system (ISHRS) in the core of a liquid metal fast breeder reactor (LMFBR) as a secondary fully inherent shutdown system for secondary loop heat exchange. All these designs were used in nuclear safety and reliability engineering for a self-actuated shutdown system. Dr. Zohuri designed a mercury heat pipe and electromagnetic pumps for large pool concepts of a LMFBR for heat rejection purposes for this reactor during 1978 and received a patent for it.

Subsequently, he was transferred to the defense division of Westinghouse, where he oversaw dynamic analysis and methods of launching and controlling MX missiles from canisters. The results were applied to MX launch seal performance and muzzle blast phenomena analysis (i.e., missile vibration and hydrodynamic shock formation). Dr. Zohuri also was involved in analytical calculations and computations in the study of non-linear ion waves in rarefying plasma. The results were applied to the propagation of so-called soliton waves and the resulting charge collector traces in the rarefaction characterization of the corona of laser-irradiated target pellets.

As part of his graduate research work at Argonne National Laboratory, he performed computations and programming of multi-exchange integrals in surface and solid-state physics. He earned various patents in areas, such as diffusion processes and diffusion furnace design, while working as a Senior Process Engineer at various semiconductor companies (e.g, Intel Corp., Varian Medical Systems, and National Semiconductor Corporation). He later joined Lockheed Martin Missile and Aerospace Corporation as Senior Chief Scientist and oversaw research and development (R&D) and the study of the vulnerability, survivability, and both radiation and laser hardening of various components of the Strategic Defense Initiative, known as Star Wars.

This work included payloads (i.e., IR sensor) for the Defense Support Program, the Boost Surveillance and Tracking System, and the Space Surveillance and Tracking Satellite against laser and nuclear threats. While at Lockheed Martin, he also performed analyses of laser beam characteristics and nuclear radiation interactions with materials, transient radiation effects in electronics, electromagnetic pulses, system-generated electromagnetic pulses, single-event upset, blast, thermo-mechanical, hardness assurance, maintenance, and device technology.

He spent several years as a consultant at Galaxy Advanced Engineering serving Sandia National Laboratories, where he supported the development of operational hazard assessments for the Air Force Safety Center in collaboration with other researchers and third parties. Ultimately, the results were included in Air Force Instructions issued specifically for directed energy weapons operational safety. He completed the first version of a comprehensive library of detailed laser tools for airborne lasers, advanced tactical lasers, tactical high-energy lasers, and mobile/tactical high-energy lasers, for example.

Dr. Zohuri also oversaw SDI computer programs in connection with Battle Management C^3I and artificial intelligence and autonomous systems. He is the author of several publications and holds several patents, such as for a laser-activated radioactive decay and results of a through-bulkhead initiator. He has published the following works: *Heat Pipe Design and Technology: A Practical Approach* (CRC Press); *Dimensional Analysis and Self-Similarity Methods for Engineering and Scientists* (Springer); *High Energy Laser (HEL): Tomorrow's Weapon in Directed Energy Weapons, Volume I* (Trafford Publishing Company); and recently the book on the subject of *Directed-Energy Weapons and Physics of High-Energy Lasers* with Springer. He has published two other books with Springer Publishing Company: *Thermodynamics in Nuclear Power Plant Systems* and *Thermal-Hydraulic Analysis of Nuclear Reactors.* Many of them can be found in most universities' technical library, can be seen on the Internet, or ordered from Amazon.com.

Presently, he holds the position of Research Associate Professor in the Department of Electrical Engineering and Computer Science at the University of New Mexico–Albuquerque, and continues his research on neural science technology and its application in super artificial intelligence. Dr. Zohuri has published a series of book in this subject as well on his research on SWs; the results of his research are presented in this book.

Chapter 1
Foundation of Electromagnetic Theory

To study the subject of a scalar wave and its physics as well as its behavior as a source driving various applications of energy, we need to have some understanding of the fundamental knowledge of electromagnetic theory; such background is essential. This chapter introduces Maxwell's equations—particularly Ampère's Law as part of his other equations. We mainly are concerned with the law's missing term as part of the complete version of Maxwell's equations. We also examine this law to show that it sometimes fails, and to find a generalization that always is valid in classical electromagnetics, whereas it fails in electrodynamics because of the missing term, which is an important factor to develop the basic scalar wave equation [1].

1.1 Introduction

Although Maxwell formulated his equations (now known as *Maxwell's equations*) more than 100 years ago, the subject of electromagnetism never has been stagnate. Production of so-called clean energy, driven by magnetic confinement of hot plasma via a controlled thermonuclear reaction between two isotopes of hydrogen—namely, deuterium (D) and tritium (T)—results in some behavior in plasma that is known as *magneto hydrodynamics* (MHD). Study of such phenomena requires knowledge of and understanding of fundamental electromagnetism and fluid dynamics combined, where the *fluid dynamics equation* and *Maxwell's equations* are joined [1].

In the study of electricity and magnetism, as part of understanding the physics of plasma, however, we need to have some knowledge of notation that may be accomplished by using vector analysis. By providing a valuable shorthand for electromagnetics (EM) and electrodynamics, vector analysis also brings to the forefront the physical ideas involved in these equations; therefore, we briefly formulate some of these vector analysis concepts and present some of their uniqueness in this chapter.

1.2 Vector Analysis

Several kinds of quantities are encountered in the study of the fundamental science of physics; in particular, we need to distinguish *vectors* and *scalars*. For our purposes, it is sufficient to define them as follows:

1. *Scalar:* A *scalar* is a quantity that is characterized completely by its magnitude. Examples of scalars are mass and volume. A simple extension of the idea of a scalar is a *scalar field*—a function of position that is entirely specified by its magnitude at all points in space.
2. *Vector:* A *vector* is a quantity that is characterized completely by its magnitude and direction. Examples of vectors are: the position from a fixed origin, velocity, acceleration, and force. The generalization to a *vector field* gives a function of position that is entirely specified by its magnitude and direction at all points in space.

Detailed review of *vector analysis* is beyond the scope of this book; thus, we briefly formulate the fundamental layout of vector analysis here for purposes of its operation for the operator developing essential electromagnetics and electrodynamics that are the foundation for understanding plasma physics.

1.2.1 Vector Algebra

Most everyone is familiar with scalar algebra from basic algebra courses; the same algebra can be applied to develop vector algebra. For the time being we use a Cartesian coordinate system to develop a three-dimensional analysis of vector algebra. The Cartesian system allows representation of a vector by its three components, denoting them with x, y, and z; or, when it is more convenient, we use notation x_1, x_2, and x_3. With respect to the Cartesian coordinate system, a vector is specified by its $x-$, $y-$, and $z-$ components. Thus, a vector $\vec{V}$ (note that the vector quantities are denoted by symbol of vector $\rightarrow$ on top) is specified by its components, V_x, V_y, and V_z, where $V_x = |\vec{V}| \cos \alpha_1, V_y = |\vec{V}| \cos \alpha_2$, and $V_z = |\vec{V}| \cos \alpha_3$. The α's are the angles between vector $\vec{V}$ and the appropriate coordinate axes of the Cartesian system.

The scalar $|\vec{V}| = \sqrt{V_x^2 + V_y^2 + V_z^2}$ is the *magnitude* of the vector or its length. On the basis of Fig. 1.1, in the case of vector fields, each of the components is to be regarded as a function of x, y, and z. It should be emphasized for the simplicity of analysis that we are using the Cartesian coordinate system, yet the similarity of these analyses applies to the other coordinates, such as cylindrical and spherical as well.

1.2 Vector Analysis

$$\vec{A} \cdot \vec{B} = 0 \tag{1.8}$$

Note that the scalar product is commutative. The length of $\vec{A}$, then, is:

$$|\vec{A}| = \sqrt{\vec{A} \cdot \vec{A}} \tag{1.9}$$

3.2 Vector Product of Two Vectors

The vector product of two vectors is a vector, which accounts for the name and alternative names: *outer product* and *cross product*. The vector product is written as $\vec{A} \times \vec{B}$. If $\vec{C}$ is the vector product of $\vec{A}$ and $\vec{B}$, then

$$\vec{C} = \vec{A} \times \vec{B} \tag{1.10}$$

or in terms of their components it can be written as:

$$\begin{aligned} C_x &= A_y B_z - A_z B_y \\ C_y &= A_z B_x - A_x B_z \\ C_z &= A_x B_y - A_y B_x \end{aligned} \tag{1.11}$$

It is important to note that the cross product depends on the order of the factors; interchanging the order of the cross product introduces a minus sign:

$$\vec{B} \times \vec{A} = -\vec{A} \times \vec{B} \tag{1.12}$$

Consequently,

$$\vec{A} \times \vec{A} = 0 \tag{1.13}$$

This definition is equivalent to the following: The vector product is the product of the magnitudes times the sine of the angle between the original vectors with the direction given by the right-hand screw rule (Fig. 1.2). Note that if we let $\vec{A}$ be rotated into $\vec{B}$ through the smallest possible angle, a right-hand screw rotated in this manner will advance in a direction perpendicular to both $\vec{A}$ and $\vec{B}$; this direction is the direction of $\vec{A} \times \vec{B}$.

The vector product may be easily expressed in terms of a determinant via the definition of unit vectors as $\hat{i}, \hat{j}$, and $\hat{k}$, which are vectors of unit magnitude in the x-, y-, and z-directions, respectively; then we can write:

$$\vec{A} \times \vec{B} = \begin{vmatrix} \hat{i} & \hat{j} & \hat{k} \\ A_x & A_y & A_z \\ B_x & B_y & B_z \end{vmatrix} \tag{1.14}$$

If this determinant is evaluated by the usual rules, the result is precisely our definition of the cross product of two vectors.

Fig. 1.2 Right-hand screw rule

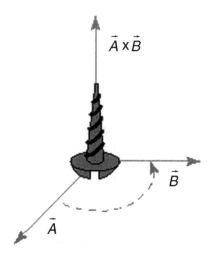

The determinant in Eq. 1.14 may be combined in many ways, and most of the results obtained are obvious; however, two triple products of sufficient importance need to be mentioned. The triple scalar product, $D = \vec{A} \cdot \vec{B} \times \vec{C}$, is easily found and given by the determinant as

$$D = \vec{A} \cdot \vec{B} \times \vec{C} = \begin{vmatrix} A_x & A_y & A_z \\ B_x & B_y & B_z \\ C_x & C_y & C_z \end{vmatrix} = -\vec{B} \cdot \vec{A} \times \vec{C} \qquad (1.15)$$

This product in Eq. 1.15 is unchanged by an exchange of dot and cross or by a cyclic permutation of the three vectors. Note that parentheses are not needed because the cross product of a scalar and a vector is undefined.

The other interesting triple product is the triple vector product, $\vec{D} = \vec{A} \times (\vec{B} \times \vec{C})$. Through repeated application of the definition of the cross product, Eqs. 1.10 and 1.11, we find:

$$\vec{D} = \vec{A} \times (\vec{B} \times \vec{C}) = \vec{B}(\vec{A} \cdot \vec{C}) - \vec{C}(\vec{A} \cdot \vec{B}) \qquad (1.16)$$

which frequently is known as the *back-cab rule*. We should bear in mind that in the cross product the parentheses are vital as part of the operation; without them the product is not well defined.

4. *Devision of Two Vectors*

At this point one might be interested in the possibly of vector division. Division of a vector by a scalar can, of course, be defined as multiplication by the reciprocal of the scalar. Division of a vector by another vector, however, is possible only if the two vectors are parallel. On the other hand, it is possible to write a general solution to vector equations and so accomplish something akin to division.

1.2 Vector Analysis

Consider this equation:

$$c = \vec{A} \cdot \vec{X} \tag{1.17}$$

where c is a known scalar, $\vec{A}$ is a known vector, and $\vec{X}$ is an unknown vector. A general solution to Eq. 1.17 is given as follows:

$$\vec{X} = \frac{c\vec{A}}{\vec{A} \cdot \vec{A}} + \vec{B} \tag{1.18}$$

where $\vec{B}$ is an arbitrary vector that is perpendicular to $\vec{A}$—that is, $\vec{A} \cdot \vec{B} = 0$. What we have done is very nearly to divide c by vector $\vec{A}$; more correctly, we have found the general form of vector $\vec{X}$ that satisfies Eq. 1.17. There is no unique solution, and this fact accounts for vector $\vec{B}$. In the same fashion, we can consider the vector equation as

$$\vec{C} = \vec{A} \times \vec{X} \tag{1.19}$$

In Eq. 1.19 both vectors $\vec{A}$ and $\vec{C}$ are known; $\vec{X}$ is an unknown vector. The general solution of this equation is then given by

$$\vec{X} = \frac{\vec{C} \times \vec{A}}{\vec{A} \cdot \vec{A}} + k\vec{A} \tag{1.20}$$

where k is an arbitrary scalar. Thus, $\vec{X}$, as defined by Eq. 1.20, is very nearly the quotient of $\vec{C}$ by $\vec{A}$; scalar k takes into account the non-uniqueness of the process. If $\vec{X}$ is required to satisfy both Eqs. 1.17 and 1.19, then the result is unique if it exists and is given by

$$\vec{X} = \frac{\vec{C} \times \vec{A}}{\vec{A} \cdot \vec{A}} + \frac{c\vec{A}}{\vec{A} \cdot \vec{A}} \tag{1.21}$$

1.2.2 Vector Gradient

Now that we have covered basic vector algebra, we pay attention to vector calculus, which extends to vector gradient, integration, vector curl, and differentiation of vectors. The simplest of these is the relation of a particular vector field to the derivative of a scalar field.

For that matter, it is convenient to introduce the idea of a *directional derivative* of a function of several variables; we leave it to the reader to find these analyses in any vector calculus book—that is, details of such a derivative are beyond the intended scope of this book. Thus, we jump to the definition of the vector gradient.

The *vector gradient* of a scalar function, φ, is a vector with a magnitude that is the maximum directional derivative at the point being considered and with a direction that is the direction of the maximum directional derivative at the point. We put this definition into some perspective using the geometry of Fig. 1.3, and it is evident that the gradient has the direction to the level surface of φ through the point, as we said that is being coinsured.

The most common mathematical symbol for gradient is $\vec{\nabla}$; in text form it is *grad*. In terms of the gradient, the directional derivative is given by

$$\frac{d\varphi}{ds} = |\operatorname{grad}\vec{\varphi}| \cos\theta \tag{1.22}$$

where θ is the angle between the direction of $d\vec{s}$ and the direction of the gradient. This result is evident immediately from Fig. 1.3. If we write $d\vec{s}$ for the vector displacement of magnitude ds, then Eq. 1.22 can be written as

$$\frac{d\varphi}{ds} = \operatorname{grad}\vec{\varphi} \cdot \frac{d\vec{s}}{ds} \tag{1.23}$$

Equation 1.23 enables us to seek the explicit form of the gradient and find it in any coordinate system in which we know the form of $d\vec{s}$. In a Cartesian or a rectangular coordinate system, we know that $d\vec{s} = \hat{i}dx + \hat{j}dy + \hat{k}dz$. We also know from differential calculus that

$$d\varphi = \frac{\partial\varphi}{\partial x}dx + \frac{\partial\varphi}{\partial y}dy + \frac{\partial\varphi}{\partial z}dz \tag{1.24}$$

From Eq. 1.22, the results are:

Fig. 1.3 Parts of two level surfaces of the function $\varphi(x, y, z)$

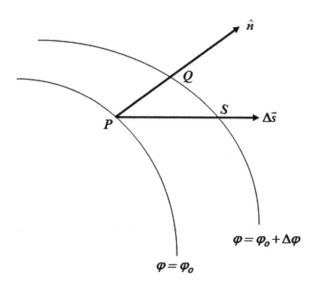

1.2 Vector Analysis

$$d\varphi = \frac{\partial \varphi}{\partial x}dx + \frac{\partial \varphi}{\partial y}dy + \frac{\partial \varphi}{\partial z}dz \qquad (1.25)$$
$$= (\mathrm{grad}\varphi)_x dx + (\mathrm{grad}\varphi)_y dy + (\mathrm{grad}\varphi)_z dz$$

Equating the coefficient of independent variables on both sides of the equation in a rectangular coordinate gives:

$$\mathrm{grad}\,\vec{\varphi} = \frac{\partial \varphi}{\partial x}\hat{i} + \frac{\partial \varphi}{\partial y}\hat{j} + \frac{\partial \varphi}{\partial z}\hat{k} \qquad (1.26)$$

In a more complicated case, the procedure is very similar as well. In spherical polar coordinates, by using Fig. 1.4 with denotation of r, θ, and ϕ, we can write Eq. 1.24 in the following form:

$$d\varphi = \frac{\partial \varphi}{\partial r}dr + \frac{\partial \varphi}{\partial \theta}d\theta + \frac{\partial \varphi}{\partial \varphi}d\phi \qquad (1.27)$$

and

$$d\vec{s} = \hat{a}_r dr + \hat{a}_\theta r d\theta + \hat{a}_\phi r \sin\theta d\phi \qquad (1.28)$$

where $\hat{a}_r$, $\hat{a}_\theta$, and $\hat{a}_\phi$ are unit vectors in the r, θ, and ϕ directions, respectively. Applying Eq. 1.23 and equating coefficients of independent variables yields:

$$\mathrm{grad}\,\vec{\varphi} = \hat{a}_r \frac{\partial \varphi}{\partial r} + \hat{a}_\theta \frac{1}{r}\frac{\partial \varphi}{\partial \theta} + \hat{a}_\phi \frac{1}{r\sin\theta}\frac{\partial \varphi}{\partial z} \qquad (1.29)$$

Equation 1.29 is established in a spherical coordinate system.

Fig. 1.4 Definition of the polar coordinates

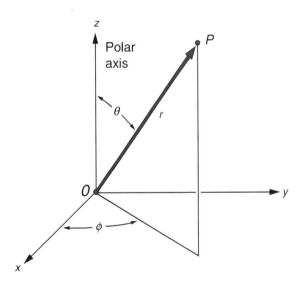

1.2.3 Vector Integration

Although there are other aspects of vector differentiation, first we need to consider vector integration. The details of such analyses are left for the reader to look up in any vector calculus book; we discuss them just briefly here. For our purposes of vector integration, we consider three kinds of integrals, according to the nature of the differential appearing in them:

1. Line integral
2. Surface integral
3. Volume integral

In either case, the integrand may be either a vector or a scalar field; however, certain combinations of integrands and differentials give rise to uninteresting integrals. Those of most interest here are the scalar line integral of a vector, the scalar surface integral of a vector, and the volume integral of both vectors and scalars.

If $\vec{F}$ is a vector field, a line integral of $\vec{F}$ is written as

$$\int_{a(C)}^{b} \vec{F}(\vec{r}) \cdot d\vec{l} \tag{1.30}$$

where C is the curve along which the integration is performed, a and b are the initial and final points on the curve, and $d\vec{l}$ is an infinitesimal vector displacement along curve C.

It is obvious that because the result of the dot product of $\vec{F}(\vec{r}) \cdot d\vec{l}$ is scalar, the result of the linear integral in Eq. 1.30 is scalar. The definition of line integral follows closely the Riemann definition of the definite integral; thus, the integral can be written as a segment of curve C between the lower and the upper bounds of a and b, respectively, and then it can be divided into a large number of small increments, $\Delta \vec{l}$. For an increment, an interior point is chosen and the value of $\vec{F}(\vec{r})$ at that point is found. In other words, Eq. 1.30 can form the following equation as

$$\int_{a(C)}^{b} \vec{F}(\vec{r}) \cdot d\vec{l} = \lim_{N \to \infty} \sum_{i=1}^{N} \vec{F}_i(\vec{r}) \cdot \Delta \vec{l} \tag{1.31}$$

It is important to emphasize that the line integral usually depends not only on the end points of a and b, but also on curve C along which the integration is to be done, because the magnitude and direction of $\vec{F}_i(\vec{r})$ and the direction of $d\vec{l}$ depend on curve C and its tangent, respectively. The line integral around a closed curve is of sufficient importance that a special notation is used for it—namely,

$$\oint_C \vec{F} \cdot d\vec{l} \tag{1.32}$$

1.2 Vector Analysis

Note that the integral around a closed curve usually is not zero. The class of vectors for which the line integral around any closed curve is zero is of considerable importance. Thus, we normally write line integrals around undesignated closed paths as

$$\oint \vec{F} \cdot d\vec{l} \tag{1.33}$$

The form of integral in Eq. 1.33 around a closed curve C is for those cases where the integral is independent of contour C within rather wide limits.

Now, paying attention to the second kind of integral—namely, surface integrals—we again can define $\vec{F}$ as a vector, and a surface integral of $\vec{F}$ is written as

$$\int_S \vec{F} \cdot \hat{n} \, da \tag{1.34}$$

where S is the surface over which the integral is taken, da is an infinitesimal area on surface S, and $\hat{n}$ is a unit vector normal to da.

There is an ambiguity of two degrees in the choice of unit vector $\hat{n}$ as far as outward or downward direction to the normal surface S is concerned if this surface is a closed one. If S is not closed and is finite, then it has a boundary, and the sense of the normal is important only with respect to the arbitrary positive sense of traversing the boundary. The positive sense of the normal is the direction in which a right-hand screw would advance if rotated in the direction of the positive sense on the bounding curve, as illustrated in Fig. 1.5. The surface integral of $\vec{F}$ over a closed surface S is sometimes denoted by

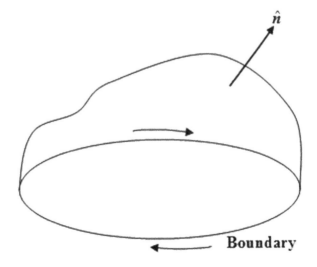

Fig. 1.5 Illustration of the relation of normal unit vector to surface and the direction of traversal of the boundary

$$\oint_S \vec{F} \cdot \hat{n}\, da \tag{1.35}$$

Comments exactly parallel to those made for the line integral can be made for the surface integral. This integral is clearly scalar, and it usually depends on surface S; cases where it does not are particularly important.

Now, we can pay attention to the third type of vector integral—namely, the volume integral—and we start with vector $\vec{F}$. Therefore, if $\vec{F}$ is a vector and φ is a scalar, then the two volume integrals in which we are interested are written:

$$J = \int_V \varphi\, dv \qquad \vec{K} = \int_V \vec{F}\, dv \tag{1.36}$$

Clearly, J is a scalar and $\vec{K}$ is a vector. The definitions of these integrals reduce quickly to just the Riemann integral in three dimensions, except that in $\vec{K}$ one must note that there is one integral for each component of $\vec{F}$. We are very familiar with these integrals, however, and require no further investigation nor any comments.

1.2.4 Vector Divergence

Another important vector operator, which plays an essential role in establishing electromagnetism equations, is a vector divergence operation; it is a derivative form. The divergence of vector $\vec{F}$, written as $\text{div}\,\vec{F}$, is defined as follows.

The divergence of a vector is the limit of its surface integral per unit volume as the volume enclosed by the surface goes to zero. This statement can be presented mathematically as follows:

$$\text{div}\,\vec{F} = \lim_{V \to 0} \oint_S \vec{F} \cdot \hat{n}\, da \tag{1.37}$$

The divergence is clearly a scalar point function; its resulting operation ends up with a scalar field, and it is defined at the limit point of the surface of integration.

A detailed proof of this concept is beyond the scope of this book, and it is left to readers to refer to any vector calculus book. Yet, the limit can be taken easily, and the divergence in rectangular coordinates is found to be:

$$\text{div}\,\vec{F} = \frac{\partial F_x}{\partial x} + \frac{\partial F_y}{\partial y} + \frac{\partial F_z}{\partial z} \tag{1.38}$$

Equation 1.38 for the vector divergence operation designated for the Cartesian coordinate and in the spherical coordinate is written in the following form:

1.2 Vector Analysis

$$\text{div}\,\vec{F} = \frac{1}{r^2}\frac{\partial}{\partial r}(r^2 F_r) + \frac{1}{r\sin\theta}\frac{\partial}{\partial \theta}(\sin\theta F_\theta) + \frac{1}{r\sin\theta}\frac{\partial F_\phi}{\partial \phi} \quad (1.39)$$

In the cylindrical coordinate it is represented by

$$\text{div}\,\vec{F} = \frac{1}{r}\frac{\partial}{\partial r}(rF_r) + \frac{1}{r}\frac{\partial}{\partial \theta}(F_\theta) + \frac{\partial}{\partial z}(F_z) \quad (1.40)$$

The method to find the explicit form of the divergence is applicable to any coordinate system, provided that the forms of the volume and the surface elements, or, alternatively, the elements of the length, are known.

Now that we have the idea behind the vector divergence operator and its operation, we can establish the *divergence theorem*. The integral of the divergence of a vector over volume V is equal to the surface integral of the normal component of the vector over the surface bounding V—that is,

$$\int_V \text{div}\,\vec{F}\,dv = \oint_S \vec{F}\cdot\hat{n}\,da \quad (1.41)$$

We leave it at that; for proof readers can refer to any vector calculus book.

1.2.5 Vector Curl

Another interesting vector differential operator is the vector *curl*. The curl of a vector, written as curl $\vec{F}$, is defined as the limit of the ratio of the integral of its cross product with the outward drawn normal, over a closed surface, to the volume enclosed by the surface as the volume goes to zero—that is,

$$\text{curl}\,\vec{F} = \lim_{V\to 0}\frac{1}{V}\oint_S \hat{n}\times\vec{F}\,da \quad (1.42)$$

Again, the details of proof are left to readers to find in a vector calculus book; we just write the final result of the curl operator in at least the rectangular coordinate, as follows:

$$\text{curl}\,\vec{F} = \begin{vmatrix} \hat{i} & \hat{j} & \hat{k} \\ \dfrac{\partial}{\partial x} & \dfrac{\partial}{\partial y} & \dfrac{\partial}{\partial z} \\ F_x & F_y & F_z \end{vmatrix} \quad (1.43)$$

Finding the form of the curl in other coordinate systems is only slightly more complicated and is left to reader to practice.

Now that we have an understanding of the vector curl operator, we can state *Stock's theorem* as follows. The line integral of a vector around a closed curve is equal to the integral of the normal component of its curl over any surface bounded by the curve—that is,

$$\oint_C \vec{F} \cdot d\hat{l} = \int_S \text{curl}\, \vec{F} \cdot \hat{n}\, da \tag{1.44}$$

where C is a closed curve that bounds surface S.

1.2.6 Vector Differential Operator

We now introduce an alternative notation for the types of vector differentiation that have been discussed—namely, gradient, divergence, and curl. This notation uses the vector differential operator, *del*, and it is identified by the symbol $\vec{\nabla}$ and written mathematically as:

$$\vec{\nabla} = \hat{i}\frac{\partial \varphi}{\partial x} + \hat{j}\frac{\partial \varphi}{\partial y} + \hat{k}\frac{\partial \varphi}{\partial z} \tag{1.45}$$

Del is a differential operator in that it is used only in front of a function of (x, y, z), which it differentiates; it is a vector in that it obeys the laws of vector algebra. It is also a vector in terms of its transformation properties, and in terms of del, Eqs. 1.25, 1.38, and 1.43 are expressed as follows:

Grad = $\vec{\nabla}$:

$$\vec{\nabla} F = \hat{i}\frac{\partial F_x}{\partial x} + \hat{j}\frac{\partial F_y}{\partial y} + \hat{k}\frac{\partial F_z}{\partial z} \tag{1.46}$$

Div = $\vec{\nabla}$:

$$\vec{\nabla} \cdot \vec{F} = \frac{\partial F_x}{\partial x} + \frac{\partial F_y}{\partial y} + \frac{\partial F_z}{\partial z} \tag{1.47}$$

Curl = $\vec{\nabla} \times$:

$$\vec{\nabla} \times \vec{F} = \begin{vmatrix} \hat{i} & \hat{j} & \hat{k} \\ \frac{\partial}{\partial x} & \frac{\partial}{\partial y} & \frac{\partial}{\partial z} \\ F_x & F_y & F_z \end{vmatrix} \tag{1.48}$$

1.3 Further Developments

The operations expressed with del are themselves independent of any special choice of coordinate system. Moreover, any identities that can be proved using the Cartesian representation hold independently of the coordinate system.

1.3 Further Developments

The first of these developments is the *Laplacian operator*, which is defined as the divergence of the gradient of a scalar field, usually written as ∇^2:

$$\vec{\nabla} \cdot \vec{\nabla} = \nabla^2 \tag{1.49}$$

In rectangular coordinates it is:

$$\nabla^2 \varphi = \frac{\partial^2 \varphi}{\partial x^2} + \frac{\partial^2 \varphi}{\partial y^2} + \frac{\partial^2 \varphi}{\partial z^2} \tag{1.50}$$

This operator is of great importance in electrostatics and will be considered in the following sections and chapters.

The curl of the gradient of any scalar field is zero. This statement is verified most easily by writing it out in rectangular coordinates. If the scalar field is φ, then we can write:

$$\vec{\nabla} \times (\vec{\nabla}\varphi) = \begin{vmatrix} \hat{i} & \hat{j} & \hat{k} \\ \frac{\partial}{\partial x} & \frac{\partial}{\partial y} & \frac{\partial}{\partial z} \\ \frac{\partial \varphi}{\partial x} & \frac{\partial \varphi}{\partial y} & \frac{\partial \varphi}{\partial z} \end{vmatrix} = i\left(\frac{\partial^2 \varphi}{\partial y \partial z} - \frac{\partial^2 \varphi}{\partial z \partial y}\right) + \cdots = 0 \tag{1.51}$$

This verifies the original statement. In operator notation it is:

$$\vec{\nabla} \times \vec{\nabla} = 0 \tag{1.52}$$

The divergence of any curl is also zero. This result is verified in rectangular coordinates by writing:

$$\vec{\nabla} \cdot (\vec{\nabla} \times \vec{F}) = \frac{\partial}{\partial x}\left(\frac{\partial F_x}{\partial y} - \frac{\partial F_y}{\partial z}\right) + \frac{\partial}{\partial y}\left(\frac{\partial F_y}{\partial z} - \frac{\partial F_z}{\partial x}\right) + \cdots = 0 \tag{1.53}$$

The two other possible second-order operations are taking the curl of the curl or the gradient of the divergence of a vector field. This is left as an exercise to show that in rectangular coordinates, the following is true as well:

Table 1.1 Differential Vector Identities

$$\vec{\nabla} \cdot \vec{\nabla} = \nabla^2 \varphi$$
$$\vec{\nabla} \cdot (\vec{\nabla} \times \vec{F}) = 0$$
$$\vec{\nabla} \times (\vec{\nabla} \varphi) = 0$$
$$\vec{\nabla} \times (\vec{\nabla} \times \vec{F}) = \vec{\nabla}(\vec{\nabla} \cdot \vec{F}) - \nabla^2 \vec{F}$$
$$\vec{\nabla}(\varphi \psi) = (\vec{\nabla} \varphi) \psi + \varphi \vec{\nabla} \psi$$
$$\vec{\nabla}(\vec{F} \cdot \vec{G}) = (\vec{F} \cdot \vec{\nabla})\vec{G} + \vec{F} \times (\vec{\nabla} \times \vec{G}) + (\vec{G} \cdot \vec{\nabla})\vec{F} + \vec{G} \times (\vec{\nabla} \times \vec{F})$$
$$\vec{\nabla} \cdot (\varphi \vec{F}) = (\vec{\nabla} \varphi) \cdot \vec{F} + \varphi \vec{\nabla} \cdot \vec{F}$$
$$\vec{\nabla} \cdot (\vec{F} \times \vec{G}) = \vec{G} \cdot (\vec{\nabla} \times \vec{F}) - \vec{F} \cdot (\vec{\nabla} \times \vec{G})$$
$$\vec{\nabla} \times (\varphi \vec{F}) = (\vec{\nabla} \varphi) \times \vec{F} + \varphi \vec{\nabla} \times \vec{F}$$
$$\vec{\nabla} \times (\vec{F} \times \vec{G}) = (\vec{\nabla} \cdot \vec{G})\vec{F} - (\vec{\nabla} \cdot \vec{F})\vec{G} + (\vec{G} \cdot \vec{\nabla})\vec{F} - (\vec{F} \cdot \vec{\nabla})\vec{G}$$
$$\nabla^2 \vec{F} = \vec{\nabla}(\vec{\nabla} \cdot \vec{F}) - \vec{\nabla} \times \vec{\nabla} \times \vec{F}$$
$$\nabla \left(\frac{1}{r} \right) = \frac{\hat{r}}{r^2}$$

$$\vec{\nabla} \times \left(\vec{\nabla} \times \vec{F} \right) = \vec{\nabla} \left(\vec{\nabla} \cdot \vec{F} \right) - \nabla^2 \vec{F} \tag{1.54}$$

This equation indicates that the Laplacian of a vector is the vector with rectangular components that are the Laplacian of those components of the original vector. In any coordinate system other than rectangular, the Laplacian of a vector is defined by Eq. 1.54. The six possible combinations of differential operators and product are listed in Table 1.1 and all can be verified easily in a rectangular coordinate system.

A derivative of a product of more than two functions, or a higher than second-order derivative of a function, can be calculated by repeated application of the identities in Table 1.1, which is therefore exhaustive. The formula can be remembered easily from the rules of vector algebra and ordinary differentiation.

Some types of function come up often enough in EM theory that it is worth mentioning their various derivatives now. For function $\vec{F} = \vec{r}$, we can write the following relationship:

$$\begin{cases} \vec{\nabla} \cdot \vec{r} = 3 \\ \vec{\nabla} \times \vec{r} = 0 \\ (\vec{G} \cdot \vec{\nabla})\vec{r} = \vec{G} \\ \nabla^2 \vec{r} = 0 \end{cases} \tag{1.55}$$

1.3 Further Developments

For a function that depends only on distance $r = |\vec{r}| = \sqrt{x^2 + y^2 + z^2}$, we can write:

$$\varphi(r) \text{ or } \vec{F}(r): \qquad \vec{\nabla} = \frac{\vec{r}}{r}\frac{d}{dr} \tag{1.56}$$

For a function that depends on scalar argument $\vec{A} \cdot \vec{r}$, where $\vec{A}$ is a constant vector,

$$\varphi(\vec{A} \cdot \vec{r}) \text{ or } \vec{F}(\vec{A} \cdot \vec{r}): \quad \vec{\nabla} = \vec{A}\left(\frac{d}{d(\vec{A} \cdot \vec{r})}\right) \tag{1.57}$$

For a function that depends on argument $\vec{R} = \vec{r} - \vec{r}'$, where $\vec{r}'$ is treated as a constant,

$$\vec{\nabla}_R = \hat{i}\frac{\partial}{\partial X} + \hat{j}\frac{\partial}{\partial Y} + \hat{k}\frac{\partial}{\partial Z} \tag{1.58}$$

where $\vec{R} = X\hat{i} + Y\hat{j} + Z\hat{k}$. If $\vec{r}$ is treated as a constant instead, it is:

$$\vec{\nabla} = -\vec{\nabla}' \tag{1.59}$$

where

$$\vec{\nabla}' = \hat{i}\frac{\partial}{\partial x'} + \hat{j}\frac{\partial}{\partial y'} + \hat{k}\frac{\partial}{\partial z'} \tag{1.60}$$

There are several possibilities for the extension of the divergence theorem and of Stokes's theorem. The most interesting of these is Green's theorem, which is:

$$\int_V (\psi \nabla^2 \varphi - \varphi \nabla^2 \psi) dv = \oint_S (\psi \vec{\nabla} \varphi - \varphi \vec{\nabla} \psi) \cdot \hat{n} da \tag{1.61}$$

This theorem follows from the application of the divergence theorem to the vector,

$$\vec{F} = \psi \vec{\nabla} \varphi - \varphi \vec{\nabla} \psi \tag{1.62}$$

Using vector $\vec{F}$ in the divergence theorem, we obtain:

$$\int_V \vec{\nabla} \cdot (\psi \nabla \varphi - \varphi \nabla \psi) dv = \oint_S (\psi \vec{\nabla} \varphi - \varphi \vec{\nabla} \psi) \cdot \hat{n} da \tag{1.63}$$

Using the identity from Table 1.1 for the divergence of scalar times a vector gives:

$$\vec{\nabla} \cdot (\psi \nabla \varphi) - \vec{\nabla} \cdot (\varphi \nabla \psi) = \psi \nabla^2 \varphi - \varphi \nabla^2 \psi \tag{1.64}$$

Combining Eqs. 1.63 and 1.64 yields Green's theorem. Some other integral theorems are listed in Table 1.2.

This section is a conclusion to our short course on vector analysis. Proof of many results are left to readers as an exercise or extra study, and the approach has been

Table 1.2 Vector Integral Theorem

$$\int_S \hat{n} \times \vec{\nabla}\varphi\, da = \oint_C \varphi\, d\vec{l}$$
$$\int_V \vec{\nabla}\varphi\, dv = \oint_S \varphi \hat{n}\, da$$
$$\int_V \vec{\nabla} \times \vec{F}\, dv = \oint_S \hat{n} \times \vec{F}\, da$$
$$\int_C (\vec{\nabla}\cdot\vec{G} + \vec{G}\cdot\vec{\nabla})\vec{F}\, dv = \oint_S \vec{F}(\vec{G}\cdot\hat{n})\, da$$
$$\int_V [\vec{F}(\nabla\cdot\vec{G}) + (\vec{G}\cdot\nabla)\vec{F}]\, dV = \oint_S \vec{F}(\vec{G}\cdot\hat{n})\, da$$

utilitarian; therefore, what one needs to understand from the view point of vector analysis has been developed to give enough tools to go on with the rest of this book.

1.4 Electrostatics

The subject of electricity is touched on briefly for rest of this chapter to provide the fundamentals of magnetism that we need in order to understand the science of plasma physics so as to go forward. We deal with the empirical concepts of charge and the force law between charges known as Coulomb's Law. We use the mathematical tools of the previous section to express this law in other, or more powerful formulations, and then extend to the basic plasma physics concept. The electric potential formulation and Gauss's Law are very important to the subsequent development of the subject.

Electric charge is a fundamental and characteristic property of the microscopic particles that make up matter. In fact, all atoms are composed of photons, neutrons, and electrons, and two of these particles bear charges. Even charge particles, the powerful electrical forces associated with these particles, however, are fairly well hidden in a macroscopic observation. The reason behind such a statement is because in nature two kinds of charges exist—namely, *positive* and *negative* charges—and an ordinary piece of matter contains approximately equal amounts of each kind.

It is understood from experimental observation that charges can neither be created nor destroyed. The total charge of a closed system cannot change. From the macroscopic point of view, charges can be regrouped and combined in different ways; nevertheless, we can state that net charge *is conserved* in a closed system [1].

1.4.1 Coulomb's Law

To establish Coulomb's Law we summarize the following three statements:

1.4 Electrostatics

1. There are two and only two kinds of electric charge, now known as positive or negative.
2. Two point charges exert on each other forces that act along the line joining them and are inversely proportional to the square of the distance between them.
3. These forces also are proportional to the product of the charges, are repulsive for like charges, and attract unlike charges.

The last two statements, with the first as preamble, together are known as *Coulomb's Law* and for point charges may be concisely formulated in the vector notation as

$$\begin{cases} \vec{F}_1 = C_u \dfrac{q_1 q_2}{r_{12}^2} \dfrac{\vec{r}_{12}}{r_{12}} \\ \vec{r}_{12} = \vec{r}_1 - \vec{r}_2 \end{cases} \quad (1.65a)$$

where $\vec{F}_1$ is the force on charge q_1, $\vec{r}_{12}$ is the vector to charge q_1 from charge q_2, r_{12} is the magnitude of vector $\vec{r}_{12}$, and C_u is a constant of proportionality that is defined to be equal to 1 in adoption with a Gaussian system of units. Figure 1.6 describes the vector $\vec{r}_{12}$ with respect to an arbitrary origin O.

In Fig. 1.6 vector $\vec{r}_{12}$ is extending from the point at the tip of vector $\vec{r}_2$ to the point at the tip of vector $\vec{r}_1$ and clearly is $\vec{r}_{12} = -\vec{r}_{21}$. Note that Coulomb's Law applies to point charges and in a macroscopic sense, a *point charge* is one with spatial dimensions that are very small compared with any other length pertinent to the problem under consideration; that is why we use the term "point charge" in this sense.

In the meter, kilogram, and/or second (MKS) system, Coulomb's Law for the force between two point charges thus can be written as

$$\vec{F}_1 = \frac{1}{4\pi\varepsilon_0} \frac{q_1 q_2}{r_{12}^2} \frac{\vec{r}_{12}}{r_{12}} \quad (1.65b)$$

If more than two point charges are present, the mutual forces are determined by the repeated application of Eqs. 1.65a and 1.65b. In particular, if a system of N charges is considered, the force on the ith charge is given by

Fig. 1.6 Representation of vector $\vec{r}_{12}$ extending between two points

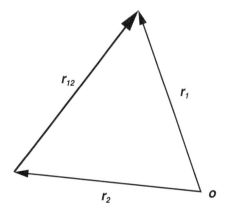

$$\vec{F} = q_i \sum_{i \neq j}^{N} \frac{q_j \vec{r}_{ij}}{4\pi\varepsilon_0 r_{ij}^3} \quad \vec{r}_{ij} = \vec{r}_i - \vec{r}_j \tag{1.66}$$

where the summation on the right-hand side (RHS) of this equation is extended over all the charges except the $i$thone. Equation 1.66 is the superposition principle for forces, which says that the total force acting on a body is the vector sum of the individual forces that act on it. Note that in an MKS unit the value of the Coulomb constant is $C = 9 \times 10^9 \, \text{N} \cdot \text{m}^2/\text{C}^2$.

There are cases (e.g., fully ionized plasma) that we may need to describe a charge distribution in terms of a *charge density function*; thus, it is defined as the limit of charge per unit volume as the volume becomes infinitesimal. Care must be taken, however, in applying this kind of description to atomic problems because in such cases only a small number of electrons are involved, and the process of taking the limit is meaningless. Nevertheless, aside from an atomic case, we can proceed as though a segment of charges might be subdivided indefinitely; thus, we can describe the charge distribution by means of point functions.

A volume charge density is defined by

$$\rho = \lim_{\Delta V \to 0} \frac{\Delta q}{\Delta V} \tag{1.67}$$

and a surface charge density is defined by

$$\sigma = \lim_{\Delta S \to 0} \frac{\Delta q}{\Delta S} \tag{1.68}$$

From the preceding statements and what has been said about point charge q, it is evident that ρ and σ are net charges, or excess charge densities. It is worthwhile to mention that in typical solid materials even a very large charge density ρ will involve a change in the local electron density of only about one part, 10^9.

Now that we have some concept of a point charge and have created Eqs. 1.60 and Eqs. 1.65a, 1.65b, and 1.66, we can extend our knowledge to more general cases. At this point, the charge is distributed through volume V with density ρ, and on surface S that bounds volume V with a surface density σ, the force exerted by this charge distribution on point charge q located at $\vec{r}$ is obtained from Eq. 1.66 by replacing q_j with $\rho_j dv'_j$ or with $\sigma_j da'_j$ and processing to the limit as follows:

$$\vec{F}_q = \frac{q}{4\pi\varepsilon_0} \int_V \frac{\vec{r} - \vec{r}'}{|\vec{r} - \vec{r}'|^3} \rho(\vec{r}') dv' \\ + \frac{q}{4\pi\varepsilon_0} \int_S \frac{\vec{r} - \vec{r}'}{|\vec{r} - \vec{r}'|^3} \sigma(\vec{r}') da' \tag{1.69}$$

Variable $\vec{r}'$ is used to locate a point within the charge distribution—that is, playing the role of source point $\vec{r}_j$ in Eq. 1.66 [1]. Equations 1.66 and 1.69 provide a ready means for obtaining an expression for the electric field because of the given

1.4 Electrostatics

Fig. 1.7 Geometry of $\vec{r}$, $\vec{r}'$, and $\vec{r} - \vec{r}'$

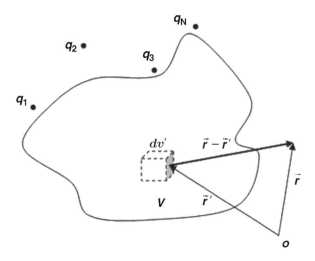

distribution of the charge as it is presented in Fig. 1.7; the electric field is discussed in the next section.

It may appear that the first integral in Eq. 1.69 will diverge if point $\vec{r}$ should fall inside the charge distribution, but that is not the case at all. In Fig. 1.7 vector $\vec{r}$ defines the observation point (i.e., field point) and $\vec{r}'$ ranges over the entire charge distribution, including point charges.

1.4.2 The Electric Field

Our first attempt to seek the *electric field* is for a point charge for the sake of simplicity. The electric field at a point is defined operationally as the limit of the force on a test charge placed at the point of the charge of the test charge, and the limit is presumed to bethe magnitude of the test charge that goes to zero. The customary symbol for electric field in an EM subject is $\vec{E}$, not to be mistaken for energy presentation, which is the case by default. Thus, we can write:

$$\vec{E} = \lim_{q \to 0} \frac{\vec{F}_q}{q} \quad (1.70)$$

The limiting process is included in the definition of the electric field to ensure that the test charge does not affect the charge distribution that produces $\vec{E}$.

Using Fig. 1.7, we let the charge distribution consist of N point $q_1, q_2, \ldots, q_N$ located at the points $\vec{r}_1, \vec{r}_2, \ldots, \vec{r}_N$, respectively; a volume distribution of charge specified by charge density $\rho(\vec{r}')$ in volume V; and a surface distribution characterized by the surface charge density $\sigma(\vec{r}')$ on surface S. If test charge q is located at point $\vec{r}$, it experiences force $\vec{F}$ given by following equation as a result of the given charge distribution:

$$\vec{F}(\vec{r}) = \frac{q}{4\pi\varepsilon_0} \sum_{i=1}^{N} q_i \frac{\vec{r}-\vec{r}_i}{|\vec{r}-\vec{r}_i|^3} + \frac{q}{4\pi\varepsilon_0} \int_V \frac{\vec{r}-\vec{r}_i}{|\vec{r}-\vec{r}_i|^3} \rho(\vec{r}')dv'$$
$$+ \frac{q}{4\pi\varepsilon_0} \int_S \frac{\vec{r}-\vec{r}_i}{|\vec{r}-\vec{r}_i|^3} \sigma(\vec{r}')da' \qquad (1.71)$$

In case of Eq. 1.71, the electric field at point $\vec{r}$ is then the limit of the ratio of this force to the test charge q. Because the ratio is independent of q, the electric field at $\vec{r}$ is just:

$$\vec{E}(\vec{r}) = \frac{1}{4\pi\varepsilon_0} \sum_{i=1}^{N} q_i \frac{\vec{r}-\vec{r}_i}{|\vec{r}-\vec{r}_i|^3} + \frac{1}{4\pi\varepsilon_0} \int_V \frac{\vec{r}-\vec{r}_i}{|\vec{r}-\vec{r}_i|^3} \rho(\vec{r}')dv'$$
$$+ \frac{1}{4\pi\varepsilon_0} \int_S \frac{\vec{r}-\vec{r}_i}{|\vec{r}-\vec{r}_i|^3} \sigma(\vec{r}')da' \qquad (1.72)$$

This equation is very general, and in most cases, one or more of the terms will not be needed.

To complete the EM foundation circle, we also quickly note the general form of the potential energy associated with an arbitrary conservative force $\vec{F}(\vec{r}')$ as the following form:

$$U(\vec{r}) = -\int_{\text{ref},\vec{r}} \vec{F}(\vec{r}') \cdot d\vec{r}' \qquad (1.73)$$

where $U(\vec{r})$ is the potential energy at $\vec{r}$ relative to the reference point at which the potential energy is arbitrarily taken to be zero. Proof is left to readers by referring them to the book by Reitz et al. [1].

1.4.3 Gauss's Law

One of the important relationships that exists between the integral of the normal component of the electric field over a closed surface, and the total charge distribution enclosed by the surface is Gauss's Law. To investigate that briefly here, we look at electric field $\vec{E}(\vec{r})$ for point charge q located at the origin; thus, we can write the following relation as before:

$$\vec{E}(\vec{r}) = \frac{q}{4\pi\varepsilon_0} \frac{\vec{r}}{r^3} \qquad (1.74)$$

1.4 Electrostatics

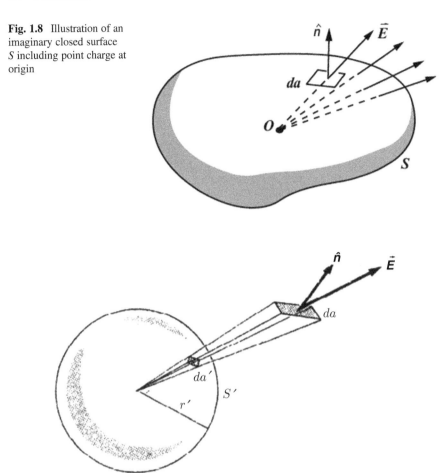

Fig. 1.8 Illustration of an imaginary closed surface S including point charge at origin

Fig. 1.9 Illustration of the construction of spherical surface S'

Consider the surface integral of the normal component of this *electric field* over a closed surface, such as shown in Fig. 1.8, that encloses the origin; consequently, charge q can be written:

$$\oint_S \vec{E} \cdot \hat{n}\, da = \frac{q}{4\pi\varepsilon_0} \oint_S \frac{\vec{r} \cdot \hat{n}}{r^3}\, da \qquad (1.75)$$

The quantity $(\vec{r}/r) \cdot \hat{n}\, da$ is the projection of da on a plane perpendicular to $\vec{r}$. This projected area divided by r^2 is the solid angle subtended by da, which is written as $d\Omega$. It is clear from Fig. 1.9 that the solid angle subtended by da is the same as the solid angle subtended by da'; it is an element of the surface area of sphere S' with a center that is at the origin and the radius of which is r'. It is then possible to write:

$$\oint_S \frac{\vec{r}\cdot\hat{n}\,da}{r^3} = \frac{q}{4\pi\varepsilon_0}\oint_S \frac{\vec{r}'\cdot\hat{n}}{r'^3}da' = 4\pi \tag{1.76}$$

which shows the following equation in the spherical case, as described before:

$$\oint_S \vec{E}\cdot\hat{n}\,da = \frac{q}{4\pi\varepsilon_0}(4\pi) = \frac{q}{\varepsilon_0} \tag{1.77}$$

Figure 1.9 illustrates the construction of spherical surface S' as an aid to evaluation of the solid angle subtended by da. If q lies outside of S, it is clear from Fig. 1.10 that S can be divided into two areas, S_1 and S_2, each of which subtends the same solid angle at charge q. For S_2, however, the direction of normal is toward q, while for S_1 it is away from q.

More details can be found in Reitz et al. [1], where readers need to go; however, in case several point charges $q_1, q_2, \ldots, q_N$ are enclosed by surface S, then the total electric field is given by the first term of Eq. 1.72. Each charge subtends a full solid angle (4π); thus, Eq. 1.72 becomes:

$$\oint_S \vec{E}\cdot\hat{n}\,da = \frac{1}{\varepsilon_0}\sum_{i=1}^{N} q_i \tag{1.78}$$

The result in Eq. 1.73 can be generalized readily for the case of a continuous distribution of charge characterized by a charge density [1].

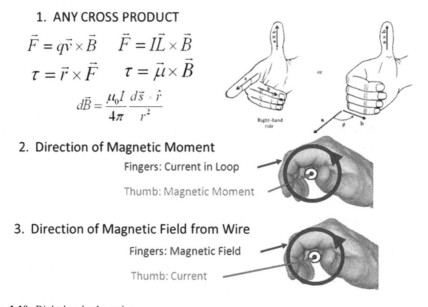

Fig. 1.10 Right-hand rule review

1.5 Solution of Electrostatic Problems

Briefly, we mention and write equations for the solution to an electrostatic problem, which is straightforward for the case in which the charge distribution is specified everywhere, as we have illustrated so far. The potential and electric fields are given as an integral form over this charge distribution as

$$\varphi(\vec{r}) = \frac{1}{4\pi\varepsilon_0} \int \frac{dq'}{|\vec{r} - \vec{r}'|} \quad (1.79)$$

$$\vec{E}(\vec{r}) = \frac{1}{4\pi\varepsilon_0} \int \frac{(\vec{r} - \vec{r}')dq'}{|\vec{r} - \vec{r}'|} \quad (1.80)$$

Yet, many of the problems that we have encountered in real practice are not of this kind. If the charge distribution is not specified in advance, it may be necessary to determine the electric field *first*, before the charge distribution can be calculated.

1.5.1 Poisson's Equation

So far, the only basic relationships we need here were developed in the preceding sections; thus, for that matter, we first write the differential form of Gauss's Law as

$$\vec{\nabla} \cdot \vec{E} = \frac{1}{\varepsilon_0} \rho \quad (1.81)$$

This equation in a purely electrostatic field $\vec{E}$ and can be expressed as minus the gradient of potential φ:

$$\vec{E} = -\vec{\nabla}\varphi \quad (1.82)$$

Combining Eqs. 1.81 and 1.82, we obtain the following relation:

$$\vec{\nabla} \cdot \vec{\nabla}\varphi = \frac{\rho}{\varepsilon_0} \quad (1.83a)$$

Using vector identity as a single differential operator, $\vec{\nabla} \cdot \vec{\nabla}$ or ∇^2, which is called the *Laplacian*, then we can state that the Laplacian is a scalar differential operator, and that Eq. 1.83a is a differential equation that is known as *Poisson's equation* and is written:

$$\nabla^2 \varphi = -\frac{\rho}{\varepsilon_0} \quad (1.83b)$$

The Laplace operator for the *Poisson equation*, in rectangular, cylindrical, and spherical coordinates, are presented here as well.

Rectangular or Cartesian Coordinate:

$$\nabla^2 \varphi \equiv \frac{\partial^2 \varphi}{\partial x^2} + \frac{\partial^2 \varphi}{\partial y^2} + \frac{\partial^2 \varphi}{\partial z^2} = -\frac{\rho}{\varepsilon_0} \quad (1.84)$$

Cylindrical Coordinate:

$$\nabla^2 \varphi \equiv \frac{1}{r}\frac{\partial}{\partial r}\left(r\frac{\partial \varphi}{\partial r}\right) + \frac{1}{r}\frac{\partial^2 \varphi}{\partial \theta} + \frac{\partial^2 \varphi}{\partial z^2} = -\frac{\rho}{\varepsilon_0} \quad (1.85)$$

Spherical Coordinate:

$$\nabla^2 \varphi \equiv \frac{1}{r^2}\frac{\partial}{\partial r}\left(r^2\frac{\partial \varphi}{\partial r}\right) + \frac{1}{r^2 \sin\theta}\frac{\partial}{\partial \theta}\left(\sin\theta\frac{\partial \varphi}{\partial \theta}\right) + \frac{1}{r^2 \sin^2\theta}\frac{\partial^2 \varphi}{\partial \phi^2} = -\frac{\rho}{\varepsilon_0} \quad (1.86)$$

For the form of the Laplacian in other, more complicated coordinated systems, readers are referred to this chapters references, which include ones on vector analysis and/or advanced calculus.

1.5.2 Laplace's Equation

For problems in electrostatics that involve conductors, all the charges either are found on the surface of the conductors or in the form of fixed-point charges. In these cases, charge density ρ is zero at most points in space, and in the absence of charge density, the Poisson equation reduces to the simpler form as follows:

$$\nabla^2 \varphi = 0 \quad (1.87)$$

Equation 1.87 is known as *Laplace's equation*.

1.6 Electrostatic Energy

From here on, without further detailed discussion and proof of various aspects of the electrostatic equation, we just write them down as basic knowledge and leave the details to the readers, referring them to various subject books on the market. Therefore, as we to go on with the subject at hand, we express that, under a static condition, the entire energy of the charge system exists as potential energy, and in this section we mainly are concerned with the potential energy that arises from the electrical interaction of the charges, so-called *electrostatic energy U*.

1.6 Electrostatic Energy

We offer that electrostatic energy U of a point charge is closely related to electrostatic potential φ at the position of point charge $\vec{r}$ per Eq. 1.73. In fact, if q is the magnitude of a particular point charge, then the work done by the force on the charge when it moves from position A to position B is given as:

$$\text{Work} = \int_A^B \vec{F} \cdot d\vec{l} = q \int_A^B \vec{E} \cdot d\vec{l}$$
$$= -q \int_A^B \vec{\nabla} \varphi \cdot d\vec{l} = -q(\varphi_B - \varphi_A) \quad (1.88)$$

Here $\vec{F}$ has been assumed to be only electric force $q\vec{E}$ at each point along the path or the *total work* is finalized to

$$W = -q(\varphi_B - \varphi_A) \quad (1.89)$$

1.6.1 Potential Energy of a Group of Point Charges

The equation for potential energy of a group of point charges can be expressed as

$$U = \sum_{j=1}^{m} W_j = \sum_{j=1}^{m} \left(\sum_{k=1}^{j-1} \frac{q_j q_k}{4\pi\varepsilon_0 r_{jk}} \right) \quad (1.90)$$

or in the summary Eq. 1.90 can be reduced to:

$$U = \frac{1}{2} \sum_{j=1}^{m} \sum_{k=1}^{m} {}' \frac{q_j q_k}{4\pi\varepsilon_0 r_{jk}} \quad (1.91)$$

Note that on the second term of summation in Eq. 1.91, where prime term $k = j$ specifically needs to be excluded, the equation can be written in a somewhat different way by noting that the final value of potential φ at the jth point charge, because of the other charges of the system, is:

$$\varphi_j = \sum_{j=1}^{m} {}' \frac{q_k}{4\pi\varepsilon_0 r_{jk}} \quad (1.92)$$

Thus, the electrostatic energy of the system is given as

$$U = \frac{1}{2} \sum_{j=1}^{m} q_j \varphi_j \quad (1.93)$$

Proof of all the preceding equations is left to the readers.

1.6.2 Electrostatic Energy of a Charge Distribution

The electrostatic energy of an arbitrary charge distribution with volume density φ and surface density can be expressed based on assembled charge distribution by bringing in charge increments δq from reference potential $\varphi_A = 0$. If the charge distribution is partially assembled, and the potential at a particular point in the system is $\varphi'(x, y, z)$, then, from Eq. 1.89, the work required to place δq at this point is written as

$$\delta W = \varphi'(x, y, z)\delta q \quad (1.94)$$

In this equation charge increment δq can be added to a volume element located at (x, y, z) so that $\delta q = \delta \rho \Delta v$, or can be added to a surface element at the point in question, in which case it is $\delta q = \delta \rho \Delta a$. The total electrostatic energy of the assembled charge distribution is obtained by summing contributions of the form in Eq. 1.94.

Let us assume at any stage of the charging process that all charge densities will be at the same fraction of their final values and are represented by symbol α; and if the final values of the charge densities are given by functions $\varphi(x, y, z)$ and $\sigma(x, y, z)$, then the charge densities at an arbitrary stage are $\alpha\varphi(x, y, z)$ and $\alpha\sigma(x, y, z)$. Furthermore, the increments in these densities are $\delta\rho = \varphi(x, y, z)d\alpha$ and $\delta\sigma = \sigma(x, y, z)d\alpha$, then the total electrostatic energy, which is obtained by summing Eq. 1.94, is given by:

$$U = \int_0^1 \delta d \int_V \varphi(x, y, z)\varphi'(x, y, z)dv \\ + \int_0^1 \delta d \int_S \sigma(x, y, z)\varphi'(x, y, z)da \quad (1.95)$$

Because all charges are at the same fraction, however, α is done readily and yields:

$$U = \frac{1}{2}\int_V \rho(\vec{r})\varphi(\vec{r})dv + \frac{1}{2}\int_S \sigma(\vec{r})\varphi(\vec{r})da \quad (1.96)$$

This equation provides the desired result for the energy of a charge distribution. If all space is filled with a single dielectric medium, except for certain conductors, the potential is then given by

$$\varphi(\vec{r}) = \frac{1}{4\pi\varepsilon}\int_V \frac{\varphi(\vec{r})dv'}{|\vec{r} - \vec{r}'|} + \frac{1}{4\pi\varepsilon}\int_V \frac{\sigma(\vec{r}')da'}{|\vec{r} - \vec{r}'|} \quad (1.97)$$

Equations 1.96 and 1.97 are the generalization of Eqs. 1.92 and 1.93 for point charges. The latter can be recovered as a special case, allowing that the following relationship is:

1.6 Electrostatic Energy

$$\rho(\vec{r}) = \sum_{j=1}^{m} q_j \delta(\vec{r} - \vec{r}_j)$$

$$\rho(\vec{r}') = \sum_{k=1}^{m'} q_j \delta(\vec{r} - \vec{r}_k) \qquad (1.98)$$

where again the prime on the second summation in Eq. 1.98, an indication of the term $k = j$, is excluded when the double sum is constructed. Note that when ρ is a continuous distribution, the vanishing of the denominator in Eq. 1.97 does not cause the integral to diverge, and it is unnecessary to exclude point $\vec{r}' = \vec{r}$.

The last integral, in part, involves integration over the surface of the conductor of interest; however, because a conductor is an equipotential region, each of these integrations can be shown as:

$$\frac{1}{2} \int_{\text{conductor} j} \sigma \varphi \, da = \frac{1}{2} Q_j \varphi_j \qquad (1.99)$$

where Q_j is the charge on the jth conductor.

Equation 1.96 for *electrostatic energy of a charge distribution*, which includes a conductor, then becomes:

$$U = \frac{1}{2} \int_V \rho \varphi \, dv + \frac{1}{2} \int_{S'} \sigma \varphi \, da + \frac{1}{2} \sum_j Q_j \varphi_j \qquad (1.100)$$

where in this equation the last summation is over all conductors, and the surface integral is restricted to non-conducting surfaces.

Furthermore, in many practical problems of interest, all the charges reside on the surfaces of a conductor. In such circumstances Eq. 1.100 reduces to the following form:

$$U = \frac{1}{2} \sum_j Q_j \varphi_j \qquad (1.101)$$

Equation 1.101 is derived based on starting with uncharged macroscopic conductors that were gradually charged by bringing in charge increments. Thus, the energy described by Eq. 1.96 includes both the interaction energy between various conductors and the self-energies of the charge on each individual conductor.

1.6.3 Forces and Torques

Thus far to some extent, we have developed several alternative procedures for calculating the electrostatic energy of a charge system. We now make an attempt to establish the force on one of the objects in the charge system that can be calculated from knowledge of this electrostatic energy.

If we are dealing with an isolated system composed of conductors, point charges, and dielectrics, we have all of the necessary items to make a small displacement $d\vec{r}$ under the influence of electrical force $\vec{F}$ acting on it. The work performed by the electrical force on the system in these circumstances is:

$$dW = \vec{F} \cdot d\vec{r} = F_x dx + F_y dy + F_z dz \qquad (1.102)$$

Because we assume the system is isolated, this work is done at the expense of electrostatic energy U. In other words, according to Eq. 1.88, we can write:

$$dW = -dU \qquad (1.103)$$

Combining Eqs. 1.102 and 1.103, the result is:

$$-dU = F_x dx + F_y dy + F_z dz \qquad (1.104)$$

and

$$\begin{aligned} F_x &= -\frac{\partial U}{\partial x} \\ F_y &= -\frac{\partial U}{\partial y} \\ F_z &= -\frac{\partial U}{\partial z} \end{aligned} \qquad (1.105)$$

Therefore, the sets in Eq. 1.105 indicate that in case $\vec{F}$ is a conservative force and $\vec{F} = -\vec{\nabla} U$. If the object under consideration is constrained to move in such a way that it rotates about an axis, then Eq. 1.102 may be replaced by the following equation:

$$dW = \vec{\tau} \cdot d\vec{\theta} \qquad (1.106)$$

where $\vec{\tau}$ is the electrical torque and $d\vec{\theta}$ is the differential angular displacement. Writing $\vec{\tau}$ and $d\vec{\theta}$ in terms of their components, $(\tau_1, \tau_2, \tau_2,)$ and $(d\theta_1, d\theta_2, d\theta_3)$, and combining Eqs. 1.103 and 1.106, we obtain the following relationships:

$$\begin{aligned} \tau_1 &= -\frac{\partial U}{\partial \theta_1} \\ \tau_2 &= -\frac{\partial U}{\partial \theta_2} \\ \tau_3 &= -\frac{\partial U}{\partial \theta_3} \end{aligned} \qquad (1.107)$$

1.6 Electrostatic Energy

This proves that our goal has been achieved and we can write:

$$\begin{cases} F_x = -\left(\dfrac{\partial U}{\partial x}\right)_Q \\ \tau_1 = -\left(\dfrac{\partial U}{\partial \theta_1}\right)_Q \end{cases} \quad (1.108\text{a})$$

$$\begin{cases} F_y = -\left(\dfrac{\partial U}{\partial y}\right)_Q \\ \tau_2 = -\left(\dfrac{\partial U}{\partial \theta_2}\right)_Q \end{cases} \quad (1.108\text{b})$$

$$\begin{cases} F_z = -\left(\dfrac{\partial U}{\partial x}\right)_Q \\ \tau_3 = -\left(\dfrac{\partial U}{\partial \theta_3}\right)_Q \end{cases} \quad (1.108\text{c})$$

where subscript Q has been added to denote that the system is isolated; thus, its total charge remains constant during the displacement of $d\vec{r}$ or $d\vec{\theta}$.

Now we are at the stage where we need to talk about EM force that is known as the *Lorentz force* here. The EM field exerts the following force on charged particles:

$$\vec{F} = q\vec{E} + q\vec{v} \times \vec{B} \quad (1.109)$$

where vector $\vec{F}$ is the force that a particle with charge q experiences, $\vec{E}$ is the electric field at the location of the particle, v is the velocity of the particle, and $\vec{B}$ is the magnetic field at the location of the particle.

The preceding equation illustrates that the Lorentz force is the sum of two vectors. One is the cross product of the velocity and the magnetic field vectors. Based on the properties of the cross product, this produces a vector that is perpendicular to both the velocity and the magnetic field vectors. The other vector goes in the same direction as the electric field. The sum of these two vectors is the Lorentz force.

Therefore, in the absence of a magnetic field, the force is in the direction of the electric field, and the magnitude of the force is dependent on the value of the charge and the intensity of the electric field. In the absence of an electric field, the force is perpendicular to the velocity of the particle and the direction of the magnetic field. If both electric and magnetic fields are present, the Lorentz force is the sum of both these vectors.

Therefore, in summary, we can express that the classical theory of electrodynamics is built on Maxwell's equations and the concepts of EM field, force, energy, and momentum, which are intimately tied together by *Poynting's theorem* and the Lorentz force law. Whereas Maxwell's macroscopic equations relate electric and magnetic fields to their material sources (i.e., charge, current, polarization, and magnetization), Poynting's theorem governs the flow of EM energy and its exchange between fields and material media, while the Lorentz Law regulates the back-and-forth transfer of momentum between the media and the fields. As it turns out, an alternative force law, first proposed in 1908 by Einstein and Laub, exists that is consistent with Maxwell's macroscopic equations and complies with conservation laws as well as with the requirements of special relativity.

Although Lorentz's Law requires the introduction of hidden energy and hidden momentum in situations where an electric field acts on magnetic material, the Einstein-Laub formulation of EM force and torque does not invoke hidden entities under such circumstances. Moreover, the total force and the total torque exerted by EM fields on any given object turn out to be independent of whether force and torque densities are evaluated using the Lorentz Law or in accordance with the Einstein-Laub formulas. Hidden entities aside, the two formulations differ only in their predicted force and torque distributions throughout material media. Such differences in distribution occasionally are measurable and could serve as a guide in deciding which formulation, if either, corresponds to physical reality [1].

Furthermore, to have some general idea about Poynting's theorem, we can say that, in electrodynamics, his theorem is a statement of conservation of energy for the EM field. Moreover, it is in the form of a partial differential equation because of British physicist John Henry Poynting. Poynting's theorem is analogous to the *work–energy theorem* in classical mechanics, and mathematically like the continuity equation because it relates the energy stored in the EM field to the work done on a charge distribution (i.e., an electrically charged object) through energy flux. The details of deriving this theorem is beyond the scope of this book; readers should refer to other classical books on electrodynamics.

In general, however, we can say that this theorem is an energy balance and the following statement does apply:

The rate of energy transfer (per unit volume) from a region of space equals the rate of work done on a charge distribution plus the energy flux leaving that region.

A second statement can also explain the theorem: "The decrease in the electromagnetic energy per unit time in a certain volume is equal to the sum of work done by the field forces and the net outward flux per unit time." Mathematically, the previous statement can be expressed, and is summarized in differential form as follows:

$$-\frac{\partial u}{\partial t} = \vec{\nabla} \cdot \vec{S} + \vec{J} \cdot \vec{E} \tag{1.110}$$

where $\vec{\nabla} \cdot \vec{S}$ is the divergence of the Poynting vector or energy flow and $\vec{J} \cdot \vec{E}$ is the rate at which the fields do work on a charged object ($\vec{J}$ is the free current density corresponding to the motion of charge, $\vec{E}$ is the electric field, and • is the dot product). The energy density u is given by

$$u = \frac{1}{2}(\vec{E} \cdot \vec{D} + \vec{B} \cdot \vec{H}) \tag{1.111}$$

In this equation $\vec{D}$ is the electric displacement filed, $\vec{B}$ is the magnetic flux density, and $\vec{H}$ is the magnetic field strength. Because only some of the charges are free to move, and the $\vec{D}$ and $\vec{H}$ fields exclude the "bound" charges and currents in the charge distribution (by their definition), one obtains free current density $\vec{J}_f$ in Poynting's theorem rather than total current density $\vec{J}$.

The integral form of Poynting's theorem can be established via utilization of the divergence theorem expressed before as:

$$-\frac{\partial u}{\partial t}\int_V u dV = \oiint_{\partial V} \vec{S} \cdot d\vec{A} + \int_V \vec{J} \cdot \vec{E} dV \qquad (1.112)$$

where ∂V is the boundary of volume V and the shape of the volume is arbitrary but is fixed for the calculation.

Summarizing of the previous two sections in this chapter, we present the perspectives shown in Fig. 1.10.

1.7 Description of Maxwell's Equations

To understand the physics and the mathematics of waves and wave equations as well as associated subjects (e.g., mechanical wave, pressure wave, transverse wave, longitudinal wave, and, consequently, scalar wave theory), we need to have some understanding of the sets of equations that are known as *Maxwell's equations*. We refer to them here briefly, and later we expand on them; the next chapter describes them in further detail.

Maxwell's equations are four vector equations:

I. Gauss's Law for electric fields
II. Gauss's Law for magnetic fields
III. Friday's Law
IV. Ampère-Maxwell Law

Each of them can be written in integral or differential form. The integral forms describe the behavior of electric and magnetic fields over surfaces or around paths, while the differential forms apply to specific locations. Both forms are relevant to EM waves, but the travel from Maxwell's equations to wave equations is somewhat more direct if you start with the differential form.

To take advantage of that shortcut travel, it is necessary to understand vector calculus and differential operators, as we have covered in previous sections earlier in this chapter—that is, called "del" (or "nabla") and written as $\vec{\nabla}$, which is a differential operator telling you to take partial spatial derivatives (e.g., $\partial/\partial x$, $\partial/\partial y$, and $\partial/\partial z$) of the function on which one is operating using a Cartesian coordinate.

At this point we are ready to introduce the importance of Maxwell's EM theory as a brief course, the so-called *displacement current*. Now, we will write all classical (i.e., non-quantum EM phenomena) that are governed by Maxwell's equations, which take the form as follows [2]:

I. Gauss's Law for Electric Fields: $\vec{\nabla} \cdot \vec{E} = \dfrac{\rho}{\varepsilon_0}$

$$\vec{\nabla} \cdot \vec{E} = \dfrac{\rho}{\varepsilon_0} \qquad \text{Also known as Coulomb's Law} \qquad (1.113)$$

Gauss's Law for electric fields states that the divergence ($\vec{\nabla}$) of the electric field ($\vec{E}$) at any location is proportional to the electric charge density (ρ) at that location. That is because electric-static field lines begin on a positive charge and end on a negative charge; thus, the field lines tend to diverge away from locations of positive charge and coverage toward locations of negative charge. The symbol ε_0 represents the electric permittivity of free space, a quantity that you will see again when we consider the phase speed and impedance of EM waves.

II. Gauss's Law for Magnetic Fields: $\vec{\nabla} \cdot \vec{B} = 0$

$$\vec{\nabla} \cdot \vec{B} = 0 \qquad \text{Also known as Gauss's Law} \qquad (1.114)$$

Gauss's Law for magnetic fields designates that the divergence ($\vec{\nabla}$) of the magnetic field ($\vec{B}$) at any location must be zero. This is true because there apparently is no isolated "magnetic charge" in the Universe, so magnetic field-lines neither diverge nor converge, and they circulate back on themselves.

III. Faraday's Law: $\vec{\nabla} \times \vec{E} = -\dfrac{\partial \vec{B}}{\partial t}$

$$\vec{\nabla} \times \vec{E} = -\dfrac{\partial \vec{B}}{\partial t} \qquad \text{Also known as Faraday's Law} \qquad (1.115)$$

Faraday's Law indicates that the *curl* ($\vec{\nabla} \times$) of the electric field ($\vec{E}$) at any location is equal to the negative of the time rate of change of the magnetic field ($\partial \vec{B}/\partial t$) at that location. That is because a changing of magnetic field produces a circulating electric field.

IV. Ampère-Maxwell Law: $\vec{\nabla} \times \vec{B} = \mu_0 \vec{J} + \mu_0 \varepsilon_0 \dfrac{\partial \vec{E}}{\partial t}$

$$\vec{\nabla} \times \vec{B} = \mu_0 \vec{J} + \mu_0 \varepsilon_0 \dfrac{\partial \vec{E}}{\partial t} \qquad \text{Also known as Ampere's Law} \qquad (1.116)$$

The Ampère-Maxwell Law, or from now on Ampère's Law as modified by Maxwell, tells us that the curl ($\vec{\nabla} \times$) of the ($\vec{B}$) at any location is proportional to the electric current density ($\vec{J}$) plus the time rate of change of the electric field ($\partial \vec{E}/\partial t$) at that location. This is the case because a circulating magnetic field is produced both by an electric current and by a changing electric field. Note that the term involving the changing electric field is the "displacement current" that was added to Ampère's Law by James Clerk Maxwell.

1.7 Description of Maxwell's Equations

The symbol μ_0 represents the magnetic permeability of free space, another quantity that you will see when we consider the speed of EM waves and electromagnetic impedance.

Notice that Maxwell's equations relate the spatial behavior of fields to the source of those fields. Those electric field sources change with density ρ, appearing in Gauss's Law for electric fields; electrical current with density $\vec{J}$, appearing in the Ampère-Maxwell Law; and changing magnetic field with time derivative $\partial \vec{E}/\partial t$, appearing in the Ampère-Maxwell Law.

All the quantities in the equations are defined as before. Here $\vec{E}(\vec{r},t)$, $\vec{B}(\vec{r},t)$, $\rho(\vec{r},t)$, and $\vec{J}(\vec{r},t)$ represent the *electric field strength*, the *magnetic field strength*, the *electrical charge density*, and the *electrical current density*, respectively. Moreover, $\varepsilon_0 = 8.8542 \times 10-12 \ C^2 N^{-1} m^{-2}$ is the *electric permittivity* of free space, whereas $\mu_0 = 4\pi \times 10^{-7} \ N \ A^{-2}$ is the *magnetic permeability* of free space.

As is well known, Eq. 1.113 is equivalent to *Coulomb's Law* for the electric fields generated by point charges. Equation 1.114 is equivalent to the statement that magnetic monopoles do not exist, which implies that magnetic field lines can never begin or end. Equation 1.115 is equivalent to Faraday's Law of *electromagnetic induction*. Finally, Eq. 1.116 is equivalent to the Biot-Savart Law for the magnetic fields generated by line currents and augmented by the induction of magnetic fields by changing electric fields.

Maxwell's equations are linear in nature. In other words, if $\rho \to \alpha\rho$ and $\vec{J} \to \alpha\vec{J}$, where α is an arbitrary spatial and temporal constant, then it is clear from Eqs. 1.113 to 1.116 that $\vec{E} \to \alpha\vec{E}$ and $\vec{B} \to \alpha\vec{B}$. The linearity of Maxwell's equations accounts for the well-known fact that the electric fields generated by point charges, as well as the magnetic fields generated by line currents, are superposable.

Taking the divergence of Eq. 1.113, and combining the resulting expression with Eq. 1.113, we obtain:

$$\frac{\partial \rho}{\partial t} + \vec{\nabla} \cdot \vec{J} = 0 \qquad (1.117)$$

In integral form, making use of the divergence theorem, this equation becomes the following:

$$\frac{d}{dt}\int_V \rho dV + \int_S \vec{J} \cdot d\vec{S} = 0 \qquad (1.118)$$

where V is a fixed volume bounded by surface S. The volume integral represents the net electric charge contained within the volume, whereas the surface integral represents the outward flux of charge across the bounding surface.

The previous equation, which states that the net rate of change of the charge contained within volume V is equal to minus the net flux of charge across bounding surface S, is clearly a statement of the *conservation of electrical charge*. Thus, Eq. 1.117 is the differential form of this conservation equation.

As is well known, point electric q moving with velocity $\vec{v}$ in the presence of electric field $\vec{E}$ and magnetic field $\vec{B}$ experiences a force that is known as a Lorentz force and was expressed before by Eq. 1.109. Likewise, a distributed of charge density ρ and current density $\vec{J}$ experiences a force density that is given as

$$\vec{f} = \rho \vec{E} + \vec{J} \times \vec{B} \tag{1.119}$$

This is the extent of our presentation for Maxwell's equations within this book; further information about these equations can be found in any classical electrodynamics book [1]. Maxwell's equations can be summarized as follows:

Maxwell's Equations

In general, In matter

$$\begin{cases} \vec{\nabla} \cdot \vec{E} = \dfrac{\rho}{\varepsilon_0} \\ \vec{\nabla} \cdot \vec{E} = -\dfrac{\partial \vec{B}}{\partial t} \\ \vec{\nabla} \cdot \vec{B} = 0 \\ \vec{\nabla} \cdot \vec{B} = \mu_0 \vec{J} + \mu_0 \varepsilon_0 \dfrac{\partial \vec{E}}{\partial t} \end{cases} \qquad \begin{cases} \vec{\nabla} \cdot \vec{D} = \rho_f \\ \vec{\nabla} \times \vec{E} = -\dfrac{\partial \vec{B}}{\partial t} \\ \vec{\nabla} \cdot \vec{B} = 0 \\ \vec{\nabla} \times \vec{H} = \vec{J}_f + \dfrac{\partial \vec{D}}{\partial t} \end{cases} \tag{1.120}$$

Auxiliary Field

Definitions Linear media

$$\begin{cases} \vec{D} = \varepsilon_0 \vec{E} + \vec{P} \\ \vec{H} = \dfrac{1}{\mu_0} \vec{B} - \vec{M} \end{cases} \qquad \begin{cases} \vec{P} = \varepsilon_0 \chi_e \vec{E}, \quad \vec{D} = \varepsilon \vec{E} \\ \vec{M} = \chi \vec{H}, \quad \vec{H} = \dfrac{1}{\mu} \vec{B}. \end{cases} \tag{1.121}$$

where by appropriate *constitutive relations*, giving $\vec{D}$ and $\vec{H}$ in terms of $\vec{E}$ and $\vec{B}$. These, of course, depend on the nature of the material and for a linear media they are given by sets of Eq. 1.121, where:

$$\begin{aligned} \varepsilon &\equiv \varepsilon_0 (1 + \chi_e) \\ \mu &\equiv \mu_0 (1 + \chi_m) \end{aligned} \tag{1.122}$$

where χ_m is a dimensionless scalar quantity that is called the *magnetic susceptibility*, while quantity χ_e is the *electric susceptibility* of materials (i.e., conductors).

Note that, although χ_m is a function of temperature, and sometimes varies quite drastically with it, generally it is safe to say that χ_m for parametric and diamagnetic materials is quite small—that is, $\chi_m \ll 1$. Identically, we can remember from our study of EMs hat $\vec{D}$ is called the *electric displacement*; that is why the second term

1.7 Description of Maxwell's Equations

in the Ampere-Maxwell equations, $\nabla \times \vec{H} = \vec{J}_f + \frac{\partial \vec{D}}{\partial t}$, came to be called the *displacement current*. In this context,

$$\vec{J}_d \equiv \frac{\partial \vec{D}}{\partial t} \tag{1.123}$$

The parameters and relationships that we need to be aware of for further use in the following chapter are:

Potentials

$$\begin{cases} \vec{E} = -\nabla V - \frac{\partial \vec{A}}{\partial t} \\ \vec{B} = \nabla \times \vec{A} \end{cases} \tag{1.124}$$

Lorentz Force Law

$$\vec{F} = q(\vec{E} + \vec{v} \times \vec{B}) \tag{1.125}$$

Energy, Momentum, and Power

$$\begin{cases} \text{Energy:} & U = \frac{1}{2} \int \left(\varepsilon_0 E^2 + \frac{1}{\mu_0} B^2 \right) \\ \text{Momentum:} & \vec{P} = \varepsilon_0 \int (\vec{E} \times \vec{B}) d\tau \\ \text{Poynting Vector:} & \vec{S} = \frac{1}{\mu_0} (\vec{E} \times \vec{B}) \\ \text{Larmor Formula:} & P = \frac{\mu_0}{6\pi c} q^2 a^2 \end{cases} \tag{1.126}$$

Here we also note the following fundamental constants and their values, in *free space* at least:

- Permittivity of free space: $\varepsilon_0 = 8.85 \times 10^{-12} C^2/Nm^2$
- Permeability of free space: $\mu_0 = 4\pi \times 10^{-7} N/A^2$
- Speed of light in vacuum: $c = 3.00 \times 10^8 m/sec$
- Charge of the electron: $e = 1.60 \times 10^{-19} C$
- Mass of the electron: $m = 9.11 \times 10^{-31} kg$

We also are expressing some of the spherical and cylindrical coordinates' mathematical relationship to the Cartesian coordinate, as a reminder to readers here as well.

Here we use Fig. 1.11 to show the spherical coordinate and Fig. 1.12 for the cylindrical coordinate.

Fig. 1.11 Spherical coordinate in respect to the Cartesian coordinate

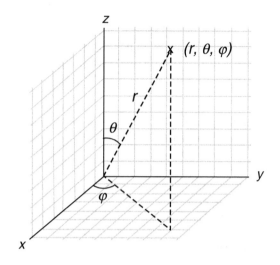

Fig. 1.12 Cylindrical coordinate in respect to the Cartesian coordinate

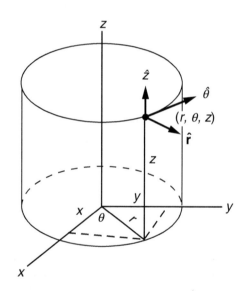

Spherical

$$\begin{cases} x = r \sin\theta \cos\varphi \\ y = r \sin\theta \sin\varphi \\ z = r \cos\theta \end{cases} \quad \begin{cases} \hat{x} = \sin\theta \cos\varphi \hat{r} + \cos\theta \cos\varphi \hat{\theta} - \sin\varphi \hat{\varphi} \\ \hat{y} = \sin\theta \sin\varphi \hat{r} + \cos\theta \sin\varphi \hat{\theta} + \cos\varphi \hat{\varphi} \\ \hat{z} = \cos\theta \hat{r} - \sin\theta \hat{\theta} \end{cases}$$

(1.127a)

$$\begin{cases} r = \sqrt{x^2+y^2+z^2} \\ \theta = \tan^{-1}\left(\sqrt{x^2+y^2}/z^2\right) \\ \varphi = \tan^{-1}(y/x) \end{cases} \qquad \begin{cases} \hat{r} = \sin\theta\cos\varphi\hat{x} + \sin\theta\sin\varphi\hat{y} + \cos\theta\hat{z} \\ \hat{\theta} = \cos\theta\cos\varphi\hat{x} + \cos\theta\sin\varphi\hat{y} - \sin\theta\hat{z} \\ \hat{\varphi} = -\sin\varphi\hat{x} + \cos\varphi\hat{y} \end{cases}$$

(1.127b)

Cylindrical

$$\begin{cases} x = r\cos\theta \\ y = r\sin\theta \\ z = z \end{cases} \qquad \begin{cases} \hat{x} = \cos\theta\hat{r} - \sin\theta\hat{\theta} \\ \hat{y} = \sin\theta\hat{r} + \cos\theta\hat{\theta} \\ \hat{z} = \hat{z} \end{cases} \quad (1.128a)$$

$$\begin{cases} r = \sqrt{x^2+y^2} \\ \theta = \tan^{-1}(y/x) \\ z = z \end{cases} \qquad \begin{cases} \hat{r} = \cos\theta\hat{x} + \sin\theta\hat{y} \\ \hat{\theta} = -\sin\theta\hat{x} + \cos\theta\hat{y} \\ \hat{z} = \hat{z} \end{cases} \quad (1.128b)$$

1.8 Time-Independent Maxwell's Equations

In this section we will take the familiar force laws of electrostatics and magnetostatics and recast them as vector field equations.

1.8.1 Coulomb's Law

According to this law, the force acting between two electric charges is radial, inverse-square, and proportional to the product of the charges. Two like charges repel one another, whereas two unlike charges attract. Suppose that two charges, q_1 and q_2, are located at position vectors $\vec{r}_1$ and $\vec{r}_2$. The electrical force acting on the second charge is written as:

$$\hat{F}_2 = \frac{q_1 q_2}{4\pi\varepsilon_0} \frac{\vec{r}_2 - \vec{r}_1}{|\vec{r}_2 - \vec{r}_1|^3} \quad (1.129)$$

Equation 1.129 is written in vector notation as illustrated in Fig. 1.13.

An equal and opposite force acts on the first charge in accordance with Newton's third law of motion. The SI unit of electrical charge is the Coulomb (C). The magnitude of the charge on an electron is 1.6022×10^{-19} C. The universal constant ε_0 is called the permittivity of free space and takes that value.

Coulomb's Law has the same mathematical form as Newton's law of gravity. Suppose that two masses, m_1 and m_2, are located at position vectors $\vec{r}_1$ and $\vec{r}_2$. The gravitational force acting on the second mass is written:

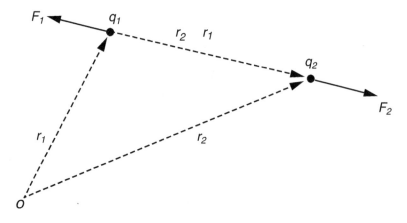

Fig. 1.13 Illustration to charges apart from each other

$$\vec{F}_2 = -Gm_1m_2 \frac{\vec{r}_2 - \vec{r}_1}{|\vec{r}_2 - \vec{r}_1|^3} \qquad (1.130)$$

This equation is written in vector notation and gravitation G takes the value of $G = 6.6726 \times 10^{-11}\,\mathrm{N\,m^2\,kg^{-2}}$.

Coulomb's Law and Newton's Law are both *inverse-square* force laws—that is,

$$|\vec{F}_2| \propto \frac{1}{|\vec{r}_2 - \vec{r}_1|^2} \qquad (1.131)$$

They differ in two crucial respects however. First, the force owing to gravity is always *attractive* (i.e., there is no such thing as a negative mass). Second, the magnitudes of the two forces are vastly different. Consider the ratio of the electrical and gravitational forces acting on two particles. This ratio is a constant, independent of the relative positions of the particles, and is given by

$$\frac{|\vec{F}_{\text{electrical}}|}{|\vec{F}_{\text{gravitational}}|} = \frac{q_1}{m_1}\frac{q_2}{m_2}\frac{1}{4\pi\varepsilon_0 G} \qquad (1.132)$$

For electrons, the charge to mass ratio is $(q/m) = 1.759 \times 10^{11}\,\mathrm{C\,kg^{-1}}$, so

$$\frac{|\vec{F}_{\text{electrical}}|}{|\vec{F}_{\text{gravitational}}|} = 4.17 \times 10^{42} \qquad (1.133)$$

This is a colossal number! Suppose we are studying a physics problem involving the motion of particles in a box under the action of two forces with the same range but differing in magnitude by a factor 10^{42}. It would seem a plausible approximation (to say the least) to start the investigation by neglecting the weaker force. Applying this reasoning to the motion of particles in the Universe, we would expect it to be

1.8 Time-Independent Maxwell's Equations

governed entirely by electrical forces. Yet, this is not the case. The force that holds us to the surface of the Earth, preventing us from floating off into space, is gravity. The force that causes the Earth to orbit the Sun is also gravity.

In fact, on astronomical length-scales gravity is the dominant force, and electrical forces are largely irrelevant. The key to understanding this paradox is that there are both positive and negative electrical charges, whereas there are only positive gravitational "charges." This means that gravitational forces are always cumulative, whereas electrical forces can cancel one another out. Suppose, for the sake of argument, that the Universe starts out with randomly distributed electrical charges. Initially, we expect electrical forces to completely dominate gravity. These forces try to make every positive charge get as far away as possible from the other positive charges, and as close as possible to the other negative charges. After a while, we expect the positive and negative charges to form close pairs. Just how close is determined by quantum mechanics; however, in general, it is very close (i.e., about 10^{-10}).

The electrical forces because of the charges in each pair effectively cancel one another out on length-scales much larger than the mutual spacing of the pair. It is only possible for gravity to be the dominant long-range force if the number of positive charges in the Universe is almost equal to the number of negative charges. In this situation, every positive change can find a negative charge to team up with, and there are virtually no charges left over. For the cancellation of long-range electrical forces to be effective, the relative difference in the number of positive and negative charges in the Universe must be incredibly small. In fact, positive and negative charges have to cancel each other out to such accuracy that most physicists believe that the net charge of the Universe is *exactly* zero.

Nonetheless, it is not enough for the Universe to start out with a zero charge. Suppose there were some elementary particle processes that did not conserve electrical charges. Even if this were to go on at a very low rate, it would not take long before the fine balance between positive and negative charges in the Universe was exhausted. So, it is important that an electrical charge is a *conserved* quantity (i.e., the net charge of the Universe can neither increase nor decrease). As far as we know, this is the case. To date, no elementary particle reactions have been discovered that create or destroy net electrical charges.

In summary, there are two long-range forces in the Universe, electromagnetism and gravity. The former is enormously stronger than the latter but is usually "hidden" away inside neutral atoms. The fine balance of forces because of negative and positive electrical charges starts to break down on atomic scales. In fact, interatomic and intermolecular forces are all electrical in nature. So, electrical forces are basically what prevent us from falling though the floor. But, this is *electromagnetism* on the microscopic or atomic scale—what is usually termed *quantum electromagnetism*. This book is about *classical electromagnetism*—that is, electromagnetism on length scales is much larger than the atomic scale.

Classical electromagnetism generally describes phenomena in which some sort of "violence" is done to matter so that the close pairing of negative and positive charges is disrupted. This allows electrical forces to manifest themselves on macroscopic length scales. Of course, very little disruption is necessary before gigantic forces are generated. It is no coincidence that the clear majority of useful machines that humankind has devised during the last century or so are electrical in nature.

Coulomb's and Newton's Laws both are examples of what are usually referred to as *action at a distance* theories. According to Eqs. 3.124 and 3.125 in Chap. 3, if the first charge or mass is moved then the force acting on the second charge or mass immediately responds. In particular, equal and opposite forces act on the two charges or masses at all times. Yet, this cannot be correct according to Einstein's theory of relativity, which implies that the maximum speed with which information can propagate through the Universe is the speed of light in a vacuum. So, if the first charge or mass is moved, then there must always be a time delay (i.e., at least the time needed for a light signal to propagate between the two charges or masses) before the second charge or mass responds.

Consider a rather extreme example. Suppose the first charge or mass is suddenly annihilated. The second charge or mass only finds out about this sometime later. During this time interval, the second charge or mass experiences an electrical or gravitational force that is as if the first charge or mass were still there. So, during this period, there is an action but no reaction, which violates Newton's third law of motion. It is clear that action at a distance is not compatible with relativity; consequently, *Newton's third law of motion* is not strictly true. Of course, Newton's third law is closely tied up with the conservation of linear momentum in the Universe.

This is a concept that most physicists are loath to abandon. It turns out that we can "rescue" momentum conservation by abandoning action at a *distance theory*, and instead adopt so-called *field theories* in which there is a medium, called a field, that transmits the force from one particle to another. In electromagnetism there are, in fact, two fields: (1) the electric field and (2) the magnetic field. Electromagnetic forces are transmitted via these fields at the speed of light, which implies that the laws of relativity are never violated. Moreover, the fields can soak up energy and momentum. This means that even when the actions and reactions acting on particles are not quite equal and opposite, momentum still is conserved. We can bypass some of the problematic aspects of action at a distance by only considering *steady-state* situations. For the moment, this is how we will proceed.

Consider N charges, q_1 through q_N, which are located at position vectors $\vec{r}_1$ through $\vec{r}_N$. Electrical forces obey what is known as the *principle of superposition*. The electrical force acting on test charge q at position vector $\vec{r}$ is simply the vector sum of all of the Coulomb Law forces from each of the N charges taken in isolation. In other words, the electrical force exerted by the ith charge (say) on the test charge is the same as if all the other charges were not there. Thus, the force acting on the test charge is given by

$$\vec{F}(\vec{r}) = q \sum_{i=1}^{N} \frac{q_i}{4\pi\varepsilon_0} \frac{\vec{r} - \vec{r}_i}{|\vec{r} - \vec{r}_i|^3} \quad (1.134)$$

1.8 Time-Independent Maxwell's Equations

It is helpful to define vector field $\vec{E}(\vec{r})$, called the *electric field*, which is the force exerted on a unit test charge located at position vector $\vec{r}$. Thus, the force on a test charge is written as

$$\vec{F} = q\vec{E} \tag{1.135}$$

and the electric field is given as

$$\vec{E}(\vec{r}) = \sum_{i=1}^{N} \frac{q_i}{4\pi\varepsilon_0} \frac{\vec{r} - \vec{r}_i}{|\vec{r} - \vec{r}_i|^3} \tag{1.136}$$

At this point, we have no reason to believe that the electric field has any real physical existence. It is just a useful device for calculating the force that acts on test charges placed at various locations.

The electric field from a single charge q located at the origin is purely radial, points outward if the charge is positive, inward if it is negative, and has magnitude of:

$$E_r(r) = \frac{q}{4\pi\varepsilon_0 r^2} \tag{1.137}$$

where $r = |\vec{r}|$.

We can represent an electric field by *field lines*. The direction of the lines indicates the direction of the local electric field, and the density of the lines perpendicular to this direction is proportional to the magnitude of the local electric field. Thus, the field of a point positive charge is represented by a group of equally spaced straight lines radiating from the charge (Fig. 1.14).

The electric field from a collection of charges is simply the vector sum of the fields from each of the charges taken in isolation. In other words, electric fields are completely *superposable*. Suppose that, instead of having discrete charges, we have

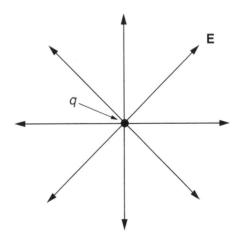

Fig. 1.14 Illustration of lines radiating from a charge

a continuous distribution of charges represented by *charge density* $\rho(\vec{r})$. Thus, the charge at position vector $\vec{r}'$ is $\rho(\vec{r}')d^3\vec{r}'$, where $d^3\vec{r}'$ is the volume element at $\vec{r}'$. It follows from a simple extension of Eq. 1.131 that the electric field generated by this charge distribution is given as

$$\vec{E}(\vec{r}) = \frac{1}{4\pi\varepsilon_0} \int \rho(\vec{r}') \frac{\vec{r} - \vec{r}'}{|\vec{r} - \vec{r}'|^3} d^3\vec{r}' \tag{1.138}$$

where the volume integral is over all space or, at least, over all space for which $\rho(\vec{r}')$ is non-zero.

1.8.2 The Electric Scalar Potential

Suppose that $\vec{r} = (x, y, z)$ and $\vec{r}' = (x', y', z')$ are in Cartesian coordinates. The x component of $(\vec{r} - \vec{r}')/|\vec{r} - \vec{r}'|^3$ is written as

$$\frac{x - x'}{\left[(x-x')^2 + (y-y')^2 + (z-z')^2\right]^{3/2}} \tag{1.139}$$

However, it is demonstrated easily that

$$\frac{x - x'}{\left[(x-x')^2 + (y-y')^2 + (z-z')^2\right]^{3/2}} = -\frac{\partial}{\partial x}\left(\frac{1}{\left[(x-x')^2 + (y-y')^2 + (z-z')^2\right]^{1/2}}\right) \tag{1.140}$$

Because there is nothing special about the x-axis, we can write:

$$\frac{\vec{r} - \vec{r}'}{|\vec{r} - \vec{r}'|^3} = -\nabla\left(\frac{1}{|\vec{r} - \vec{r}'|}\right) \tag{1.141}$$

where $\nabla \equiv (\partial/\partial x, \partial/\partial y, \partial/\partial z)$ is a differential operator that involves the components of $\vec{r}$ but not those of $\vec{r}'$.

It follows from Eq. 1.133 that:

$$\vec{E} = -\nabla\phi(\vec{r}) \tag{1.142}$$

1.8 Time-Independent Maxwell's Equations

where

$$\phi(\vec{r}) = \frac{1}{4\pi\varepsilon_0} \int \frac{\rho(\vec{r}')}{|\vec{r}-\vec{r}'|} d^3\vec{r}' \qquad (1.143)$$

Thus, the electric field generated by a collection of fixed charges can be written as the gradient of a scalar potential, and this potential can be expressed as a simple volume integral involving the charge distribution.

The scalar potential generated by charge q located at the origin is:

$$\phi(r) = \frac{q}{4\pi\varepsilon_0 r} \qquad (1.144)$$

According to Eq. 1.136, the scalar potential generated by a set of N discrete charges q_i, located at $\vec{r}_i$, is:

$$\phi(\vec{r}) = \sum_{i=1}^{N} \phi_i(\vec{r}) \qquad (1.145)$$

where

$$\phi_i(\vec{r}) = \frac{q_i}{4\pi\varepsilon_0 |\vec{r}-\vec{r}_i|} \qquad (1.146)$$

Thus, the scalar potential is just the sum of the potentials generated by each of the charges taken in isolation.

Suppose that a particle of charge q is taken along some path from point A to point B. The net work done on the particle by electrical forces is:

$$W = \int_A^B \vec{F} \cdot d\vec{l} \qquad (1.147)$$

where $\vec{F}$ is the electrical force and $d\vec{l}$ is a line element along the path from point A to point B. Making use of Eqs. 1.130 and 1.137, we obtain:

$$W = q \int_A^B \vec{E} \cdot d\vec{l} = -q \int_A^B \nabla\phi \cdot d\vec{l} = -q[\phi(B) - \phi(A)] \qquad (1.148)$$

Thus, the work done on the particle is simply minus its charge times the difference in electric potential between the end point and the beginning point. This quantity clearly is independent of the path taken between A and B. So, an electric field generated by stationary charges is an example of a conservative field. In fact, this result follows immediately from vector field theory once we are told, in Eq. 1.142, that the electric field is the gradient of a scalar potential. The work done on the particle when it is taken around a closed loop is zero so:

$$\oint_C \vec{E} \cdot d\vec{l} = 0 \qquad (1.149)$$

for any closed loop C. This implies, from Stokes's theorem, that:

$$\vec{\nabla} \times \vec{E} = 0 \qquad (1.150)$$

for any electric field generated by stationary charges. Equation 1.150 also follows directly from Eq. 1.138 because $\vec{\nabla} \times \vec{\nabla}\phi = 0$ for any scalar potential ϕ. The SI unit of electric potential is the volt (V), which is equivalent to a joule (J) per Coulomb. Thus, according to Eq. 1.148, the electrical work done on a particle when it is taken between two points is the product of its charge and the voltage difference between the points.

We are familiar with the idea that a particle moving in a gravitational field possesses potential energy as well as kinetic energy. If the particle moves from point A to a lower point B, then the gravitational field does work on the particle causing its kinetic energy to increase. The increase in kinetic energy of the particle is balanced by an equal decrease in its potential energy so that the overall energy of the particle is a conserved quantity. Therefore, the work done on the particle as it moves from A to B is *minus* the difference in its gravitational potential energy between points B and A. Of course, it only makes sense to talk about gravitational potential energy because the gravitational field is conservative. Thus, the work done in taking a particle between two points is path-independent and, therefore, well defined. This means that the difference in potential energy of the particle between the beginning and end points also is well defined.

We already have seen that an electric field generated by stationary charges is a conservative field. It follows that we can define an electrical potential energy of a particle moving in such a field. By analogy with gravitational fields, the work done in taking a particle from point A to point B is equal to minus the difference in potential energy of the particle between points B and A. It follows from Eq. 1.148 that the potential energy of the particle at a general point B, relative to some reference point A (where the potential energy is set to zero), is given by

$$\mathcal{E}(B) = q\phi(B) \qquad (1.151)$$

Free particles try to move down gradients of potential energy in order to attain a minimum potential energy state. Thus, free particles in the Earth's gravitational field tend to fall downward. Likewise, positive charges moving in an electric field tend to migrate toward regions with the most negative voltage and vice versa for negative charges.

The scalar electric potential is undefined to an additive constant. So, the transformation is given by

$$\phi(\vec{r}) \rightarrow \phi(\vec{r}) + c \qquad (1.152)$$

which leaves the electric field unchanged according to Eq. 1.142. The potential can be fixed unambiguously by specifying its value at a single point. The usual

1.8 Time-Independent Maxwell's Equations

convention is to say that the potential is zero at infinity. This convention is implicit in Eq. 1.138, where it can be seen that $\phi \to 0$ as $|\vec{r}| \to \infty$, provided that total charge $\int \rho(\vec{r}')d^3\vec{r}'$ is finite.

1.8.3 Gauss's Law

Consider a single charge located at the origin. The electric field generated by such a charge is given by Eq. 1.137. Suppose that we surround the charge by concentric spherical surface S of radius r (Fig. 1.15). The flux of the electric field through this surface is given by

$$\oint_S \vec{E} \cdot d\vec{S} = \oint_S E_r dS_r = E_r(r)4\pi r^2 = \frac{q}{4\pi\varepsilon_0 r^2} 4\pi r^2 = \frac{q}{\varepsilon_0} \quad (1.153)$$

because what is normal to the surface is always parallel to the local electric field. Nevertheless, we also know from Gauss's theorem that

$$\oint_S \vec{E} \cdot d\vec{S} = \int_V \vec{\nabla} \cdot \vec{E} d^3\vec{r} \quad (1.154)$$

where V is the volume enclosed by surface S. Let us evaluate $\vec{\nabla} \cdot \vec{E}$ directly. In Cartesian coordinates, the field is written as

$$\vec{E} = \frac{q}{4\pi\varepsilon_0}\left(\frac{x}{r^3}, \frac{y}{r^3}, \frac{z}{r^3}\right) \quad (1.155)$$

where $r^2 = x^2 + y^2 + z^2$; thus, we have

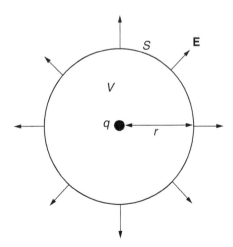

Fig. 1.15 Illustration of a concentric spherical shape

$$\frac{\partial E_x}{\partial x} = \frac{q}{4\pi\varepsilon_0}\left(\frac{1}{r^3} - \frac{3xx}{r^4\,r}\right) = \frac{q}{4\pi\varepsilon_0}\frac{r^2 - 3x^2}{r^5} \qquad (1.156)$$

Here, we have used the following relation:

$$\frac{\partial r}{\partial x} = \frac{x}{r} \qquad (1.157)$$

Formulas analogous to Eq. 1.156 can be obtained for $\partial E_y/\partial y$ and $\partial E_z/\partial z$. The divergence of the field thus is given by

$$\vec{\nabla} \cdot \vec{E} = \frac{\partial E_x}{\partial x} + \frac{\partial E_y}{\partial y} + \frac{\partial E_z}{\partial z} = \frac{q}{4\varepsilon_0}\frac{3r^2 - 3x^2 - 3y^2 - 3z^2}{r^5} = 0 \qquad (1.158)$$

This is a puzzling result! We know from Eqs. 1.153 and 1.149 that:

$$\int_V \vec{\nabla} \cdot \vec{E}\, d^3\vec{r} = \frac{q}{\varepsilon_0} \qquad (1.159)$$

and yet we have just proved that $\vec{\nabla} \cdot \vec{E} = 0$. This paradox can be resolved after a close examination of Eq. 1.158. At the origin ($r = 0$) we find that $\vec{\nabla} \cdot \vec{E} = 0/0$, which means that $\vec{\nabla} \cdot \vec{E}$ can take any value at this point.

Thus, Eqs. 1.158 and 1.159 can be reconciled if $\vec{\nabla} \cdot \vec{E}$ is some sort of "spike" function (i.e., it is zero everywhere except arbitrarily close to the origin, where it becomes very large). This must occur in such a manner that the volume integral over the spike is finite.

Using Fig. 1.16, let us examine how we might construct a one-dimensional spike function. Consider the "box-car" function:

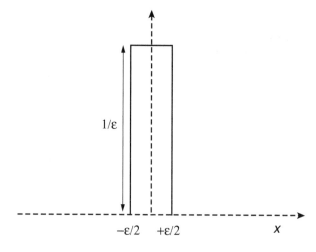

Fig. 1.16 Illustration of the spike function "box-car"

1.8 Time-Independent Maxwell's Equations

$$g(x, \varepsilon) = \begin{cases} 1/\varepsilon & \text{for } |x| < \varepsilon/2 \\ 0 & \text{otherwise} \end{cases} \quad (1.160)$$

From Fig. 1.16 and the relation in Eq. 1.155, it is clear that:

$$\int_{-\infty}^{+\infty} g(x, \varepsilon) dx = 1 \quad (1.161)$$

Now consider the function

$$\delta(x) \lim_{\varepsilon \to 0} g(x, \varepsilon) \quad (1.162)$$

This is zero everywhere except arbitrarily close to $x = 0$. According to Eq. 1.156, it also possesses a finite integral:

$$\int_{-\infty}^{+\infty} \delta(x) dx = 1 \quad (1.163)$$

Thus, $\delta(x)$ has all of the required properties of a spike function. The one-dimensional spike function $\delta(x)$ is called the *Dirac delta-function* after the Cambridge physicist Paul Dirac who invented it in 1927 while investigating quantum mechanics. The *delta-function* is an example of what mathematicians call a *generalized function*; it is not clearly defined at $x = 0$, but its integral is nevertheless well defined. Consider the integral

$$\int_{-\infty}^{+\infty} f(x) \delta(x) dx \quad (1.164)$$

where $f(x)$ is a function that is well behaved in the vicinity of $x = 0$. Since the delta-function is zero everywhere apart from very close to $x = 0$, it is clear that:

$$\int_{-\infty}^{+\infty} f(x) \delta(x) dx = f(0) \int_{-\infty}^{+\infty} \delta(x) dx = f(0) \quad (1.165)$$

where use has been made of Eq. 1.158. The preceding equation, which is valid for any well-behaved function, $f(x)$, is effectively the definition of a delta-function. A simple change of variables allows us to define $\delta(x - x_0)$, which is a spike function centered on $x = x_0$. Equation 1.160 gives

$$\int_{-\infty}^{+\infty} f(x) \delta(x - x_0) dx = f(x_0) \quad (1.166)$$

Actually, we want a three-dimensional spike function—that is, a function that is zero everywhere apart from arbitrarily close to the origin and with a volume integral that is unity. If we denote this function by $\delta(\vec{r})$, then it can be easily seen that the three-dimensional delta-function is the product of three one-dimensional delta-functions:

$$\delta(\vec{r}) = \delta(x)\delta(y)\delta(z) \tag{1.167}$$

This function is clearly zero everywhere except at the origin. But is its volume integral unity? Let us integrate over a cube of dimensions $2a$ that is centered on the origin and aligned along the Cartesian axes. This volume integral obviously is separable, so

$$\int \delta(\vec{r}) d^3\vec{r} = \int_{-a}^{+a} \delta(x)dx \int_{-a}^{+a} \delta(y)dy \int_{-a}^{+a} \delta(z)dz \tag{1.168}$$

The integral can be turned into an integral over all space by taking the limit $a \to \infty$. Still, we know that for one-dimensional delta-functions $\int_{-\infty}^{+\infty} \delta(x)dx = 1$, it follows from the previous equation that:

$$\int \delta(\vec{r}) d^3\vec{r} = 1 \tag{1.169}$$

which is the desired result. A simple generalization of previous arguments yields the following:

$$\int f(\vec{r})\delta(\vec{r}) d^3\vec{r} = f(0) \tag{1.170}$$

where $f(\vec{r})$ is any well-behaved scalar field. Finally, we can change variables and write:

$$\delta(\vec{r} - \vec{r}') = \delta(x - x')\delta(y - y')\delta(z - z') \tag{1.171}$$

which is a three-dimensional spike function on $\vec{r} = \vec{r}'$. It is easily demonstrated that:

$$\int f(\vec{r})\delta(\vec{r} - \vec{r}') d^3\vec{r} = f(\vec{r}') \tag{1.172}$$

Up to now, we have only considered volume integrals taken over all space. Nevertheless, it should be obvious that the preceding result also holds for integrals over any finite volume V that contains point $\vec{r} = \vec{r}'$. Likewise, the integral is zero if V does not contain $\vec{r} = \vec{r}'$.

Let us now return to the problem at hand. The electric field generated by charge q located at the origin has $\vec{\nabla} \cdot \vec{E} = 0$ everywhere apart from the origin, and also satisfies

1.8 Time-Independent Maxwell's Equations

$$\int_V (\vec{\nabla} \cdot \vec{E}) d^3\vec{r} = \frac{q}{\varepsilon_0} \tag{1.173}$$

for spherical volume V centered on the origin. These two facts imply that:

$$\vec{\nabla} \cdot \vec{E} = \frac{q}{\varepsilon_0} \delta(\vec{r}) \tag{1.174}$$

where use has been made of Eq. 1.169.

At this stage, vector field theory has yet to show its worth. After all, we have just spent an inordinately long time proving something using vector field theory which we previously proved in one line (see Eq. 1.148) using conventional analysis. It is time to demonstrate the power of vector field theory. Consider, again, charge q at the origin surrounded by spherical surface S, which is centered on the origin. Suppose that we now displace surface S so that it is no longer centered on the origin. What is the flux of the electric field out of S? This is not a simple problem for conventional analysis because the normal to the surface is no longer parallel to the local electric field. Using vector field theory, however, this problem is no more difficult than the previous one. We have the following form:

$$\oint_S \vec{E} \cdot d\vec{S} = \int_V (\vec{\nabla} \cdot \vec{E}) d^3\vec{r} \tag{1.175}$$

from Gauss's theorem, plus Eq. 1.169. From these equations, it is clear that the flux of $\vec{E}$ out of S is q/ε_0 for a spherical surface displaced from the origin. Still, the flux becomes zero when the displacement is sufficiently large that the origin is no longer enclosed by the sphere.

It is possible to prove this through conventional analysis, but it is certainly not easy. Suppose that surface S is not spherical but is instead highly distorted. What is the flux of $\vec{E}$ out of S now? This is a virtually impossible problem with conventional analysis, but it still is easy using vector field theory. Gauss's theorem and Eq. 1.169 tell us that the flux is q/ε_0 provided that the surface contains the origin, and that the flux is zero otherwise. This result is completely independent of the shape of S.

Considering N charges q_i located at $\vec{r}_i$, a simple generalization of Eq. 1.174 gives:

$$\vec{\nabla} \cdot \vec{E} = \sum_{i=1}^{N} \frac{q_i}{\varepsilon_0} \delta(\vec{r} - \vec{r}_i) \tag{1.176}$$

Thus, Gauss's theorem, Eq. 1.175 implies that:

$$\oint_S \vec{E} \cdot d\vec{S} = \int_V (\vec{\nabla} \cdot \vec{E}) d^3\vec{r} = \frac{Q}{\varepsilon_0} \tag{1.177}$$

where Q is the total charge enclosed by surface S. This result is called Gauss's Law and does not depend on the shape of the surface.

Suppose, finally, that instead of having a set of discrete charges, we have a continuous charge distribution described by charge density $\rho(\vec{r})$. The charge contained in a small rectangular volume of dimensions dx, dy, and dz located at position $\vec{r}$ is $Q = \rho(\vec{r})dxdydz$. If we integrate $\vec{\nabla} \cdot \vec{E}$ over this volume element, however, we obtain

$$(\vec{\nabla} \cdot \vec{E})dxdydz = \frac{Q}{\varepsilon_0} = \frac{\rho dxdydz}{\varepsilon_0} \tag{1.178}$$

where Eq. 1.177 has been used. Here, the volume element is assumed to be sufficiently small that $\vec{\nabla} \cdot \vec{E}$ does not vary significantly across it. Thus, we obtain

$$\vec{\nabla} \cdot \vec{E} = \frac{\rho}{\varepsilon_0} \tag{1.179}$$

This is the first of four field equations, called Maxwell's equations, which together form a complete description of electromagnetism. Of course, our derivation of Eq. 1.179 is only valid for electric fields generated by stationary charge distributions. In principle, additional terms might be required to describe fields generated by moving charge distributions. It turns out that this is not the case, however, and that Eq. 1.179 is universally valid.

Equation 1.179 is a differential equation describing the electric field generated by a set of charges. We already know the solution to this equation when the charges are stationary; it is given by Eq. 1.137:

$$\vec{E}(\vec{r}) = \frac{1}{4\pi\varepsilon_0} \int \rho(\vec{r}\,') \frac{\vec{r} - \vec{r}\,'}{|\vec{r} - \vec{r}\,'|^3} d^3\vec{r}\,' \tag{1.180}$$

Equations 1.179 and 1.180 can be reconciled provided that:

$$\vec{\nabla} \cdot \left(\frac{\vec{r} - \vec{r}\,'}{|\vec{r} - \vec{r}\,'|^3}\right) = -\nabla^2 \left(\frac{1}{|\vec{r} - \vec{r}\,'|}\right) = 4\pi\delta(\vec{r} - \vec{r}\,') \tag{1.181}$$

where use has been made of Eq. 1.140. It follows that:

$$\begin{aligned} \vec{\nabla} \cdot \vec{E}(\vec{r}) &= \frac{1}{4\pi\varepsilon_0} \int \rho(\vec{r}\,') \vec{\nabla} \cdot \left(\frac{\vec{r} - \vec{r}\,'}{|\vec{r} - \vec{r}\,'|^3}\right) d^3\vec{r}\,' \\ &= \int \frac{\rho(\vec{r}\,')}{\varepsilon_0} \delta(\vec{r} - \vec{r}\,') d^3\vec{r}\,' = \frac{\rho(\vec{r})}{\varepsilon_0} \end{aligned} \tag{1.182}$$

which is the desired result. The most general form of Gauss's Law, Eq. 1.182, is obtained by integrating Eq. 1.179 over volume V surrounded by surface S and making use of Gauss's theorem:

1.8 Time-Independent Maxwell's Equations

$$\oint_S \vec{E} \cdot d\vec{S} = \frac{1}{\varepsilon_0} \int_V \rho(\vec{r}) d^3 \vec{r} \qquad (1.183)$$

One particularly interesting application of Gauss's Law is *Earnshaw's theorem*, which states that it is impossible for a collection of charged particles to remain in static equilibrium solely under the influence of electrostatic forces. For instance, consider the motion of the ith particle in electric field $\vec{E}$ generated by all the other static particles. The equilibrium position of the ith particle corresponds to some point $\vec{r}_i$, where $\vec{E}(\vec{r}_i) = 0$. By implication $\vec{r}_i$ does not correspond to the equilibrium position of any other particle. Nevertheless, for $\vec{r}_i$ to be a *stable* equilibrium point, the particle must experience a *restoring force* when it is moved a small distance away from $\vec{r}_i$ in *any* direction.

Assuming that the ith particle is positively charged, this means that the electric field must point radially toward $\vec{r}_i$ at all neighboring points. Therefore, if we apply Gauss's Law to a small sphere centered on $\vec{r}_i$, then there must be a negative flux of $\vec{E}$ through the surface of the sphere, implying the presence of a negative charge at $\vec{r}_i$. However, there is no such charge at $\vec{r}_i$. Thus, we conclude that $\vec{E}$ cannot point radially toward $\vec{r}_i$ at all neighboring points. In other words, there must be some neighboring points at which $\vec{E}$ is directed *away* from $\vec{r}_i$. Consequently, a positively charged particle placed at $\vec{r}_i$ can always escape by moving to such points. One corollary of *Earnshaw's theorem* is that classical electrostatics cannot account for the stability of atoms and molecules.

1.8.4 Poisson's Equation

We have seen that the electric field generated by a set of stationary charges can be written as the gradient of a scalar potential so that:

$$\vec{E} = -\vec{\nabla}\phi \qquad (1.184)$$

This equation can be combined with field Eq. 1.179 to give a partial differential equation for the scalar potential:

$$\nabla^2 \phi = -\frac{\rho}{\varepsilon_0} \qquad (1.185)$$

This is an example of a very famous type of partial differential equation known as *Poisson's equation*.

In its most general form, Poisson's equation is written:

$$\nabla^2 u(\vec{r}) = v(\vec{r}) \qquad (1.186)$$

where $u(\vec{r})$ is some scalar potential that is to be determined and $v(\vec{r})$ is a known "source function." The most common boundary condition applied to this equation is that potential u is zero at infinity. The solutions to Poisson's equation are completely superposable. Thus, if u_1 is the potential generated by source function v_1, and u_2 is the potential generated by source function v_2, we have:

$$\begin{cases} \nabla^2 u_1 = v_1 \\ \nabla^2 u_2 = v_2 \end{cases} \quad (1.187)$$

Then, the potential generated by $v_1 + v_2$ is $u_1 + u_2$, because

$$\nabla^2(u_1 + u_2) = \nabla^2 u_1 + u_2 = v_1 + v_2 \quad (1.188)$$

Poisson's equation has this property because it is *linear* in both the potential and the source terms.

The fact that the solutions to Poisson's equation are superposable suggests a general method for solving this equation. Suppose that we could construct all the solutions generated by point sources. Of course, these solutions must satisfy the appropriate boundary conditions. Any general source function can be built out of a set of suitably weighted point sources, so the general solution of Poisson's equation must be expressible as a weighted sum over the point source solutions. Thus, once we know all the point source solutions, we can construct any other solution. In mathematical terminology, we require the solution to be

$$\nabla^2 G(\vec{r}, \vec{r}\,') = \delta(\vec{r}, \vec{r}\,') \quad (1.189)$$

which goes to zero as $|\vec{r}| \to \infty$.

The function $G(\vec{r}, \vec{r}\,')$ is the solution generated by a unit point source located at position $\vec{r}\,'$. This function is known to mathematicians as a *Green's function*. The solution generated by a general source function $v(\vec{r})$ is simply the appropriately weighted sum of all the Green's function solutions:

$$u(\vec{r}) = \int G(\vec{r}, \vec{r}\,') v(\vec{r}\,') d^3 \vec{r}\,' \quad (1.190)$$

We easily can demonstrate that this is the correct solution, as follows:

$$\nabla^2 u(\vec{r}) = \int [\nabla^2 G(\vec{r}, \vec{r}\,')] v(\vec{r}\,') d^3 \vec{r}\,' = \int \delta(\vec{r} - \vec{r}\,') v(\vec{r}) d^3 \vec{r}\,' = v(\vec{r}) \quad (1.191)$$

Let us return to Eq. 1.185 as:

$$\nabla^2 \phi = -\frac{\rho}{\varepsilon_0} \quad (1.192)$$

Green's function for this equation satisfies Eq. 1.189 with $|G| \to \infty$ as $|\vec{r}| \to 0$. It follows from Eq. 1.192 that

1.8 Time-Independent Maxwell's Equations

$$G(\vec{r}, \vec{r}') = -\frac{1}{4\pi} \frac{1}{|\vec{r}, \vec{r}'|} \qquad (1.193)$$

Note, from Eq. 1.146, that Green's function has the same form as the potential generated by a point charge. This is hardly surprising given the definition of a Green's function. It follows from Eqs. 1.190 and 1.193 that the general solution to *Poisson's equation* 1.192, is written as:

$$\phi(\vec{r}) = \frac{1}{4\pi\varepsilon_0} \int \frac{\rho(\vec{r}')}{|\vec{r} - \vec{r}'|} d^3\vec{r}' \qquad (1.194)$$

In fact, we already have obtained this solution by another method (see Eq. 1.143).

1.8.5 Ampère's Experiments

As legend has it, in 1820 the Danish physicist Hans Christian Orsted was giving a lecture demonstration of various electrical and magnetic effects. Suddenly, much to his surprise, he noticed that the needle of a compass he was holding was deflected when he moved it close to a current-carrying wire. Up until then, magnetism had been thought of as solely a property of some rather unusual rocks called loadstones. Word of this discovery spread quickly along the scientific grapevine, and the French physicist Andre Marie Ampère immediately decided to investigate further. Ampère's apparatus consisted (essentially) of a long straight wire carrying an electric current I.

Ampère quickly discovered that the needle of a small compass maps out a series of concentric circular loops in the plane perpendicular to a current-carrying wire (Fig. 1.17). The direction of circulation around these magnetic loops conventionally is taken to be the direction in which the *North* pole of the compass needle points. Using this convention, the circulation of the loops is given by a right-hand rule—that

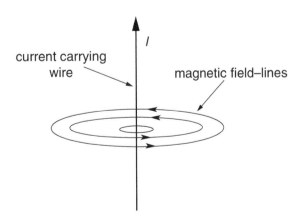

Fig. 1.17 Illustration of concentric circular loops

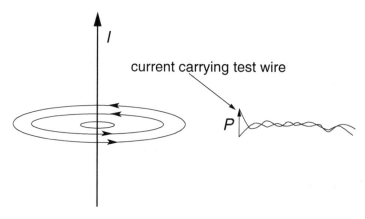

Fig. 1.18 Twisted feed wires illustration

is, if the thumb of the right-hand points along the direction of the current then the fingers of the right hand circulate in the same way as the magnetic loops.

Ampère's next series of experiments involved bringing a short test wire, carrying current I', close to the original wire, and investigating the force exerted on the test wire (Fig. 1.18). This experiment is not quite as clear-cut as Coulomb's experiment because, unlike electrical charges, electric currents cannot exist as point entities—they have to flow in complete circuits. We must imagine that the circuit that connects with the central wire is sufficiently far away so that it has no appreciable influence on the outcome of the experiment. The circuit that connects with the test wire is more problematic. Fortunately, if the feed wires are twisted around each other, as shown in Fig. 1.18, then they effectively cancel one another out, and do not influence the outcome of the experiment.

Ampère discovered that the force exerted on the test wire is directly proportional to its length. He also made the following observations:

- If the current in the test wire (i.e., the test current) flows parallel to the current in the central wire then the two wires attract one another.
- If the current in the test wire is reversed, then the two wires repel one another.
- If the test current points radially toward the central wire (and the current in the central wire flows upward), then the test wire is subject to a downward force.
- If the test current is reversed, then the force is upward. If the test current is rotated in a single plane so that it starts parallel to the central current and ends up pointing radially toward it, then the force on the test wire is of constant magnitude and is always at right angles to the test current.
- If the test current is parallel to a magnetic loop, then there is no force exerted on the test wire.
- If the test current is rotated in a single plane so that it starts parallel to the central current and $\theta = 0$ corresponds to the case where the test current is parallel to the central current, and its direction is again always at right angles to the test current.

1.8 Time-Independent Maxwell's Equations

Finally, Ampère was able to establish that the attractive force between two parallel current-carrying wires is proportional to the product of the two currents and falls off like the inverse of the perpendicular distance between the wires.

This rather complicated force law can be summed up succinctly in vector notation provided that we define vector field $\vec{B}$, called the *magnetic field,* with a direction that is always parallel to the loops mapped out by a small compass. The dependence of the force per unit length, $\vec{F}$, acting on a test wire with the different possible orientations of the test current is described by

$$\vec{F} = \vec{I} \times \vec{B} \tag{1.195}$$

where $\vec{I}$ is a vector with a direction and magnitude that are the same as those of the test current. Incidentally, the SI unit of electrical current is the ampere (A), which is the same as a Coulomb per second. The SI unit of magnetic field strength is the tesla (T), which is the same as a Newton per ampere per meter.

The variation of the force per unit length acting on a test wire with the strength of the central current and the perpendicular distance r to the central wire is summed up by saying that the magnetic field strength is proportional to I and inversely proportional to r. Thus, defining cylindrical polar coordinates aligned along the axis of the central current, we have

$$B_\theta = \frac{\mu_0 I}{2\pi r} \tag{1.196}$$

with $B_r = B_z = 0$. The constant of proportionality μ_0 is called the *permeability of free space* and takes the value:

$$\mu_0 = 4\pi \times 10^{-7}\,\text{NA}^{-2} \tag{1.197}$$

The concept of a magnetic field allows the calculation of the force on a test wire to be conveniently split into two parts. In the first part, we calculate the magnetic field generated by the current flowing in the central wire. This field circulates in the plane normal to the wire; its magnitude is proportional to the central current and inversely proportional to the perpendicular distance from the wire. In the second part, we use Eq. 1.195 to calculate the force per unit length acting on a short current-carrying wire located in the magnetic field generated by the central current. This force is perpendicular to both the magnetic field and the direction of the test current. Note that, at this stage, we have no reason to suppose that the magnetic field has any real physical existence. It is introduced merely to facilitate the calculation of the force exerted on the test wire by the central wire.

1.8.6 The Lorentz Force

The flow of an electric current down a conducting wire is ultimately attributable to the motion of electrically charged particles (in most cases, electrons) through the conducting medium. It seems reasonable, therefore, that the force exerted on the wire when it is placed in a magnetic field is really the result of the forces exerted on these moving charges. Let us suppose that this is the case.

Let A be the (uniform) cross-sectional area of the wire and let n be the number density of mobile charges in the conductor. Assume that the mobile charges each have a charge q and a velocity $\vec{v}$. We must accept that the conductor also contains stationary charges of charge $-q$ and a number density n (say) so that the net charge density in the wire is zero. In most conductors, the mobile charges are electrons and the stationary charges are atomic nuclei.

The magnitude of the electric current flowing through the wire is simply the number of Coulombs per second that flow past a given point. In 1 second (s), a mobile charge moves a distance of v, so all the charges contained in a cylinder of cross-sectional area A and length v flow past a given point. Thus, the magnitude of the current is $qnAv$. The direction of the current is the same as the direction of motion of the charges, so the vector current is $\vec{I} = qnA\vec{v}$.

According to Eq. 1.190, the force per unit length acting on the wire is:

$$\vec{f} = qnA\vec{v} \times \vec{B} \tag{1.198}$$

Yet, a unit length of the wire contains nA moving charges. So, assuming that each charge is subject to an equal force from the magnetic field (we have no reason to suppose otherwise), the force acting on an individual charge is:

$$\vec{F} = q\vec{v} \times \vec{B} \tag{1.199}$$

We can combine this with Eq. 1.130 to give the force acting on charge q moving with velocity $\vec{v}$ in electric field $\vec{E}$ and magnetic field $\vec{B}$:

$$\vec{F} = q\vec{E} + q\vec{v} \times \vec{B} \tag{1.200}$$

This is called the *Lorentz force law* after the Dutch physicist Hendrik Antoon Lorentz who first formulated it. The electrical force on a charged particle is parallel to the local electric field. The magnetic force, however, is perpendicular to both the local magnetic field and the particle's direction of motion. No magnetic force is exerted on a stationary charged particle.

The equation of motion of a free particle of charge q and mass m moving in electric and magnetic fields is:

$$m\frac{d\vec{v}}{dt} = q\vec{E} + q\vec{v} \times \vec{B} \tag{1.201}$$

according to the Lorentz force law. This equation of motion was first verified in a famous experiment carried out by the Cambridge physicist J. J. Thompson in 1897. Thompson was investigating *cathode rays*, a then mysterious form of radiation emitted by a heated metal element held at a large negative voltage (i.e., a cathode) with respect to another metal element (i.e., an anode) in an evacuated tube.

1.8 Time-Independent Maxwell's Equations

German physicists held that cathode rays were a form of EM radiation, while British and French physicists suspected that they were, in reality, a stream of charged particles. Thompson was able to demonstrate that the latter view was correct. In Thompson's experiment, the cathode rays passed through a region of "crossed" electric and magnetic fields (still in a vacuum). The fields were perpendicular to the original trajectory of the rays and were also mutually perpendicular.

Let us analyze Thompson's experiment. Suppose that the rays were originally traveling in the x-direction and were subject to a uniform electric field $\vec{E}$ in the z-direction and a uniform magnetic field $\vec{B}$ in the $-y$-direction. Assume, as Thompson did, that cathode rays are a stream of particles of mass m and charge q. The equation of motion of the particles in the z-direction is:

$$m\frac{d^2z}{dt^2} = q(\vec{E} - v\vec{B}) \tag{1.202}$$

where v is the velocity of the particles in the x-direction. Thompson started off his experiment by only turning on the electric field in his apparatus and measuring deflection d of the ray in the z-direction after it had traveled a distance of l through the electric field. It is clear from the equation of motion that:

$$d = \frac{q}{m}\frac{Et^2}{2} = \frac{q}{m}\frac{El^2}{2v^2} \tag{1.203}$$

where the "time of flight" t is replaced by l/v. This formula is valid only if $d \ll l$, which is assumed to be the case.

Next, Thompson turned on the magnetic field in his apparatus and adjusted it so that the cathode ray was no longer deflected. The lack of deflection implies that the net force on the particles in the z-direction was zero. In other words, the electric and magnetic forces balanced exactly. It follows from Eq. 1.197 that with a properly adjusted magnetic field strength:

$$v = \frac{E}{B} \tag{1.204}$$

Thus, Eqs. 1.203 and 1.204 can be combined and rearranged to give the charge to mass ratio of the particles in terms of measured quantities:

$$\frac{q}{m} = \frac{2Ed}{l^2B^2} \tag{1.205}$$

Finally, if a particle is subject to force $\vec{f}$ and moves a distance of $\delta\vec{r}$ in time interval δt, then the work done on the particle by the force is:

$$\delta W = \vec{f} \cdot \delta\vec{r} \tag{1.206}$$

The power input to the particle from the force fields is:

$$\vec{P} = \lim_{\delta t \to 0} \frac{\delta W}{\delta t} = \vec{f} \cdot \vec{v} \qquad (1.207)$$

where $\vec{v}$ is the particle's velocity. It follows from the Lorentz force law of Eq. 1.200 that the power input to a particle moving in the electric and magnetic fields is:

$$P = q\vec{v} \cdot \vec{E} \qquad (1.208)$$

Note that a charged particle can gain (or lose) energy from an electric field, but not from a magnetic field. This is because the magnetic force always is perpendicular to the particle's direction of motion; therefore, it does no work on the particle (see Eq. 1.206). Thus, in particle accelerators, magnetic fields are often used to guide particle motion (e.g., in a circle) but the actual acceleration is performed by electric fields.

1.8.7 Ampère's Law

Magnetic fields, like electric fields, are completely superposable. So, if field $\vec{B}_1$ is generated by current I_1 flowing through some circuit, and field $\vec{B}_2$ is generated by current I_2 flowing through another circuit, then when currents I_1 and I_2 flow through both circuits simultaneously, the generated magnetic field is $\vec{B}_1 + \vec{B}_2$.

Consider two parallel wires separated by a perpendicular distance of r and carrying electric currents of I_1 and I_2, respectively (Fig. 1.19). The magnetic field strength at the second wire because of the current flowing in the first wire is $B = \mu_0 I_1 / 2\pi r$. This field is oriented at right angles to the second wire, so the force per unit length exerted on the second wire is:

$$F = \frac{\mu_0 I_1 I_2}{2\pi r} \qquad (1.209)$$

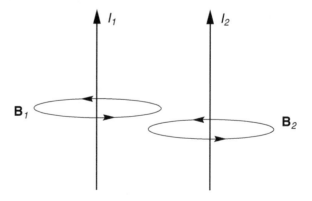

Fig. 1.19 Illustration of two parallel current-carrying wires

1.8 Time-Independent Maxwell's Equations

This follows from Eq. 1.195, which is valid for continuous wires as well as short test wires. The force acting on the second wire is directed radially inward toward the first wire. The magnetic field strength at the first wire owing to the current flowing in the second wire is $B = \mu_0 I_2 / 2\pi r$. This field is oriented at right angles to the first wire, so the force per unit length acting on the first wire is equal and opposite to that acting on the second wire, according to Eq. 1.190. Equation 1.203 is sometimes called *Ampère's Law* and is clearly another example of an action at a distance law—that is, if the current in the first wire is suddenly changed, then the force on the second wire immediately adjusts. In reality, there should be a short time delay, at least as long as the propagation time for a light signal between the two wires. Clearly, Ampère's Law is not strictly correct. Nevertheless, if we restrict our investigations to *steady* currents, it is perfectly adequate.

1.8.8 Magnetic Monopoles

Suppose that we have an infinite straight wire carrying an electric current of I. Let the wire be aligned along the z-axis. The magnetic field generated by such a wire is written as

$$\vec{B} = \frac{\mu_0 I}{2\pi} \left(\frac{-y}{r^2}, \frac{x}{r^2}, 0 \right) \tag{1.210}$$

In Cartesian coordinates, where $r = \sqrt{x^2 + y^2}$, the divergence of this field is:

$$\vec{\nabla} \cdot \vec{B} = \frac{\mu_0 I}{2\pi} \left(\frac{2yx}{r^4} - \frac{2xy}{r^4} \right) = 0 \tag{1.211}$$

where use has been made of $(\partial r / \partial x) = (x/r)$, and so on. We saw in Sect. 1.8.3 that the divergence of the electric field appeared, at first sight, to be zero. But, in reality, it was a delta-function because the volume integral of $\vec{\nabla} \cdot \vec{E}$ was non-zero. Does the same sort of thing happen for the divergence of the magnetic field? Well, if we could find a closed surface S for which $\oint_S \vec{B} \cdot d\vec{S} \neq 0$ then, according to Gauss's theorem, $\int_V \vec{\nabla} \cdot \vec{B} dV \neq 0$, where V is the volume enclosed by S. This would certainly imply that $\vec{\nabla} \cdot \vec{B}$ is some sort of delta-function. So, can we find such a surface? The short answer is "no." Consider a cylindrical surface aligned with the wire. The magnetic field is tangential to the outward surface element everywhere, so this surface certainly has zero magnetic flux coming out of it. In fact, it is impossible to invent any closed surface for which $\oint_S \vec{B} \cdot d\vec{S} \neq 0$ with $\vec{B}$ given by Eq. 1.204 (if you do not believe this, try it yourself). This suggests that the divergence of a magnetic field generated by steady electric currents really is zero. Admittedly, we have only proved

Fig. 1.20 Illustration of magnetic poles

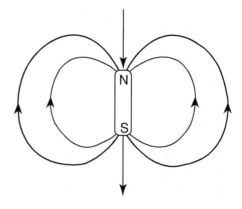

this for infinite straight currents, but, as will be demonstrated presently, it is true in general.

If $\vec{\nabla} \cdot \vec{B} = 0$, then $\vec{B}$ is a *solenoidal* vector field. In other words, field-lines of $\vec{B}$ never begin or end. This is certainly the case in Eq. 1.204, where the field-lines are a set of concentric circles centered on the z-axis. What about magnetic fields generated by permanent magnets (i.e., the modern equivalent of loadstones)? Do they also never begin or end? Well, we know that a conventional bar magnet has both a North and a South magnetic pole (like the Earth). If we track the magnetic field-lines with a small compass, they all emanate from the South Pole, spread out, and eventually reconverge on the North Pole (Fig. 1.20). It appears likely (but we cannot prove it with a compass) that the field-lines inside the magnet connect from the North to the South Pole to form closed loops that never begin or end.

Can we produce an isolated North or South magnetic pole—for instance, by snapping a bar magnet in two? A compass needle would always point toward an isolated North Pole, so this would act like a negative "magnetic charge." Likewise, a compass needle would always point away from an isolated South Pole, so this would act like a positive "magnetic charge."

It is clear from Fig. 1.21 that if we take a closed surface S containing an isolated magnetic pole, which is usually termed a *magnetic monopole*, then $\oint_S \vec{B} \cdot d\vec{S} \neq 0$; the flux will be positive for an isolated South Pole and negative for an isolated North Pole. It follows from Gauss's theorem that if $\oint_S \vec{B} \cdot d\vec{S} \neq 0$, then $\vec{\nabla} \cdot \vec{B} \neq 0$. Thus, the statement that magnetic fields are solenoidal, or that $\vec{\nabla} \cdot \vec{B} = 0$, is equivalent to the statement that there are *no magnetic monopoles*. It is not clear a priori that this is a true statement. In fact, it is quite possible to formulate electromagnetism to allow for magnetic monopoles. As far as we know, however, there are no magnetic monopoles in the Universe. At least, if there are any, they are all hiding from us!

We know that if we try to make a magnetic monopole by snapping a bar magnet in two, we just end up with two smaller bar magnets. If we snap one of these smaller magnets in two, we end up with four even smaller bar magnets. We can continue this

1.8 Time-Independent Maxwell's Equations

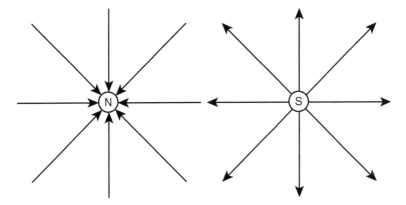

Fig. 1.21 Illustration of magnetic charges

process down to the atomic level without ever producing a magnetic monopole. In fact, permanent magnetism is generated by electric currents circulating on the atomic scale, so this type of magnetism is not fundamentally different from the magnetism generated by macroscopic currents.

In conclusion, *all* steady magnetic fields in the Universe are generated by circulating electric currents of some description. Such fields are solenoidal; that is, they never begin or end and satisfy the field equation:

$$\vec{\nabla} \cdot \vec{B} = 0 \tag{1.212}$$

This, incidentally, is the second of Maxwell's equations. Essentially, it says that there is no such thing as a magnetic monopole. We have only proved that $\vec{\nabla} \cdot \vec{B} = 0$ for steady magnetic fields, but in fact this is also the case for time-dependent fields.

1.8.9 Ampère's Circuital Law

Consider, again, an infinite straight wire aligned along the z-axis and carrying a current of I. The field generated by such a wire is written:

$$B_\theta = \frac{\mu_0 I}{2\pi r} \tag{1.213}$$

in cylindrical polar coordinates. Consider a circular loop C in the $x - y$ plane which is centered on the wire. Suppose that the radius of this loop is r. Let us evaluate line integral $\oint_C \vec{B} \cdot d\vec{l}$. This integral is easy to perform because the magnetic field is always parallel to the line element; thus, we have:

$$\oint_C \vec{B} \cdot d\vec{l} = \oint B_\theta r d\theta = \mu_0 I \tag{1.214}$$

Yet, we know from Stokes' theorem that

$$\oint_S \vec{B} \cdot d\vec{l} = \int_S \vec{\nabla} \times \vec{B} \cdot d\vec{S} \tag{1.215}$$

where S is any surface attached to loop C.

Let us evaluate $\vec{\nabla} \times \vec{B}$ directly. According to Eq. 1.210, we have:

$$\left(\vec{\nabla} \times \vec{B}\right)_x = \frac{\partial B_z}{\partial y} - \frac{\partial B_y}{\partial z} = 0 \tag{1.216}$$

$$\left(\vec{\nabla} \times \vec{B}\right)_y = \frac{\partial B_x}{\partial z} - \frac{\partial B_z}{\partial x} = 0 \tag{1.217}$$

$$\left(\vec{\nabla} \times \vec{B}\right)_z = \frac{\partial B_y}{\partial x} - \frac{\partial B_x}{\partial y} = \frac{\mu_0 I}{2\pi}\left(\frac{1}{r^2} - \frac{2x^2}{r^4} + \frac{1}{r^2} - \frac{2y^2}{r^4}\right) = 0 \tag{1.218}$$

where use has been made of $\partial r/\partial x = x/r$, and so on. We now have a problem. Equations 1.208 and 1.209 imply that

$$\int_S \vec{\nabla} \times \vec{B} \cdot d\vec{S} = \mu_0 I \tag{1.219}$$

But, we have just demonstrated that $\vec{\nabla} \times \vec{B} = 0$. This problem is very reminiscent of the difficulty we had earlier with $\vec{\nabla} \cdot \vec{E}$. Recall that $\int_V \vec{\nabla} \cdot \vec{E} dV = q/\varepsilon_0$ for volume V contains discrete charge q, but that $\vec{\nabla} \cdot \vec{E} = 0$ at a general point. We got around this problem by saying that $\vec{\nabla} \cdot \vec{E}$ is a three-dimensional delta-function with a spike that is coincident with the location of the charge.

Likewise, we can get around our present difficulty by saying that $\vec{\nabla} \times \vec{B}$ is a two-dimensional delta-function. A three-dimensional delta-function is a singular (but integrable) *point* in space, whereas a two-dimensional delta-function is a singular *line* in space. It is clear from an examination of Eqs. 1.216 through 1.218 that the only component of $\vec{\nabla} \times \vec{B}$ that can be singular is the z - component, and that this can only be singular on the z-axis (i.e., $r = 0$). Thus, the singularity coincides with the location of the current, and we can write:

$$\vec{\nabla} \times \vec{B} = \mu_0 I \delta(x) \delta(y) \hat{z} \tag{1.220}$$

The preceding equation certainly gives $\left(\vec{\nabla} \times \vec{B}\right)_x = \left(\vec{\nabla} \times \vec{B}\right)_y = 0$ and $\left(\vec{\nabla} \times \vec{B}\right)_z = 0$ everywhere apart from the z-axis, in accordance with Eqs. 1.216 through 1.218. Suppose that we integrate over a plane surface of S connected to loop C. The surface element is $d\vec{S} = d(x)d(y)\hat{z}$, so

1.8 Time-Independent Maxwell's Equations

$$\int_S \vec{\nabla} \times \vec{B} \cdot d\vec{S} = \mu_0 I \iint \delta(x) \delta(y) dx dy \quad (1.221)$$

where the integration is performed over the region $\sqrt{x^2 + y^2} \leq r$. Because the only part of S that actually contributes to the surface integral is the bit that lies infinitesimally close to the z-axis, we can integrate overall x and y without changing the result. Thus, we obtain

$$\int_S \vec{\nabla} \times \vec{B} \cdot d\vec{S} = \mu_0 I \int_{-\infty}^{+\infty} \delta(x) dx \int_{-\infty}^{+\infty} \delta(y) dy = \mu_0 I \quad (1.222)$$

which is in agreement with Eq. 1.219.

Nonetheless, why have we gone to so much trouble to prove something using vector field theory that can be demonstrated in one line via conventional analysis (see Eq. 1.208)? The answer, of course, is that the vector field result is easily generalized, whereas the conventional result is just a special case. For instance, it is clear that Eq. 1.216 is true for *any* surface attached to loop C, not just a plane surface. Moreover, suppose that we distort simple circular loop C so that it is no longer circular or even lies in one plane. What is the line integral of $\vec{B}$ around the loop now? This is no longer a simple problem for conventional analysis because the magnetic field is not parallel to a line element of the loop. Nonetheless, according to Stokes's theorem,

$$\oint_C \vec{B} \cdot d\vec{l} = \int_S \vec{\nabla} \times \vec{B} \cdot d\vec{S} \quad (1.223)$$

with $\vec{\nabla} \times \vec{B}$ given by Eq. 1.220. Note that the only part of S that contributes to the surface integral is an infinitesimal region centered on the z-axis. So, as long as S actually intersects with the z-axis, it does not matter what shape the rest the surface is; we always get the same answer for the surface integral—namely,

$$\oint_C \vec{B} \cdot d\vec{l} = \int_S \vec{\nabla} \times \vec{B} \cdot d\vec{S} = \mu_0 I \quad (1.224)$$

Thus, provided curve C circulates the z-axis and, therefore, any surface S attached to C intersects the z-axis line integral $\oint_C \vec{B} \cdot d\vec{l}$ is equal to $\mu_0 I$. Of course, if C does not circulate the z-axis, then an attached surface S does not intersect the z-axis and $\oint_C \vec{B} \cdot d\vec{l}$ is zero.

There is one more proviso. The line integral $\oint_C \vec{B} \cdot d\vec{l}$ is $\mu_0 I$ for a loop that circulates the z-axis in a clockwise direction (looking up the z-axis). Still, if the loop circulates in a counterclockwise direction, then the integral is $-\mu_0 I$. This follows because in the latter case the z-component of surface element dS is directed opposite to the current flow at the point where the surface intersects the wire.

Let us now consider N wires directed along the z-axis, with coordinates (x_i, y_i) in the $x - y$ plane, each carrying a current of I_i in the positive z-direction. It is fairly obvious that Eq. 1.214 generalizes to

$$\vec{\nabla} \times \vec{B} = \mu_0 \sum_{i=1}^{N} I_i \delta(x - x_i) \delta(y - y_i) \hat{z} \tag{1.225}$$

If we integrate the magnetic field around some closed curve C, which can have any shape and does not necessarily lie in one plane, then Stokes's theorem and the previous equation imply that

$$\oint_C \vec{B} \cdot d\vec{l} = \int_S \vec{\nabla} \times \vec{B} \cdot d\vec{S} = \mu_0 \mathcal{J} \tag{1.226}$$

where $\mathcal{J}$ is the total current enclosed by the curve. Again, if the curve passes around the ith wire in a clockwise direction (looking down the direction of current flow), the wire contributes I_i to the aggregate current $\mathcal{J}$. On the other hand, if the curve passes around the wire in a counterclockwise direction then the wire contributes $-I_i$. Finally, if the curve does not pass around the wire at all, the wire contributes nothing to $\mathcal{J}$.

Equation 1.219 is a field equation describing how a set of z-directed current-carrying wires generate a magnetic field. These wires have zero thickness, which implies that we are trying to squeeze a finite amount of current into an infinitesimal region. This accounts for the delta-functions on the RHS of the equation. Likewise, we obtained delta-functions in Sect. 1.8.3 because we were dealing with point charges. Let us now generalize to the more realistic case of *diffuse* currents.

Suppose that the z-current flowing through a small rectangle in the $x - y$ plane, centered on coordinates (x, y) and of dimensions dx and dy, is $j_x(y, z)dydz$. Here, j_x is termed the current density in the z-direction. Let us integrate $(\vec{\nabla} \times \vec{B}_z)$ over this rectangle, which is assumed to be sufficiently small so that $(\vec{\nabla} \times \vec{B}_z)$ does not vary appreciably across it. According to Eq. 1.220, this integral is equal to μ_0 times the total z-current flowing through the rectangle. Thus,

$$(\vec{\nabla} \times \vec{B}_z)dxdy = \mu_0 j_x dxdy \tag{1.227}$$

which implies that

$$(\vec{\nabla} \times \vec{B})_z = \mu_0 j_z \tag{1.228}$$

Of course, there is nothing special about the z-axis. Suppose we have a set of diffuse currents flowing in the x-direction. The current flowing through a small rectangle in the $y - z$ plane, centered on coordinates (y, z) and of dimensions dy and dz, is given by $j_x(y, z)dydz$, where j_x is the current density in the x-direction. It is fairly, obvious that we can write:

1.8 Time-Independent Maxwell's Equations

$$(\vec{\nabla} \times \vec{B})_x = \mu_0 j_x \tag{1.229}$$

with a similar equation for diffuse currents flowing along the y - axis. We can combine these equations with Eq. 1.228 to form a single vector field equation that describes how electric currents generate magnetic fields, as follows:

$$\vec{\nabla} \times \vec{B} = \mu_0 \vec{J} \tag{1.230}$$

where $\vec{J} = (j_x, j_y, j_z)$ is the vector current density. This is the third of Maxwell's equations.

The electric current flowing through small area $d\vec{S}$ located at position $\vec{r}$ is $\vec{J}(\vec{r}) \cdot d\vec{S}$. Suppose that the space is filled with particles of charge q, number density $n(\vec{r})$, and velocity $v(\vec{r})$. The charge density is given by $\rho(\vec{r}) = qn$. The current density is given by $\vec{J}(\vec{r}) = qn\vec{v}$ and is obviously a proper vector field (velocities are proper vectors because they are derived from displacements ultimately).

If we form the line integral of $\vec{B}$ around some general closed curve C, making use of Stokes's theorem and the field Eq. 1.223, we obtain

$$\oint_C \vec{B} \cdot d\vec{l} = \mu_0 \oint_S \vec{j} \cdot d\vec{S} \tag{1.231}$$

In other words, the line integral of the magnetic field around any closed loop C is equal to μ_0 times the flux of the current density through C. This result is called *Ampère's circuital law*. If the currents flow in zero-thickness wires, then this law reduces to Eq. 1.226.

The flux of the current density through C is evaluated by integrating $\vec{J} \cdot d\vec{S}$ over any surface $\vec{s}$ attached to C. Suppose that we take two different surfaces, S_1 and S_1. It is clear that if Ampère's circuital law is to make any sense, then surface integral $\int_{S_1} \vec{J} \cdot d\vec{S}$ had better be equal integral $\int_{S_2} \vec{J} \cdot d\vec{S}$. That is, when we work out the flux of current though C using two different attached surfaces, we had better get the same answer; otherwise, Eq. 1.230 is wrong (because the left-hand side, LHS, is clearly independent of the surface spanning C). We saw at the beginning of this chapter that if the integral of a vector field A over some surface attached to a loop depends only on the loop, and is independent of the surface which spans it, then this implies that $\vec{\nabla} \cdot \vec{A} = 0$.

The flux of the current density through any loop C is calculated by evaluating integral $\int_S \vec{J} \cdot d\vec{S}$ for any surface S that spans the loop. According to Ampère's circuital law, this integral depends only on C and is completely independent of S (i.e., it is equal to the line integral of $\vec{B}$ around C, which depends on C but not on S). This implies that $\vec{\nabla} \cdot \vec{J} = 0$. In fact, we can obtain this relationship directly from the field in Eq. 1.230. We know that the divergence of a *curl* is automatically zero, so taking the divergence of Eq. 1.129, we obtain

$$\vec{\nabla} \cdot \vec{J} = 0 \qquad (1.232)$$

We have shown that if Ampère's circuital law is to make any sense, we need $\vec{\nabla} \cdot \vec{J} = 0$. Physically, this implies that the net current flowing through any closed surface S is zero. Up to now, we have only considered stationary charges and steady currents. It is clear that if all charges are stationary and all currents are steady, then there can be no net current flowing through a closed surface S because this would imply a buildup of charge in volume V enclosed by S. In other words, as long as we restrict our investigation to stationary charges and steady currents, we expect $\vec{\nabla} \cdot \vec{J} = 0$, so Ampère's circuital law makes sense. In spite of this, suppose that we now relax this restriction.

Presume that some of the charges in a volume V decide to move outside V. Clearly, there will be a non-zero net flux of electric current through the bounding surface S while this is happening. This implies, from Gauss's theorem, that $\vec{\nabla} \cdot \vec{J} \neq 0$. Under these circumstances Ampère's circuital law collapses in a heap. We will see later that we can rescue this law by adding an extra term involving a time derivative to the RHS of the field in Eq. 1.224. For steady-state situations (i.e., $\partial/\partial t$), this extra term can be neglected. Thus, field equation $\vec{\nabla} \times \vec{B} = \mu_0 \vec{J}$ is in fact only two-thirds of Maxwell's third equation—there is a term missing on the RHS.

We now have derived two field equations involving magnetic fields (actually, we have only derived one and two-thirds):

$$\vec{\nabla} \cdot \vec{B} = 0 \qquad (1.233)$$
$$\nabla \times \vec{B} = \mu_0 \vec{J} \qquad (1.234)$$

We obtained these equations by looking at the fields generated by infinitely long, straight, steady currents. This, of course, is a rather special class of currents.

We should go back and repeat the process for general currents now. In fact, if we did this we would find that the preceding field equations still hold (provided that the currents are steady). Unfortunately, this demonstration is rather messy and extremely tedious. There is a better approach. Let us *assume* that the field equations are valid for any set of steady currents. We then can use, with relatively little effort, these equations to generate the correct formula for the magnetic field induced by a general set of steady currents, thus proving that our assumption is correct. More of this later.

1.8.10 Helmholtz's Theorem

Let us embark on a slight mathematical digression. Up to now, we have only studied the electric and magnetic fields generated by stationary charges and steady currents. We have found that these fields are describable in terms of four field equations:

1.8 Time-Independent Maxwell's Equations

$$\vec{\nabla} \cdot \vec{E} = \frac{\rho}{\varepsilon_0} \tag{1.235}$$

$$\vec{\nabla} \times \vec{E} = 0 \tag{1.236}$$

for electric fields, and

$$\vec{\nabla} \cdot \vec{B} = 0 \tag{1.237}$$

$$\vec{\nabla} \times \vec{B} = \mu_0 \vec{J} \tag{1.238}$$

for magnetic fields. There are no other field equations. This clearly suggests that if we know the divergence and the curl of a vector field, then we know everything there is to know about the field. In fact, this is the case. There is a mathematical theorem, which sums this up, called *Helmholtz's theorem* after the German polymath Hermann Ludwig Ferdinand von Helmholtz.

Let us start with scalar fields. Field equations are a type of differential equation—that is, they deal with the infinitesimal differences in quantities between neighboring points. The question is: Which differential equation completely specifies a scalar field? This is easy. Suppose that we have scalar field ϕ and a field equation that tells us the gradient of this field at all points; something like

$$\vec{\nabla}\phi = \vec{A}(\vec{r}) \tag{1.239}$$

where $\vec{A}(\vec{r})$ is a vector field. Note that we need $\vec{\nabla} \times \vec{A} = 0$ for self-consistency because the curl of a gradient is automatically zero. The preceding equation specifies ϕ completely once we are given the value of the field at a single point, P say. Thus,

$$\phi(Q) = \phi(P) + \int_P^Q \vec{\nabla}\phi \cdot d\vec{l} = \phi(P) + \int_P^Q \vec{A} \cdot d\vec{l} \tag{1.240}$$

where Q is a general point. The fact that $\vec{\nabla} \times \vec{A} = 0$ means that $\vec{A}$ is a conservative field, which guarantees that the equation, gives a unique value for ϕ at a general point in space.

Suppose that we have vector field $\vec{F}$. How many differential equations do we need to specify this field completely? Hopefully, we only need two: one giving the divergence of the field, and one giving its curl. Let us test this hypothesis. Suppose that we have two field equations:

$$\vec{\nabla} \cdot \vec{F} = D \tag{1.241}$$

and

$$\vec{\nabla} \times \vec{F} = \vec{C} \tag{1.242}$$

where D is a scalar field and $\vec{C}$ is a vector field. For self-consistency, we need

$$\vec{\nabla} \cdot \vec{C} = 0 \tag{1.243}$$

because the divergence of a curl is automatically zero. The question is: Do these two field equations plus some suitable boundary conditions completely specify $\vec{F}$? Suppose that we write

$$\vec{F} = -\nabla U + \vec{\nabla} \times \vec{W} \tag{1.244}$$

In other words, we are saying that general field $\vec{F}$ is the sum of a conservative field, ∇U, and a solenoidal field, $\vec{\nabla} \times \vec{W}$. This sounds plausible, but it remains to be verified. Let us start by taking the divergence of Eq. 1.244 and making use of Eq. 1.235. We get

$$\nabla^2 U = -D \tag{1.245}$$

Note that vector field $\vec{W}$ does not figure into this equation because the divergence of a curl is automatically zero.

Let us take the curl of Eq. 1.244 now:

$$\vec{\nabla} \times \vec{F} = \vec{\nabla} \times \vec{\nabla} \times \vec{W} = \vec{\nabla}(\vec{\nabla} \cdot \vec{W}) - \nabla^2 \vec{W} = -\nabla^2 \vec{W} \tag{1.246}$$

Here, we assume that the divergence of $\vec{W}$ is zero. This is another thing that remains to be proved. Note that scalar field U does not figure into this equation because the curl of a divergence is automatically zero. Using Eq. 1.242, we get

$$\nabla^2 W_x = -C_x \tag{1.247}$$
$$\nabla^2 W_y = -C_y \tag{1.248}$$
$$\nabla^2 W_z = -C_z \tag{1.249}$$

So, we have transformed our problem into four differential equations, Eqs. 1.244 through 1.249, which we need to solve. Let us look at them. We immediately notice that they all have exactly the same form. In fact, they are all versions of Poisson's equation. We can make use of a principle made famous by *Richard P. Feynman*: "[T]he same equations have the same solutions." Recall that earlier we came across the following equation:

$$\nabla^2 \phi = -\frac{\rho}{\varepsilon_0} \tag{1.250}$$

where ϕ is the electrostatic potential and ρ is the charge density. We verified that the solution to this equation, with the boundary condition that ϕ goes to zero at infinity, is:

$$\phi(\vec{r}) = \frac{1}{4\pi\varepsilon_0} \int \frac{\rho(\vec{r}')}{|\vec{r} - \vec{r}'|} d^3\vec{r}' \tag{1.251}$$

1.8 Time-Independent Maxwell's Equations

Well, if the same equations have the same solutions, and Eq. 1.251 is the solution to Eq. 1.250, then we immediately can write down the solutions to Eq. 1.245 and Eqs. 1.247 through 1.249. We get

$$U(\vec{r}) = \frac{1}{4\pi} \int \frac{D(\vec{r}')}{|\vec{r} - \vec{r}'|} d^3\vec{r}' \tag{1.252}$$

and

$$W_x(\vec{r}) = \frac{1}{4\pi} \int \frac{C_x(\vec{r}')}{|\vec{r} - \vec{r}'|} d^3\vec{r}' \tag{1.253}$$

$$W_y(\vec{r}) = \frac{1}{4\pi} \int \frac{C_y(\vec{r}')}{|\vec{r} - \vec{r}'|} d^3\vec{r}' \tag{1.254}$$

$$W_z(\vec{r}) = \frac{1}{4\pi} \int \frac{C_z(\vec{r}')}{|\vec{r} - \vec{r}'|} d^3\vec{r}' \tag{1.255}$$

The last three equations can be combined to form a single vector equation:

$$\vec{W}(\vec{r}) = \frac{1}{4\pi} \int \frac{\vec{C}(\vec{r}')}{|\vec{r} - \vec{r}'|} d^3\vec{r}' \tag{1.256}$$

We assumed earlier that $\vec{\nabla} \cdot \vec{W}(\vec{r}) = 0$. Let us check to see whether this is true. Note that

$$\frac{\partial}{\partial x}\left(\frac{1}{|\vec{r} - \vec{r}'|}\right) = -\frac{x - x'}{|\vec{r} - \vec{r}'|^3} = \frac{x - x'}{|\vec{r} - \vec{r}'|^3} = -\frac{\partial}{\partial x'}\left(\frac{1}{|\vec{r} - \vec{r}'|}\right) \tag{1.257}$$

which implies that

$$\nabla\left(\frac{1}{|\vec{r} - \vec{r}'|}\right) = -\nabla'\left(\frac{1}{|\vec{r} - \vec{r}'|}\right) \tag{1.258}$$

where ∇' is the operator $(\partial/\partial x', \partial/\partial y', \partial/\partial z')$. Taking the divergence of Eq. 1.250, and making use of the previous relation, we obtain:

$$\vec{\nabla} \cdot \vec{W}(\vec{r}) \frac{1}{4\pi} \int \vec{C}(\vec{r}') \cdot \nabla\left(\frac{1}{|\vec{r} - \vec{r}'|}\right) d^3\vec{r}'$$

$$= -\frac{1}{4\pi} \int \vec{C}(\vec{r}') \cdot \nabla'\left(\frac{1}{|\vec{r} - \vec{r}'|}\right) d^3\vec{r}' \tag{1.259}$$

Now,

$$\int_{-\infty}^{+\infty} g \frac{\partial f}{\partial x} dx = [gf]_{-\infty}^{+\infty} f \frac{\partial g}{\partial x} dx \qquad (1.260)$$

if $gf \to 0$ as $x \to \pm\infty$, however, we can neglect the first term on the RHS of this equation and write:

$$\int_{-\infty}^{+\infty} g \frac{\partial f}{\partial x} dx = -\int_{-\infty}^{+\infty} f \frac{\partial g}{\partial x} dx \qquad (1.261)$$

A simple generalization of this result yields:

$$\int \vec{g} \cdot \nabla f d^3 \vec{r} = -\int f \nabla \cdot \vec{g} d^3 \vec{r} \qquad (1.262)$$

provided that $g_x f \to 0$ as $|\vec{r}| \to \infty$, and so on. Thus, we can deduce from Eq. 1.259 that

$$\vec{\nabla} \cdot \vec{W}(\vec{r}) = \frac{1}{4\pi} \int \frac{\vec{\nabla}' \cdot \vec{C}(\vec{r}')}{|\vec{r} - \vec{r}'|} d^3 \vec{r}' \qquad (1.263)$$

provided $|\vec{C}(\vec{r})|$ is bounded as $|\vec{r}| \to \infty$. Although we have already shown that $\vec{\nabla} \cdot \vec{C}(\vec{r}) = 0$ from self-consistency arguments, the preceding equation implies that $\vec{\nabla} \cdot \vec{W}(\vec{r}) = 0$, which is the desired result.

We have constructed vector field $\vec{F}$ that satisfies Eqs. 1.235 and 1.236 and behaves sensibly at infinity (i.e., $|\vec{F}| \to 0$ as $|\vec{r}| \to \infty$). But, is our solution the only possible one for Eqs. 1.235 and 1.236 with sensible boundary conditions at infinity? Another way of posing this question is to ask whether there are any solutions for

$$\begin{cases} \nabla^2 U = 0 \\ \nabla^2 W_i = 0 \end{cases} \qquad (1.264)$$

where i denotes x, y, or z, which are bounded at infinity.

If there are, we are in trouble because we can take our solution and add to it an arbitrary amount of a vector field with zero divergence and zero curl, thereby obtaining another solution that also satisfies physical boundary conditions. This would imply that our solution is not unique. In other words, it is not possible to unambiguously reconstruct a vector field given its divergence, its curl, and its physical boundary conditions. Fortunately, the equation

$$\nabla^2 \phi = 0 \qquad (1.265)$$

which is called *Laplace's equation*, has a very nice property: Its solutions are *unique*. That is, if we can find a solution to Laplace's equation that satisfies the boundary conditions, then we are guaranteed that this is the *only* solution. We shall prove this

1.8 Time-Independent Maxwell's Equations

later on in this book. Well, let us invent some solutions to Eq. 1.264 that are bounded at infinity. How about

$$U = W_i = 0 \qquad (1.266)$$

These solutions certainly satisfy Laplace's equation and are well behaved at infinity. Because the solutions to Laplace's equation are unique, we know that Eq. 1.260 is the only solution to Eq. 1.258. This means that there is no vector field that satisfies physical boundary equations at infinity and has zero divergence and zero curl. In other words, our solution to Eqs. 1.235 and 1.236 are the *only* solutions. Thus, we have unambiguously reconstructed vector field **F** given its divergence, its curl, and its sensible boundary conditions at infinity. This is *Helmholtz's theorem*.

We have just proved several very useful and important points. First, according to Eq. 1.238, a general vector field can be written as the sum of a conservative field and a solenoidal field. Thus, we ought to be able to write electric and magnetic fields in this form. Second, a general vector field that is zero at infinity is completely specified once its divergence and its curl are given. Thus, we can guess that the laws of electromagnetism can be written as four field equations:

$$\vec{\nabla} \cdot \vec{E} = \text{something} \qquad (1.267)$$

$$\vec{\nabla} \times \vec{E} = \text{something} \qquad (1.268)$$

$$\vec{\nabla} \cdot \vec{B} = \text{something} \qquad (1.269)$$

$$\vec{\nabla} \times \vec{B} = \text{something} \qquad (1.270)$$

without knowing the first thing about electromagnetism (other than the fact that it deals with two vector fields). Of course, Eqs. 1.235 through 1.238 are of exactly this form. We also know that there are only four field equations because the preceding ones are sufficient to completely reconstruct both $\vec{E}$ and $\vec{B}$.

Furthermore, we know that we can solve the field equations without even knowing what the RHS look like. After all, we solved Eqs. 1.241 and 1.242 for completely general right-hand sides. (Actually, the RHSs must go to zero at infinity; otherwise, integrals like Eq. 1.252 blow up.) We also know that any solutions we find are unique. In other words, there only is one possible steady electric and one magnetic field that can be generated by a given set of stationary charges and steady currents.

The third thing, which we proved, is that if the RHSs of the previous field equations are all zero, then the only physical solution is $\vec{E} = \vec{B} = 0$. This implies that steady electric and magnetic fields cannot generate themselves. Instead, they must be generated by stationary charges and steady currents. So, if we come across a steady electric field, we know that if we trace the field lines back we will eventually find a charge. Likewise, a steady magnetic field implies that there is a steady current flowing somewhere. All these results follow from vector field theory (i.e., from the general properties of fields in three-dimensional space) prior to any investigation of electromagnetism.

1.8.11 The Magnetic Vector Potential

Electric fields generated by stationary charges obey

$$\vec{\nabla} \times \vec{E} = 0 \tag{1.271}$$

This immediately allows us to write:

$$\vec{E} = -\vec{\nabla}\phi \tag{1.272}$$

because the curl of a gradient is automatically zero. In fact, whenever we come across an irrotational vector field in physics, we can always write it as the gradient of some scalar field. This is clearly a useful thing to do because it enables replacement of a vector field by a much simpler scalar field. Quantity ϕ in Eq. 1.272 is known as the *electric scalar potential*.

Magnetic fields generated by steady currents (and unsteady currents, for that matter) satisfy

$$\vec{\nabla} \cdot \vec{B} = 0 \tag{1.273}$$

This immediately allows us to write:

$$\vec{B} = \vec{\nabla} \times \vec{A} \tag{1.274}$$

because the divergence of a curl is automatically zero. In fact, whenever we come across a solenoidal vector field in physics, we can always write it as the curl of some other vector field. This obviously is not a useful thing to do, however, because it only allows us to replace one vector field by another. Nevertheless, Eq. 1.272 is one of the most useful equations we will come across in this section. Quantity $\vec{A}$ is known as the *magnetic vector potential*.

We know from Helmholtz's theorem that a vector field is specified fully by its divergence and its curl. The curl of the vector potential gives us the magnetic field using Eq. 1.268. The divergence of $\vec{A}$, however, has no physical significance. In fact, we are completely free to choose $\vec{\nabla} \cdot \vec{A}$ to be whatever we like. Note that, according to Eq. 1.272, the magnetic field is invariant under this transformation:

$$\vec{A} \rightarrow \vec{A} - \nabla \psi \tag{1.275}$$

In other words, the vector potential is undetermined in the gradient of a scalar field. This is just another way of saying that we are free to choose $\vec{\nabla} \cdot \vec{A}$. Recall that the electric scalar potential is undetermined in an arbitrary additive constant because the transformation

1.8 Time-Independent Maxwell's Equations

$$\phi \to \phi + c \tag{1.276}$$

leaves the electric field invariant in Eq. 1.266. The transformations of Eqs. 1.275 and 1.276 are examples of what mathematicians call *gauge transformations*.

The choice of a specific function ψ or a particular constant c is referred to as a *choice of the gauge*. We are free to set the gauge to be whatever we like. The most sensible choice is the one that makes equations as simple as possible. The usual gauge for scalar potential ϕ is such that $\phi \to 0$ at infinity. The usual gauge for $\vec{A}$ is such that

$$\vec{\nabla} \cdot \vec{A} = 0 \tag{1.277}$$

This choice is known as the *Coulomb gauge*.

It is obvious that we can always add a constant to ϕ to make it zero at infinity. But it is not at all obvious that we can always perform a gauge transformation to make $\vec{\nabla} \cdot \vec{A}$ zero. Suppose that we have found some vector field $\vec{A}$ with a curl that gives the magnetic field but the divergence of which is non-zero. Let,

$$\vec{\nabla} \cdot \vec{A} = v(\vec{r}) \tag{1.278}$$

The question is: Can we find scalar field ψ such that after we perform the gauge transformation in Eq. 1.275 we are left with $\vec{\nabla} \cdot \vec{A} = 0$?

Taking the divergence of Eq. 1.275, it is clear that we need to find function ψ that satisfies

$$\nabla^2 \psi = v \tag{1.279}$$

Nonetheless, this is just Poisson's equation. We know that we can always find a unique solution for this equation (see Sect. 1.8.10). This proves that, in practice, we can always set the divergence of $\vec{A}$ equal to zero.

Let us consider an infinite straight wire directed along the z-axis and carrying current I again. The magnetic field generated by such a wire is written:

$$\vec{B} = \frac{\mu_0 I}{2\pi} \left(\frac{-y}{r^2}, \frac{x}{r^2}, 0 \right) \tag{1.280}$$

We want to find vector potential $\vec{A}$ with a curl that is equal to the preceding magnetic field, and a divergence that is zero. It is not difficult to see that

$$\vec{A} = -\frac{\mu_0 I}{4\pi} \left(0, 0 \ln \left[x^2 + y^2 \right] \right) \tag{1.281}$$

fits the bill. Note that the vector potential is parallel to the direction of the current. This would seem to suggest that there is a more direct relationship between the vector potential and the current than there is between the magnetic field and the current. The potential is not very well behaved on the z-axis, but this is just because we are dealing with an infinitely thin current.

Let us take the curl of Eq. 1.274. We find from Table 1.1 that

$$\vec{\nabla} \times \vec{B} = \vec{\nabla} \times \vec{\nabla} \times \vec{A} = \nabla(\vec{\nabla} \cdot \vec{A}) - \nabla^2 \vec{A} = -\nabla^2 \vec{A} \tag{1.282}$$

where the Coulomb gauge condition of Eq. 1.277 has been used. We can combine the previous relation with the field in Eq. 1.230 to give

$$\nabla^2 \vec{A} = -\mu_0 \vec{J} \tag{1.283}$$

Writing this in component form, we obtain:

$$\nabla^2 A_x = -\mu_0 j_x \tag{1.284}$$

$$\nabla^2 A_y = -\mu_0 j_y \tag{1.285}$$

$$\nabla^2 A_z = -\mu_0 j_z \tag{1.286}$$

But then again, this is just *Poisson's equation* three times over. We immediately can write the unique solutions to the preceding equations:

$$A_x(\vec{r}) = \frac{\mu_0}{4\pi} \int \frac{j_x(\vec{r}')}{|\vec{r} - \vec{r}'|} d^3\vec{r}' \tag{1.287}$$

$$A_y(\vec{r}) = \frac{\mu_0}{4\pi} \int \frac{j_y(\vec{r}')}{|\vec{r} - \vec{r}'|} d^3\vec{r}' \tag{1.288}$$

$$A_z(\vec{r}) = \frac{\mu_0}{4\pi} \int \frac{j_z(\vec{r}')}{|\vec{r} - \vec{r}'|} d^3\vec{r}' \tag{1.289}$$

These solutions can be recombined to form a single vector solution:

$$\vec{A}(\vec{r}) = \frac{\mu_0}{4\pi} \int \frac{\vec{J}(\vec{r}')}{|\vec{r} - \vec{r}'|} d^3\vec{r}' \tag{1.290}$$

Of course, we have seen an equation like this before:

$$\phi(\vec{r}) = \frac{\mu_0}{4\pi} \int \frac{\rho(\vec{r}')}{|\vec{r} - \vec{r}'|} d^3\vec{r}' \tag{1.291}$$

Equations 1.290 and 1.291 are the unique solutions (given the arbitrary choice of gauge) to field Eqs. 1.234 through 1.237; they specify the magnetic vector and electric scalar potentials generated by a set of stationary charges, of charge density $\rho(\vec{r})$, and a set of steady currents, of current density $\vec{j}(\vec{r})$. Incidentally, we can prove that Eq. 1.290 satisfies the gauge condition $\vec{\nabla} \cdot \vec{A} = 0$ by repeating the analysis of

1.8 Time-Independent Maxwell's Equations

Eqs. 1.260 through 1.262 (with $\vec{W}(\vec{r}) \to \vec{A}(\vec{r})$ and $\vec{C}(\vec{r}) \to \mu_0 \vec{j}(\vec{r})$) and using the fact that $\vec{\nabla} \cdot \vec{j}(\vec{r}) = 0$ for steady currents.

1.8.12 The Biot-Savart Law

According to Eq. 1.266, we can obtain an expression for the electric field generated by stationary charges by taking minus the gradient of Eq. 1.290. This yields:

$$\vec{E}(\vec{r}) = \frac{1}{4\pi\varepsilon_0} \int \rho(\vec{r}\,') \frac{\vec{r}-\vec{r}\,'}{|\vec{r}-\vec{r}\,'|^3} d^3\vec{r}\,' \tag{1.292}$$

which we recognize as Coulomb's Law written for a continuous charge distribution. According to Eq. 1.274, we can obtain an equivalent expression for the magnetic field generated by steady currents by taking the curl of Eq. 1.296. This gives:

$$\vec{B}(\vec{r}) = \frac{\mu_0}{4\pi} \int \frac{\vec{j}(\vec{r}\,') \times (\vec{r}-\vec{r}\,')}{|\vec{r}-\vec{r}\,'|^3} d^3\vec{r}\,' \tag{1.293}$$

where the vector identity $\vec{\nabla} \times (\phi \vec{A}) = \phi \vec{\nabla} \times \vec{A} + \vec{\nabla}\phi \times \vec{A}$ has been used. Equation 1.287 is known as the *Biot-Savart Law* after the French physicists Jean Baptiste Biot and Felix Savart; it completely specifies the magnetic field generated by a steady (but otherwise quite general) distributed current.

Let us reduce our distributed current to an idealized zero thickness wire. We can do this by writing

$$\vec{j}(\vec{r}) d^3\vec{r} = \vec{I}(\vec{r}) dl \tag{1.294}$$

where $\vec{I}(\vec{r})$ is the vector current (i.e., its direction and magnitude specify the direction and magnitude of the current) and dl is an element of length along the wire. Equations 1.287 and 1.288 can be combined to give:

$$\vec{B}(\vec{r}) = \frac{\mu_0}{4\pi} \int \frac{\vec{I}(\vec{r}\,') \times (\vec{r}-\vec{r}\,')}{|\vec{r}-\vec{r}\,'|^3} dl \tag{1.295}$$

which is the form in which the Biot-Savart Law most usually is written.

This law is to magnetostatics (i.e., the study of magnetic fields generated by steady currents) what Coulomb's Law is to electrostatics (i.e., the study of electric fields generated by stationary charges). Furthermore, it can be verified experimentally given a set of currents, a compass, a test wire, and a great deal of skill and patience. This justifies the earlier assumption that field Eqs. 1.227 and 1.228 are valid for general current distributions. (Recall that we derived them by studying the fields generated by infinite, straight wires.) Note that both Coulomb's Law and the

Fig. 1.22 Illustration of an infinite wire

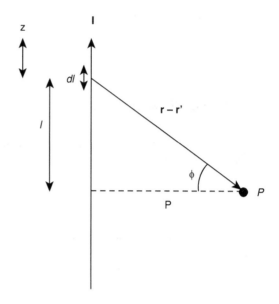

Biot-Savart Law are *gauge independent*—that is, they do not depend on a particular choice of gauge.

Consider an infinite straight wire, directed along the z-axis, and carrying current $\vec{I}(\vec{r})$ (Fig. 1.22). Let us reconstruct the magnetic field generated by the wire at point P using the Biot-Savart Law. Suppose that the perpendicular distance to the wire is ρ. It easily can be seen that

$$\vec{I} \times (\vec{r} - \vec{r}') = I\rho\hat{\theta} \tag{1.296}$$

$$l = \rho \tan \phi \tag{1.297}$$

$$dl = \frac{\rho}{\cos^2 \phi} d\phi \tag{1.298}$$

$$|\vec{r} - \vec{r}'| = \frac{\rho}{\cos \phi} \tag{1.299}$$

Thus, according to Eq. 1.295, we have:

$$\begin{aligned} B_\theta &= \frac{\mu_0}{4\pi} \int_{-\pi/2}^{+\pi/2} \frac{I\rho}{\rho^3 (\cos \phi)^{-3}} \frac{\rho}{\cos^2 \phi} d\phi \\ &= \frac{\mu_0 I}{4\pi\rho} \int_{-\pi/2}^{+\pi/2} \cos \phi d\phi = \frac{\mu_0 I}{4\pi\rho} [\sin \phi]_{-\pi/2}^{+\pi/2} \end{aligned} \tag{1.300}$$

which gives the familiar result

$$B_\theta = \frac{\mu_0 I}{2\pi\rho} \tag{1.301}$$

1.8.13 Electrostatics and Magnetostatics

We have now completed the theoretical investigation of electrostatics and magnetostatics. Our next task is to incorporate time variation into the analysis. Before we start this, however, let us briefly review the progress so far. We have found that the electric fields generated by stationary charges, and the magnetic fields generated by steady currents, are describable in terms of four field equations:

$$\vec{\nabla} \cdot \vec{E} = \frac{\rho}{\varepsilon_0} \tag{1.302}$$

$$\vec{\nabla} \times \vec{E} = 0 \tag{1.303}$$

$$\vec{\nabla} \cdot \vec{B} = 0 \tag{1.304}$$

$$\vec{\nabla} \times \vec{B} = \mu_0 \vec{J} \tag{1.305}$$

The boundary conditions are that the fields are zero at infinity, if the generating charges and currents are localized to some region in space. According to Helmholtz's theorem, the preceding field equations, plus the boundary conditions, are sufficient to *uniquely* specify the electric and magnetic fields. The physical significance of this is that divergence and curl are the only *rotationally invariant* first-order differential properties of a general vector field—that is, the only quantities that do not change their physical characteristics when the coordinate axes are rotated. Because physics does not depend on the orientation of the coordinate axes (which is, after all, quite arbitrary), divergence and curl are the *only* quantities, which can appear in first-order differential field equations, that claim to describe physical phenomena.

The field equations can be integrated to give:

$$\oint_S \vec{E} \cdot d\vec{S} = \frac{1}{\varepsilon_0} \int_V \rho \, dV \tag{1.306}$$

$$\oint_C \vec{E} \cdot d\vec{l} = 0 \tag{1.307}$$

$$\oint_S \vec{B} \cdot d\vec{S} = 0 \tag{1.308}$$

$$\oint_C \vec{B} \cdot d\vec{l} = \mu_0 \int_{S'} \vec{J} \cdot d\vec{S} \tag{1.309}$$

Here S is a closed surface enclosing volume V. Also, C is a closed loop, and S' is some surface attached to this loop. Field Eqs. 1.296 through 1.299 can be deduced from Eqs. 1.300 through 1.302 using Gauss's and Stokes's theorems.

Equation 1.300 is called *Gauss's Law* and says that the flux of the electric field out of a closed surface is proportional to the enclosed electrical charge. Equation 1.302 has no particular name and says that there is no such thing as a magnetic monopole. Equation 1.303 is called *Ampère's circuital law* and says that the line integral of the magnetic field around any closed loop is proportional to the flux of the current through the loop. Finally, Eqs. 1.301 and 1.303 are incomplete: each acquires an extra term on the RHS in time-dependent situations.

Field Eq. 1.297 is satisfied automatically if we write:

$$\vec{E} = -\vec{\nabla}\phi \tag{1.310}$$

Likewise, field Eq. 1.298 is satisfied automatically if we write:

$$\vec{B} = \vec{\nabla} \times \vec{A} \tag{1.311}$$

Here ϕ is the electric scalar potential, and $\vec{A}$ is the magnetic vector potential. The electric field clearly is unchanged if we add a constant to the scalar potential:

$$\vec{E} \to \vec{E} \quad \text{as} \quad \phi \to \phi + c \tag{1.312}$$

Similarly, the magnetic field is unchanged if we add the gradient of a scalar field to the vector potential:

$$\vec{B} \to \vec{B} \quad \text{as} \quad \vec{A} \to \vec{A} + \nabla\psi \tag{1.313}$$

The preceding transformations, which leave the $\vec{E}$ and $\vec{B}$ fields invariant, are called *gauge transformations*. We are free to choose c and ψ to be whatever we like—that is, we are free to select the gauge. The most sensible gauge is the one that makes the equations as simple and symmetrical as possible. This corresponds to this choice:

$$\phi(\vec{r}) \to 0 \quad \text{as} \quad |\vec{r}| \to \infty \tag{1.314}$$

and

$$\vec{\nabla} \cdot \vec{A}(\vec{r}) = 0 \tag{1.315}$$

The latter convention is known as the *Coulomb gauge*.

1.8 Time-Independent Maxwell's Equations

Taking the divergence of Eq. 1.304 and the curl of Eq. 1.311, and making use of the Coulomb gauge, we find that the four field Eqs. 1.302 through 1.305 can be reduced to Poisson's equation written four times over:

$$\nabla^2 \phi = -\frac{\rho}{\varepsilon_0} \tag{1.316}$$

$$\nabla^2 \vec{A} = -\mu_0 \vec{J} \tag{1.317}$$

Poisson's equation is just about the simplest *rotationally invariant* second-order partial differential equation it is possible to write.

Note that ∇^2 is clearly rotationally invariant because it is the divergence of a gradient, and both divergence and gradient are rotationally invariant. We always can construct the solution to Poisson's equation given the boundary conditions. Furthermore, we have a *uniqueness theorem* that tells us that the solution is the only possible one. Physically, this means that there is only one electric and one magnetic field that is consistent with a given set of stationary charges and steady currents. This sounds like an obvious, almost trivial, statement. Nonetheless, there are many areas of physics (e.g., fluid mechanics and plasma physics) where we also believe, for physical reasons, that for a given set of boundary conditions the solution should be unique.

The difficulty is that in most cases when we reduce the problem to a partial differential equation, we end up with something far worse than Poisson's equation. In general, we cannot solve this equation. In fact, we usually cannot even prove that it possesses a solution for general boundary conditions, let alone a solution that is unique. So, we are very fortunate indeed that in electrostatics and magnetostatics the problem boils down to solving a nice partial differential equation. When physicists make statements to the effect that "electromagnetism is the best understood theory in physics," which they often do, what they really are saying is that the partial differential equations that crop up in this theory are soluble and have good properties.

Poisson's equation,

$$\nabla^2 u = v \tag{1.318}$$

is *linear*, which means that its solutions are superposable. We can exploit this fact to construct a general solution to this equation. Suppose that we can find the solution to

$$\nabla^2 G(\vec{r}, \vec{r}\,') = \delta(\vec{r}, \vec{r}\,') \tag{1.319}$$

which satisfies the boundary conditions. This is the solution driven by a unit amplitude point source located at position vector $\vec{r}\,'$. Because any general source can be built out of a weighted sum of point sources, it follows that a general solution to Poisson's equation can be built out of a weighted superposition of point source solutions. Mathematically, we can write:

$$u(\vec{r}) = \int G(\vec{r}, \vec{r}\,') v(\vec{r}\,') d^3 \vec{r}\,' \tag{1.320}$$

Function G is called the *Green's function*. This function for Poisson's equation is:

$$G(\vec{r}, \vec{r}') = -\frac{1}{4\pi} \frac{1}{|\vec{r} - \vec{r}'|} \quad (1.321)$$

Note that Green's function is proportional to the scalar potential of a point charge located at $\vec{r}'$; this is hardly surprising given the definition of a Green's function.

According to Eqs. 1.316 through 1.321, the scalar and vector potentials generated by a set of stationary charges and steady currents take the form:

$$\phi(\vec{r}) = \frac{1}{4\pi\varepsilon_0} \int \frac{\rho(\vec{r}')}{|\vec{r} - \vec{r}'|} d^3\vec{r}' \quad (1.322)$$

$$\vec{A}(\vec{r}) = \frac{\mu_0}{4\pi} \int \frac{\vec{J}(\vec{r}')}{|\vec{r} - \vec{r}'|} d^3\vec{r}' \quad (1.323)$$

Making use of Eqs. 1.310, 1.311, 1.322, and 1.323, we obtain the fundamental force laws for electric and magnetic fields—Coulomb's Law,

$$\vec{E}(\vec{r}) = \frac{1}{4\pi\varepsilon_0} \int \rho(\vec{r}') \frac{\vec{r} - \vec{r}'}{|\vec{r} - \vec{r}'|^3} d^3\vec{r}' \quad (1.324)$$

and the *Biot-Savart Law*,

$$\vec{B}(\vec{r}) = \frac{\mu_0}{4\pi} \int \frac{\vec{J}(\vec{r}') \times (\vec{r} - \vec{r}')}{|\vec{r} - \vec{r}'|^3} d^3\vec{r}' \quad (1.325)$$

Of course, both of these laws are examples of action-at-a-distance laws and, therefore, violate the theory of relativity. This is not a problem, however, as long as we restrict ourselves to fields generated by *time-independent* charges and current distributions.

Now, the question is: By how much is the scheme that we have just worked out going to be disrupted when we take time variation into account? The answer, somewhat surprisingly, is by very little indeed. Therefore, in Eqs. 1.312 through 1.325 we already can discern the basic outline of classical electromagnetism. Let us continue our investigation.

1.9 Time-Dependent Maxwell's Equations

In this section, we will take the time independent set Maxwell's equations, derived in the previous section, and generalize them to the full set of Maxwell's time-dependent equations.

1.9.1 Faraday's Law

Let us first consider Faraday's experiments, having put them in their proper historical context. Prior to 1830, the only known way to make an electric current flow through a conducting wire was to connect the ends of the wire to the positive and negative terminals of a battery. We measure a battery's ability to push current down a wire in terms of its *voltage*, by which we mean the voltage difference between its positive and negative terminals.

What does voltage correspond to in physics? Well, volts are the units used to measure electric scalar potential, so when we talk about a 6-V battery, what we are really saying is that the difference in electric scalar potential between its positive and negative terminals is six volts. This insight allows us to write:

$$V = \phi(\oplus) - \phi(\odot) = -\int_{\oplus}^{\odot} \vec{\nabla}\phi \cdot d\vec{l} = \int_{\oplus}^{\odot} \vec{E} \cdot d\vec{l} \tag{1.326}$$

where V is the battery voltage, $\oplus$ denotes the positive terminal, $\odot$ the negative terminal, and $d\vec{l}$ is an element of length along the wire. Of course, this equation is a direct consequence of $\vec{E} = -\vec{\nabla}\phi$. Clearly, a voltage difference between two ends of a wire attached to a battery implies the presence of an electric field that pushes charges through the wire.

This field is directed from the positive terminal of the battery to the negative terminal and, therefore, as such it forces electrons to flow through the wire from the negative to the positive terminal. As expected, this means that there is a net positive current flow from the positive to the negative terminal. The fact that $\vec{E}$ is a conservative field ensures that voltage difference V is independent of the path of the wire. In other words, two different wires attached to the same battery develop identical voltage differences.

Now we consider a closed loop of wire (with no battery). The voltage around such a loop, which is sometimes called the *electromotive force*, or e.m.f., is:

$$V = \oint \vec{E} \cdot d\vec{l} = 0 \tag{1.327}$$

This is a direct consequence of field equation $\vec{\nabla} \times \vec{E} = 0$. So, because $\vec{E}$ is a conservative field the electromotive force around a closed loop of wire is automatically zero, and no current flows around the wire. This all seems to make sense.

Nevertheless, in 1830 Michael Faraday threw a spanner (i.e., wrench) in the works! He discovered that a changing magnetic field can cause a current to flow around a closed loop of wire (in the absence of a battery). Well, if current flows through a wire then there must be an electromotive force. So,

$$V = \oint \vec{E} \cdot d\vec{l} \neq 0 \tag{1.328}$$

which immediately implies that $\vec{E}$ is not a conservative field and that $\vec{\nabla} \times \vec{E} \neq 0$. Clearly, we are going to have to modify some of our ideas regarding electric fields.

Faraday continued his experiments and found that another way of generating an electromotive force around a loop of wire was to keep the magnetic field constant and move the loop. Eventually, Faraday was able to formulate a law that explained all his experiments. The e.m.f. generated around a loop of wire in a magnetic field is proportional to the rate of change of the flux of the magnetic field through the loop. So, if the loop is denoted C, and S is some surface attached to the loop, then Faraday's experiments can be summed up by writing:

$$V = \oint_C \vec{E} \cdot d\vec{l} = A \frac{\partial}{\partial t} \int_S \vec{B} \cdot d\vec{S} \tag{1.329}$$

where A is a constant of proportionality. Thus, the changing flux of the magnetic field through the loop creates an electric field directed around the loop. This process is known as *magnetic induction*.

The International System of Units (SI) has been chosen carefully to make $|A = 1|$ in the previous equation. The only thing we now must decide is whether $A = +1$ or $A = -1$. In other words, which way around the loop does the induced e.m.f. want to drive the current? We possess a general principle that allows us to decide questions like this—*LeChatelier's principle*. According to this principle, every change generates a reaction that tries to minimize the change. Essentially, this means that the Universe is stable to small perturbations. When LeChatelier's principle is applied to the special case of magnetic induction, it usually is called *Lenz's Law*.

According to Lenz's Law, the current induced around a closed loop is always such that the magnetic field it produces tries to counteract the change in magnetic flux that generates the electromotive force. From Fig. 1.23, it is clear that if magnetic field $\vec{B}$ is increasing and current I circulates clockwise (as shown before) then it generates field $\vec{B}'$, which opposes the increase in magnetic flux through the loop in accordance with Lenz's Law. The direction of the current is opposite to the sense of current loop C (assuming that the flux of $\vec{B}$ through the loop is positive), so this implies that $A = -1$ in Eq. 1.323. Thus, Faraday's Law takes the form:

$$\oint_C \vec{E} \cdot d\vec{l} = -\frac{\partial}{\partial t} \int_S \vec{B} \cdot d\vec{S} \tag{1.330}$$

Experimentally, Faraday's Law has been found to correctly predict the e.m.f. (i.e., $\oint \vec{E} \cdot d\vec{l}$) generated in any wire loop, irrespective of the position or shape of the loop. It is reasonable to assume that the same e.m.f. would be generated in the absence of the wire (of course, no current would flow in this case). Thus, Eq. 1.330 is valid for

1.9 Time-Dependent Maxwell's Equations

any closed loop C. If Faraday's Law is to make any sense, it also must be true for any surface S attached to loop C.

Clearly, if the flux of the magnetic field through the loop depends on the surface on which it is evaluated then Faraday's Law is going to predict different e.m.f.s for unlike surfaces. Because there is no preferred surface for a general non-coplanar loop, this would not make very much sense. The condition for the flux of the magnetic field, $\int_S \vec{B} \cdot d\vec{S}$, to depend only on loop C to which surface S is attached, and not on the nature of the surface itself, is:

$$\int_S \vec{B} \cdot d\vec{S} = 0 \qquad (1.331)$$

for any closed surface S'.

Faraday's Law, Eq. 1.330, can be converted into a field equation using Stokes's theorem. We can obtain:

$$\vec{\nabla} \times \vec{E} = -\frac{\partial \vec{B}}{\partial t} \qquad (1.332)$$

This is the final one of Maxwell's equations. It describes how a changing magnetic field can generate, or induce, an electric field. *Gauss's theorem* applied to Eq. 1.325 yields the familiar field equation,

$$\vec{\nabla} \cdot \vec{B} = 0 \qquad (1.333)$$

This ensures that the magnetic flux through a loop is a well-defined quantity.

The divergence of Eq. 1.326 yields:

$$\frac{\partial \vec{\nabla} \cdot \vec{B}}{\partial t} = 0 \qquad (1.334)$$

Thus, field Eq. 1.326 actually demands that the divergence of the magnetic field be constant in time for self-consistency (i.e., the flux of the magnetic field through a loop need not be a well-defined quantity, as long as its time derivative is well defined). A constant non-solenoidal magnetic field, however, can be generated only by magnetic monopoles, and magnetic monopoles do not exist (as far as this author is aware); thus, $\vec{\nabla} \cdot \vec{B} = 0$.

The absence of magnetic monopoles is an observational fact; it cannot be predicted by any theory. If magnetic monopoles were discovered tomorrow, this would not cause physicists any problems. We know how to generalize Maxwell's equations to include both magnetic monopoles and currents of magnetic monopoles. In this generalized formalism, Maxwell's equations are completely symmetrical with respect to electric and magnetic fields and $\vec{\nabla} \cdot \vec{B} \neq 0$. Nevertheless, an extra term (involving the current of magnetic monopoles) must be added to the RHS of Eq. 1.326 in order to make it self-consistent.

1.9.2 Electric Scalar Potential

Now we have a problem. We can write only the electric field in terms of a scalar potential (i.e., $\vec{E} = -\vec{\nabla}\phi$) provided that $\vec{\nabla} \cdot \vec{E} \neq 0$. We have just found, however, that in the presence of a changing magnetic field the curl of the electric field is non-zero. In other words, $\vec{E}$ is not, in general, a conservative field. Does this mean that we must abandon the concept of electric scalar potential? Fortunately, no; it is still possible to define a scalar potential that is physically meaningful.

Let us start from the equation

$$\vec{\nabla} \cdot \vec{B} = 0 \tag{1.335}$$

which is valid for both time-varying and non-time-varying magnetic fields. Because the magnetic field is solenoidal, we can write it as the curl of a vector potential:

$$\vec{B} = \vec{\nabla} \times \vec{A} \tag{1.336}$$

So, there is no problem with the vector potential in the presence of time-varying fields. Let us substitute Eq. 1.336 into the field in Eq. 1.332. We obtain

$$\vec{\nabla} \times \vec{E} = -\frac{\partial(\vec{\nabla} \times \vec{A})}{\partial t} \tag{1.337}$$

which can be written:

$$\vec{\nabla} \times \left(\vec{E} + \frac{\partial \vec{A}}{\partial t}\right) = 0 \tag{1.338}$$

We know that a curl-free vector field can always be expressed as the gradient of a scalar potential, so we can write:

$$\vec{E} + \frac{\partial \vec{A}}{\partial t} = -\vec{\nabla}\phi \tag{1.339}$$

or

$$\vec{E} = -\vec{\nabla}\phi + \frac{\partial \vec{A}}{\partial t} \tag{1.340}$$

This is a very nice equation! It tells us that scalar potential ϕ only describes the conservative electric field generated by electrical charges. The electric field induced by time-varying magnetic fields is non-conservative and is described by *magnetic vector potential* $\vec{A}$.

1.9 Time-Dependent Maxwell's Equations

1.9.3 Gauge Transformations

Electric and magnetic fields can be written in terms of scalar and vector potentials, as follows:

$$\vec{E} = -\vec{\nabla}\phi + \frac{\partial \vec{A}}{\partial t} \quad (1.341)$$

$$\vec{B} = \vec{\nabla} \times \vec{A} \quad (1.342)$$

Nonetheless, this prescription is not unique. There are many different potentials that can generate the same fields. We have come across this problem before. It is called *gauge invariance*. The most general transformation that leaves the $\vec{E}$ and $\vec{B}$ fields unchanged in Eqs. 1.335 and 1.336 is:

$$\phi \rightarrow \phi + \frac{\partial \psi}{\partial t} \quad (1.343)$$

$$\vec{A} \rightarrow \vec{A} - \vec{\nabla}\psi \quad (1.344)$$

This is clearly a generalization of the gauge transformation that we found earlier for static fields:

$$\phi \rightarrow \phi + c \quad (1.345)$$

$$\vec{A} \rightarrow \vec{A} - \vec{\nabla}\psi \quad (1.346)$$

where c is a constant. In fact, if $\psi(\vec{r},t) \rightarrow \psi(\vec{r}) + ct$, then Eqs. 1.343 and 1.338 reduce to Eqs. 1.345 and 1.346.

We are free to choose the gauge to make the equations as simple as possible. As before, the most sensible gauge for the scalar potential is to cause it to go to zero at infinity:

$$\phi(\vec{r}) \rightarrow 0 \quad \text{as} \quad |\vec{r}| \rightarrow \infty \quad (1.347)$$

For steady fields, we found that the optimum gauge for the vector potential was the so-called Coulomb gauge:

$$\vec{\nabla} \cdot \vec{A} = 0 \quad (1.348)$$

We can still use this gauge for non-steady fields.

The argument that we gave earlier (see Sect. 1.8.11), that it is always possible to transform away the divergence of a vector potential, remains valid. One of the nice features of the Coulomb gauge is that when we write the electric field as,

Fig. 1.23 Illustration of magnetic field in a current loop

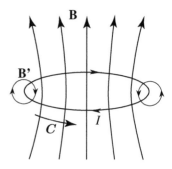

$$\vec{E} = -\nabla\phi - \frac{\partial \vec{A}}{\partial t} \qquad (1.349)$$

we find that the part generated by charges (i.e., the first term on the RHS) is conservative, and the part induced by magnetic fields (i.e., the second term on the RHS) is purely solenoidal.

Earlier on we proved mathematically that a general vector field can be written as the sum of a conservative field and a solenoidal field (see Sect. 1.8.10). Now we are finding that when we split up the electric field in this manner, the two fields have diverse physical origins—that is, the conservative part of the field emanates from electrical charges, whereas the solenoidal part is induced by magnetic fields.

Equation 1.349 can be combined with the field equation to give:

$$\vec{\nabla} \cdot \vec{E} = \frac{\rho}{\varepsilon_0} \qquad (1.350)$$

(which remains valid for non-steady fields) to give:

$$-\nabla^2\phi - \frac{\partial(\vec{\nabla} \cdot \vec{A})}{\partial t} = \frac{\rho}{\varepsilon_0} \qquad (1.351)$$

With the Coulomb gauge condition, $\vec{\nabla} \cdot \vec{A} = 0$, the preceding expression reduces to

$$\nabla^2\phi = -\frac{\rho}{\varepsilon_0} \qquad (1.352)$$

which is simply *Poisson's equation*. Thus, we can write down an expression for the scalar potential generated by non-steady fields immediately. It is exactly the same as our previous expression for the scalar potential generated by steady fields—namely,

1.9 Time-Dependent Maxwell's Equations

$$\phi(\vec{r},t) = \frac{1}{4\pi\varepsilon_0} \int \frac{\rho(\vec{r}',t)}{|\vec{r}-\vec{r}'|} d^3\vec{r}' \qquad (1.353)$$

Nevertheless, this apparently simple result is *extremely* deceptive. Equation 1.353 is a typical action-at-a-distance law. If the charge density changes suddenly at $\vec{r}'$, then the potential at **r** responds *immediately*. Still, we will see later that the full time-dependent Maxwell's equations only allow information to propagate at the speed of light (i.e., they do not violate relativity). How can these two statements be reconciled? The crucial point is that the scalar potential cannot be measured directly; it can only be inferred from the electric field. In the time-dependent case, there are two parts to the electric field: the part that comes from the scalar potential and the part that comes from the vector potential (see Eq. 1.343). Thus, if the scalar potential responds immediately to some distance rearrangement of charge density, it does not necessarily follow that the electric field also has an immediate response.

What really happens is that the change in the part of the electric field that comes from the scalar potential is balanced by an equal and opposite change in the part that comes from the vector potential, so the overall electric field remains unchanged. This persists at least until sufficient time has elapsed for a light signal to travel from the distant charges to the region in question. Thus, relativity is not violated because it is the electric field, not the scalar potential, that carries physically accessible information.

It is clear that the apparent action at a distance nature of Eq. 1.347 is highly misleading. This suggests, very firmly, that the Coulomb gauge is not the optimum gauge in the time-dependent case. A more sensible choice is the so-called *Lorentz gauge*:

$$\vec{\nabla} \cdot \vec{A} = -\varepsilon_0 \mu_0 \frac{\partial \phi}{\partial t} \qquad (1.354)$$

It can be shown, by analogy with earlier arguments (see Sect. 1.8.11), that it is always possible to make a gauge transformation at a given instance in time, such that the preceding equation is satisfied. Substituting the Lorentz gauge condition into Eq. 1.345, we obtain:

$$\varepsilon_0 \mu_0 \frac{\partial^2 \phi}{\partial t^2} - \nabla^2 \phi = \frac{\rho}{\varepsilon_0} \qquad (1.355)$$

It turns out that this is a three-dimensional wave equation in which information propagates at the speed of light. But, more of this later. Note that the magnetically induced part of the electric field (i.e., $-\partial \vec{A}/\partial t$) is not purely solenoidal in the Lorentz gauge. This is a slight disadvantage of this gauge with respect to the Coulomb gauge. This disadvantage, however, is more than offset by other advantages that will become apparent presently. Incidentally, the fact that the part of the electric field, which we ascribe to magnetic induction, changes when we change the gauge suggests that the separation of the field into magnetically induced and charge-induced components is not unique in the general time-varying case (i.e., it is a convention).

1.9.4 The Displacement Current

Michael Faraday revolutionized physics in 1830 by showing that electricity and magnetism were interrelated phenomena. He achieved this breakthrough by careful experimentation. Between 1864 and 1873, James Clerk Maxwell achieved a similar breakthrough by pure thought. Of course, this was only possible because he was able to take the experimental results of Faraday, Ampère, and others as his starting point.

Prior to 1864, the laws of electromagnetism were written in integral form. Thus, Gauss's Law was (in SI units) the "flux of the electric field through a closed surface equals the total enclosed charge, divided by ε_0." The no magnetic monopole law was the "flux of the magnetic field through any closed surface is zero." Faraday's Law was the "electromotive force generated around a closed loop equals minus the rate of change of the magnetic flux through the loop." Finally, Ampère's circuital law was the "line integral of the magnetic field around a closed loop equals the total current flowing through the loop, multiplied by μ_0."

Maxwell's first great achievement was to realize that these laws could be expressed as a set of first-order partial differential equations. Of course, he wrote his equations out in component form because modern vector notation did not come into vogue until about the time of the First World War. In modern notation, Maxwell first wrote:

$$\vec{\nabla} \cdot \vec{E} = \frac{\rho}{\varepsilon_0} \qquad (1.356)$$

$$\vec{\nabla} \cdot \vec{B} = 0 \qquad (1.357)$$

$$\vec{\nabla} \times \vec{E} = -\frac{\partial \vec{B}}{\partial t} \qquad (1.358)$$

$$\vec{\nabla} \times \vec{B} = \mu_0 \vec{J} \qquad (1.359)$$

Maxwell's second great achievement was to realize that these equations were wrong.

We can see that there is something slightly unusual about Eqs. 1.356 through 1.359. They are very unfair to electric fields! After all, time-varying magnetic fields can induce electric fields, but electric fields apparently cannot affect magnetic fields in any way. Nevertheless, there is a far more serious problem associated with these equations, which we alluded to earlier. Consider the integral form of the last of Maxwell's equations (i.e., Ampère's circuital law):

$$\oint_C \vec{B} \cdot d\vec{l} = \mu_0 \int_S \vec{J} \cdot d\vec{S} \qquad (1.360)$$

1.9 Time-Dependent Maxwell's Equations

This says that the line integral of the magnetic field around closed loop C is equal to μ_0 times the flux of the current density through the loop.

The problem is that the flux of the current density through a loop is not, in general, a well-defined quantity. For the flux to be well defined, the integral of $\vec{J} \cdot d\vec{S}$ over some surface S attached to loop C must depend on C but not on the details of S. This is only the case if

$$\vec{\nabla} \cdot \vec{J} = 0 \qquad (1.361)$$

Unfortunately, this condition is only satisfied for non-time-varying fields.

Why do we say that, in general, $\vec{\nabla} \cdot \vec{J} \neq 0$? Well, consider the flux of $\vec{J}$ out of some closed surface S enclosing volume V. This clearly is equivalent to the rate at which charge flows out of S. If charge is a conserved quantity (we certainly believe it is), however, then the rate at which charge flows out of S must equal the rate of decrease of the charge contained in volume V. Thus,

$$\oint_S \vec{J} \cdot d\vec{S} = -\frac{\partial}{\partial t} \int_V \rho \, dV \qquad (1.362)$$

Making use of *Gauss's theorem*, this yields:

$$\vec{\nabla} \cdot \vec{J} = -\frac{\partial \rho}{\partial t} \qquad (1.363)$$

Thus, $\vec{\nabla} \cdot \vec{J} = 0$ is only true in a steady state (i.e., when $\partial/\partial t \equiv 0$).

The problem with Ampère's circuital law is competently illustrated by the following very famous example. Consider a long straight wire interrupted by a parallel plate capacitor. Suppose that C is some loop that circles the wire. In the non-time-dependent situation, the capacitor acts like a break in the wire, so no current flows and no magnetic field is generated. There is clearly no problem with Ampère's Law in this case. In the time-dependent situation, however, a transient current flow in the wire as the capacitor charges up, or charges down, and so a transient magnetic field is generated. Thus, the line integral of the magnetic field around C is (transiently) non-zero. According to Ampère's circuital law, the flux of the current through any surface attached to C also should be (transiently) non-zero. Let us consider two such surfaces.

The first surface, S_1, intersects the wire. This surface causes no problem because the flux of $\vec{J}$ though the surface is clearly non-zero (because it intersects a current-carrying wire). The second surface, S_2, passes between the plates of the capacitor; therefore, it does not intersect the wire at all. Clearly, the flux of the current through this surface is zero. The current fluxes through surfaces S_1 and S_2 are obviously different. Both surfaces, however, are attached to the same loop C, so the fluxes should be the same, according to Ampère's Law—Eq. 1.354.

It would appear that Ampère's circuital law is about to disintegrate! But, we notice that although surface S_2 does not intersect any electric current, it does pass through a region of a strong changing electric field as it threads between the plates of the charging (or discharging) capacitor. Perhaps, if we add a term involving $\partial \vec{E}/\partial t$ to the RHS of Eq. 1.353, we can somehow fix Ampère's circuital law. This is, essentially, how Maxwell reasoned more than 100 years ago.

Let us try this scheme. Suppose that we write

$$\vec{\nabla} \times \vec{B} = \mu_0 \vec{J} + \lambda \frac{\partial \vec{E}}{\partial t} \tag{1.364}$$

instead of Eq. 1.359. Here λ is some constant. Does this resolve the problem? We want the flux of the RHS of the preceding equation through some loop C to be well defined; that is, it should only depend on C, not the particular surface S (which spans C) on which it is evaluated. This is another way of saying that we want the divergence of the RHS to be zero.

In fact, we can see that this is necessary for self-consistency because the divergence of the LHS is automatically zero. So, taking the divergence of Eq. 1.358, we obtain

$$0 = \mu_0 \vec{\nabla} \cdot \vec{J} + \lambda \frac{\partial (\nabla \cdot \vec{E})}{\partial t} \tag{1.365}$$

But, we know that

$$\vec{\nabla} \cdot \vec{E} = \frac{\rho}{\varepsilon_0} \tag{1.366}$$

so combining the previous two equations we arrive at

$$\mu_0 \vec{\nabla} \cdot \vec{J} + \frac{\lambda}{\varepsilon_0} \frac{\partial \rho}{\partial t} = 0 \tag{1.367}$$

Now, the charge conservation law in Eq. 1.357 can be written:

$$\vec{\nabla} \cdot \vec{J} + \frac{\partial \rho}{\partial t} = 0 \tag{1.368}$$

The previous two equations are in agreement provided $\lambda = \varepsilon_0 \mu_0$. Therefore, if we modify the final one of Maxwell's equations such that it reads:

$$\vec{\nabla} \times \vec{B} = \mu_0 \vec{J} + \varepsilon_0 \mu_0 \frac{\partial \vec{E}}{\partial t} \tag{1.369}$$

we find that the divergence of the RHS is zero because of charge conservation. The extra term is called the *displacement current* (this name was created by Maxwell). In summary, we have shown that although the flux of the real current through a loop is *not* adequately defined, if we form the sum of the real current and the displacement current, the flux of this new quantity through a loop *is* thoroughly defined.

1.9 Time-Dependent Maxwell's Equations

Of course, the displacement current is not a current at all. It in fact is associated with the generation of magnetic fields by time-varying electric fields. Maxwell came up with this rather curious name because many of his ideas regarding electric and magnetic fields were completely wrong. For instance, Maxwell believed in the "other," and he thought that electric and magnetic fields were some sort of stresses in this medium. He also thought that the displacement current was associated with displacements of the other (thus, the name). The reason that these misconceptions did not invalidate his equations is quite simple. Maxwell based his equations on the results of experiments, and he added in his extra term to make these equations mathematically self-consistent. Both steps are valid irrespective of the existence or non-existence of the other.

"But, hang on a minute," you might say; "you cannot go around adding terms to laws of physics just because you feel like it!" The field Eqs. 1.356 through 1.359 are derived directly from the results of famous nineteenth-century experiments. If there is a new term involving the time derivative of the electric field that needs to be added into these equations, how come there is no corresponding nineteenth-century experiment that demonstrates this? We have Faraday's Law that shows that changing magnetic fields generate electric fields. Why is there no *Joe Blogg's Law* that says that changing electric fields generate magnetic fields? This is a perfectly reasonable question. The answer is that the new term describes an effect that is far too small to have been observed in nineteenth-century experiments. Let us demonstrate this.

First, we will show that it is comparatively easy to detect the induction of an electric field by a changing magnetic field in a desktop laboratory experiment. The Earth's magnetic field is about 1 gauss (i.e., 10^{-4} T). Magnetic fields generated by electromagnets (that will fit on a laboratory desktop) are typically about 100 times larger than this. Let us, therefore, consider a hypothetical experiment in which a 100-G magnetic field is switched on suddenly. Suppose that the field ramps up in one-tenth of a second. Which electromotive force is generated in a 10-centimeter square loop of wire located in this field?

Faraday's Law is written:

$$V = -\frac{\partial}{\partial t} \oint \vec{B} \cdot d\vec{S} \sim \frac{BA}{t} \tag{1.370}$$

where $B = 0.01$ T is the field-strength, $A = 0.01$ m² is the area of the loop, and $t = 0.1$ is the ramp time. It follows that $V \sim 1$ millivolt (mV). Well, 1 mV is easily detectable. Most hand-held laboratory voltmeters in fact are calibrated in millivolts. It is clear that we would have no difficulty whatsoever detecting the magnetic induction of electric fields in a nineteenth-century style laboratory experiment.

We now consider the electric induction of magnetic fields. Suppose that the electric field is generated by a parallel plate capacitor of a spacing of 1 cm that is charged to 100 V. This gives a field of 10^4 V per meter (m). Suppose, further, that the capacitor is discharged in one-tenth of a sec. The law of electric induction is obtained by integrating Eq. 1.363 and neglecting the first term on the RHS. Thus,

$$\oint \vec{B} \cdot d\vec{l} = \varepsilon_0 \mu_0 \frac{\partial}{\partial t} \int \vec{E} \cdot d\vec{S} \tag{1.371}$$

Let us consider a loop 10^2 cm. What is the magnetic field generated around this loop (we could try to measure this with a Hall probe). Very approximately, we find that

$$lB \sim \varepsilon_0 \mu_0 \frac{El^2}{t} \tag{1.372}$$

where $l = 0.1$ m is the dimension of the loop, B is the magnetic field-strength, $E = 10^4$ V/m is the electric field, and $t = 0.1$ sec is the decay time of the field. We find that $B \sim 10^{-9}$ G. Modern technology is unable to detect such a small magnetic field, so we cannot really blame Faraday for not noticing electric induction in 1830.

So, you might say, "Why did you bother mentioning this displacement current thing in the first place if it is undetectable?" Again, a perfectly fair question. The answer is that the displacement current *is* detectable in some experiments. Suppose that we take an FM radio signal, amplify it so that its peak voltage is 100 V, and then apply it to the parallel plate capacitor in the previous hypothetical experiment. What size of magnetic field would this generate? Well, a typical FM signal oscillates at 10^{-9} Hz, so t in the previous example changes from 0.1 to 10^{-9}. Thus, the induced magnetic field is about 10^{-1} G. This certainly is detectable by modern technology.

Therefore, it appears that if the electric field is oscillating fast, then electric induction of magnetic fields is an observable effect. In fact, there is a virtually infallible rule for deciding whether the displacement current can be neglected in Eq. 1.369. If *EM radiation* is important, then the displacement current must be included. On the other hand, if EM radiation is unimportant, the displacement current can be safely neglected. Clearly, Maxwell's inclusion of the displacement current in Eq. 1.369 was a vital step in his later realization that his equations allowed propagating wave-like solutions. These solutions are, of course, EM waves. But, more on this later.

We are now in a position to write out Maxwell's equations in all their glory! We get

$$\vec{\nabla} \cdot \vec{E} = \frac{\rho}{\varepsilon_0} \tag{1.373}$$

$$\vec{\nabla} \cdot \vec{B} = 0 \tag{1.374}$$

$$\vec{\nabla} \times \vec{E} = -\frac{\partial \vec{B}}{\partial t} \tag{1.375}$$

$$\vec{\nabla} \times \vec{B} = \mu_0 \vec{J} + \varepsilon_0 \mu_0 \frac{\partial \vec{E}}{\partial t} \tag{1.376}$$

These four partial differential equations constitute a *complete* description of the behavior of electric and magnetic fields. The first equation describes how electric fields are induced by charges. The second says that there is no such thing as a

1.9 Time-Dependent Maxwell's Equations

magnetic monopole. The third equation describes the induction of electric fields by changing magnetic fields, and the fourth describes the generation of magnetic fields by electrical currents and the induction of magnetic fields by changing electric fields.

Note that with the inclusion of the displacement current these equations treat electric and magnetic fields on an equal footing—that is, electric fields can induce magnetic fields and vice versa. Equations 1.373 through 1.376 sum up the experimental results of Coulomb, Ampère, and Faraday very succinctly; they are called *Maxwell's equations* because James Clerk Maxwell was the first to write them down (in component form). Maxwell also fixed them up so that they made mathematical sense.

1.9.5 Potential Formulation

We have seen that Eqs. 1.357 and 1.358 are satisfied automatically if we write the electric and magnetic fields in terms of potentials:

$$\vec{E} = -\nabla \phi - \frac{\partial \vec{A}}{\partial t} \tag{1.377}$$

$$\vec{B} = \vec{\nabla} \times \vec{A} \tag{1.378}$$

This prescription is not unique, but we can make it unique by adopting the following conventions:

$$\phi(\vec{r}) \to 0 \quad \text{as} \quad |\vec{r}| \to \infty \tag{1.379}$$

$$\vec{\nabla} \cdot \vec{A} = -\varepsilon_0 \mu_0 \frac{\partial \phi}{\partial t} \tag{1.380}$$

These equations can be combined with Eq. 1.356 to give:

$$\varepsilon_0 \mu_0 \frac{\partial^2 \phi}{\partial t^2} - \nabla^2 \phi = \frac{\rho}{\varepsilon_0} \tag{1.381}$$

Let us now consider Eq. 1.376. Substitution of Eqs. 1.377 and 1.378 into this formula yields:

$$\vec{\nabla} \times \vec{\nabla} \times \vec{A} \equiv \vec{\nabla}(\vec{\nabla} \cdot \vec{A}) - \nabla^2 \vec{A} = \mu_0 \vec{J} - \varepsilon_0 \mu_0 \frac{\partial(\nabla \phi)}{\partial t} - \varepsilon_0 \mu_0 \frac{\partial^2 \vec{A}}{\partial t^2} \tag{1.382}$$

or

$$\varepsilon_0\mu_0 \frac{\partial^2 \vec{A}}{\partial t^2} - \nabla^2 \vec{A} = \mu_0 \vec{J} - \nabla\left(\vec{\nabla}\cdot\vec{A} + \varepsilon_0\mu_0 \frac{\partial \phi}{\partial t}\right) \quad (1.383)$$

We can see quite clearly where the Lorentz gauge condition of Eq. 1.354 comes from. This equation is, in general, very complicated because it involves both vector and scalar potentials. But, if we adopt the Lorentz gauge, the last term on the RHS becomes zero, and the equation simplifies considerably, such that it only involves the vector potential.

Thus, we find that Maxwell's equations reduce to the following:

$$\varepsilon_0\mu_0 \frac{\partial^2 \phi}{\partial t^2} - \nabla^2 \phi = \frac{\rho}{\varepsilon_0} \quad (1.384)$$

$$\varepsilon_0\mu_0 \frac{\partial^2 \vec{A}}{\partial t^2} - \nabla^2 \vec{A} = \mu_0 \vec{J} \quad (1.385)$$

This is the same (scalar) equation written four times over. In a steady state (i.e., $\partial/\partial t$), it reduces to Poisson's equation, which we know how to solve. With the $\partial^2/\partial t^2$ terms included, it becomes a slightly more complicated equation—in fact, a driven three-dimensional wave equation.

1.9.6 Electromagnetic Waves

This is an appropriate point at which to demonstrate that Maxwell's equations possess propagating wave-like solutions. Let us start with them in free space (i.e., with no charges and no currents):

$$\vec{\nabla}\cdot\vec{E} = 0 \quad (1.386)$$

$$\vec{\nabla}\cdot\vec{B} = 0 \quad (1.387)$$

$$\vec{\nabla}\times\vec{E} = -\frac{\partial \vec{B}}{\partial t} \quad (1.388)$$

$$\vec{\nabla}\times\vec{B} = \varepsilon_0\mu_0 \frac{\partial \vec{E}}{\partial t} \quad (1.389)$$

Note that these equations exhibit a nice symmetry between the electric and magnetic fields.

There is an easy way to show that the preceding equations possess wave-like solutions, and a hard way. The easy way is to assume that the solutions are going to be wave-like beforehand. Specifically, let us search for plane-wave solutions of the form:

1.9 Time-Dependent Maxwell's Equations

$$\vec{E}(\vec{r},t) = \vec{E}_0 \cos(\vec{k} \cdot \vec{r} - \omega t) \tag{1.390}$$

$$\vec{B}(\vec{r},t) = \vec{B}_0 \cos(\vec{k} \cdot \vec{r} - \omega t + \phi) \tag{1.391}$$

Here $\vec{E}_0$ and $\vec{B}_0$ are constant vectors, $\vec{k}$ is called the wave vector, and ω is the angular frequency. The frequency in Hz, f, is related to the angular frequency using $\omega = 2\pi f$. The frequency is conventionally defined to be positive. Quantity ϕ is a phase difference between the electric and the magnetic fields.

Actually, it is more convenient to write:

$$\vec{E} = \vec{E}_0 e^{i(\vec{k} \cdot \vec{r} - \omega t)} \tag{1.392}$$

$$\vec{B} = \vec{B}_0 e^{i(\vec{k} \cdot \vec{r} - \omega t)} \tag{1.393}$$

where, by convention, the physical solution is the *real part* of the equations. Phase difference ϕ is absorbed into constant vector $\vec{B}_0$ by allowing it to become complex; thus, $\vec{B}_0 \to \vec{B}_0 e^{i\phi}$. In general, vector $\vec{E}_0$ is also complex.

A wave maximum of the electric field satisfies

$$\vec{k} \cdot \vec{r} = \omega t + n2\pi + \phi \tag{1.394}$$

where n is an integer and ϕ is some phase angle. The solution to this equation is a set of equally spaced parallel planes (one plane for each possible value of n), with a normal that lies in the direction of wave-vector $\vec{k}$, and that propagate in this direction with phase velocity:

$$v = \frac{\omega}{k} \tag{1.395}$$

The spacing between adjacent planes (i.e., the wavelength) shown in Fig. 1.24 is given by

$$\lambda = \frac{2\pi}{k} \tag{1.396}$$

Fig. 1.24 The spacing between adjacent planes

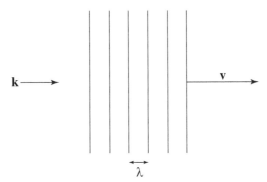

Consider a general plane-wave vector field:

$$\vec{A} = \vec{A}_0 e^{i(\vec{k}\cdot\vec{r}-\omega t)} \tag{1.397}$$

What is the divergence of $\vec{A}$? This is easy to evaluate. We have

$$\begin{aligned}\vec{\nabla}\cdot\vec{A} &= \frac{\partial A_x}{\partial x} + \frac{\partial A_y}{\partial y} + \frac{\partial A_z}{\partial z} \\ &= (A_{0x} ik_x + A_{0y} ik_y + A_{0z} ik_z) \\ &= i\vec{k}\cdot\vec{A}\end{aligned} \tag{1.398}$$

How about the curl of $\vec{A}$? This is slightly more difficult. We have

$$\begin{aligned}(\vec{\nabla}\times\vec{A})_x &= \frac{\partial A_z}{\partial y} - \frac{\partial A_y}{\partial z} = (ik_x A_z - ik_z A_y) \\ &= i(\vec{k}\times\vec{A})_x\end{aligned} \tag{1.399}$$

This is easily generalized to:

$$\vec{\nabla}\times\vec{A} = i\vec{k}\times\vec{A} \tag{1.400}$$

We can see that vector field operations on a plane wave simplify to replacing the ∇ operator with $i\vec{k}$.

The first of Maxwell's equations, Eq. 1.380, reduces to:

$$i\vec{k}\cdot\vec{E}_0 = 0 \tag{1.401}$$

using the assumed electric and magnetic fields in Eqs. 1.392 and 1.393 as well as Eq. 1.398. Thus, the electric field is perpendicular to the direction of propagation of the wave. Likewise, the second of Maxwell's equations gives

$$i\vec{k}\cdot\vec{B}_0 = 0 \tag{1.402}$$

implying that the magnetic field also is perpendicular to the direction of propagation. Clearly, the wave-like solutions of Maxwell's equations are a type of *transverse wave* (TW). The third of Maxwell's equations gives

$$i\vec{k}\times\vec{E}_0 = i\omega\vec{B}_0 \tag{1.403}$$

where Eq. 1.394 has been used.

Dotting this equation with $\vec{E}_0$ yields:

$$\vec{E}_0\cdot\vec{B}_0 = \frac{\vec{E}_0\cdot\vec{k}\times\vec{E}_0}{\omega} = 0 \tag{1.404}$$

1.9 Time-Dependent Maxwell's Equations

Thus, the electric and magnetic fields are mutually perpendicular. Dotting Eq. 1.403 with $\vec{B}_0$ yields:

$$\vec{B}_0 \cdot \vec{k} \times \vec{E}_0 = \omega B_0^2 > 0 \tag{1.405}$$

Thus, vectors $\vec{E}_0$, $\vec{B}_0$, and $\vec{k}$ are mutually perpendicular and form a right-handed set. The final one of Maxwell's equations gives

$$i\vec{k} \times \vec{B}_0 = -i\varepsilon_0 \mu_0 \omega \vec{E}_0 \tag{1.406}$$

Combining this with Eq. 1.403 yields:

$$\vec{k} \times (\vec{k} \times \vec{E}_0) = (\vec{k} \cdot \vec{E}_0)\vec{k} - k^2 \vec{E}_0 = k^2 \vec{E}_0 = -\varepsilon_0 \mu_0 \omega^2 \vec{E}_0 \tag{1.407}$$

or

$$k^2 = \varepsilon_0 \mu_0 \omega^2 \tag{1.408}$$

where use has been made of Eq. 1.401. Still, we know from Eq. 1.395 that phase-velocity c is related to the magnitude of the wave-vector and the angular wave frequency via $c = \omega/k$. Thus, we obtain

$$c = \frac{1}{\sqrt{\varepsilon_0 \mu_0}} \tag{1.409}$$

So, we have found TW solutions of the free-space in Maxwell's equations, propagating at some phase-velocity c, which is given by a combination of ε_0 and μ_0. The constants ε_0 and μ_0 are easily measurable. The former is related to the force acting between stationary electrical charges, and the latter to the force acting between steady electrical currents. Both constants were fairly well known in Maxwell's time. Incidentally, he was the first person to look for wave-like solutions to his equations and, thus, to derive Eq. 1.409. The modern values of ε_0 and μ_0 are:

$$\varepsilon_0 = 8.8542 \times 10^{-12} \text{ C}^2 \text{N}^{-1} \text{ m}^{-2} \tag{1.410}$$

$$\mu_0 = 4\pi \times 10^{-7} \text{ NA}^{-2} \tag{1.411}$$

Let us use these values to find the phase velocity of "electromagnetic waves." So, we obtain:

$$c = \frac{1}{\sqrt{\varepsilon_0 \mu_0}} = 2.998 \times 10^8 \text{ m s}^{-1} \tag{1.412}$$

Of course, we immediately recognize this as the velocity of light. Maxwell also made this connection back in the 1870s. He conjectured that light, with a nature that had previously been unknown, was a form of EM radiation. This was a remarkable

Table 1.3 The Electromagnetic Spectrum

Radiation type	Wavelength range (m)
Gamma rays	$<10^{-11}$
X-rays	$10^{-11}-10^{-9}$
Ultraviolet	$10^{-9}-10^{-7}$
Visible	$10^{-7}-10^{-6}$
Infrared	$10^{-6}-10^{-4}$
Microwave	$10^{-4}-10^{-1}$
TV – FM	$10^{-1}-10^{1}$
Radio	$>10^{1}$

extrapolation. After all, Maxwell's equations were derived from the results of benchtop laboratory experiments involving charges, batteries, coils, and currents, which apparently had nothing whatsoever to do with light.

Maxwell was able to make another remarkable prediction. The wavelength of light was well known in the late nineteenth century from studies of diffraction through slits, and so on. Actually, visible light occupies a surprisingly narrow wavelength range. The shortest wavelength, blue light, that is visible has $\lambda = 0.4$ microns (μ)—1 μ is 10^{-6} m. The longest wavelength, red light, that is visible has $\lambda = 0.76\,\mu$. There is nothing in our analysis, however, that suggests that this particular range of wavelengths is special. Electromagnetic waves can have any wavelength. Maxwell concluded that visible light was a small part of a vast spectrum of previously undiscovered types of EM radiation.

Since Maxwell's time, virtually all the non-visible parts of the EM spectrum have been observed. Table 1.3 gives a brief guide to that spectrum. Electromagnetic waves are particularly important because they are our only source of information regarding the Universe around us. Radio waves and microwaves, which are comparatively hard to scatter, have provided much of our knowledge about the center of our Galaxy. This is completely unobservable in visible light, which is strongly scattered by interstellar gas and dust lying in the galactic plane.

For the same reason, the spiral arms of our Galaxy can be mapped out only by using radio waves. Infrared (IR) radiation is useful for detecting protostars, which are not yet hot enough to emit visible radiation. Of course, visible radiation is still the mainstay of astronomy. Satellite-based ultraviolet (UV) observations have yielded invaluable insights into the structure and distribution of distant galaxies. Finally, x-ray and γ-ray astronomy usually concentrates on exotic objects in the Galaxy (e.g., pulsars and supernova remnants).

Equations 1.402, 1.403, and the relation $c = \omega/k$ imply that

$$B_0 = \frac{E_0}{c} \tag{1.413}$$

Thus, the magnetic field associated with an EM wave is smaller in magnitude than the electric field by a factor of c. Consider a free charge interacting with an EM wave. The force exerted on the charge is given by the *Lorentz formula*:

1.9 Time-Dependent Maxwell's Equations

$$\vec{F} = q(\vec{E} + \vec{v} \times \vec{B}) \qquad (1.414)$$

The ratio of the electric and magnetic forces is:

$$\frac{F_{\text{magnetic}}}{F_{\text{electric}}} \sim \frac{vB_0}{E_0} \sim \frac{v}{c} \qquad (1.415)$$

So, unless the charge is relativistic, the electric force greatly exceeds the magnetic force. Clearly, in most terrestrial situations EM waves are an essentially *electric* phenomenon (as far as their interaction with matter goes). For this reason, EM waves usually are characterized by their wave vector, which specifies direction of propagation and wavelength and the plane of polarization (i.e., the plane of oscillation) of the associated electric field. For a given wave-vector $\vec{k}$, the electric field can have any direction in the plane normal to $\vec{k}$. There are, however, only two *independent* directions in a plane (i.e., we can define only two linearly independent vectors). This implies that there are only two independent polarizations of an EM wave once its direction of propagation is specified.

Let us now derive the velocity of light from Maxwell's equations the hard way. Suppose that we take the curl of the fourth of Maxwell's equations, Eq. 1.399. We obtain:

$$\vec{\nabla} \times \vec{\nabla} \times \vec{B} = \vec{\nabla}(\vec{\nabla} \cdot \vec{B}) - \nabla^2 \vec{B} = -\nabla^2 \vec{B} = \varepsilon_0 \mu_0 \frac{\partial(\vec{\nabla} \times \vec{E})}{\partial t} \qquad (1.416)$$

Here, we have used the fact that $\vec{\nabla} \cdot \vec{B} = 0$. The third of Maxwell's equations, Eq. 1.398, yields

$$\left(\nabla^2 - \frac{1}{c^2}\frac{\partial^2}{\partial t^2}\right)\vec{B} = 0 \qquad (1.417)$$

where Eq. 1.412 has been used. A similar equation can be obtained for the electric field by taking the curl of Eq. 1.388:

$$\left(\nabla^2 - \frac{1}{c^2}\frac{\partial^2}{\partial t^2}\right)\vec{E} = 0 \qquad (1.418)$$

We have found that electric and magnetic fields both satisfy equations of the form

$$\left(\nabla^2 - \frac{1}{c^2}\frac{\partial^2}{\partial t^2}\right)\vec{A} = 0 \qquad (1.419)$$

in free space. As is easily verified, the most general solution to this equation (with a positive frequency) is

$$A_x = F_x(\vec{k} \cdot \vec{r} - kct) \tag{1.420}$$

$$A_y = F_y(\vec{k} \cdot \vec{r} - kct) \tag{1.421}$$

$$A_z = F_z(\vec{k} \cdot \vec{r} - kct) \tag{1.422}$$

where $F_x(\phi)$, $F_y(\phi)$, and $F_z(\phi)$ are one-dimensional scalar functions. Looking along the direction of the wave-vector so that $r = (\vec{k}/k)r$, we find that

$$A_x = F_x[(k(r - ct)] \tag{1.423}$$

$$A_y = F_y[(k(r - ct)] \tag{1.424}$$

$$A_z = F_z[(k(r - ct)] \tag{1.425}$$

The x-component of this solution is shown schematically in Fig. 1.25. It clearly propagates in r with velocity c. If we look along a direction that is perpendicular to $\vec{k}$ then $\vec{k} \cdot \vec{r} = 0$, and there is no propagation. Thus, the components of $\vec{A}$ are arbitrarily shaped pulses that propagate, without changing shape, along the direction of $\vec{k}$ with velocity c. These pulses can be related to the sinusoidal plane-wave solutions we found earlier by *Fourier transformation*. Thus, any arbitrarily shaped pulse propagating in the direction of $\vec{k}$ with velocity c can be broken down into lots of sinusoidal oscillations propagating in the same direction with the same velocity.

The operator

$$\nabla^2 - \frac{1}{c^2}\frac{\partial^2}{\partial t^2} \tag{1.426}$$

is called the *d'Alembertian*. It is the four-dimensional equivalent of the Laplacian. Recall that the Laplacian is invariant under rotational transformation. The d'Alembertian goes one better than this because it is both rotationally invariant and *Lorentz invariant*. The d'Alembertian conventionally is denoted $\square^2$. Thus, EM waves in free space satisfy these wave equations:

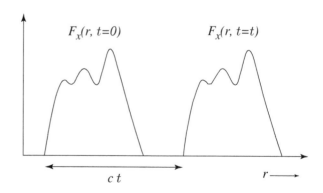

Fig. 1.25 Illustration of wave propagation

1.9 Time-Dependent Maxwell's Equations

$$\Box^2 \vec{E} = 0 \tag{1.427}$$

$$\Box^2 \vec{B} = 0 \tag{1.428}$$

When written in terms of the vector and scalar potentials, Maxwell's equations reduce to:

$$\Box^2 \phi = -\frac{\rho}{\varepsilon_0} \tag{1.429}$$

$$\Box^2 \vec{A} = -\mu_0 \vec{J} \tag{1.430}$$

These clearly are driven wave equations. Our next task is to find the solutions to these equations.

1.9.7 Green's Function

Earlier on in this chapter, we had to solve Poisson's equation,

$$\nabla^2 u = v \tag{1.431}$$

where $v(\vec{r})$ is denoted the source function. The potential $u(\vec{r})$ satisfies the boundary condition

$$u(\vec{r}) \to 0 \quad \text{as} \quad |\vec{r}| \to \infty \tag{1.432}$$

provided that the source function is reasonably localized. The solutions to Poisson's equation are superposable (because the equation is linear). This property is exploited in the Green's function method of solving this equation. Green's function $G(\vec{r}, \vec{r}\,')$ is the potential, which satisfies the appropriate boundary conditions, generated by a unit amplitude point source located at $\vec{r}\,'$. Thus,

$$\nabla^2 G(\vec{r}, \vec{r}\,') = \delta(\vec{r} - \vec{r}\,') \tag{1.433}$$

Any source function $v(\vec{r})$ can be represented as a weighted sum of point sources:

$$v(\vec{r}) = \int \delta(\vec{r} - \vec{r}\,') v(\vec{r}\,') d^3 \vec{r}\,' \tag{1.434}$$

It follows from superposability that the potential generated by source $v(\vec{r})$ can be written as the weighted sum of point source-driven potentials (i.e., Green's function):

$$u(\vec{r}) = \int G(\vec{r}, \vec{r}\,')v(\vec{r})d^3\vec{r}\,' \tag{1.435}$$

We found earlier that the Green's function for Poisson's equation is

$$G(\vec{r}, \vec{r}\,') = -\frac{1}{4\pi}\frac{1}{|\vec{r} - \vec{r}\,'|} \tag{1.436}$$

It follows that the general solution to Eq. 1.431 is written:

$$u(\vec{r}) = -\frac{1}{4\pi}\int \frac{v(\vec{r}\,')}{|\vec{r} - \vec{r}\,'|}d^3\vec{r}\,' \tag{1.437}$$

Note that the point source-driven potential of Eq. 1.436 is perfectly sensible. It is spherically symmetrical about the source and falls off smoothly with increasing distance from it.

We now need to solve the wave equation:

$$\left(\nabla^2 - \frac{1}{c^2}\frac{\partial^2}{\partial t^2}\right)u(\vec{r}, t) = v(\vec{r}, t) \tag{1.438}$$

where $v(\vec{r}, t)$ is a time-varying source function. The potential $u(\vec{r}, t)$ satisfies the boundary conditions,

$$u(\vec{r}) \to 0 \quad \text{as} \quad |\vec{r}| \to \infty \quad \text{and} \quad |t| \to \infty \tag{1.439}$$

The solutions to Eq. 1.438 are superposable (because the equation is linear), so a Green's function method of solution is again appropriate. Green's function $G(\vec{r}, ; \vec{r}\,', t, t')$ is the potential generated by a point *impulse* located at position $\vec{r}\,'$ and applied at time t'. Thus,

$$\left(\nabla^2 - \frac{1}{c^2}\frac{\partial^2}{\partial t^2}\right)G(\vec{r}, ; \vec{r}\,', t, t') = \delta(\vec{r} - \vec{r}\,')\delta(t - t') \tag{1.440}$$

Of course, the Green's function must satisfy the correct boundary conditions. A general source $v(\vec{r}, t)$ can be built up from a weighted sum of point impulses:

$$v(\vec{r}, t) = \iint \delta(\vec{r} - \vec{r}\,')\delta(t - t')v(\vec{r}\,', t')d^3\vec{r}\,'dt' \tag{1.441}$$

It follows that the potential generated by $v(\vec{r}, t)$ can be written as the weighted sum of point impulse-driven potentials:

$$u(\vec{r}, t) = \iint G(\vec{r}, ; \vec{r}\,', t, t')v(\vec{r}\,', t')d^3\vec{r}\,'dt' \tag{1.442}$$

1.9 Time-Dependent Maxwell's Equations

So, how do we find the Green's function?
Consider

$$G(\vec{r},;\vec{r};',t,t') = \frac{F(t - t' - |\vec{r} - \vec{r}'|/c)}{|\vec{r} - \vec{r}'|} \tag{1.443}$$

where $F(\phi)$ is a general scalar function. Let us try to prove the following theorem:

$$\left(\nabla^2 - \frac{1}{c^2}\frac{\partial^2}{\partial t^2}\right)G(\vec{r},;\vec{r};',t,t') = -4\pi F\left(t - t'\right)\delta(\vec{r} - \vec{r}') \tag{1.444}$$

At a general point, $\vec{r} \neq \vec{r}'$, the expression reduces to

$$\left(\nabla^2 - \frac{1}{c^2}\frac{\partial^2}{\partial t^2}\right)G(\vec{r},;\vec{r};',t,t') = 0 \tag{1.445}$$

So, we basically have to show that G is a valid solution of the free-space wave equation. We can easily show that

$$\frac{\partial |\vec{r} - \vec{r}'|}{\partial x} = \frac{x - x'}{|\vec{r} - \vec{r}'|} \tag{1.446}$$

It follows by simple differentiation that

$$\frac{\partial^2 G}{\partial x^2} = \left(\frac{3(x - x')^2 - |\vec{r} - \vec{r}'|}{|\vec{r} - \vec{r}'|^5}\right)F$$
$$+ \left(\frac{3(x - x')^2 - |\vec{r} - \vec{r}'|}{|\vec{r} - \vec{r}'|^4}\right)\frac{F'}{c} \tag{1.447}$$
$$+ \left(\frac{(x - x')^2}{|\vec{r} - \vec{r}'|^3}\right)\frac{F''}{c^2}$$

where $F'(\phi) = dF(\phi)/d\phi$. We can derive analogous equations for $\partial^2 G/\partial y^2$ and $\partial^2 G/\partial z^2$. Thus,

$$\nabla^2 G = \frac{\partial^2 G}{\partial x^2} + \frac{\partial^2 G}{\partial y^2} + \frac{\partial^2 G}{\partial z^2} = \frac{F''}{|\vec{r} - \vec{r}'|c^2} = \frac{1}{c^2}\frac{\partial^2 G}{\partial t^2} \tag{1.448}$$

giving

$$\left(\nabla^2 - \frac{1}{c^2}\frac{\partial^2}{\partial t^2}\right)G(\vec{r},;\vec{r};',t,t') = 0 \tag{1.449}$$

which is the desired result.

Consider, now, the region around $\vec{r} = \vec{r}'$. It is clear from Eq. 1.438 that the dominant term on the RHS as $|\vec{r} - \vec{r}'| \to 0$ is the first one, which is essentially $F\partial^2\left(|\vec{r} - \vec{r}'|^{-1}\right)/\partial x^2$. It is also clear that $(1/c^2)(\partial^2 G/\partial t^2)$ is negligible compared to this term. Thus, as $|\vec{r} - \vec{r}'| \to 0$ we find that

$$\left(\nabla^2 - \frac{1}{c^2}\frac{\partial^2}{\partial x^2}\right)G \to F\left(t - t'\right)\nabla^2\left(\frac{1}{|\vec{r} - \vec{r}'|}\right) \tag{1.450}$$

According to Eqs. 1.433 and 1.436, however,

$$\nabla^2\left(\frac{1}{|\vec{r} - \vec{r}'|}\right) = -4\pi\delta(\vec{r} - \vec{r}') \tag{1.451}$$

We conclude that

$$\left(\nabla^2 - \frac{1}{c^2}\frac{\partial^2}{\partial x^2}\right)G = -4\pi F\left(t - t'\right)\delta(\vec{r} - \vec{r}') \tag{1.452}$$

which is the desired result.

Let us now make the special choice:

$$F(\phi) = -\frac{\delta(\phi)}{4\pi} \tag{1.453}$$

It follows from Eq. 1.452 that

$$\left(\nabla^2 - \frac{1}{c^2}\frac{\partial^2}{\partial x^2}\right)G = \delta(\vec{r} - \vec{r}')\delta\left(t - t'\right) \tag{1.454}$$

Thus,

$$G(\vec{r} - \vec{r}'; t, t') = \frac{1}{4\pi}\frac{\delta(t - t' - |\vec{r} - \vec{r}'|/c)}{|\vec{r} - \vec{r}'|} \tag{1.455}$$

The equation is the Green's function for the driven wave of Eq. 1.438.

The time-dependent Green's function, Eq. 1.455, is the same as the steady-state Green's function, Eq. 1.436, apart from the delta-function appearing in the former. What does this delta-function do? Well, consider an observer at point $\vec{r}$. Because of the delta-function, our observer only measures a non-zero potential at one specific time:

$$t = t' + \frac{|\vec{r} - \vec{r}'|}{c} \tag{1.456}$$

1.9 Time-Dependent Maxwell's Equations

It is clear that this is the time the impulse was applied at position $\vec{r}'$ (i.e., t') *plus* the time taken for a light signal to travel between points $\vec{r}'$ and $\vec{r}$. At time $t > t'$, the locus of all the points at which the potential is non-zero is:

$$|\vec{r} - \vec{r}'| = c(t - t') \qquad (1.457)$$

In other words, it is a sphere centered on $\vec{r}'$ with a radius that is the distance traveled by light in the time interval since the impulse was applied at position $\vec{r}'$. Thus, Green's function, Eq. 1.446, describes a spherical wave that emanates from position $\vec{r}'$ at time t' and propagates at the speed of light. The amplitude of the wave is inversely proportional to the distance from the source.

1.9.8 Retarded Potentials

We are now able to solve Maxwell's equations. Recall that in a steady state, Maxwell's equations reduce to:

$$\nabla^2 \phi = -\frac{\rho}{\varepsilon_0} \qquad (1.458)$$

$$\nabla^2 \vec{A} = -\mu_0 \vec{J} \qquad (1.459)$$

The solutions to these equations are easily found using the Green's function for Poisson's Eq. 1.436:

$$\phi(\vec{r}) = \frac{1}{4\pi\varepsilon_0} \int \frac{\rho(\vec{r}')}{|\vec{r} - \vec{r}'|} d^3\vec{r}' \qquad (1.460)$$

$$\vec{A}(\vec{r}) = \frac{\mu_0}{4\pi} \int \frac{\vec{J}(\vec{r}')}{|\vec{r} - \vec{r}'|} d^3\vec{r}' \qquad (1.461)$$

The time-dependent Maxwell's equations reduce to:

$$\Box^2 \phi = -\frac{\rho}{\varepsilon_0} \qquad (1.462)$$

$$\Box^2 \vec{A} = -\mu_0 \vec{J} \qquad (1.463)$$

We can solve these equations using the time-dependent Green's function, Eq. 1.455. From this equation we find that

$$\phi(\vec{r}, t) = \frac{1}{4\pi\varepsilon_0} \iint \frac{\delta(t - t' - |\vec{r} - \vec{r}'|/c) \rho(\vec{r}', t')}{|\vec{r} - \vec{r}'|} d^3\vec{r}' dt' \qquad (1.464)$$

with a similar equation for $\vec{A}$. Using the well-known property of delta-functions, these equations reduce to:

$$\phi(\vec{r},t) = \frac{1}{4\pi\varepsilon_0} \int \frac{\rho(\vec{r}',t-|\vec{r}-\vec{r}'|/c)}{|\vec{r}-\vec{r}'|} d^3\vec{r}' \qquad (1.465)$$

$$\vec{A}(\vec{r},t) = \frac{\mu_0}{4\pi} \int \frac{\vec{J}(\vec{r}',t-|\vec{r}-\vec{r}'|/c)}{|\vec{r}-\vec{r}'|} d^3\vec{r}' \qquad (1.466)$$

These are the general solutions to Maxwell's equations. Note that the time-dependent solutions, Eqs. 1.465 and 1.466, are the same as the steady-state solutions, Eqs. 1.460 and 1.461, apart from the weird way in which time appears in the former.

According to Eqs. 1.465 and 1.466, if we want to work out the potentials at position $\vec{r}$ and time t, we have to perform integrals of the charge density and current density over all space (just like in the steady-state situation). Yet, when we calculate the contribution of charges and currents at position $\vec{r}'$ to these integrals, we do not use the values at time t; instead, we use the values at some earlier time, $t - |\vec{r} - \vec{r}'|/c$.

What is this earlier time? It is simply the latest time at which a light signal emitted from position $\vec{r}'$ would be received at position $\vec{r}$ before time t. This is called *retarded time*. Likewise, the potentials Eqs. 1.465 and 1.466 are called *retarded potentials*. It is often useful to adopt the following notation:

$$\vec{A}(\vec{r}',t-|\vec{r}-\vec{r}'|/c) \equiv \left[\vec{A}(\vec{r}',t)\right] \qquad (1.467)$$

The square brackets denote retardation (i.e., using retarded time instead of real time). Using this notation, Eqs. 1.465 and 1.466 become:

$$\phi(\vec{r}) = \frac{1}{4\pi\varepsilon_0} \int \frac{[\rho(\vec{r}')]}{|\vec{r}-\vec{r}'|} d^3\vec{r}' \qquad (1.468)$$

$$\phi(\vec{r}) = \frac{\mu_0}{4\pi} \int \frac{[\vec{J}(\vec{r}')]}{|\vec{r}-\vec{r}'|} d^3\vec{r}' \qquad (1.469)$$

The time dependence in these equations is accepted as they read.

We are now in a position to understand electromagnetism at its most fundamental level. Charge distribution $\rho(\vec{r},t)$ can be thought of as built out of a collection, or series, of charges that instantaneously come into existence at some point $\vec{r}'$ and some time t' and then disappear again. Mathematically, this is written:

$$\rho(\vec{r},t) = \iint \delta(\vec{r}-\vec{r}')\delta(t-t')\rho(\vec{r}',t')d^3\vec{r}'dt' \qquad (1.470)$$

1.9 Time-Dependent Maxwell's Equations

Likewise, we can think of current distribution $\vec{J}(\vec{r},t)$ as built out of a collection or series of currents that instantaneously appear and then disappear:

$$\vec{J}(\vec{r},t) = \iint \delta(\vec{r}-\vec{r}')\delta(t-t')\vec{J}(\vec{r}',t')d^3\vec{r}'dt' \tag{1.471}$$

Each of these ephemeral charges and currents excites a spherical wave in the appropriate potential. Thus, the charge density at $\vec{r}'$ and t' sends out a wave in the scalar potential:

$$\phi(\vec{r},t) = \frac{\rho(\vec{r}',t')}{4\pi\varepsilon_0}\frac{\delta(t-t'-|\vec{r}-\vec{r}'|/c)}{|\vec{r}-\vec{r}'|} \tag{1.472}$$

Similarly, the current density at $\vec{r}'$ and t' sends out a wave in the vector potential:

$$\vec{A}(\vec{r},t) = \frac{\mu_0 \vec{J}(\vec{r}',t')}{4\pi}\frac{\delta(t-t'-|\vec{r}-\vec{r}'|/c)}{|\vec{r}-\vec{r}'|} \tag{1.473}$$

These waves can be thought of as messengers that inform other charges and currents about the charges and currents present at position $\vec{r}'$ and time t'. These messengers travel at a finite speed however (i.e., the speed of light). So, by the time they reach other charges and currents their message is a little out of date.

Every charge and every current in the Universe emits these spherical waves. The resultant scalar and vector potential fields are given by Eqs. 1.468 and 1.469. Of course, we can turn these fields into electric and magnetic fields using Eqs. 1.377 and 1.378. We then can evaluate the force exerted on charges using the Lorentz formula.

We can see that we have now escaped from the apparent action at a distance nature of Coulomb's Law and the Biot-Savart Law. The EM information is carried by spherical waves in the vector and scalar potentials, therefore, travels at the velocity of light. Thus, if we change the position of a charge, a distant charge can respond only after a time delay sufficient for a spherical wave to propagate from the former to the latter charge.

Let us compare the steady-state law,

$$\phi(\vec{r}) = \frac{1}{4\pi\varepsilon_0}\int \frac{\rho(\vec{r}')}{|\vec{r}-\vec{r}'|}d^3\vec{r}' \tag{1.474}$$

with the corresponding time-dependent law,

$$\phi(\vec{r}) = \frac{1}{4\pi\varepsilon_0}\int \frac{\rho(\vec{r}')}{|\vec{r}-\vec{r}'|}d^3\vec{r}' \tag{1.475}$$

These two formulas look very similar indeed, but there is an important difference. We can imagine (rather pictorially) that every charge in the Universe is continuously performing the integral in the preceding equation and also is performing a similar integral to find the vector potential.

After evaluating both potentials, the charge can calculate the fields; and, using the Lorentz force law, it can then work out its equation of motion. The problem is that the information the charge receives from the rest of the Universe is carried by spherical waves and is always slightly out of date (because the waves travel at a finite speed). As the charge considers more and more distant charges or currents, its information gets more and more out of date. (Similarly, when astronomers look out to more and more distant galaxies in the Universe, they are also looking backward in time. In fact, the light we receive from the most distant observable galaxies was emitted when the Universe was only about one-third of its present age.) So, what does our electron do? It simply uses the most up-to-date information about distant charges and currents that it possesses. So, instead of incorporating charge density $\rho(\vec{r},t)$ into its integral, the electron uses *retarded* charge density $\rho(\vec{r},t)$ (i.e., the density evaluated at the retarded time). This is effectively what Eq. 1.466 says.

Consider a thought experiment in which charge q appears at position $\vec{r}_0$ at time t_1, persists for a while, and then disappears at time t_2. What is the electric field generated by such a charge? Using Eq. 1.466, we find that

$$\phi(\vec{r}) = \frac{q}{4\pi\varepsilon_0 |\vec{r}-\vec{r}_0|} \quad \text{for} \quad t_1 \le t - |\vec{r}-\vec{r}_0|/c \le t_2$$
$$= 0 \qquad \text{otherwise} \tag{1.476}$$

Now, $\vec{E} = -\nabla \phi$ (because there are no currents, therefore no vector potential is generated), so

$$\vec{E}(\vec{r}) = \frac{q}{4\pi\varepsilon_0 |\vec{r}-\vec{r}_0|^3} \quad \text{for} \quad t_1 \le t - |\vec{r}-\vec{r}_0|/c \le t_2$$
$$= 0 \qquad \text{otherwise} \tag{1.477}$$

This solution is shown pictorially in Fig. 1.26. We can see that the charge effectively emits a Coulomb electric field that propagates radially away from the charge at the

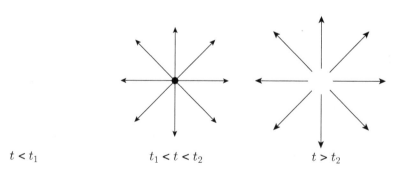

Fig. 1.26 Illustration of a Coulomb electric field

1.9 Time-Dependent Maxwell's Equations

speed of light. Likewise, it is easy to show that a current-carrying wire effectively emits an Ampèrian magnetic field at the speed of light.

We can now appreciate the essential difference between time-dependent electromagnetism and the action at a distance laws of Coulomb and Biot and Savart. In the latter theories, the field lines act rather like rigid wires attached to charges (or circulating around currents). If the charges (or currents) move, so do the field lines, leading inevitably to unphysical action at a distance-type behavior. In the time-dependent theory, charges act rather like water sprinklers—that is, they spray out the Coulomb field in all directions at the speed of light. Similarly, current-carrying wires throw out magnetic field loops at the speed of light. If we move a charge (or current), field lines emitted beforehand are not affected, so the field at a distant charge (or current) only responds to the change in a position after a time delay sufficient for the field to propagate between the two charges (or currents) at the speed of light.

In Coulomb's Law and the Biot-Savart Law, it is not entirely obvious that the electric and magnetic fields have a real existence. After all, the only measurable quantities are the forces acting between charges and currents. We can describe the force acting on a given charge or current because of the other charges and currents in the Universe, in terms of the local electric and magnetic fields; however, we have no way of knowing whether these fields persist when the charge or current is not present. That is, we could argue that electric and magnetic fields are just a convenient way of calculating forces; yet, in reality, the forces are transmitted directly between charges and currents by some form of magic. Nevertheless, it is patently obvious that electric and magnetic fields have a real existence in the time-dependent theory.

Consider the following thought experiment. Suppose that charge q_1 comes into existence for a period of time, emits a Coulomb field, and then disappears. Presume that a distant charge q_2 interacts with this field but is sufficiently far from the first charge that by the time the field arrives the first charge has already disappeared. The force exerted on the second charge is ascribable only to the electric field; it cannot be ascribed to the first charge because this charge no longer exists by the time the force is exerted. The electric field clearly transmits energy and momentum between the two charges. Anything that possesses energy and momentum is "real" in a physical sense. Later in this book, we will demonstrate that electric and magnetic fields conserve energy and momentum.

Now we consider a moving charge. Such a charge continually is emitting spherical waves in the scalar potential, and the resulting wavefront pattern is sketched in Fig. 1.27. Clearly, the wavefronts are more closely spaced in front of the charge than they are behind it, suggesting that the electric field in front is larger than the field behind. In a medium, such as water or air, where waves travel at a finite speed, c (say), it is possible to get a very interesting effect if the wave source travels at some velocity v that exceeds the wave speed. This is illustrated in Fig. 1.28.

The locus of the outermost wavefront is now a cone instead of a sphere. The wave intensity on the cone is extremely large—that is, a *shock wave*! The half-angle θ of the shock wave cone is simply $\cos^{-1}(c/v)$. In water, shock waves are produced by fast-moving boats. We call these *bow waves*. In air, shock waves are produced by speeding bullets and supersonic jets. In the latter case, we call these *sonic booms*. Is

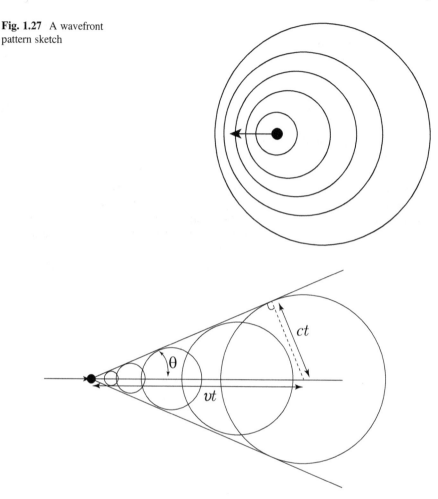

Fig. 1.27 A wavefront pattern sketch

Fig. 1.28 Illustration of wave source traveling

there any such thing as an EM shock wave? At first sight, the answer to this question would appear to be "no." After all EM waves travel at the speed of light, and no wave source (i.e., an electrically charged particle) can travel faster than this velocity. This is a rather disappointing conclusion.

Nevertheless, when an EM wave travels through matter, a remarkable thing happens. The oscillating electric field of the wave induces a slight separation of the positive and negative charges in the atoms that make up the material. We call separated positive and negative charges an *electric dipole*. Of course, the atomic dipoles oscillate in sympathy with the field that induces them. Still, an oscillating electric dipole radiates EM waves. Amazingly, when we add the original wave to these induced waves, it is exactly as if the original wave propagates through the material in question at a velocity *slower* than the velocity of light in a vacuum.

1.9 Time-Dependent Maxwell's Equations

Suppose, now, that we shoot a charged particle through the material faster than the slowed-down velocity of EM waves. This is possible because the waves are traveling slower than the velocity of light in vacuum. In practice, the particle must be traveling pretty close to the velocity of light (i.e., it has to be relativistic), but modern particle accelerators produce copious amounts of such particles. Today, we can get an EM shock wave. We expect an intense cone of emission, just like the bow wave produced by a fast ship. In fact, this type of radiation has been observed. It is called *Cherenkov radiation*, and it is very useful in high-energy physics.

Cherenkov radiation typically is produced by surrounding a particle accelerator with Perspex blocks. Relativistically charged particles emanating from the accelerator pass through the Perspex traveling faster than the local velocity of light, therefore emitting Cherenkov radiation. We know the velocity of light (c_*, say) in *Perspex* (this can be worked out from the refractive index), so if we can measure half-angle θ of the radiation cone emitted by each particle, then we can evaluate the speed of particle v via geometric relation $\cos\theta = c_*/v$.

1.9.9 Advanced Potentials

We have defined the retarded time,

$$t_r = t - |\vec{r} - \vec{r}'|/c \tag{1.478}$$

as the latest time at which a light signal emitted from position $\vec{r}'$ would reach position $\vec{r}$ before time t. We also have shown that a solution to Maxwell's equations can be written in terms of retarded potentials:

$$\phi(\vec{r},t) = \frac{1}{4\pi\varepsilon_0} \int \frac{\rho(\vec{r}',t_r)}{|\vec{r} - \vec{r}'|} d^3\vec{r}' \tag{1.479}$$

and so on. But, is this the most general solution? Suppose that we define the *advanced time*,

$$t_a = t + |\vec{r} - \vec{r}'|/c \tag{1.480}$$

This is the time a light signal emitted at time t from position r would reach position r'. It turns out that we also can write a solution to Maxwell's equations in terms of *advanced potentials*:

$$\phi(\vec{r},t) = \frac{1}{4\pi\varepsilon_0} \int \frac{\rho(\vec{r}',t_a)}{|\vec{r} - \vec{r}'|} d^3\vec{r}' \tag{1.481}$$

and so on. In fact, this is just as good a solution to Maxwell's equations as the one involving retarded potentials. To get some idea what is going on, let us examine the Green's function corresponding to the retarded potential solution:

$$\phi(\vec{r},t) = \frac{\rho(\vec{r}\,',t')}{4\pi\varepsilon_0} \frac{\delta(t-t'-|\vec{r}-\vec{r}\,'|/c)}{|\vec{r}-\vec{r}\,'|} \tag{1.482}$$

with a similar equation for the vector potential. This says that the charge density present at position $\vec{r}\,'$ and time t' emits a spherical wave in the scalar potential that *propagates forward in time*. The *Green's function* corresponding to our advanced potential solution is:

$$\phi(\vec{r},t) = \frac{\rho(\vec{r}\,',t')}{4\pi\varepsilon_0} \frac{\delta(t-t'+|\vec{r}-\vec{r}\,'|/c)}{|\vec{r}-\vec{r}\,'|} \tag{1.483}$$

This says that the charge density present at position $\vec{r}\,'$ and time t' emits a spherical wave in the scalar potential that *propagates backward in time*. "But, hang on a minute," you might say; "everybody knows that EM waves can't travel backward in time. If they did then causality would be violated." Well, *you* know that EM waves do not propagate backward in time, and we know that they do not propagate backward in time, but the question is do Maxwell's equations know this? Consider the wave equation for the scalar potential:

$$\left(\nabla^2 - \frac{1}{c^2}\frac{\partial^2}{\partial t^2}\right)\phi = -\frac{\rho}{c} \tag{1.484}$$

This equation is manifestly symmetrical in time (i.e., it is invariant under transformation $t \to -t$). Thus, backward-traveling waves are just as good a solution to this equation as forward-traveling waves. The equation also is symmetrical in space (i.e., it is invariant under transformation $x \to -x$). So, why do we adopt the Green's function, Eq. 1.473, which is symmetric in space (i.e., it is invariant under $x \to -x$) but asymmetrical in time (i.e., it is not invariant under $t \to -t$)? Would it not be better to use the completely symmetrical *Green's function*:

$$\phi(\vec{r},t) = \frac{\rho(\vec{r}\,',t')}{4\pi\varepsilon_0}\frac{1}{2}\left(\frac{\delta(t-t'-|\vec{r}-\vec{r}\,'|/c)}{|\vec{r}-\vec{r}\,'|} + \frac{\delta(t-t'-|\vec{r}-\vec{r}\,'|/c)}{|\vec{r}-\vec{r}\,'|}\right) \tag{1.485}$$

In other words, a charge emits half of its waves running forward in time (i.e., retarded waves), and the other half running backward in time (i.e., advanced waves). This sounds completely crazy! In the 1940s, however, *Richard P. Feynman* and *John A. Wheeler* pointed out that under certain circumstances this prescription gives the right answer. Consider a charge interacting with "the rest of the Universe," where the "rest of the Universe" denotes all the distant charges in it and is, by implication, an awful long way away from the original charge. Suppose that the "rest of the Universe" is a perfect reflector of advanced waves and a perfect absorber of retarded waves. The waves emitted by the charge can be written schematically as

1.9 Time-Dependent Maxwell's Equations

$$F = \frac{1}{2}(\text{redarded}) + \frac{1}{2}(\text{advanced}) \tag{1.486}$$

The response of the rest of the Universe is written:

$$R = \frac{1}{2}(\text{retarded}) + \frac{1}{2}(\text{advanced}) \tag{1.487}$$

This is illustrated in the spacetime diagram in Fig. 1.29. Here A and R denote the advanced and retarded waves emitted by the charge, respectively. The advanced wave travels to "the rest of the Universe" and is reflected—that is, the distant charges oscillate in response to the advanced wave and emit a retarded wave a, as shown. Retarded wave a is a spherical wave that converges on the original charge, passes through the charge, and then diverges again. The divergent wave is denoted by $a\ a$. Note that a looks like a negative advanced wave emitted by the charge, whereas $a\ a$ looks like a positive retarded wave emitted by the charge. This is essentially what Eq. 1.487 says. Retarded waves R and $a\ a$ are absorbed by "the rest of the Universe."

If we add the waves emitted by the charge to the response of "the rest of the Universe" we obtain:

$$F' = F + R = (\text{retarded}) \tag{1.488}$$

Thus, charges *appear* to emit only retarded waves, which agrees with our everyday experience. Clearly, in this model we have sidestepped the problem of a time asymmetrical Green's function by adopting time asymmetrical boundary conditions to the Universe; that is, the distant charges in the Universe absorb retarded waves and reflect advanced waves.

This is possible because the absorption takes place at the end of the Universe (i.e., at the "big crunch," or whatever), and the reflection takes place at the beginning of

Fig. 1.29 The spacetime diagram

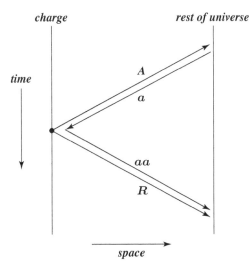

the Universe (i.e., at the "Big Bang"). It is quite plausible that the state of the Universe (thus, its interaction with EM waves) is completely different at these two times. It should be pointed out that the Feynman-Wheeler model runs into trouble when one tries to combine electromagnetism with quantum mechanics. These difficulties have yet to be resolved, so at present the status of this model is that it is "an interesting idea," but it is still not fully accepted into the canon of physics.

1.9.10 Retarded Fields

We know the solution to Maxwell's equations in terms of retarded potentials. Let us now construct the associated electric and magnetic fields using,

$$\vec{E} = -\vec{\nabla}\phi - \frac{\partial \vec{A}}{\partial t} \tag{1.489}$$

$$\vec{B} = \vec{\nabla} \times \vec{A} \tag{1.490}$$

It is helpful to write:

$$\vec{R} = \vec{r} - \vec{r}' \tag{1.491}$$

where $R = |\vec{r} - \vec{r}'|$. The retarded time becomes $t_r = t - R/c$, and a general retarded quantity is written $[F(\vec{r}, t_r)] \equiv F(\vec{r}, t_r)$. Thus, we can write the retarded potential solutions of Maxwell's equations in an especially compact form:

$$\phi = \frac{1}{4\pi\varepsilon_0} \int \frac{[\rho]}{R} dV' \tag{1.492}$$

$$\vec{A} = \frac{\mu_0}{4\pi} \int \frac{[\vec{J}]}{R} dV' \tag{1.493}$$

where $dV' = d^3\vec{r}'$.

It can be easily seen that

$$\begin{aligned}\vec{\nabla}\phi &= \frac{1}{4\pi\varepsilon_0} \int \left([\rho]\vec{\nabla}(R^{-1}) + \frac{[\partial\rho/\partial t]}{R}\vec{\nabla}t_r \right) dV' \\ &= -\frac{1}{4\pi\varepsilon_0} \int \left(\frac{[\rho]}{R^3}\vec{R} + \frac{[\partial\rho/\partial t]}{cR^2}\vec{\nabla}t_r \right) dV'\end{aligned} \tag{1.494}$$

where this equation has been used:

1.9 Time-Dependent Maxwell's Equations

$$\nabla R = \frac{\vec{R}}{R}$$

$$\nabla (R^{-1}) = -\frac{\vec{R}}{R^3} \tag{1.495}$$

$$\nabla t_r = -\frac{\vec{R}}{cR}$$

Likewise,

$$\vec{\nabla} \times \vec{A} = \frac{\mu_0}{4\pi} \int \left(\nabla(R^{-1}) \times [\vec{J}] + \frac{\nabla t_r \times [\partial \vec{J}/\partial t]}{R} \right)$$

$$= -\frac{\mu_0}{4\pi} \int \left(\frac{\vec{R} \times [\vec{J}]}{R^3} + \frac{\vec{R} \times [\partial \vec{J}/\partial t]}{cR^2} \right) dV' \tag{1.496}$$

Equations 1.481, 1.489, 1.490, and 1.496 can be combined to give,

$$\vec{E} = \frac{1}{4\pi\varepsilon_0} \int \left([\rho]\frac{\vec{R}}{R^3} + \left[\frac{\partial\rho}{\partial t}\right]\frac{\vec{R}}{cR^2} - \frac{[\partial\vec{J}/\partial t]}{c^2 R} \right) dV' \tag{1.497}$$

which is the time-dependent generalization of Coulomb's Law, and

$$\vec{B} = \frac{\mu_0}{4\pi} \int \left(\frac{[\vec{J}] \times \vec{R}}{R^3} + \frac{[\partial\vec{J}/\partial t] \times \vec{R}}{c^2 R} \right) dV' \tag{1.498}$$

Suppose that the typical variation time scale of the charges and currents is t_0. Let us define $R_0 = ct_0$, which is the distance a light ray travels in time t_0. We can evaluate Eqs. 1.488 and 1.489 in two asymptotic limits: *near-field* region $R \ll R_0$ and *far-field* region $R \gg R_0$. In the near-field region,

$$\frac{|t - t_r|}{t_0} = \frac{R}{R_0} \ll 1 \tag{1.499}$$

so the difference between retarded time and standard time is relatively small. This allows us to expand retarded quantities in a *Taylor series*. Thus,

$$[\rho] \simeq \rho + \frac{\partial \rho}{\partial t}(t_r - t) + \frac{1}{2}\frac{\partial^2 \rho}{\partial t^2}(t_r - t)^2 + \cdots \tag{1.500}$$

giving

$$[\rho] \simeq \rho - \frac{\partial \rho}{\partial t}\frac{R}{c} + \frac{1}{2}\frac{\partial^2 \rho}{\partial t^2}\frac{R^2}{c^2} + \cdots \tag{1.501}$$

Expansion of the retarded quantities in the near-field region yields:

$$\vec{E} \simeq \int \left(\frac{\rho \vec{R}}{R^3} - \frac{1}{2} \frac{\partial^2 \rho}{\partial t^2} \frac{\vec{R}}{c^2 R} - \frac{\partial \vec{J}/\partial t}{c^2 R} + \cdots \right) dV' \qquad (1.502)$$

$$\vec{B} \simeq \int \left(\frac{\vec{J} \times \vec{R}}{R^3} - \frac{1}{2} \frac{(\partial^2 \vec{J}/\partial t^2) \times \vec{R}}{c^2 R} + \cdots \right) dV' \qquad (1.503)$$

In Eq. 1.502 the first term on the RHS corresponds to Coulomb's Law, the second term is the correction owing to retardation effects, and the third term corresponds to Faraday induction. In Eq. 1.503, the first term on the RHS is the *Biot-Savart Law*, and the second term is the correction because of retardation effects. Note that the retardation corrections are only of the order $(R/R_0)^2$. We might suppose, from looking at Eqs. 1.497 and 1.498, that the corrections should be of the order R/R_0.

All the order R/R_0 terms, however, canceled out in the previous expansion. Suppose, then, that we have a direct current (DC) circuit sitting on a laboratory benchtop. Let the currents in the circuit change on a typical time scale of one-tenth of a second. In this time, light can travel about 3×10^7 m, so $R_0 \sim 30,000$ km. The length scale of the experiment is about 1 m, so $R = 1$ m. Thus, the retardation corrections are of order $(3 \times 10^7)^{-2} \sim 10^{-15}$. It is clear that we are fairly safe just using Coulomb's Law, *Faraday's Law*, and the *Biot-Savart Law* to analyze the fields generated by this type of circuit.

In the far-field region, $R \gg R_0$, Eqs. 1.497 and 1.498 are dominated by the terms that vary like R^{-1}, so

$$\vec{E} \simeq -\frac{1}{4\pi\varepsilon_0} \int \frac{[\partial \vec{J}_\perp/\partial t]}{c^2 R} dV' \qquad (1.504)$$

$$\vec{B} \simeq \int \frac{[\partial \vec{J}_\perp/\partial t] \times \vec{R}}{cR^2} dV' \qquad (1.505)$$

where

$$\vec{J}_\perp = \vec{J} - \frac{(\vec{J} \cdot \vec{R})}{R^2} \vec{R} \qquad (1.506)$$

Here, use has been made of $[\partial \rho/\partial t] = -[\vec{\nabla} \cdot \vec{J}]$ and $[\vec{\nabla} \cdot \vec{J}] = -[\partial \vec{J}/\partial t] \cdot (\vec{R}/cR) + O(1/R^2)$. Suppose that the charges and currents are localized to some region in the vicinity of $\vec{r}' = \vec{r}_*$. Let $\vec{R}_* = \vec{r} - \vec{r}_*$, with $R_* = |\vec{r} - \vec{r}_*|$. Suppose that the extent of the current- and charge-containing regions is much less than R_*. It follows that retarded quantities can be written:

$$[\rho(\vec{r},t)] \simeq \rho(\vec{r}, t - R_*/c) \qquad (1.507)$$

and so on. Thus, the electric field reduces to

1.9 Time-Dependent Maxwell's Equations

$$\vec{E} \simeq -\frac{1}{4\pi\varepsilon_0} \frac{\left[\int (\partial \vec{J}_\perp/\partial t) dV'\right] \times \vec{R}_*}{c^2 R_*} \tag{1.508}$$

whereas the magnetic field is given by

$$\vec{B} \simeq -\frac{1}{4\pi\varepsilon_0} \frac{\left[\int (\partial \vec{J}_\perp/\partial t) dV'\right] \times \vec{R}_*}{c^3 R_*^2} \tag{1.509}$$

Note that

$$\frac{E}{B} = c \tag{1.510}$$

and

$$\vec{E} \cdot \vec{B} = 0 \tag{1.511}$$

This configuration of electric and magnetic fields is characteristic of an EM wave (see Sect. 1.9.6). Thus, Eqs. 1.508 and 1.509 describe a wave propagating *radially* away from the charge- and current-containing regions. Note that the wave is driven by time-varying electric currents. Now, charges moving with a constant velocity constitute a steady current, so a non-steady current is associated with *accelerating charges*. We conclude that accelerating electrical charges emit EM waves. The wave fields, Eqs. 1.508 and 1.509, fall off like the inverse of the distance from the wave source.

This behavior should be contrasted with that of the Coulomb or Biot-Savart fields, which fall off like the inverse square of the distance from the source. The fact that wave fields attenuate fairly gently with increasing distance from the source is what makes astronomy possible. If wave fields obeyed an inverse square law, then no appreciable radiation would reach us from the rest of the Universe.

In conclusion, electric and magnetic fields look simple in the near-field region (they are just Coulomb fields, etc.) and also in the far-field region (they are just EM waves). Only in the intermediate region, $R \sim R_0$, do things start getting really complicated; thus, we generally do not look in this region!

1.9.11 Summary

This marks the end of our theoretical investigation of Maxwell's equations. Here we summarize what we have learned so far. The field equations that govern electric and magnetic fields are written:

$$\vec{\nabla} \cdot \vec{E} = \frac{\rho}{\varepsilon_0} \tag{1.512}$$

$$\vec{\nabla} \cdot \vec{B} = 0 \tag{1.513}$$

$$\vec{\nabla} \times \vec{E} = -\frac{\partial \vec{B}}{\partial t} \tag{1.514}$$

$$\vec{\nabla} \times \vec{B} = \mu_0 \vec{J} + \frac{1}{c^2} \frac{\partial \vec{E}}{\partial t} \tag{1.515}$$

These equations can be integrated to give

$$\oint_S \vec{E} \cdot d\vec{S} = \frac{1}{\varepsilon_0} \int_V \rho \, dV \tag{1.516}$$

$$\oint_S \vec{B} \cdot d\vec{S} = 0 \tag{1.517}$$

$$\oint_C \vec{E} \cdot d\vec{l} = -\frac{\partial}{\partial t} \int_S \vec{B} \cdot d\vec{S} \tag{1.518}$$

$$\oint_C \vec{B} \cdot d\vec{l} = \mu_0 \int_S \vec{J} \cdot d\vec{S} + \frac{1}{c^2} \frac{\partial}{\partial t} \int_S \vec{E} \cdot d\vec{S} \tag{1.519}$$

Equations 1.513 and 1.514 are satisfied automatically by writing

$$\vec{E} = -\vec{\nabla} \phi - \frac{\partial \vec{A}}{\partial t} \tag{1.520}$$

$$\vec{B} = \vec{\nabla} \times \vec{A} \tag{1.521}$$

This prescription is not unique (there are many choices of ϕ and $\vec{A}$ that generate the same fields), but we can make it unique by adopting the following conventions:

$$\phi(\vec{r}) \to 0 \quad \text{as} \quad |\vec{r}| \to \infty \tag{1.522}$$

and

$$\frac{1}{c^2} \frac{\partial \phi}{\partial t} + \vec{\nabla} \cdot \vec{A} = 0 \tag{1.523}$$

Equations 1.502 and 1.505 reduce to

$$\Box^2 \phi = -\frac{\rho}{\varepsilon_0} \tag{1.524}$$

$$\Box^2 \vec{A} = -\mu_0 \vec{J} \tag{1.525}$$

These are driven wave equations of the general form:

$$\Box^2 u \equiv \left(\nabla^2 - \frac{1}{c^2} \frac{\partial^2}{\partial t^2} \right) u = v \tag{1.526}$$

Green's function for this equation, which satisfies the boundary conditions and is consistent with causality, is:

$$G(\vec{r},;\vec{r}\,';t,t') = -\frac{1}{4\pi}\frac{\delta(t-t'-|\vec{r}-\vec{r}\,'|/c)}{|\vec{r}-\vec{r}\,'|} \tag{1.527}$$

Thus, the solutions to Eqs. 1.524 and 1.525 are:

$$\phi(\vec{r},t) = \frac{1}{4\pi\varepsilon_0}\int\frac{[\rho]}{R}dV' \tag{1.528}$$

$$\vec{A}(\vec{r},t) = \frac{\mu_0}{4\pi}\int\frac{[\vec{J}]}{R}dV' \tag{1.529}$$

where $R = |\vec{r}-\vec{r}\,'|$ and $dV' = d^3\vec{r}\,'$, with $[A] \equiv A(\vec{r}\,',t-R/c)$. These solutions can be combined with Eqs. 1.510 and 1.511 to give:

$$\vec{E}(\vec{r},t) = \frac{1}{4\pi\varepsilon_0}\int\left([\rho]\frac{\vec{R}}{R^3} + \left[\frac{\partial\rho}{\partial t}\right]\frac{\vec{R}}{cR^2} - \frac{[\partial\vec{J}/\partial t]}{c^2R}\right)dV' \tag{1.530}$$

$$\vec{B}(\vec{r},t) = \frac{\mu_0}{4\pi}\int\left(\frac{[\vec{J}]\times\vec{R}}{R^3} + \frac{[\partial\vec{J}/\partial t]\times\vec{R}}{cR^2}\right)dV' \tag{1.531}$$

Equations 1.512 through 1.531 constitute the complete theory of classical electromagnetism. We can express the same information in terms of field equations (Eqs. 1.512 and 1.515), integrated field equations (Eqs. 1.516 through 1.519), retarded EM potentials (Eqs. 1.528 and 1.529), and retarded EM fields (Eqs. 1.530 and 1.531). Chapter 2 considers the applications of this theory [3].

References

1. B. Zohuri, *Plasma Physics and Controlled Thermonuclear Reactions Driven Fusion Energy* (Springer Publishing Company, Cham, 2016)
2. D. Fleisch, L. Kinnaman, *A Student's Guide to Waves*, 4th edn. (Cambridge University Press, Cambridge, 2016)
3. R. Fitzpatrick, *Maxwell's Equations and the Principles of Electromagnetism* (Infinity Science Press LLC, New Delhi, 2008)

Chapter 2
Maxwell's Equations—Generalization of Ampère-Maxwell's Law

Ampère's Law, relating a steady electric current to a circulating magnetic field, was well known by the time James Clerk Maxwell started his research in a similar field in the 1850s. Although Ampère's Law was known to apply only to static situations involving steady currents, it was Maxwell's effort to add another source term—a change of electric flux—that extended the applicability of Ampère's Law to time-dependent conditions. More important, it was the presence of this term in Ampère's equation that led to it being known as Ampère-Maxwell's Law. It allowed Maxwell to distinguish the electromagnetic nature of light and to develop a comprehensive theory of electromagnetism [1].

2.1 Introduction

From our knowledge of college electromagnetics, we have learned that the integral form of the Ampère Law is the magnetic field because a current distribution satisfies the following relationship:

$$\oint_C \vec{H} \times d\vec{l} = \int_S \vec{J} \cdot \hat{n} \, da \tag{2.1}$$

This equation can be derived with the aid of Stock's theorem; we leave the deviation of it to the readers and refer them to a reference book such as Reitz et al. [2].

Nevertheless, we briefly show how Eq. 2.1 can be derived using the *Biot-Savart Law*. Magnetic field $\vec{B}$ in terms of this law is defined as Eq. 2.2:

$$\vec{B}(\vec{r}) = \frac{\mu_0}{4\pi} \int_V \vec{J}(\vec{r}') \times \frac{(\vec{r} - \vec{r}')}{|\vec{r} - \vec{r}'|^3} dV' \tag{2.2}$$

© Springer International Publishing AG, part of Springer Nature 2019
B. Zohuri, *Scalar Wave Driven Energy Applications*,
https://doi.org/10.1007/978-3-319-91023-9_2

The permeability, μ_0, indictates the strength of the magnetic field given current distribution.

Note the equivalent to Eq. 2.2 is known as Biot-Savart's magneto-static field produced by its source $\vec{J}$, also known as *current density*; in addition, there exists Coulomb's electrostatic field produced by its source ρ, also known as *charge density*. In the absence of $\dot{\rho} = 0$ and $\dot{\vec{J}} = 0$, in statics this is defined as:

$$\vec{E}(\vec{r}) = \frac{1}{4\pi\varepsilon_0} \int_v \frac{\rho(\vec{r}')(\vec{r}-\vec{r}')}{|\vec{r}-\vec{r}'|^3} dV' \qquad (2.3)$$

Note also the notation of charge density ρ, and current density $\vec{J}$, arise from *one* reality, which is the very existence of *electrically* charged particles in motion:

$$\rho(\vec{r},t) = \sum q_i \delta(\vec{r} - r_i(t)) \qquad (2.4)$$

and

$$\vec{J}(\vec{r},t) = \sum q_i \frac{d\vec{r}_i(t)}{dt} \delta(\vec{r} - r_i(t)) \qquad (2.5)$$

As with Coulomb's Law, we can apply mathematics to the Biot-Savart Law to obtain another of Maxwell's equations. Nevertheless, the essential physics already is inherent in the Biot-Savart Law. Note that the Biot-Savart Law, like Coulomb's Law, is incomplete because it also implies an instantaneous response of the magnetic field to a reconfiguration of the currents.

The generalized version of the Biot-Savart Law, another of Jefimenko's equations, incorporates the fact that electromagnetic (EM) waves travel at the speed of light. Ironically, Gauss's Law for magnetic fields and Maxwell's version of Ampère's Law, derived from the Biot-Savart Law, hold perfectly whether the currents are steady or vary in time. The Jefimenko equations, analogs of Coulomb and Biot-Savart Laws, also embody Faraday's Law, the only of Maxwell's equations that cannot be derived from the usual forms of Coulomb's and Biot-Savart laws.

To continue the discussion of ways to derive Ampère's Law in Eq. 2.1, we can take advantage of Fig. 2.1 and write the following by merely taking the inversion of the Biot-Savart Law of Eq. 2.2, so that $\vec{J}$ appears by itself, unfettered by integrals or the like [3].

Thus, we can write the inversion of the Biot-Savart Law by taking the *curl* (i.e., $\vec{\nabla} \times$) of Eq. 2.2 over variable vector $\vec{r}$ as follows:

$$\nabla \times \vec{B}(\vec{r}) = \frac{\mu_0}{4\pi} \int_v \nabla_r \times \vec{J}(\vec{r}') \times \frac{(\vec{r}-\vec{r}')}{|\vec{r}-\vec{r}'|^3} dV' \qquad (2.6)$$

Now, if we take the last row of vector identity of Table 1.1 in Chap. 1 and apply it to Eq. 2.6, while noting that $\vec{J}(\vec{r}')$ does not depend on $\vec{r}$ so that only two terms survive, the $\nabla \times \vec{B}(\vec{r})$ then becomes

2.1 Introduction

Fig. 2.1 Illustration of Ampère's Law

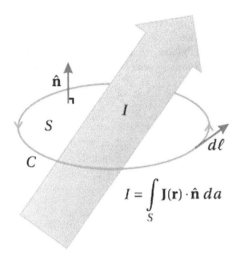

$$\nabla \times \vec{B}(\vec{r}) = \frac{\mu_0}{4\pi} \int_V \left\{ \vec{J}(\vec{r}') \left[\nabla_r \cdot \frac{(\vec{r} - \vec{r}')}{|\vec{r} - \vec{r}'|} - (\vec{J}(\vec{r}') \cdot \nabla_r) \frac{(\vec{r} - \vec{r}')}{|\vec{r} - \vec{r}'|} \right] \right\} dV' \quad (2.7)$$

Taking advantage of modern mathematical language, however, the vector expression in the integral on right-hand side (RHS) of Eq. 2.7 is a three-dimensional Dirac delta-function; thus, we can reduce the first term in the integral to $4\pi \vec{J}(\vec{r}') \delta^3(\vec{r}' - \vec{r})$, which is easily integrated. Note that for a derivation of Gauss's Law from Coulomb's Law it does not rely directly on the Dirac delta-function (see Jackson [4]).

To make progress on the second term, we observe that the gradient can be changed to operate on the primed variables without affecting the final result (i.e., $\nabla_r \to -\nabla_{r'}$). In addition, we can take advantage of a vector integral theorem (see last term of vector integral identity in Table 1.1) to arrive at the following equation:

$$\nabla \times \vec{B}(\vec{r}) = \mu_0 \vec{J}(\vec{r}) - \frac{\mu_0}{4\pi} \int_V \frac{(\vec{r} - \vec{r}')}{|\vec{r} - \vec{r}'|} [\nabla_{r'} \cdot \vec{J}(\vec{r}')] dV'$$

$$+ \frac{\mu_0}{4\pi} \oint_S \frac{(\vec{r} - \vec{r}')}{|\vec{r} - \vec{r}'|} [\vec{J}(\vec{r}') \cdot \hat{n}] da' \quad (2.8)$$

The last term in Eq. 2.8 vanishes if we assume that current density $\vec{J}$ is completely contained within volume V, so that it is zero at surface S. Thus, the expression for the $\nabla \times \vec{B}(\vec{r})$ reduces to the following form:

$$\nabla \times \vec{B}(\vec{r}) = \mu_0 \vec{J}(\vec{r}) - \frac{\mu_0}{4\pi} \int_V \frac{(\vec{r} - \vec{r}')}{|\vec{r} - \vec{r}'|^3} [\nabla_r \cdot \vec{J}(\vec{r}')] dV' \quad (2.9)$$

The latter term in Eq. 2.9 vanishes if we consider it approximately a steady state, as can be seen in the following equation:

$$\nabla \cdot \vec{J}(\vec{r}') \cong 0 \tag{2.10}$$

in which case, we successfully have isolated $\vec{J}$ and obtained Ampère's Law.

Without Maxwell's correction, Ampère's Law is

$$\nabla \times \vec{B} = \mu_0 \vec{J} \tag{2.11}$$

which only applies to quasi steady-state situations. The physical interpretation of Ampère's Law is more apparent in integral form. We integrate both sides of Eq. 2.11 over open surface S bounded by contour C and apply Stokes's theorem as Eq. 2.12 to the left-hand side (LHS):

Stokes's theorem

$$\oint_C \vec{F} \cdot d\vec{l} = \int_S (\nabla \times \vec{F}) \cdot \hat{n} \, da \tag{2.12}$$

As we said, if Eq. 2.12 is applied to the LHS of Eq. 2.11, we obtain the following result that is analogous to Eq. 2.1. This demonstrates the Ampère-Maxwell equation in materials, along with the magnetic permeability of $\mu_0 = 4\pi \times 10^{-7}$ volt-seconds/ Ampère meters (Vs/Am) in an SI unit, or sometimes given as Newtons/square Ampère (N/A^2) in free space (or, vacuum permeability). We have used what is shown in Fig. 2.1; that is,

$$\oint_C \vec{B}(r) \cdot d\vec{l} = \mu_0 \int_S \vec{J}(\vec{r}) \cdot \hat{n} \, da \equiv \mu_0 I \tag{2.13(a)}$$

or

$$\oint_C \frac{\vec{B}(r)}{\mu_0} \cdot d\vec{l} = \int_S \vec{J}(\vec{r}) \cdot \hat{n} \, da \equiv I \tag{2.13(b)}$$

This law says that the line integral of $\vec{B}$ around closed loop C is proportional to the total current flowing through the loop (see Fig. 2.1). The unit of $\vec{J}$ is current per area, so the surface integral containing $\vec{J}$ yields current I in the unit of charge per time. To put Eq. 2.13 in perspective, we can write what is presented in Fig. 2.2, which is an expanded view of the Ampère-Maxwell Law.

You can use this law to determine the circulation of a magnetic field if you are given information about the enclosed current or the change in electric flux. Furthermore, in highly symmetric situations you may be able to extract magnetic field $\vec{B}$ from the dot product and the integral to determine the magnitude of the magnetic field, as shown in the following equation:

$$\oint_C \vec{B} \cdot d\vec{l} \qquad \text{The Magnetic Field Circulation} \tag{2.14}$$

The Ampère-Maxwell Law tells you that this quantity is proportional to the enclosed current and rate of change of electric flux through any surface bounded

2.2 Permeability of Free Space μ_0

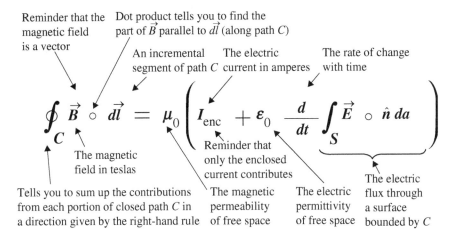

Fig. 2.2 Expansion of the Ampère-Maxwell Law in free space

by your path of integration (C). Nonetheless, if you hope to use this law to determine the value of the magnetic field, you will need to fetch magnetic field $\vec{B}$ out of the dot product and out of the integral. That means you will have to choose your path around the wire very carefully. Just as you had to choose a "special Gaussian surface" to extract the electric field from Gauss's Law, you will need a "special Ampèrian loop" to determine the magnetic field.

2.2 Permeability of Free Space μ_0

The constant of proportionality between the magnetic circulation on the LHS of the Ampère-Maxwell Law and the enclosed current and rate of flux change on the RHS is 10, the permeability of free space. Just as electric permittivity characterizes the response of a dielectric to an applied electrical field, the magnetic permeability determines a material's response to an applied magnetic field. The permeability in the Ampère-Maxwell Law is that of free space (or "vacuum permeability"), which is why it carries the subscript zero.

As in the case of electric permittivity in Gauss's Law for electric fields, the presence of this quantity does not mean that the Ampère-Maxwell Law applies only to sources and fields in a vacuum. This form of the Ampère-Maxwell law is general as long as you consider all currents (bound as well as free). In the Appendix, you will find a version of this law that is more useful when dealing with currents and fields in magnetic materials [1].

One interesting difference between the effect of dielectrics on electric fields and the effect of magnetic substances on magnetic fields is that the magnetic field really is stronger than the applied field within many magnetic materials. The reason for this

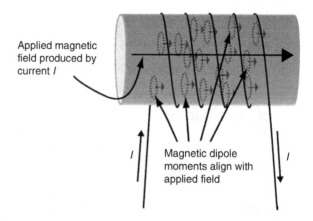

Fig. 2.3 Effect of magnetic core on field inside a solenoid

is that these materials become magnetized when exposed to an external magnetic field, and the induced magnetic field is in the same direction as the applied field, as shown in Fig. 2.3.

The permeability of a magnetic material often is expressed as the relative permeability, which is the factor by which the material's permeability exceeds that of free space. Relative permeability, μ_r, is given by the following equation:

$$\mu_r = \frac{\mu}{\mu_0} \qquad (2.15)$$

Note, however, that materials are classified as diamagnetic, paramagnetic, or ferromagnetic on the basis of relative permeability.

Diamagnetic materials have μ_r slightly less than 1.0 because the induced field weakly opposes the applied field. Examples of diamagnetic materials include gold and silver, which have a μ_r of approximately 0.99997. The induced field within paramagnetic materials weakly reinforces the applied field, so these materials have a μ_r slightly greater than 1.0. One example of a paramagnetic material is aluminum, with a μ_r of 1.00002 [1].

The situation is more complex for ferromagnetic materials, for which the permeability depends on the applied magnetic field. Typical maximum values of permeability range from several hundred for nickel and cobalt to more than 5000 for reasonably pure iron. As we may recall from a basic course in general physics, the inductance, L, of a long solenoid is given by the expression:

$$L = \frac{\mu N^2 A}{\ell} \qquad (2.16)$$

where μ is the magnetic permeability of the material within the solenoid, N is the number of turns, A is the cross-sectional area, and ℓ is the length of the coil. As this expression makes clear, adding an iron core to a solenoid may increase the inductance by a factor of 5000 or more.

2.2 Permeability of Free Space μ_0

Like electrical permittivity, the magnetic permeability of any medium is a fundamental parameter in the determination of the speed with which an EM wave propagates through that medium. This makes it possible to determine the speed of light in a vacuum simply by measuring μ_0 and ε_0 using an inductor and a capacitor; an experiment for which, to paraphrase Maxwell, the only use of light is to see the instruments.

Note that magnetic field is a force that is created by moving electric charges (i.e., electric currents) and magnetic dipoles and exerting a force on other nearby moving charges and magnetic dipoles. At any given point, it has a *direction* and a *magnitude* (or strength), so it is represented by a vector field. The term is used for two distinct, but closely related, fields denoted by the symbols $\vec{B}$ and $\vec{H}$, where in the International System of Units (SI), $\vec{H}$ is measured in units of Am per meter (m) and $\vec{B}$ is measured in tesla's (T) or Newtons per m per Am. In a vacuum $\vec{B}$ and $\vec{H}$ are the same aside from units; however, in a material with magnetization (denoted by the symbol $\vec{M}$), $\vec{B}$ is solenoidal (having no divergence in its spatial dependence) while $\vec{H}$ is irrotational (curl-free) (Fig. 2.4).

In absence of $\vec{M}$ we can write $\vec{H} = \vec{B}/\mu_0$ for linear media, and, then by substituting this value into Eq. 2.13(b), the result will reduce to Eq. 2.1, which is the Ampère-Maxwell Law of materials.

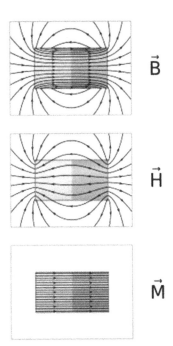

Fig. 2.4 Comparison of $\vec{B}$, $\vec{H}$, and $\vec{M}$ inside and outside of a cylindrical bar magnet

2.3 Generalization of Ampère's Law with Displacement Current

It can be easy to examine and show that the Ampère Law sometimes fails and to find a more generalizable aspect of Eq. 2.1 that always is valid.

To establish this matter, we take the circuit that is shown in Fig. 2.5 into consideration; it consists of a small parallel-plate capacitor being charged by a constant current I. We do not have to be worried about what causes the current at this point. If we apply the Ampère Law in Eq. 2.1 to contour C, and surface S_1, which is surrounded by the contour, we conclude that

$$\oint_C \vec{H} \cdot d\vec{l} = \int_{S_1} \vec{J} \cdot \hat{n}\, da = I \tag{2.17}$$

If, on the other hand, Ampère's Law is applied to contour C and surrounding surface S_2, then current density, $\vec{J}$, is *zero* at all points on surface S_2 and then, from Eq. 2.1, we conclude:

$$\oint_C \vec{H} \times d\vec{l} = \int_{S_2} \vec{J} \cdot \hat{n}\, da = 0 \tag{2.18}$$

Analyzing both Eqs. 2.17 and 2.18, we can see that they contradict each other, and thus cannot both be correct. If contour C is imagined to be a great distance from the capacitor, it is clear that the situation is not substantially different from the standard Ampère's Law cases, which we have considered before in a previous section and know about it so far.

This led us to think that Eq. 2.17 is correct because it is not dependent on the new feature—namely, the capacitor. Equation 2.18, on t he other hand, requires consideration of the capacitor for its deduction. It would appear, then, that it requires some

Fig. 2.5 The contour and two surfaces, S_1 and S_2

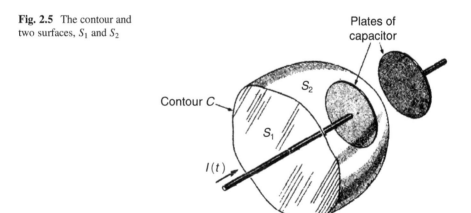

2.3 Generalization of Ampère's Law with Displacement Current

sort of modification. Because these equations arise by integrating *magnetic intensity* $\vec{H}$, that is defined as follows:

$$\vec{\nabla} \times \vec{H} = \vec{J} \tag{2.19}$$

The preceding arguments also forces Eq. 2.19 to be modified [2].

The proper modification can be made by noting that Eqs. 2.17 and 2.18 give dissimilar results because the integrals on the right-hand-sides are different. Thus, to phrase it mathematically, we can write the following conclusion:

$$\int_{S_2} \vec{J} \cdot \hat{n}_2 da - \int_{S_1} \vec{J} \cdot \hat{n}_1 da \neq 0 \tag{2.20}$$

Both surfaces S_1 and S_2 are together for a closed surface that joins at C; considering Fig. 2.5, however, unit vector $\hat{n}_2$ is outward drawn, while unit vector $\hat{n}_1$ is inward drawn. If this fact is taken into account, then Eq. 2.20 may be written in the following form:

$$\oint_{S_1+S_2} \vec{J} \cdot \hat{n} da \neq 0 \tag{2.21}$$

To state this physically, the net transport current through closed surface $S_1 + S_2$ does not vanish because charge is piling up on the plate of the condenser enclosed by the surface. Charge conservation requires, according to Eq. 2.13(b) and its modified version in Eq. 2.22, an arbitrary *closed* surface, $S = S_1 + S_2$. Here we can write the electric current entering volume V, where this volume V also is enclosed by the same surface, S, as:

$$I = -\oint_S \vec{J} \cdot \hat{n} da = -\int_V (\vec{\nabla} \cdot \vec{J}) dV \tag{2.22}$$

Note that the last integral being obtained through the use of the divergence theorem and the minus sign in Eq. 2.22 comes into consideration because unit surface vector $\hat{n}$ is the outward normal, and we want to keep current I positive when the net flow of charge is from the outside of volume V to within. Nonetheless, current I is equal to the rate at which charge is transported into V and can be written as the following equation:

$$I = \frac{dQ}{dt} = \frac{d}{dt} \int_V \rho dV \tag{2.23(a)}$$

where ρ is the charge density within volume V.

Because we are looking at a fixed volume V, the time derivative operates only on the function ρ. But ρ is a function of position as well as of time, so the time derivative becomes the partial derivative with respect to time when it is moved inside the integral. Thus, we can write the new version of Eq. 2.23(a) as:

$$I = \int_V \frac{\partial \rho}{\partial t} dV \qquad (2.23(b))$$

Equations 2.22 and 2.23(b) combined result in the following form by equating to each other, then we have:

$$\int_V \left(\frac{\partial \rho}{\partial t} + \vec\nabla \cdot \vec{J}\right) dV = 0 \qquad (2.24)$$

But, volume V is completely arbitrary, and the only way that Eq. 2.24 holds its validity for an *arbitrary volume segment* of the medium is for the integrand to vanish at each point. Thus, the equation of continuity is developed:

$$\frac{\partial \rho}{\partial t} + \vec\nabla \cdot \vec{J} = 0 \qquad (2.25)$$

Later, we will use Eq. 2.25 for further analysis as part of a scalar wave (SW) study. Now, going back out to the discussion on the generalization of Ampère's Law with current density $\vec{J}$, we can state that charge conservation requires, according to Eqs. 2.22 and 2.23, the following form:

$$\oint_{S=S_1+S_2} \vec{J} \cdot \hat{n}\, da = -\int_V \frac{\partial \rho}{\partial t} dV \qquad (2.26)$$

because inside volume V, enclosed by $S = S_1 + S_2$, charge density ρ is changing with time on the condenser plate. In differential form Eq. 2.26 is expressed by the equation of continuity as it was established in Eq. 2.25:

$$\begin{cases} \vec\nabla \cdot \vec{J} + \dfrac{\partial \rho}{\partial t} = 0 \\ \vec\nabla \cdot \vec{J} = -\dfrac{\partial \rho}{\partial t} \end{cases} \qquad (2.27)$$

By now, it is very clear, what is wrong with Eq. 2.19. Taking its divergence, we have:

$$\begin{cases} \vec\nabla \cdot (\vec\nabla \times \vec{H}) = 0 = \vec\nabla \cdot \vec{J} \\ \vec\nabla \cdot \vec{J} = 0 \end{cases} \qquad (2.28)$$

As can be seen from both Eqs. 2.27 and 2.8, once again a significant inconsistency does exist. Divergence $\vec{J}$ cannot be both zero and $-\partial \varphi/\partial t$ simultaneously. There is no apparent problem with momentum in Eq. 2.27, or similarly Eq. 2.25; it is difficult to imagine a way of modifying it to remove the inconsistency. What appears to be needed and essential is a modification of Eq. 2.19 that will change the RHS into

2.3 Generalization of Ampère's Law with Displacement Current

a vector with zero divergence. To do such a maneuver is to use Gauss's Law in a manner such as dielectric, as it is given in Chap. 1, in the form of $\vec{\nabla} \cdot \vec{D} = \rho$ [2].

This allows us to replace charge density ρ in Eq. 2.27 (i.e., the momentum equation) by $\vec{\nabla} \times \vec{D}$, thus a new form of Eq. 2.27 can be produced (i.e., a new momentum equation form) as:

$$\vec{\nabla} \cdot \left[\vec{J} + \frac{\partial \vec{D}}{\partial t} \right] = 0 \tag{2.29}$$

where, we have assumed that $\vec{D}$ is a sufficiently continuous function of the space and time variable so that the order of the derivatives can be interchanged. It is now clear that if $\partial \vec{D}/\partial t$ were added to the RHS of Eq. 2.19, the inconsistency would no longer exist, and it will disappear—that is, the divergence of either side would be zero.

With this logical argument, we are therefore in position to *revise* Ampère's Law and write it in the following form:

$$\vec{\nabla} \times \vec{H} = \vec{J} + \frac{\partial \vec{D}}{\partial t} \tag{2.30}$$

This equation is referring to the time derivative of $\vec{D}$ as the *displacement field* or in some texts you find it named *displacement current*.

Introduction of the displacement field or current demonstrated in Eq. 2.30 makes EM waves possible; this will lead us to SW analyses, which we present in the next chapters. We will expand them to different energy applications as well.

To establish EM wave analyses, either in free space or media, we should note that for a good conductor, a metal, the conductivity is of the order 10^8 S/m for frequencies below the far infrared (IR). Thus, the conduction current is of the order $10^8 \vec{E}$. The magnitude of the displacement field or current is dominated by the factor $\varepsilon \omega$, where again ε is *permittivity*, and for free space it is $\varepsilon_0 = 8.854 \times 10^{-12} C^2/N \cdot m$. All the while ω is the angular frequency and is equal to $2\pi \nu$, where ν is the frequency of the wave. Free space permittivity ε_0, however, is small except at very high frequencies (i.e., far IR), where the simple discussion breaks down for other reasons. For frequencies up to 10^{11} Hertz (Hz) the displacement field in metals can be ignored. In the case of a good dielectric material, the conduction current is very small or zero, therefore the displacement current can never be ignored. Even at 60 Hz, all the current passing through a capacitor in an alternating circuit (AC) is a displacement current. It is not necessary to consider the displacement field or current explicitly, simply because the time-varying fields in the interior of a capacitor are not going to be examined when AC circuits are analyzed [2].

We now turn attention to the examination of the full set of Maxwell's equations and their implications in this section. The entire set of Maxwell's equations is shown in Eq. 2.30, and three other equations, briefly detailed in Chap. 1, are rewritten here as well:

$$\vec{\nabla} \times \vec{E} = -\frac{\partial \vec{B}}{\partial t} \qquad \text{Faraday's Law} \qquad (2.31)$$

$$\vec{\nabla} \cdot \vec{D} = \rho \qquad \text{Gauss's Law} \qquad (2.32)$$

$$\vec{\nabla} \cdot \vec{B} = 0 \qquad \text{Coulomb's Law} \qquad (2.33)$$

$$\vec{\nabla} \times \vec{H} = \vec{J} + \frac{\partial \vec{D}}{\partial t} \qquad \text{Ampere-Maxwell's Law} \qquad (2.34)$$

All the preceding equations are expressed in their differential forms, and it is clear that the set of Maxwell's equations represent mathematical expressions of certain experimental results. Although it appears that they cannot be proved, the applicability to any situation can be verified.

Coupling these sets of equations with Lorentz force equation $\vec{F} = q(\vec{E} + \vec{v} \times \vec{B})$, which describes the action of the fields on a charged particle, gives a complete set of laws of classical electromagnetics; they describe Maxwell's equation and their empirical basis for interacting particles.

As we have seen earlier, the displacement field or displacement current introduces necessity in order to have charge conservation and that, when it is included in Maxwell's equations, they imply continuity equations as we saw in first part of Eq. 2.27, so the latter need not be added to the set of fundamental equations. Maxwell equations have two more interesting consequences, and they are used in developing EM *energy* and *wave* equations as well.

2.4 Electromagnetic Induction

The induction of the electromotive force by changing magnetic flux was first observed by Faraday and Henry in the early nineteenth century. Electromotive force, also callede.m.f. (denoted by symbol $\mathcal{E}$ and measured in volts), is the voltage developed by any source of electrical energy such as a battery or dynamo. It generally is defined as the electrical potential for a source in a circuit. A device that converts other forms of energy to electrical energy supplies an e.m.f. to a circuit. The word "force" in this case is not used to mean mechanical force, measured in newtons, but a potential, or energy per unit of charge, measured in volts (V).

In EM induction, e.m.f. can be defined around a closed loop as the electromagnetics work that would be done on a charge if it travels once around that loop. (While the charge travels around the loop, it can simultaneously lose the energy gained via resistance into thermal energy.) For a time-varying magnetic flux linking a loop, the electrical potential scalar field is not defined because of the circulating electric vector field; nevertheless, an e.m.f. does work and can be measured as a virtual electric potential around that loop.

2.4 Electromagnetic Induction

The equation that characterizes electrostatic from Maxwell's equations is

$$\vec{\nabla} \times \vec{E} = 0 \qquad (2.35)$$

or in integral form

$$\oint \vec{E} \cdot d\vec{l} = 0 \qquad (2.36)$$

where $d\vec{l}$ is an element of the path. The sets in Eqs. 2.35 and 2.36 are immediate, following Coulomb's Law, and they are not downgraded by the magnetic force because of a steady current. They do not hold for more general time-dependent fields, however, and these cases are what this section will now consider.

Inside a source of e.m.f. that is open-circuited, the conservative electrostatic field created by the separation of charge exactly cancels the forces producing it. Thus, the e.m.f. has the same value but an opposite sign as the integral of the electric field aligned with an internal path between two terminals, A and B, of a source of e.m.f. in an open-circuit condition; that is, the path is taken from the negative terminal to the positive terminal to yield a positive e.m.f., indicating work done on the electrons moving in the circuit [2]. Mathematically, we define this as:

$$\mathcal{E} = -\int_A^B \vec{E}_{cs} \times d\vec{l} \qquad (2.37)$$

where $\vec{E}_{cs}$ is the conservative electrostatic field created by the charge separation associated with the e.m.f., and this time $d\vec{l}$ is an element of the path from terminal A to terminal B.

This equation applies only to locations A and B that are terminals and does not apply to paths between points A and B with portions outside the source of e.m.f.. It involves the electrostatic electric field because of charge separation $\vec{E}_{cs}$ and does not involve, for example, any non-conservative component of an electric field because of Faraday's Law of induction.

In the case of a closed path in the presence of a varying magnetic field, the integral of the electric field around a closed loop may be non-zero; one common application of the concept of e.m.f., known as "induced e.m.f." is the voltage generated in such a loop. The induced e.m.f., or the *electromotive force*, around a stationary or a circuit, closed path C is:

$$\mathcal{E} = \oint_C \vec{E} \cdot d\vec{l} \qquad (2.38)$$

where now $\vec{E}$ is the entire electric field around path C, either conservative or non-conservative, and the integral is around an arbitrary but stationary closed curve of this path, for which there is a varying magnetic field, with static E and Bfields; this e.m.f. is always zero.

The electrostatic field does not contribute to the net e.m.f. around a circuit because the electrostatic portion of the electric field is conservative (i.e., the work done against the field around a closed path is zero). In other words, we take up cases where it is not zero because this E-field cannot be defined from Coulomb's Law. It is legitimate to ask what does define it then. It is defined, so that the Lorentz force:

$$\vec{F} = q(\vec{E} + \vec{v} \times \vec{B}) \tag{2.39}$$

is *always* the EM force on a test particle charge q. Thus, the definition can be extended to an arbitrary source of e.m.f. and moving paths such as C:

$$\begin{aligned}\mathcal{E} = &\oint_C [\vec{E} + \vec{v} \times \vec{B}] d\vec{l} \\ &+ \frac{1}{q}\oint_C \text{effective chemical force} \cdot d\vec{l} \\ &+ \frac{1}{q}\oint_C \text{effective chemical force} \cdot d\vec{l}\end{aligned} \tag{2.40}$$

which is a conceptual equation, mainly because the determination of the "effective forces" is difficult.

The results of a large number of experiments can be summarized by associating an e.m.f., such as $\mathcal{E} = d\Phi/dt$, with a change in magnetic flux Φ through a circuit. This result, which is known as Faraday's Law of EM induction, is found to be independent of the way in which the flux is changed and the value of $\vec{B}$ at various points inside the circuit may be changed in any way. It is very important to accept that the relationship of $\mathcal{E} = d\Phi/dt$ does represent an independent experimental law, thus it cannot be derived from other experimental laws. In addition, it certainly is not, as is sometimes stated, a consequence of conservation of energy applied to the energy balance of the current in magnetic fields.

Because by definition we have the following sets of equation, such as Eq. 2.28, and these:

$$\varepsilon = \frac{d\Phi}{dt} \tag{2.41}$$

and

$$\Phi = \int_S \vec{B} \cdot \hat{n} \, da \tag{2.42}$$

then, Eq. 2.41, with help from Eq. 2.28, produces

$$\oint_C \vec{E} \cdot d\vec{l} = -\frac{d}{dt} \oint_S \vec{B} \cdot \hat{n} \, da \tag{2.43}$$

2.5 Electromagnetic Energy and the Poynting Vector

Yet, if the circuit is a rigid stationary one, the time derivative can be taken inside the integral where it becomes a partial time derivative. Furthermore, Stokes's theorem can be used to transform the line integral of $\vec{E}$ into the surface integral of $\vec{\nabla} \times \vec{E}$. The result of these transformations is:

$$\int_S \vec{\nabla} \times \vec{E} \cdot \hat{n}\, da = -\int_S \frac{\partial \vec{B}}{\partial t} \cdot \hat{n}\, da \tag{2.44}$$

If this must be true for all fixed surfaces, S, it follows that

$$\vec{\nabla} \times \vec{E} = -\frac{\partial \vec{B}}{\partial t} \tag{2.45}$$

This equation is the differential form of Faraday's Law, and it requires generalization of $\vec{\nabla} \times \vec{E} = 0$, which holds for static fields.

2.5 Electromagnetic Energy and the Poynting Vector

As we have learned from our knowledge of classical mechanics, many problems in that field can be reduced to a simple problem significantly by means of energy consideration because of its nature of being conserved. By the same argument, when the mechanical behavior of an electrical system is to be taken under consideration, it may be advantageous to use the energy method as well. By the nature of energy being conserved, the energy of a system of charges, like that of any other mechanical system, may be divided into its *potential* and *kinetic* contributions or a general combination of them.

Here we study EM energy, both electrostatically and electrodynamically, similar to a mechanical system. Under static conditions, however, the entire energy of the charge system presents itself in the form of potential energy, and we are especially concerned with the potential energy that arises from the electric interaction of the charges—called *electrostatic energy*. On the other hand, considering Fig. 2.6, the fundamental problem electrodynamics hopes to solve, the energy related problem, when the electric charges $q_1, q_2, q_3, \ldots, q_i$; thus, we may say they are *source charges* and *test charges*, Q, in motion.

To identify and compute the electrostatic potential energy given the depiction in Fig. 2.6 as a system of charges, where charges $q_1, q_2, q_3, \ldots, q_i$ are seated at arbitrary distances, $r_1, r_2, r_3, \ldots, r_i$, from test charge Q, all we need to do to derive this energy is compute the work done by electrical field $\vec{E}$ as it acts on these point charges (i.e., $q_1, q_2, q_3, \ldots, q_i$) in the neighborhood of Q. Thus, we compute this work done by field $\vec{E}$ on system point charges by moving from point A to point B, which is defined as the *electric potential difference*, $U = U_A - U_B$, between points A and B. Here we assume that test charge Q moves from infinity (i.e., point B is set at infinity) and places it at point A, then the work done would be written:

Fig. 2.6 Illustration of the combination of source and test charges

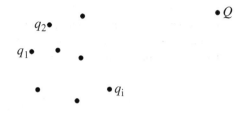

"Source" charges   "Test" charge

$$W_{\infty \to A} = Q(U_A - U_\infty) = U$$
$$= Q\left(\frac{kq_1}{r_1} + \frac{kq_2}{r_2} + \frac{kq_3}{r_3} + \cdots + \frac{kq_i}{r_i}\right) \quad (2.46)$$
$$= (Qk)\sum_{i=1}^{i}\frac{q_i}{r_i}$$

The total electric potential energy of this system of charges—namely, the work needed to bring them to their current positions—can be calculated as follows. First bring q_1 (zero work because there is no charge around it yet), then in the field of q_1 bring q_2, then in the fields of q_1 and q_2 bring q_3. Add all the effort needed to compute the total work. The result would be:

$$U = \sum_{i=1}^{i} W_i = \sum_{i=1}^{i}\left(\sum_{j=1}^{i-1}\frac{q_i q_j}{4\pi\varepsilon_0 r_{ij}}\right) \quad (2.47)$$

where $\sum_{i=1}^{i} W_i$ is a sum of all the work by $q_1, q_2, q_3, \ldots, q_i$ charges in system S as depicted in Fig. 2.6. If we abbreviate Eq. 2.36, this will result in the total electrostatic energy U:

$$U = \sum_{i=1}^{i}\sum_{j=1}^{j-1} W_{ij} \quad (2.48)$$

To find the electric field from electric potential, we simply can argue that the component of electric field $\vec{E}$ in any direction is the negative of the rate of change of the potential with distance in that direction, and mathematically could be expressed as

$$\vec{E} = -\vec{\nabla}U = -\left(\frac{\partial U}{\partial x}\hat{i} + \frac{\partial U}{\partial y}\hat{j} + \frac{\partial U}{\partial z}\hat{k}\right) \quad (2.49)$$

Basically, we can express that the electric field is the gradient of electric potential, and electric field-lines are always perpendicular to the equipment's surfaces.

2.5 Electromagnetic Energy and the Poynting Vector

If the charge distribution bears zero net charge, then the potential field at a great distance (i.e., almost infinity) acts like some multipole and falls off *more* rapidly than r^{-1}. Again, the contribution from system S' within volume V that it includes system S (see Fig. 2.5) may be seen to vanish; thus, for electrostatic energy, one can have the following general formation:

$$U = \frac{1}{2}\int_V (\vec{D} \cdot \vec{E}) dv \qquad (2.50)$$

where $\vec{D}$ is a field vector and the integral is taken over the volume of the system external to the conductors (i.e., over the various dielectrics in it). The integration may, of course, be extended to include all space because electric field $\vec{E}$ is equal to zero inside a conductor. If this formation is applied to fields that are produced, in part by point charges, it is essential to subtract their infinite "self-energy" explicitly [2].

In a similar way, we can calculate the magnetic energy of a current system in terms of circuit parameters by expressing which could be produced directly from Faraday's Law of induction. If a source of voltage $\mathcal{V}$ is applied to a circuit, then, in general, the current through the circuit can be presented by the following equation:

$$\mathcal{V} + \mathcal{E} = IR \qquad (2.51)$$

In this equation the symbol $\mathcal{E}$ is the induced electromotive force (e.m.f.) and R is the resistance of the current circuit. The work done by source voltage $\mathcal{V}$ in moving charge increment $dq = Idt$ through the circuit is then:

$$\begin{aligned}\mathcal{V}dq = \mathcal{V}Idt &= -\mathcal{E}Idt + I^2 Rdt \\ &= Id\Phi + I^2 Rdt\end{aligned} \qquad (2.52)$$

Equation 2.52 is obtained with the help of Faraday's Law. The term $I^2 Rdt$ represents the irreversible conservation of electric energy into heat by the circuit, but this term absorbs the entire work input *only* in cases where the flux change is zero. The additional term, $Id\Phi$, is the work done against the induced e.m.f. in the circuit; it is that part of the work done by $\mathcal{V}$ that is effective in altering the magnetic field structure. Ignoring the $I^2 Rdt$ term, however, we can write the following relationship:

$$dW_b = Id\Phi \qquad (2.53)$$

where subscript b indicates that this is work performed by external electrical energy sources. The work increment given in Eq. 2.53 may be either positive or negative. It is positive when flux change $d\Phi$ through the circuit is in the same direction as the flux produced by current I. For a rigid stationary circuit that is showing no energy losses, other than Joule heat loss (i.e., no hysteresis[1]), the term dW_b is equal to the

---

[1] The phenomenon in which the value of a physical property lags behind changes in the effect causing it; for instance, when magnetic induction lags behind the magnetizing force.

change in *magnetic energy* of the circuit. Hysteresis loss is discussed by Reitz et al. [2]; however, our attention here is restricted to the reversal magnetic system.

As part of the magnetic energy of coupled circuits, we derive an expression for the it of a system interacting with current circuits. If there are n circuits, then, according to Eq. 2.53, the electrical work done against the induced e.m.f.s is given by

$$dW_b = \sum_{i=1}^{n} I_i d\Phi_i \tag{2.54}$$

This equation is a very generalized form of the preceding expression, and it is valid independently of how flux increment $d\Phi_i$ is produced. Particularly, we are interested in the case where $d\Phi_i$ is produced by current change in the n circuits themselves. Under these circumstances, the flux changes of $d\Phi_i$ are correlated directly with changes in these currents as

$$d\Phi_i = \sum_{j=1}^{n} \frac{d\Phi_{ij}}{dI_j} dI_j = \sum_{j=1}^{n} M_{ij} dI_i \tag{2.55}$$

Under this condition, however, where the circuits are rigid and stationary, no mechanical work takes place that is associated with the flux changes of $d\Phi_i$, and dW_b is just equal to the change in magnetic energy dU of the system. Note again, that our attention is restricted to just stationary circuits so that magnetic energy can be calculated as a work term. Nevertheless, magnetic energy U of a system of n rigid stationary circuits can be obtained by integration of the Eq. 2.54 interval, from zero flux condition, corresponding to all $I_i = 0$ to the final set of flux values [2].

For a group of *rigid circuits* continuing, or located in *linear magnetic media*, the Φ_i is linearly related to the currents in the circuits, and the magnetic energy is independent of the way in which these currents are brought to their final set of values. If we concentrate on the linear case of a rigid circuit, and because the final energy is independent of the order in which the currents are varied, we may choose a scenario for which electrical work W can be calculated easily. This scenario is the case that all currents, and thus all fluxes, are brought to their final values stage in a concert mode. In other words, at any moment in time all currents and fluxes will be at the same fraction of their final values.

If we call this fraction quantity α, then the final values of the current are given by the symbols:

$$I_1, I_2, \cdots I_n \tag{2.56}$$

Then at any stage $I'_i = \alpha I_i$, and furthermore the flux change, will be presented by $d\Phi_i = \alpha_i d\alpha$, and integrating Eq. 2.54, will result in the following form:

2.5 Electromagnetic Energy and the Poynting Vector

$$\int dW_b = \int_0^1 d\alpha \sum_{i=1}^n I'_i \Phi_i = \sum_{i=1}^n I_i \Phi_i \int_0^1 \alpha \, d\alpha$$

$$= \frac{1}{2} \sum_{i=1}^n I_i \Phi_i \qquad (2.57)$$

Thus, magnetic energy U is given as

$$U = \frac{1}{2} \sum_{i=1}^n I_i \Phi_i \qquad (2.58)$$

With the use of Eq. 2.57, which for a rigid-circuit linear system may be integrated directly, and the magnetic energy may be expressed in the following form:

$$U = \frac{1}{2} \sum_{i=1}^n \sum_{j=1}^n M_{ij} I_i I_j$$

$$= \frac{1}{2} L_1 I_1^2 + \frac{1}{2} L_2 I_2^2 + \cdots + \frac{1}{2} L_n I_n^2$$

$$+ M_{12} I_1 I_2 + M_{13} I_1 I_3 + \cdots M_{1n} I_1 I_n \qquad (2.59)$$

$$+ M_{23} I_2 I_3 + \cdots M_{n-1,n} I_{n-1} I_n$$

(rigid circuit, linear media)

Here, we have used the results and the notation of $M_{ij} = M_{ji}$ and $L_i \equiv M_{ii}$. In case of two coupled circuits, Eq. 2.59 reduces to:

$$U = \frac{1}{2} L_1 I_1^2 M I_1 I_2 + \frac{1}{2} L_2 I_2^2 \qquad (2.60)$$

For a single circuit, we can also write the following relationship for flux Φ:

$$\Phi = LI \qquad (2.61)$$

Thus, we can write magnetic energy U in terms of flux Φ as:

$$U = \frac{1}{2} I \Phi = \frac{1}{2} L I^2 = \frac{1}{2} \frac{\Phi^2}{L} \qquad (2.62)$$

Equation 2.59 provides the magnetic energy of a current system in terms of circuit parameters such as currents and inductance. Such information is particularly useful simply because these parameters are capable of direct experimental measurement. On the other hand, an alternative formulation of the magnetic energy in terms of field vector $\vec{B}$ (magnetic field) and $\vec{H}$ (magnetic intensity) is of considerable interest because it provides a picture in which energy is stored in the magnetic field itself. We can show how energy moves through the EM field in non-stationary processes [2].

Now, as part of energy density in the magnetic field, we consider a group of rigid current-carrying circuits, none of which extends to infinity, immersed in a medium with linear magnetic properties. The energy of such a system is initiated from Eq. 2.58, and, for purposes of this argument, it is convenient to assume that each circuit consists of only a single loop, then flux Φ_i may be expressed as:

$$\Phi_i = \int_{S_i} \vec{B} \cdot \hat{n} \, da = \oint_{C_i} \vec{A} \cdot d\vec{l}_i \tag{2.63}$$

where $\vec{A}$ is the local vector potential. Substitution of this result into Eq. 2.58 would yield the following:

$$U = \frac{1}{2} \sum_i \oint_{C_i} I_i \vec{A} \cdot d\vec{l} \tag{2.64}$$

Note that C_i is the contiguous circuits loop.

At this point, however, we want to make a more general form of Eq. 2.64; for that we suppose that we do not have current circuits defined by wires, but instead each "circuit" is a closed path in the medium, which we assume to be conducting and follows a line of current density. Then Eq. 2.64 may be made to approximate this situation very closely by choosing a large number of contiguous circuits (C_i); replacing $I_i d\vec{l}_i \rightarrow \vec{J} dv$; and, finally, by the substitution of $\int_V$ for $\sum_i \oint_{C_i}$, we can write Eq. 2.64 in a new form:

$$U = \frac{1}{2} \int_V \vec{J} \cdot \vec{A} \, dv \tag{2.65}$$

The equation may be further transformed by using field equation $\vec{\nabla} \times \vec{H} = \vec{J}$, and vector identity of the following form:

$$\vec{\nabla} \cdot (\vec{A} \times \vec{H}) = \vec{H} \cdot \vec{\nabla} \times \vec{A} - \vec{A} \cdot \vec{\nabla} \times \vec{H} \tag{2.66}$$

Thus, we can write,

$$U = \frac{1}{2} \int_V \vec{H} \cdot \vec{\nabla} \times \vec{A} \, dv - \frac{1}{2} \int_S \vec{A} \cdot \vec{H} \cdot \hat{n} \, da \tag{2.67}$$

where S is the surface that bounds volume V. Furthermore, we assumed none of the current "circuits" extends to infinity; it is convenient to move surface S out to a very great distance so that all parts of this surface are far from the currents. Of course, the volume of the system must be increased accordingly.

Now, magnetic intensity $\vec{H}$ falls off at least as fast as $1/r^2$, where r is the distance from an origin near the middle of the current distribution to a characteristic point on surface S; local vector potential $\vec{A}$ falls off at least as fast as $1/r$; and the surface area is proportional to r^2. Thus, the contribution from the surface integral in Eq. 2.67 falls

2.5 Electromagnetic Energy and the Poynting Vector

off as per $1/r$ or even faster, and if S is moved out to infinity, this contribution vanishes [2].

By dropping the surface integral in Eq. 2.67 and extending the volume term to include all space, and because $\vec{B} = \vec{\nabla} \times \vec{A}$ then we obtain the following result for magnetic energy:

$$U = \frac{1}{2} \int_v \vec{H} \cdot \vec{B} \, dv \qquad (2.68)$$

The result in this equation is analogous to the one expressed for electrostatic energy, as represented by Eq. 2.50.

Equation 2.68 is restricted to systems containing linear magnetic media because it was derived from Eq. 2.58. With little reasoning provided by Reitz et al. [2], we are led to the concept of energy density u in the magnetic field as:

$$u = \frac{1}{2} \vec{H} \cdot \vec{B} \qquad (2.69)$$

which, for the case of isotropic, linear, magnetic materials reduces to

$$u = \frac{1}{2} \mu H^2 = \frac{1}{2} \frac{B^2}{\mu} \qquad (2.70)$$

Now, the question for the applicability of the expressions in Eq. 2.58 and Eq. 2.68 to non-static situations arises.

If we take the scalar product of Eq. 2.34 with electric field $\vec{E}$, and take the result of that operation to subtract from the scalar product of Eq. 2.32 with magnetic intensity $\vec{H}$, then the resulting equation is provided as follows:

$$\vec{H} \cdot \vec{\nabla} \times \vec{E} - \vec{E} \cdot \vec{\nabla} \times \vec{H} = -\vec{H} \cdot \frac{\partial \vec{B}}{\partial t} - \vec{E} \cdot \frac{\partial \vec{D}}{\partial t} - \vec{E} \cdot \vec{J} \qquad (2.71)$$

The LHS of this equation can be converted into a divergence by using the following vector identity:

$$\vec{\nabla} \cdot (\vec{F} \times \vec{G}) = \vec{G} \cdot \vec{\nabla} \times \vec{F} - \vec{F} \cdot \vec{\nabla} \times \vec{G} \qquad (2.72)$$

Thus, by using the vector identity in Eq. 2.73, we obtain:

$$\vec{\nabla} \cdot (\vec{E} \times \vec{H}) = -\vec{H} \cdot \frac{\partial \vec{B}}{\partial t} - \vec{E} \cdot \frac{\partial \vec{D}}{\partial t} - \vec{E} \cdot \vec{J} \qquad (2.73)$$

If the medium for which Eq. 2.73 will be used and applied is linear and non-dispersive—that is, if electric displacement $\vec{D}$ is proportional to electric field $\vec{E}$ and magnetic field $\vec{B}$ is proportional to magnetic intensity $\vec{H}$—then the time derivatives on the RHS of Eq. 2.73 can be modified to the following terms:

$$\vec{E} \cdot \frac{\partial \vec{D}}{\partial t} = \vec{E} \cdot \frac{\partial}{\partial t} \varepsilon \vec{E} = \frac{1}{2} \varepsilon \frac{\partial \vec{E}^2}{\partial t} = \frac{1}{2} \frac{\partial}{\partial t} (\vec{E} \cdot \vec{D}) \qquad (2.74)$$

and

$$\vec{H} \cdot \frac{\partial \vec{B}}{\partial t} = \vec{H} \cdot \frac{\partial}{\partial t} (\mu \vec{H}) = \frac{1}{2} \mu \frac{\partial}{\partial t} (\vec{H}^2) = \frac{\partial}{\partial t} \left(\frac{1}{2} \vec{H} \cdot \vec{B} \right) \qquad (2.75)$$

Using the relationship in Eqs. 2.74 and 2.75 by substituting them into Eq. 2.73, it reduces to the form:

$$\vec{\nabla} \cdot (\vec{E} \times \vec{H}) = -\frac{\partial}{\partial t} \left(\frac{1}{2} \right) [\vec{E} \cdot \vec{D} + \vec{B} \cdot \vec{H}] - (\vec{J} \cdot \vec{E}) \qquad (2.76)$$

where $\vec{J}$ is presenting the displacement current in a linear and non-dispersive medium, as before.

The first term on the RHS of Eq. 2.76 is totally the time derivative of the sum of the electric and magnetic energy densities, while, in many cases, the second term is just the negative of Joule heating rate per unit volume, if $\vec{J} = g\vec{E}$. Integration of Eq. 2.76, over a fixed volume V, which is bounded by the surface S, results in:

$$\int_v \vec{\nabla} \cdot (\vec{E} \times \vec{H}) dv = -\frac{d}{dt} \int_v \frac{1}{2} [\vec{E} \cdot \vec{D} + \vec{B} \cdot \vec{H}] dv - \int_v (\vec{J} \cdot \vec{E}) dv \qquad (2.77)$$

Applying the divergence theorem to the LHS of Eq. 2.77, we get the following result:

$$\oint_S \vec{E} \times \vec{H} \cdot \hat{n} \, da = -\frac{d}{dt} \int_{v} \frac{1}{2} (\vec{E} \cdot \vec{D} + \vec{B} \cdot \vec{H}) - \int_V \vec{J} \cdot \vec{E} dv \qquad (2.78)$$

Rearranging Eq. 2.78, we can then rewrite it as:

$$-\int_V \vec{J} \cdot \vec{E} dv = -\frac{d}{dt} \int_v \frac{1}{2} (\vec{E} \cdot \vec{D} + \vec{B} \cdot \vec{H}) + \oint_S \vec{E} \times \vec{H} \cdot \hat{n} \, da \qquad (2.79)$$

From this equation, clearly, we can see that the $\vec{J} \cdot \vec{E}$ term is comprised of two parts: (1) the rate of change of EM energy stored in volume V and (2) a surface integral. Furthermore, the LHS of Eq. 2.79 is the power transferred *into* the EM field through the motion of a free charge in volume V.

If there are no sources of e.m.f. in volume V, then the LHS of Eq. 2.79 is negative and equal to minus the Joule heat production per unit of time, as was indicated in the preceding. Under certain circumstances, however, the LHS of Eq. 2.79 may be positive. Let us suppose that charged particle q moves with constant velocity $\vec{v}$ under the Lorentz force, which is combination of the influence of mechanical, electrical, and magnetic forces, then the rate at which the mechanical force performs work on the particle is given by:

2.5 Electromagnetic Energy and the Poynting Vector

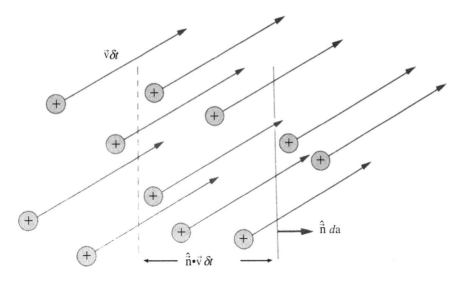

Fig. 2.7 The drift motion of charge carriers across the plane *da* in time *δt*

$$\vec{F}_m \cdot \vec{v} = -q(\vec{E} + \vec{v} \times \vec{B}) \cdot \vec{v} = -q\vec{E} \cdot \vec{v} \tag{2.80}$$

Additionally, we need to establish current density $\vec{J}$ as the summation of current over the various carrier types for the drift motion of charges across plane *da* in time *δt*, as depicted in Fig. 2.7.

For this purpose, we consider a conducting medium that has only one type of charge carrier of particle charge q type. The number of these carriers per unit volume will be denoted by N; meanwhile, we ignore any random thermal motions of these particles and assign the same drift velocity $\vec{v}$ to each carrier. We now calculate the current through an element of area *da*, during time *δt*, where each carrier moves a distance of $\vec{v}\delta t$. From Fig. 2.7 it is obvious that charge δQ, which moves across surface element *da* during time *δt*, is q times the sum of all charge carriers in volume $\vec{v} \cdot \hat{n} \,\delta t da$, where $\hat{n}$ is a unit vector normal to element area *da*.

If current I is defined as the rate at which charge is transported, through a given surface in a conducting system for δQ and duration of time δt, we can claim that:

$$I = \frac{\delta Q}{\delta t} \tag{2.81}$$

Then, we can write:

$$I = \frac{\delta Q}{\delta t} = \frac{qN\vec{v} \cdot \hat{n}\, \delta t da}{\delta t}$$
$$= Nq\vec{v} \cdot \hat{n}\, da \tag{2.82}$$

Still, if more than one kind of charge carrier exists, then there will be a contribution of the form presented in Eq. 2.82 from each type of carrier. In general, we can write:

$$dI = \left[\sum_i N_i q_i \vec{v}_i\right] \cdot \hat{n} \, da \tag{2.83}$$

This equation presents of the current going through element area da, and thus the summation is over the various carrier types. The quantity in brackets in Eq. 2.83 is a vector that has dimensions of current per unit area; this quantity is called the current density and is illustrated by symbol $\vec{J}$ as in the following form:

$$\vec{J} = \sum_i N_i q_i \vec{v}_i \tag{2.84}$$

Now comparing Eq. 2.80, with Eq. 2.84, which is current density $\vec{J}$, the rate at which mechanical work is done per unit volume is given by:

$$\sum_i N_i \vec{F}_m \cdot \vec{v}_i = -\vec{E} \cdot \vec{J} \tag{2.85}$$

Thus, this equation is the power density that is transferred into the EM field.

Because the surface integral in Eq. 2.79 involves only the electric and magnetic fields, it is feasible to interpret this term as the rate of energy flow across the surface. Equation 2.79 thus expresses the conservation of energy in a fixed volume V. Now, if we take the corresponding differential, Eq. 2.76, which will express the local conservation of energy at a certain point, we can make the following designations:

$$\vec{S} = \vec{E} \times \vec{H} \tag{2.86}$$

and

$$u = \frac{1}{2}\left(\vec{E} \cdot \vec{D} + \vec{B} \cdot \vec{H}\right) \tag{2.87}$$

Thus, Eq. 2.76, implies that at any point we can write:

$$\vec{\nabla} \cdot \vec{S} + \frac{\partial u}{\partial t} = -\vec{J} \cdot \vec{E} \tag{2.88}$$

From Eq. 2.88, without a doubt, we can see that $\vec{J} \cdot \vec{E}$ is the work done by the local field on charged particles per unit volume. Here u was interpreted as the energy density of the electric and magnetic fields. If $\vec{\nabla} \cdot \vec{S} = 0$, then Eq. 2.88 would express local conservation of energy. The rate of change of field energy equals power dissipation per unit volume at each point. If, on the other hand, $\vec{\nabla} \cdot \vec{S} \neq 0$, but $\vec{J} \cdot \vec{E} = 0$ (e.g., in case of a non-conducting medium), then Eq. 2.88 reduces to this form:

$$\vec{\nabla} \cdot \vec{S} + \frac{\partial u}{\partial t} = 0 \tag{2.89}$$

This equation has exactly the mathematical form of the equation of continuity, as presented in Eq. 2.27 for charge, except that energy density u takes the place of charge density ρ. If Eq. 2.89 is validating the description of conservation of energy, then the relationship of $\vec{\nabla} \cdot \vec{S}$ must represent the divergence of an energy current density or, in other words, of a rate of energy flow per unit area.

We now, by the definition of electrodynamics, treat Eq. 2.86 as some quantity that is known as the *Poynting vector*, as the local energy flow per unit area; however, note that there is a continuing controversy over this point [5]. We will use these interpretations of u and $\vec{S}$, while recognizing that it is only their time derivative and divergence, respectively, the interpretations of which are directly required by Maxwell's equations. It is usually only the latter that are physically measurable anyway. In any case Eq. 2.88 expresses energy conservation locally, as does Eq. 2.79 in its integral form [2].

Note that in the linear media the relationship between magnetic intensity $\vec{H}$ and magnetic field $\vec{B}$ is given by a factor known as the *permeability* of free space—$\mu_0 = 4\pi \times 10^{-7}$ N/A^2. Thus, we have:

$$\vec{H} = \frac{1}{\mu_0} \vec{B} \quad (2.90)$$

Using Eq. 2.90, substituting it into Eq. 2.86, we have a new form of this equation:

$$\vec{S} = \frac{1}{\mu_0} (\vec{E} \times \vec{B}) \quad (2.91)$$

which is the general form of the flux vector or another version of the Poynting vector. It is independent of the choice of coordinate systems.

In Chap. 3 we show further details about usage of the Poynting vector in applications, when we derive the energy flux and energy density of EM fields in a vacuum and for a plane wave as well.

2.6 Simple Classical Mechanics Systems and Fields

This section quickly reviews the classical mechanical systems largely from the point of view of *Lagrangian dynamics*, consequently, introducing the corresponding *Hamiltonian dynamic* associated with Lagrangian and momentum p of the particle.

To demonstrate this matter, we start with the Lagrangian of a discrete system, like a single particle q, as:

$$\mathcal{L}(x, \dot{x}) = T - V \quad (2.92)$$

where quantity T and V are associated with the kinetic and potential energy of the particle under consideration, and $\dot{x}$ is the time derivative of the coordinate in which the single particle is moving. Thus, the Lagrange's equations are written:

$$\frac{d}{dt}\left(\frac{\partial \mathcal{L}}{\partial \dot{x}}\right) - \frac{\partial \mathcal{L}}{\partial x_i} = 0 \qquad (2.93)$$

where x_i are the coordinates of the particle. Equation 2.93 is derivable from the relationship that is known as the *Principle of Least Action*. You will see more detail about it in Sect. 2.7, Lagrangian and Hamiltonian of Relativistic Mechanics. Thus, we write the following relation:

$$\delta \int_{t_1}^{t_2} \mathcal{L}(x_i, \dot{x}_i)\, dt = 0 \qquad (2.94)$$

where in this integral the coefficient of δ corresponds to some constant. Similarly, we can define the Hamiltonian as follows:

$$\mathcal{H}(x_i, p_i) = \sum_i p_i \dot{x}_i - \mathcal{L} \qquad (2.95)$$

where p_i are the momenta conjugate to coordinates x_i and are presented by the following form:

$$p_i = \frac{\partial \mathcal{L}}{\partial \dot{x}_i} \qquad (2.96)$$

For continuous system, like a string, the Lagrangian is an integral of Lagrangian density $\mathscr{L}$ function as:

$$\mathcal{L} = \int \mathscr{L}\, dx \qquad (2.97)$$

In case of the string example, the factor $\mathscr{L}$ will be in the form:

$$\mathscr{L} = \frac{1}{2}\left[\mu \dot{\eta}^2 - Y\left(\frac{\partial \eta}{\partial x}\right)^2\right] \qquad (2.98)$$

where Y is the *Young's module* for the material of the string and μ is the mass density. The Euler-Lagrange equation for a continuous system is also a derivative from the Principle of Least Action states as previously discussed. For the string, this would be:

$$\frac{\partial}{\partial x}\left(\frac{\partial \mathscr{L}}{\partial (\partial \eta/\partial x)}\right) + \frac{\partial}{\partial t}\left(\frac{\partial \mathscr{L}}{\partial (\partial \eta/\partial t)}\right) - \frac{\partial \mathscr{L}}{\partial \eta} = 0 \qquad (2.99)$$

Recall that the Lagrangian is a function of parameter η, and it is space and time derivative.

The Hamiltonian density can be computed from the Lagrangian density and is a function of the coordinate η; it is conjugate momentum and is written:

$$\mathcal{H} = \dot{\eta}\left(\frac{\partial \mathcal{L}}{\partial \dot{\eta}}\right) - \mathcal{L} \qquad (2.100)$$

In this example of a string, $\eta(x,t)$ is a simple *scalar field*. The string has a displacement at each point along it that varies as a function of time.

If we apply the Euler-Lagrange, Eq. 2.99, we obtain the following differential equation that the string's displacement will satisfy:

$$\mathcal{L} = \frac{1}{2}\left[\mu\dot{\eta}^2 - Y\left(\frac{\partial \eta}{\partial x}\right)^2\right]$$

$$\frac{\partial}{\partial x}\left(\frac{\partial \mathcal{L}}{\partial(\partial \eta/\partial x)}\right) + \frac{\partial}{\partial t}\left(\frac{\partial \mathcal{L}}{\partial(\partial \eta/\partial t)}\right) - \frac{\partial \mathcal{L}}{\partial \eta} = 0$$

$$\frac{\partial \mathcal{L}}{\partial(\partial \eta/\partial x)} = 0$$

$$\frac{\partial \mathcal{L}}{\partial(\partial \eta/\partial t)} = \mu\dot{\eta} \qquad (2.101)$$

$$-Y\frac{\partial^2 \eta}{\partial x^2} + \mu\ddot{\eta} + 0 = 0$$

$$\ddot{\eta} = \frac{Y}{\mu}\frac{\partial^2 \eta}{\partial x^2}$$

This equation is the wave equation for the string. There are easier ways to get to this wave equation; thus, as we move away from simple classical mechanical systems, a formal way of proceeding will be very desirable.

2.7 Lagrangian and Hamiltonian of Relativistic Mechanics

To be able to study motion, and consequently, the energy and momentum of charge particle q, we need to have some understanding of a principle known as the *Principle of Least Action*, where it leads us to the establishment of Lagrangian and Hamiltonian statements for the motion of material particles, eventually, yielding to energy and momentum of those particles [6].

The principle's definition states that for each mechanical system, there exists a certain integral $\mathcal{A}$, which is called an action that has a minimum value for the actual motion, in a way that its variation $\delta\mathcal{A}$ is zero. Strictly speaking, the principle indicates that integral $\boldsymbol{a}$ must be a minimum only for infinitesimal lengths of the path of integration where the integral takes place. For paths of an arbitrary length, we can express that only integral $\boldsymbol{a}$ must be a extremum, not necessarily a minimum.

To determine the action integral for a free material particle that is not under the influence of any external force, we can see that this integral must not depend on the choice of any reference system, which means it must be invariant under any Lorentz force transformation. Then, the integral follows that its dependency should be of a scalar form. Furthermore, mathematically, it is clear that the integrand element under the integral must be a first-order differential form. The only scalar of this kind that one can build on for a free particle motion, however, is interval $d\boldsymbol{a}$ or $\alpha d\boldsymbol{a}$, where a coefficient of α is some constant. Thus, for a free particle the action must keep to the following form of integration for a Principle of Least Action:

$$\boldsymbol{a} = -a \int_a^b d\boldsymbol{a} \tag{2.102}$$

where $\int_a^b d\boldsymbol{a}$ is an integral along the world line of the particle between the two particular events of the arrival of it at initial position a and at final position b of two given world points. It needs to be at defined times associated with these two points at time t_1 and t_2, respectively, where α is some constant characterizing the particle, and the value of it has to be a positive quantity and should be a maximum along the straight world line for all particles [6].

By integrating along a curved world line, we can make the integral arbitrarily small. According to Minkowski, for a geometrical shape (see Fig. 2.8), he said: "Henceforth space by itself, and time by itself are fading away into mere shadows according to the Principle of Relativity" [7]. This way then integral $\int_a^b d\boldsymbol{a}$ with the positive sign cannot have a minimum; with the opposite sign it clearly has a minimum value along the straight world line. We will further explain the details about Fig. 2.8 under Lorentz transformation in Sect. 2.8, and it is described in Appendix A as well.

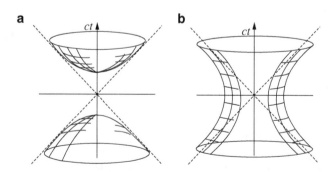

Fig. 2.8 (**a**) Hyperboloid of two sheets and (**b**) hyperboloid of one sheet

2.7 Lagrangian and Hamiltonian of Relativistic Mechanics

The action integral then can be represented as an integral with respect to time as:

$$a = \int_{t_1}^{t_2} \mathcal{L} dt \qquad (2.103)$$

In this equation the symbol $\mathcal{L}$ as a coefficient of dt is a representation of the *Lagrange function* of the mechanical system. Furthermore, to determine the Lagrangian $\mathcal{L}$ for some velocity v of the material particle, in a proper time, we need to define and calculate this proper time within a certain inertial reference system, where we can observe clocks that may be moving relative to us in an arbitrary manner.

To achieve such a goal, we suppose that at each differential moment of time, this motion can be considered as uniform as possible. Thus, at each moment of time, we can introduce a coordinate system rigidly linked to the moving clocks, which with the clocks constitutes an initial reference system. In the course of an infinitesimal time interval, dt, as it reads by a clock in our frame at rest, however, the moving clock goes a distance of $\sqrt{dx^2 + dy^2 + dz^2}$.

Now let us ask: What time interval dt' is indicated for this period by moving clocks? In a system of coordinates linked to the moving clocks, the latter are at rest (i.e., $dx' = dy' = dz' = 0$). Because of the invariance of intervals, we can write the following sets of equations:

$$da^2 = c^2 dt^2 - dx^2 - dy^2 - dz^2 = c^2 dt'^2 \qquad (2.104)$$

From which

$$dt' = \left[\sqrt{1 - \frac{dx^2 + dy^2 + dz^2}{c^2 dt^2}}\right] dt \qquad (2.105)$$

But, we know that

$$v^2 = \frac{dx^2 + dy^2 + dz^2}{dt^2} \qquad (2.106)$$

where v is the velocity of the moving clocks; therefore, we can write the following result:

$$dt' = \frac{da}{c} = \left[\sqrt{1 - \frac{v^2}{c^2}}\right] dt \qquad (2.107)$$

We need to use this equation as an aid to Eq. 2.103; we find that:

$$a = -\int_{t_1}^{t_2} \alpha c \sqrt{1 - \frac{v^2}{c^2}} dt \qquad (2.108)$$

where again in Eq. 2.108, the parameter v is the velocity of the material particle. Comparing the integrand $\mathcal{L}$ in Eq. 2.103 and the one in Eq. 2.108, we conclude that the following is valid for the Lagrangian for the particle:

$$\mathcal{L} = -\alpha c \sqrt{1 - (v^2/c^2)} \qquad (2.109)$$

Again, as we discussed in the preceding, quantity α is a symbol that characterizes the particle. In classical mechanics each particle is characterized by mass, m, as we know from Newtonian mechanics.

Now, let us find the relationship between these two characterization symbols—namely, α and m. It can be shown from the fact that in the limit as speed of light, $c \to \infty$, our expression for Lagrangian function $\mathcal{L}$ must go over into classical in a transition from relative expression as $\mathcal{L} = mv^2/2$. To carry out this transition, we expand $\mathcal{L}$ using the *Taylor expansion method* in powers of v/c, neglecting terms of higher order; we find that

$$\mathcal{L} = -\alpha c \sqrt{1 - \frac{v^2}{c^2}} \approx -\alpha c + \frac{\alpha v^2}{2c} \qquad (2.110)$$

Constant terms in Lagrangian do not affect the equation of motion, and it can be omitted. Omitting the constant αc in $\mathcal{L}$ and showing that $\mathcal{L} = \alpha v^2/2c$, and comparing it the classical expression of $\mathcal{L} = mv^2/2$, we can conclude that $\alpha = mc$. Thus, the action for a free material point, using Eq. 2.102, is:

$$a = -mc \int_a^b da \qquad (2.111)$$

By the same statement, using Eq. 2.109, Lagrangian function $\mathcal{L}$ is given by:

$$\mathcal{L} = -mc^2 \sqrt{1 - \frac{v^2}{c^2}} \qquad (2.112)$$

2.7.1 Four-Dimensional Velocity

From the ordinary three-dimensional velocity vector (see Appendix C, Definition of Three-Dimensional Vector), one can develop a 4-vector form. This four-dimensional velocity vector (*four-velocity*) of a particle is this vector:

$$u^i = \frac{dx^i}{da} \qquad (2.113)$$

where da is structured by Eq. 2.104; thus, according to this equation, we can find the following relation, as well as all the components in Eq. 2.111, so we can write:

2.7 Lagrangian and Hamiltonian of Relativistic Mechanics

$$da = cdt\sqrt{1 - \frac{v^2}{c^2}} \qquad (2.114)$$

where, in this equation, the parameter of v is the ordinary three-dimensional velocity of the particle. Thus, we can write the following relations:

$$u^1 = \frac{dx^1}{da} = \frac{dx}{cdt\sqrt{1-\frac{v^2}{c^2}}} = \frac{v_x}{c\sqrt{1-\frac{v^2}{c^2}}} \qquad (2.115(a))$$

Or, in general, we can write it:

$$u^i = \left(\frac{1}{\sqrt{1-\frac{v^2}{c^2}}}, \frac{\vec{v}}{c\sqrt{1-\frac{v^2}{c^2}}}\right) \qquad (2.115(b))$$

Note that the four-velocity is a dimensionless quantity.

The components of the four-velocity are not independent. Noting that $dx_i dx^i = ds^2$ (see Appendix C again), we have the following form for the velocity vector:

$$u^i u_i = 1 \qquad (2.116)$$

Geometrically, u^i is a unit 4-vector tangent to the world line of the particle under consideration for purposes of this analysis. Similar to the definition of the four-velocity, the second derivative is:

$$u^i = \frac{d^2 x^i}{dx^2} = \frac{du^i}{da} \qquad (2.117)$$

This can be called the *four-acceleration*. Differentiating Eq. 2.116, we find:

$$u_i w^i = 0 \qquad (2.118)$$

Note that, also, the 4-vectors of velocity and acceleration are mutually perpendicular.

2.7.2 Energy and Momentum in Relativistic Mechanics

By the definition of momentum of a particle, we mean the vector $\vec{p} = \partial L/\partial \vec{v}$, where the components of this momentum vector $\vec{p}$ are the derivatives of Lagrangian function $\mathcal{L}$ with respect to the corresponding components of vector velocity. Using Eq. 2.112, we find:

$$\vec{p} = \frac{m\vec{v}}{\sqrt{1 - \frac{v^2}{c^2}}} \tag{2.119}$$

In this equation, for a small velocity of $v \ll c$, or in the limit as $c \to \infty$, it approaches the classical mechanics form as $\vec{p} = m\vec{v}$, and for $v = c$, naturally, the momentum becomes an infinite value.

The time derivative of the momentum, mathematically, indicates that is the force acting on the particle. Now, if we suppose that the velocity of the particle changes only in a direction perpendicular to the direction of force, we can write [7]:

$$\frac{d\vec{p}}{dt} = \frac{m}{\sqrt{1 - \frac{v^2}{c^2}}} \frac{d\vec{v}}{dt} \tag{2.120}$$

If the velocity changes only in magnitude—that is, if the force is parallel to the velocity—then we can write:

$$\frac{d\vec{p}}{dt} = \frac{m}{\left(1 - \frac{v^2}{c^2}\right)^{\frac{3}{2}}} \frac{d\vec{v}}{dt} \tag{2.121}$$

From both Eqs. 2.120 and 2.121, we clearly can see that the ratio of force to acceleration is different in both cases as well.

Energy $\mathcal{E}$ of the particle is defined as the following relationship:

$$\mathcal{E} = \vec{p} \cdot \vec{v} - \mathcal{L} \tag{2.122}$$

Now, if we substitute for $\mathcal{L}$ from Eq. 2.112 and for $\vec{p}$ from Eq. 2.119 into Eq. 2.108, we obtain the following result:

$$\mathcal{E} = \frac{mc^2}{\sqrt{1 - \frac{v^2}{c^2}}} \tag{2.123}$$

This equation is a very important formula and shows, in particular, that in relativistic mechanics the energy of a free particle does not approach a zero value for the velocity of $v = 0$; rather, it takes on a finite value, which is same as for the famous Einstein formula of relativity, for the *rest energy* of the particle as follows:

$$\mathcal{E} = mc^2 \tag{2.124}$$

For small velocities of $v/c \ll 1$, however, we can establish a different form of Eq. 2.106 by expanding the denominator via Taylor's series in a power of v/c and obtain the following result:

$$\mathcal{E} \approx mc^2 + \frac{mv^2}{2} \tag{2.125}$$

2.7 Lagrangian and Hamiltonian of Relativistic Mechanics

which, except for the rest energy in relativistic mechanics, also is included in the equation—the classical mechanics expression for the kinetic energy term of a free particle.

Note that, although we talk about a "particle" in the preceding discussion, we have now made use of a particle that is "elementary." Thus, the equations and formulas to any composite body consisting of many particles, where by mass m we mean the total mass of the body, and by velocity v of that body's motion we mean a whole. In particular, Eq. 2.110 is valid for any composite body that is at rest as a whole [7].

If we pay attention to the fact that in relativistic mechanics the energy of a free composite body of any closed system is a completely definite quantity, it is always positive and is directly related to the mass of the body. In this scenario, we can recall from our knowledge of classical mechanics that the energy of a body is defined only within an arbitrary constant, and it can be either positive or negative [7].

The energy of a composite body at rest contains, in addition to the rest energies of its constituent particles, the kinetic energy of the particles and the energy of their interactions with one another. In other words, mc^2 is not equal to $\sum m_a c^2$, where m_a is the representation of the masses of the particles as a whole, so m is not equal to $\sum m_a c^2$. Thus, in relativistic mechanics the law of conservation of mass does not hold, which means that the mass of a composite body is not equal to the sum of the masses of its parts. Instead, only the law of conservation of energy, in which the rest energies of the particles are included, is valid [7].

In addition, by squaring both Eqs. 2.119 and 2.122 and comparing the results, we obtain the following relation between energy and momentum in relativistic momentum of a particle:

$$\frac{\mathcal{E}^2}{c^2} = p^2 + m^2 c^2 \qquad (2.126)$$

The energy expressed in terms of momentum is called the Hamiltonian function, $\mathcal{H}$, and it is written as:

$$\mathcal{H} = c\sqrt{p^2 + m^2 c^2} \qquad (2.127)$$

For low velocities, where $p \ll mc$, we can obtain an approximate value for Hamiltonian function as follows:

$$\mathcal{H} \approx mc^2 + \frac{p^2}{2m} \qquad (2.128)$$

As we can see in Eq. 2.128, except for the rest energy in relativistic mechanics, we included the familiar classical mechanics expression for the Hamiltonian as well. Further investigation of Eqs. 2.119 and 2.122 reveals that we are able to get the following relation between the energy, momentum, and velocity of a free particle:

$$\vec{p} = \mathcal{E}\frac{\vec{v}}{c^2} \tag{2.129}$$

For a condition where $v = c$, the momentum and energy of the particle become infinite. This reveals that a particle with mass m differs from zero and cannot move with the velocity of light. Nevertheless, in relativistic mechanics, there are particles of zero mass (e.g., light quanta and neutrinos) that can move with equal velocity to light; thus, for such particles Eq. 2.129 provides the following result [7]:

$$p = \frac{\mathcal{E}}{c} \tag{2.130}$$

Similarly, the same formula holds approximately for particles with non-zero mass in the so-called *ultra-relativistic* case, when particle energy $\mathcal{E}$ is large compared to its rest mass energy, mc^2.

Now, if we take advantage of 4-vector (see Appendix C for details of 4-vector scenario), we can write all the equations in four-dimensional form. According to the Principle of Least Action, as:

$$\delta \mathcal{A} = -mc\delta \int_a^b da = 0 \tag{2.131}$$

Moreover, to set up the expression for $\delta\mathcal{A}$, we note that $da = \sqrt{dx_i dx^i}$; therefore, we can write the following new form of Eq. 2.125:

$$\delta \mathcal{A} = -mc \int_a^b \frac{dx_i dx^i}{\sqrt{da}} = -mc \int_a^b u_i d\delta x^i \tag{2.132}$$

To carry on the integration in this equation, we use the method of integration by part and obtain the following result:

$$\delta \mathcal{A} = -mcu_i \delta x^i \Big|_a^b + mc \int_a^b \delta x^i \frac{du_i}{da} da \tag{2.133}$$

Therefore, to get the equation of motion, we compare different trajectories between the same two points—that is, at the limits $(\delta x^i)_a = (\delta x^i)_b = 0$. The actual trajectory is then determined from condition $\delta \mathcal{A} = 0$ [7]. From Eq. 2.127, we thus obtain the equation $(du_i/da) = 0$, which is a constant velocity for the free particle in four-dimensional form.

To be able to determine the variation of the action as a function of the coordinates, one must consider point a as a fixed point so that $(\delta x^i)_a = 0$. The second point b is to be considered as variable, but only actual trajectories are allowed (i.e., those that satisfy the equations of motion). Therefore, the integral of Eq. 2.127 for $\delta \mathcal{A} = 0$ and in place of $(\delta x^i)_b$ for the upper limit of the first term on the RHS of the integral of Eq. 2.127, may be written simply as δx^i, thus obtaining the following relationship:

2.7 Lagrangian and Hamiltonian of Relativistic Mechanics

$$\delta a = -mcu_i \delta x^i \tag{2.134}$$

The 4-vector provides the following form of momentum p_i:

$$p_i = -\frac{\delta a}{\partial x^i} \tag{2.135}$$

From a relativistic mechanics point of view, however, the derivatives of $\delta a/\partial x$, $\delta a/\partial y, \delta a/\partial z$, are the three components of momentum vector $\vec{p}$ of the particle, while the time derivative of $-\delta a/\partial t$ is particle energy $\mathcal{E}$. Thus, the covariant components of the four-momentum are a function of $\mathcal{E}/c$ and vector momentum $\vec{p}$; it is shown here as $p_i = (\mathcal{E}/c, \vec{p})$, while the contravariant components are:

$$p_i = (\mathcal{E}/c, \vec{p}) \tag{2.136}$$

Note that the *contravariant* components are related to the corresponding three-dimensional vectors $\vec{r}$ for x', $\vec{p}$ for p^i with the "right" positive sign.

From Eq. 2.132 we can see that the components of the four-momentum of a free particle are:

$$p^i = mcu^i \tag{2.137}$$

By substituting the components of the four-velocity from Eq. 2.115(a, b), we can see that we get Eqs. 2.119 and 2.123 for momentum $\vec{p}$ and energy $\mathcal{E}$, respectively. Thus, in relativistic mechanics, momentum and energy are the components of a single 4-vector. From this argument, we immediately get the formulas for transformation of momentum and energy from one intertidal frame of reference to another. By substitution of Eq. 2.136 into general formulas established in Eq. 2.115(a, b) for transformation of 4-vectors, we find:

$$p_x = \frac{p'_x + \frac{v}{c^2}\mathcal{E}'}{\sqrt{1 - \frac{v^2}{c^2}}} \qquad p_y = p'_y \qquad p_z = p'_z \qquad p_x = \frac{\mathcal{E}' + vp'_x}{\sqrt{1 - \frac{v^2}{c^2}}} \tag{2.138}$$

where p_x, p_y, p_z are the components of three-dimensional vector $\vec{p}$.

From Eq. 2.137 of the four-momentum and the identity $u^i u_i = 1$, for the square of the four-momentum of a free particle, we have:

$$p_i p^i = m^2 c^2 \tag{2.139}$$

By substituting Eq. 2.136, we obtain the result of Eq. 2.126 repeatedly. If we analyze this with the usual definition of force, the force 4-vector is defined as the derivative:

$$g^i = \frac{dp^i}{da} = me\left(\frac{du^i}{da}\right) \tag{2.140}$$

The components of Eq. 2.140 satisfy the identity of $g_i u^i = 0$. The components of this 4-vector are expressed in terms of the usual three-dimensional force vector $\vec{f} = d\vec{p}/dt$:

$$g_i = \left(\frac{\vec{f} \cdot \vec{v}}{c^2 \sqrt{1 - \frac{v^2}{c^2}}}, \frac{\vec{f}}{c \sqrt{1 - \frac{v^2}{c^2}}} \right) \qquad (2.141)$$

The time component in $\vec{f} = d\vec{p}/dt$ is related to the work done by the force term.

The relativistic *Hamilton-Jacobi equation* is obtained by the first derivative of the *Principle of Least Action*, $\mathcal{A}$, in respect to x^i as a form of $\partial \mathcal{A}/\partial x^i$ for momentum that is given by Eq. 2.137 [6]:

$$\frac{\partial \mathcal{A}}{\partial x_i} \frac{\partial \mathcal{A}}{\partial x^i} = g^{ik} \frac{\partial \mathcal{A}}{\partial x^i} \frac{\partial \mathcal{A}}{\partial x^j} m^2 c^2 \qquad (2.142)$$

Or, by writing the sum explicitly as follows:

$$\frac{1}{c^2} \left(\frac{\partial \mathcal{A}}{\partial t} \right)^2 - \left(\frac{\partial \mathcal{A}}{\partial x} \right)^2 - \left(\frac{\partial \mathcal{A}}{\partial y} \right)^2 - \left(\frac{\partial \mathcal{A}}{\partial z} \right)^2 = m^2 c^2 \qquad (2.143)$$

The transition to the limitation case of classical mechanics in Eq. 2.142 is made using the following considerations and concepts. First, we notice that just as in the corresponding transition with Hamiltonian function $\mathcal{H}$ in Eq. 2.127, the energy of a particle in *relativistic* mechanics contains the term mc^2, which it obviously does not have in classical mechanics. As much as the Principle of Least Action, $\mathcal{A}$, is related to energy $\mathcal{E}$ by $\mathcal{E} = -(\partial \mathcal{A}/\partial t)$, in making the transition to classical mechanics, we must, in place of $\mathcal{A}$, substitute a new action, $\mathcal{A}'$, according to the relation [6]:

$$\mathcal{A} = \mathcal{A}' - mc^2 t \qquad (2.144)$$

By substituting Eq. 2.144 into Eq. 2.143, we get the following result [6]:

$$\frac{1}{2m} \left[\left(\frac{\partial \mathcal{A}'}{\partial x} \right)^2 + \left(\frac{\partial \mathcal{A}'}{\partial y} \right)^2 + \left(\frac{\partial \mathcal{A}'}{\partial x} \right)^2 \right] - \frac{1}{2mc^2} \left(\frac{\partial \mathcal{A}'}{\partial t} \right)^2 + \frac{\partial \mathcal{A}'}{\partial t} = 0 \qquad (2.145)$$

In the limit $c \to \infty$ this equation goes over into the classical Hamilton-Jacobi equation.

2.8 Lorentz versus Galilean Transformation

Lorentz transformation is explained to some degree in Appendix A of this book, however, in this brief summary, we refer to the Galilean transformation versus Lorentz transformation for comparison purposes.

2.8 Lorentz versus Galilean Transformation

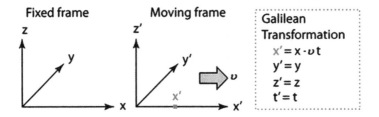

Fig. 2.9 The nature of the Galilean transformation

Fig. 2.10 Illustration of fixed frame versus moving frame of reference

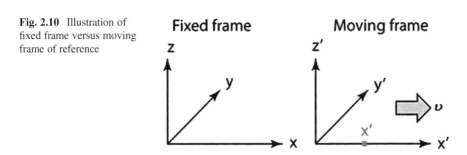

Using Fig. 2.9, we depict the nature of Galilean transformation and its connection to a fixed frame of reference versus a moving frame of reference, say in x-axis; thus, show the relationship in Cartesian coordinate system. In the figure the following presumptions are valid:

- The primed frame on the RHS moves with velocity v in the direction of x with respect to the fixed reference frame.
- The reference frames coincide at $t = t' = 0$.
- The point x' is moving with the primed frame.
- The Galilean transformation gives the coordinates of the point as measured from the fixed frame in terms of its location in the moving reference frame.

The Galilean transformation is the commonsense relationship that agrees with our everyday experience in the Euclidian space in which we live. It has embedded within it the presumption that the passage of time is the same for every observer.

Under Lorentz transformation, along with the illustration in Fig. 2.10, we are able to establish it between fixed and moving frames of reference in the x-axis direction. The primed frame moves with velocity v in the x-direction with respect to the fixed reference frame. The reference frames coincide at $t = t' = 0$. Point x' is moving with the primed frame as shown by the following sets of equations:

$$\begin{cases} x' = \dfrac{x - vt}{\sqrt{1 - \dfrac{v^2}{c^2}}} \\ y' = y \\ z' = z \\ t' = \dfrac{t - \dfrac{vx}{c^2}}{\sqrt{1 - \dfrac{v^2}{c}}} \end{cases} \quad (2.146)$$

The reverse transformation is then given by:

$$\begin{cases} x = \dfrac{x' + vt'}{\sqrt{1 - \dfrac{v^2}{c^2}}} \\ t = \dfrac{t' + vx'}{\sqrt{1 - \dfrac{v^2}{c^2}}} \end{cases} \quad (2.147)$$

2.9 The Structure of Spacetime, Interval, and Diagram

The postulates of relativity, taken together, lead to a description of spacetime in which the notions of simultaneity, time duration, and spatial distance are well defined in each inertial reference frame; however, their values, for a given pair of events, can vary from one reference frame to another. In particular, objects evolve more slowly and are contracted along their direction of motion when observed in a reference frame relative to which they are in motion.

In *special relativity* we only can use inertial frames to assign coordinates to events. There are many different types of inertial frames. Still, it is convenient to adhere to those with *standard coordinates*. That is, spatial coordinates that are right-handed rectilinear Cartesians based on a standard unit of length and time-scales based on a standard unit of time. We will continue to assume that we are employing standard coordinates.

As indicated previously in this chapter and later in Appendix C, for defining the Lorentz transformation using Fig. 2.9, we considered two Cartesian frames of reference, S and S', with coordinates (x, y, z, t) and (x', y', z', t') in a *standard configuration* in which moving frame S' moves in the x-direction of fixed frame S with a designated uniform speed v; the corresponding axes of S and S' remain parallel throughout the motion, while coordinating at time $t = t' = 0$. Under these constraints, we are assuming that the same units of distance and time are adopted in both frames,

2.9 The Structure of Spacetime, Interval, and Diagram

and we wrote the set in Eqs. C.1 through C.4 (see Appendix C) for the Lorentz factor $\gamma = \gamma(v) = 1/(1 - v^2/c^2)^{1/2}$, which approaches unity for $c \to \infty$ or can refer to the following set of equations:

$$\begin{cases} x' = x - vt \\ y' = y \\ z' = z \\ t' = t \end{cases} \quad (2.148)$$

Now recall that a "point in space" is called an "event" that is happening at an instant of time and point in space, with infinitesimal time duration and spatial extension. We now consider two events, such as A and B, taking place in neighboring areas to each other with coordinates dx, dy, dz, dt in frame system S for event A, and coordinate elements of dx', dy', dz', dt' in frame S' for event B, so we can write the following. For an observer in frame S for $dt > 0$, we can express:

$$dx^2 + dy^2 + dz^2 - c^2 dt^2 = 0 \quad (2.149)$$

Similarly, for an observer in frame S' for $dt' > 0$, we can express:

$$dx'^2 + dy'^2 + dz'^2 - c^2 dt'^2 = 0 \quad (2.150)$$

Any event near A with coordinates that satisfy either Eq. 2.149 or Eq. 2.150 is illuminated by the flash from A, therefore, its coordinates must satisfy both Eqs. 2.147 and 2.150. Now, no matter what form the transformation between coordinates in the two inertial frames takes, the transformation between differentials at any fixed event A is linear and homogeneous. In other words, if:

$$x' = F(x, y, z, t) \quad (2.151)$$

where symbol F is a general function, then we can write the transformation matrix as follows:

$$dx' = \frac{\partial F}{\partial x} dx + \frac{\partial F}{\partial y} dy + \frac{\partial F}{\partial z} dz + \frac{\partial F}{\partial t} dt \quad (2.152)$$

In index notation Eq. 2.152 is summarized thus:

$$dx' = \sum_\lambda \frac{\partial F}{\partial x^\lambda} dx^\lambda \quad (2.153)$$

Relations provided by Eq. 2.152 allows us to get the following result:

$$dx'^2 + dx'^2 + dx'^2 - c^2 dt'^2 = adx^2 + bdy^2 + cdz^2 + ddt^2$$
$$+ gdxdt + hdydt + kdzdt \qquad (2.154)$$
$$+ ldydz + mdxdz + ndxdy$$

where parameters a, b, c, $\cdots$, and so on are functions of x, y, z, and t. We know that the RHS of the preceding expression vanishes for all real values of the differentials that satisfy Eq. 2.149. It follows that the right side is a multiple of the quadratic in Eq. 2.149; that is,

$$dx'^2 + dy'^2 + dz'^2 - c^2 dt'^2 = K\left[dx^2 + dy^2 + dz^2 - c^2 dt^2\right] \qquad (2.155)$$

where K is a function of x, y, z, and t.

Note that K at A also is independent of the choice of standard coordinates in S and S'. Because the frames are Euclidian the values of $dx^2 + dy^2 + dz^2$ and $dx'^2 + dy'^2 + dz'^2$ are relevant to A and B independent of the choice of axes. Furthermore, the values of dt^2 and dt'^2 are independent of the choice of the origins of time. Thus, without affecting the value of K at A, we can choose coordinates such that $A = (0, 0, 0, 0)$ in both S and S'. Because the orientations of the axes in S and S' are, at present, arbitrary; and because inertial frames are isotropic, the relation of S and S' relative to each other, to the event A and to the locus of possible event B, is now completely symmetric. Thus, in addition to Eq. 2.155, we can write the following:

$$dx^2 + dy^2 + dz^2 - c^2 dt^2 = K\left[dx'^2 + dy'^2 + dz'^2 - c^2 dt'^2\right] \qquad (2.156)$$

It follows that $K = \pm 1$. The quantity of $K = -1$, however, can be dismissed immediately because the intervals $dx^2 + dy^2 + dz^2 - c^2 dt^2$ and $dx'^2 + dy'^2 + dz'^2 - c^2 dt'^2$ must coincide exactly when there is no motion of frame S' relative to S. Thus, we have:

$$dx'^2 + dy'^2 + dz'^2 - c^2 dt'^2 = dx^2 + dy^2 + dz^2 - c^2 dt^2 \qquad (2.157)$$

This equation implies that the transformation equations between primed and unprimed coordinates must be *linear*. The linearity of the transformation allows the coordinate axes in the two frames to be oriented so as to give the *standard configuration* mentioned earlier [9].

Under a linear transformation, the finite coordinate differences satisfy the same transformation equations as the differentials. It follows from Eq. 2.157, assuming that the events $(0, 0, 0, 0)$ coincide in both frames, that for any event with coordinates (x, y, z, t) in S and (x', y', z', t') in S', the following relationship holds:

$$x^2 + y^2 + z^2 - c^2 t^2 = x'^2 + y'^2 + z'^2 - c^2 t'^2 \qquad (2.158)$$

By hypothesis, coordinate planes $y = 0$ and $y' = 0$ coincide permanently. Thus, $y = 0$ must imply $y' = 0$, which suggests that:

2.9 The Structure of Spacetime, Interval, and Diagram

$$y' = Ay \tag{2.159}$$

where A is a constant. We can reverse the directions of the x- and z-axes in S and S', which has the effect of interchanging the roles of these frames. This procedure does not affect Eq. 2.156, but by symmetry we also have:

$$y = Ay' \tag{2.160}$$

It is clear that $A = \pm 1$. The negative sign can again be dismissed because $y = y'$ when there is no motion between S and S'. The argument for z is similar. Thus, we have:

$$y' = y \tag{2.161}$$

and

$$z' = z \tag{2.162}$$

as in the Galilean transformation.

As noted in the opening statement of this section, we will continue to assume that we are employing standard coordinates. From now on, however, we intend to make no assumptions about the relative configuration of the two sets of spatial axes, and the origins of time, when dealing with two inertial frames. Thus, the most general transformation between two inertial frames consists of a Lorentz transformation in the standard configuration plus a translation; this includes a translation in time and a rotation of the coordinate axes. The resulting transformation is called a *general Lorentz transformation*, as opposed to a Lorentz transformation in the standard configuration, which hereafter will be termed a *standard Lorentz transformation*.

As per our description in the preceding and what we proved as a finding and conclusion in Eq. 2.158, between two frame of reference presented is a fixed system S and a moving system S' as inertial frames; for their corresponding differential between these frames, we can continue with our discussion on the subject of *spacetime schema*. Thus, the expectation is that the relationship in Eq. 2.158 will remain invariant under a general Lorentz transformation; because such a transformation is *linear*, it follows that:

$$\begin{aligned}(x_2 - x_1)^2 + (y_2 - y_1)^2 + (y_2 - y_1)^2 - c^2(t_2 - t_1)^2 = \\ (x'_2 - x'_1)^2 + (y'_2 - y'_1)^2 + (z'_2 - z'_1)^2 - c^2(t'_2 - t'_1)^2\end{aligned} \tag{2.163}$$

In this equation the corresponding coordinates of (x_1, y_1, z_1, t_1) and (x_2, y_2, z_2, t_2) are the coordinates of any two events in system S and primed symbols denote the corresponding coordinates in system S'. Then, it is convenient to write the following:

$$ds^2 = -dx^2 - dy^2 + c^2 dt^2 \tag{2.164}$$

and

$$s^2 = -(x_2 - x_1)^2 - (y_2 - y_1)^2 - (z_2 - z_1)^2 + c^2(t_2 - t_1)^2 \qquad (2.165)$$

The differential ds, or the finite number s, defined by these equations is called the *interval* between the corresponding events. Equations 2.164 and 2.165 express the fact that the *interval between two events is invariant*, in the sense that it has the same value in all inertial frames. In other words, the interval between two events is invariant under a general Lorentz transformation.

We are now in position to establish a 4-vector tensor rank again in a spacetime infrastructure, then take steps toward a *spacetime diagram* that is also known as a *Minkowski diagram*. Let us consider entities defined in terms of four variables:

$$x^1 = x, \quad x^2 = y, \quad x^3 = z, \quad x^4 = ct \qquad (2.166)$$

and that transform as tensors under a general Lorentz transformation. From now on such entities will be referred to as *4-tensors* [9].

If we introduce non-singular tensor $g_{\mu\nu}$, it is called a *fundamental tensor*, which is used to define the operation of raising and lowering suffixes from a tensor analysis point of view. Thus, the fundamental tensor is usually introduced using the metric $ds^2 = g_{\mu\nu} dx^\mu dx^\nu$, where ds^2 is a differential invariant, which we established in Eq. 2.164, and with a tensor analysis perspective, this equation can be written in a different form as:

$$\begin{aligned} ds^2 &= -dx^2 - dy^2 - dz^2 + c^2 dt^2 \\ &= -(dx^1)^2 - (dx^2)^2 - (dx^3)^2 + (dx^4)^2 \\ &= g_{\mu\nu} dx^\mu dx^\nu \end{aligned} \qquad (2.167)$$

where the Greek suffixes μ and ν contain values from 1 to 4 that are conventionally and commonly used in 4-tensor analysis and theory.

Furthermore, in some textbooks the Roman suffixes are reserved for tensors in three-dimensional Euclidian space, so-called *3-tensors*. Per this establishment, the 4-tensor $g_{\mu\nu}$ has the components $g_{11} = g_{22} = g_{33} = -1$ and $g_{44} = 1$, and $g_{\mu\nu} = 0$ is a valid equality when $\mu \neq \nu$ is in all permissible coordinate frames. For the time being for the purposes of spacetime structure, we treat the quantity $g_{\mu\nu}$ as a non-singular 4-tensors, thus $g_{\mu\nu}$ can be thought of as the *metric tensor* of the space with points that are the events, such as (x^1, x^2, x^3, x^4), and that is why this space is called or usually referred to as *space-time* (i.e., *spacetime*) for obvious reasons by now.

The distribution of signs in the metric ensures that time coordinate x^4 is not on the same footing as the three-space coordinates. Thus, spacetime has a non-isotropic nature that is quite unlike Euclidian space, with its positive definite metric. According to the relativity principle, all physical laws are expressible as interrelationships between 4-tensors in spacetime [9].

A tensor of rank one is called a *4-vector*. We also will have occasion to use ordinary vectors in three-dimensional Euclidian space. Such vectors are called *3-vectors* and are conventionally represented by boldface symbols. We will use

2.9 The Structure of Spacetime, Interval, and Diagram

the Latin suffixes i, j, k, and so on to denote the components of a 3-vector; these suffixes are understood to range from 1 to 3. Thus, $\vec{u} = u^i = dx^i/dt$ denotes a velocity vector. For 3-vectors, we shall use the notation $u^i = u_i$ interchangeably—that is, the level of the suffix has no physical significance [9].

When tensor transformations from one frame to another actually has to be computed, we usually find it possible to choose coordinates in the standard configuration, so the standard Lorentz transformation applies. Under such a transformation, any contravariant 4-vector, T^μ, transforms according to the same scheme as the difference in coordinates $x_2^\mu - x_1^\mu$ between two points in spacetime. Thus, it follows that

$$T^{1'} = \gamma(T^1 - \beta T^4) \tag{2.168}$$

$$T^{2'} = T^2 \tag{2.169}$$

$$T^{3'} = T^3 \tag{2.170}$$

$$T^{4'} = \gamma(T^4 - \beta T^{-1}) \tag{2.171}$$

In the preceding set of equations the symbol $\beta = v/c$ is used, while γ is a Lorentz factor as defined before. Higher ranked 4-tensors transform according to the rules defined in the form of *covariant tensor* and *contravariant tensor* as well as a *mixed tensor*; the formal definition of tensors is as follows:

1. An entity having components $A_{ij...k}$ in the x^i system and $A_{i'j'...k'}$ in the $x^{i'}$ system is said to behave as a covariant tensor under the transformation $x^i \to x^{i'}$ if

$$A_{i'j'...k'} = A_{ij...k} p_{i'}^i, p_{j'}^j, \cdots p_{k'}^k \tag{2.172}$$

2. Similarly, $A^{ij...k}$ is said to behave as a contravariant tensor under $x^i \to x^{i'}$ if

$$A^{i'j'...k'} = A^{ij...k} p_i^{i'}, p_j^{j'}, \cdots p_k^{k'} \tag{2.173}$$

3. Finally, $A_{k...l}^{i...j}$ is said to behave as a mixed tensor (contravariant in $i \cdots j$ and a covariant in $k \cdots l$) under $x^i \to x^{i'}$ if

$$A_{k'...l'}^{i'...j'} = A_{k...l}^{i...j} p_i^{i'} \cdots p_j^{j'} p_{k'}^k \cdots p_{l'}^l \tag{2.174}$$

When an entity is described as a tensor, it is generally understood that it behaves as a tensor under *all* non-singular differentiable transformations of the relevant coordinates [9]. Taking Eqs. 2.172, 2.173, and 2.174 into consideration under the rules, the higher ranked 4-tensors transform accordingly, and the transformation coefficients take the form of the following tensors:

$$p^{\mu}_{\ \mu'} = \begin{pmatrix} +\gamma & 0 & 0 & -\gamma\beta \\ 0 & 1 & 0 & 0 \\ 0 & 1 & 0 & 0 \\ -\gamma\beta & 0 & 0 & +\gamma \end{pmatrix} \tag{2.175}$$

and

$$p^{\ \mu}_{\mu'} = \begin{pmatrix} +\gamma & 0 & 0 & +\gamma\beta \\ 0 & 1 & 0 & 0 \\ 0 & 0 & 1 & 0 \\ +\gamma\beta & 0 & 0 & +\gamma \end{pmatrix} \tag{2.176}$$

Often, the first three components of a 4-vector coincide with the components of a 3-vector. For example, the x^1, x^2, x^3 in $R^\mu = (x^1, x^2, x^3, x^4)$ are the components of $\vec{r}$, the position 3-vector of the point at which the event occurs. In such cases, we adopt the notation exemplified by $R^\mu = (\vec{r}, c, t)$. The covariant form of such a vector is simply $R^\mu = (-\vec{r}, c, t)$. The squared magnitude of the vector is $(R)^2 = R_\mu R^\mu = -r^2 + c^2 t^2$. The inner product $g_{\mu\nu} R^\mu Q^\nu = R_\mu Q^\mu$ of R^μ with a similar vector $Q^\mu = (\vec{q}, k)$ is given by $R_\mu Q^\mu = -\vec{r} \cdot \vec{q} + ctk$. The vectors R^μ and Q^μ are said to be *orthogonal* if $R_\mu Q^\mu = 0$.

Because a general Lorentz transformation is a *linear* transformation, the partial derivative of a 4-tensor is also a 4-tensor:

$$\frac{\partial A^{\nu\sigma}}{\partial x^\mu} = A^{\nu\sigma}_{\ \ \mu} \tag{2.177}$$

Obviously, a general 4-tensor acquires an extra covariant index after partial differentiation with respect to the contravariant coordinate x^μ. It is helpful to define a covariant derivative operator, as follows:

$$\partial_\mu \equiv \frac{\partial}{\partial x^\mu} = \left(\nabla, \frac{1}{c}\frac{\partial}{\partial t} \right) \tag{2.178}$$

where

$$\partial_\mu A^{\nu\sigma} \equiv A^{\nu\sigma}_{\ \ \mu} \tag{2.179}$$

There is a corresponding contravariant derivative operator :

$$\partial^\mu \equiv \frac{\partial}{\partial x_\mu} = \left(-\nabla, \frac{1}{c}\frac{\partial}{\partial t} \right) \tag{2.180}$$

where

2.9 The Structure of Spacetime, Interval, and Diagram

$$\partial^\mu A^{\nu\sigma} \equiv g^{\mu\tau} A^{\nu\sigma}_\tau \tag{2.181}$$

The four-divergence of a 4-vector, $A^\mu = (\vec{A}, A^0)$, is the invariant and is written:

$$\partial^\mu A_\mu = \partial_\mu A^\mu = \vec{\nabla} \cdot \vec{A} + \frac{1}{c}\frac{\partial A^0}{\partial t} \tag{2.182}$$

The four-dimensional Laplacian operator, or *d'Alembertian*, is equivalent to the invariant contraction as follows:

$$W \equiv \partial_\mu \partial^\mu = -\nabla^2 + \frac{1}{c^2}\frac{\partial^2}{\partial t^2} \tag{2.183}$$

We still need to prove, from Eq. 2.157 or Eq. 2.167, the invariance of the differential metric,

$$ds^2 = dx'^2 + dy'^2 + dz'^2 - cdt'^2 = dx^2 + dy^2 + dz^2 - cdt^2 \tag{2.184}$$

between two general inertial frames implies that the coordinate transformation between such frames is necessarily linear. To put it another way, we need to demonstrate that a transformation that transforms a metric, $g_{\mu\nu}dx^\mu dx^\nu$, with constant coefficients into metric $g_{\mu'\nu'}dx^{\mu'}dx^{\nu'}$ with constant coefficients must be linear. Now,

$$g_{\mu\nu} = g_{\mu'\nu'} p_\mu^{\mu'} p_\nu^{\nu'} \tag{2.185}$$

Taking the differential of Eq. 2.185 in respect to x^σ, we obtain:

$$g_{\mu'\nu'} p_{\mu\sigma}^{\mu'} p_\nu^{\nu'} + g_{\mu'\nu'} p_\mu^{\mu'} p_{\nu\sigma}^{\nu'} = 0 \tag{2.186}$$

where

$$p_{\mu\sigma}^{\mu'} = \frac{\partial p_\mu^{\mu'}}{\partial x^\sigma} = \frac{\partial^2 x^{\mu'}}{\partial x^\mu \partial x^\sigma} = p_{\sigma\mu}^{\mu'} \tag{2.187}$$

and so on. Interchanging the indices μ and σ yields:

$$g_{\mu'\nu'} p_{\mu\sigma}^{\mu'} p_\nu^{\nu'} + g_{\mu'\nu'} p_\sigma^{\mu'} p_{\nu\mu}^{\nu'} = 0 \tag{2.188}$$

Interchanging the indices ν and σ gives the following:

$$g_{\mu'\nu'} p_\sigma^{\mu'} p_{\nu\mu}^{\nu'} + g_{\mu'\nu'} p_\mu^{\mu'} p_{\nu\sigma}^{\nu'} = 0 \tag{2.189}$$

where indices μ' and ν' have been interchanged in the first term. It follows from Eqs. 2.186, 2.187, and 2.188 that this relationship is valid:

$$g_{\mu'\nu'} p_{\mu\sigma}^{\mu'} p_\nu^{\nu'} = 0 \tag{2.190}$$

Multiplying Eq. 2.190 by $p_{\sigma'}^{\nu}$ will yield to the following result:

$$g_{\mu'\nu'}p_{\mu\sigma}^{\mu'}p_{\nu}^{\nu'}p_{\sigma'}^{\nu'} = g_{\mu'\sigma'}p_{\mu\sigma}^{\mu'} = 0 \qquad (2.191)$$

Finally, multiplication of Eq. 2.190 by $g^{\nu'\sigma'}$ results in the following relation:

$$g_{\mu'\sigma'}g^{\nu'\sigma'}p_{\mu\sigma}^{\mu'} = p_{\mu\sigma}^{\nu'} = 0 \qquad (2.192)$$

This proves that the coefficients $p_{\mu}^{\nu'}$ are constants, thus, that the transformation is linear. Consequently, these fact-finding results prove that the following differential metric also is a linear form between two general inertial frames, S and S', as well and that is [9]:

$$ds^2 = dx'^2 + dy'^2 + dz'^2 - c^2dt'^2 = dx^2 + dy^2 + dz^2 - c^2dt^2 \qquad (2.193)$$

2.9.1 Spacetime or the Minkowski Diagram

If we want to represent the motion of a particle graphically by using a typical Cartesian coordinate system, we normally plot a position-versus-time coordinate; x is a *vertical axis* while coordinate time t is the *horizonal axis* from basic physics and calculus of classical mechanics facts. In such a graph the velocity can be read as the slope of the curve. In relativity mechanics, however, the convention of position-versus-time is reversed, where the time t-axis is vertical while the position x-axis is horizonal; velocity is then given by the *reciprocal* of the slop for the equation motion defined by $x^0 = ct$.

Under this convention, a particle at rest is represented by a vertical line; a photon, traveling at the speed of light, is described by a 45° line, and a rocket going at some intermediate speed follows a line of slope $(c/v) = (1/\beta)$, as is demonstrated in Fig. 2.11. We call such a plot in this figure and associated plots a *Minkowski diagram*.

The trajectory of a particle on a Minkowski diagram is called a *world line*. To have a better understanding of this line, suppose we set out from the origin at time $t = 0$; and, because no material object per special relativity theory and concept, can travel faster than the speed of light, our worldline can never have a slope less than 1. Thus, the motion is restricted to the wedge-shaped region bounded by the two 45° lines, as depicted in Fig. 2.12.

In this figure we call this "your future," in the sense that it is the locus of all points, as Griffiths [8] describes it. As time goes on and you move along the chosen worldline, your options progressively narrow; thus, "your future" at any moment is the forward "wedge" constructed at whatever point you find yourself. Meanwhile, the *backward* wedge represents your "past," in the sense that it is the locus of all points from which you might have come. As for the rest of the region outside the

Fig. 2.11 Illustration of Minkowski diagram

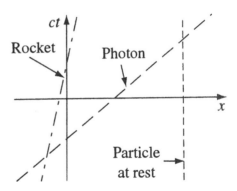

Fig. 2.12 Illustration of 45° world line

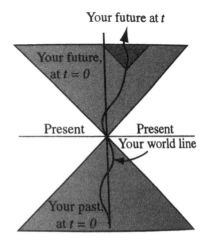

forward and backward wedges, this is the generalized "present." You cannot *get* there, and you did not *come* from there. In fact, there is no way you can influence any event in the present, where the information would have to travel faster than the speed of light, which is a vast expanse of spacetime that is absolutely inaccessible to you [8].

Now, if you start thinking about the idea of spacetime, we can recall, from the earlier discussion in this section, that a *point in space* is called an *event*. This is something happening at an instant of time at a point in space, with infinitesimal time duration and spatial extension. For an example, tap the tip of a pencil once on a table top, or click your fingers.

If we include a *y*-axis coming out of the page, the "wedges" become cones, as illustrated in Fig. 2.13, with an undrawable *z*-axis (i.e., hypercones). Because their boundaries are the trajectories of light rays, we call them the *forward light cone* and the *backward light cone*. Thus, in other words, your future lies within your forward light cone, while your past is within the backward light cone. Note that the slope of the line connecting two events on a spacetime diagram tells you immediately

Fig. 2.13 Description of a world line along past and present for a light cone

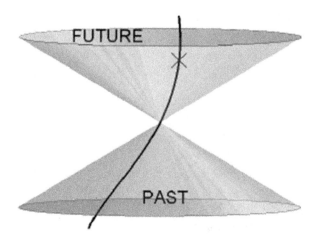

whether the displacement between them is a time-like slope greater than 1, a space-like slope less than 1, or a light-like slope of 1. See definition of time-like, space-like, and light-like in the preceding, as well as the next section.

A *particle* is a physical object of infinitesimal spatial extent that can exist for some extended period of time. The line of events that gives the location of the particle as a function of time, as we described in the preceding, is called its *world line* (see Fig. 2.13). The figure presents a spacetime, or Minkowski diagram, showing a world line and light cone (past and future branches). The cross (X) marks an example event. The apex of the cone is another event. Again, as we discussed previously, if two events have coordinates (t_1, x_1, y_1, z_1) and (t_2, x_2, y_2, z_2) in some reference frame, then the quantity

$$s^2 = -c^2(t_2 - t_1)^2 + (x_2 - x_1)^2 + (y_2 - y_1)^2 + (z_2 - z_1)^2 \qquad (2.194)$$

is called the *squared spacetime interval* between them. Note the crucial minus sign in front of the first term. We emphasize it by writing Eq. 2.194 in different format as follows:

$$s^2 = -c^2 \Delta t^2 + \Delta x^2 + \Delta y^2 + \Delta z^2 \qquad (2.195)$$

If $s^2 < 0$, then the time between the events is sufficiently long so that a particle or other signal moving at speeds less than the speed of light, c, could move from one event to another. Such a pair of events is said to be separated by a *time-like interval*. If $s^2 > 0$, then the time between the events is too short for any physical influence to move between them. This is called a *space-like interval*. If $s = 0$, then we have a *null interval*; this means that a light pulse or other light-speed signal could move directly from one event to another. More details are given in the next section.

Although the parts t_i, x_i, y_i, z_i needed to calculate an interval will vary from one reference frame to another, we already find that the net result, s^2, is independent of a

2.9 The Structure of Spacetime, Interval, and Diagram

reference frame; all reference frames agree on the value of this quantity. This is similar to the fact that the length of a vector is unchanged by rotations of the vector. A quantity with a value that is the same in all reference frames is called a *Lorentz invariant* (or *Lorentz scalar*). Lorentz invariants play a *central role in special relativity*.

The set of events with a null spacetime interval from any given event lie on a cone called the *light cone*. The part (or "branch") of this cone extending into the past is made up of the world lines of photons that form a spherical pulse of light collapsing onto the event, and the part extending into the future is made up of the world lines of photons that form a spherical pulse of light emitted by the event. The cone is an abstraction—the incoming and outgoing light pulses do not have to be there. The past part of the light cone surface of any event A is called the *past light cone* of A; the future part of the surface is called the *future light cone* of A. The whole of the future cone (i.e., the body of the cone as well as the surface) is called the *absolute future* of A; it consists of all events that could possibly be influenced by A (in view of the *light speed postulate*). The whole of the past cone is called the *absolute past* of A, and it consists of all possible events.

2.9.2 Time Dilation

Consider a light tray that leaves a bulb and shines on the floor of a moving train car, directly below by striking it, as depicted in Fig. 2.14.

Now, we are interested in knowing how long it takes for the light beam to travel from the bulb to the floor of the train car from the point of view of an observer on the train. The answer is an easy one and can be presented as a simple relationship; if the height of celling from the floor is h, then the time of travel is:

$$\Delta t' = \frac{h}{c} \tag{2.196}$$

This equation is valid so long as the observer is on the train; on the other hand, what would be the time of travel if it is observed from the ground for the same beam. Naturally, because the train is moving, the same beam must travel farther. From Fig. 2.14, it can be observed easily that the distance is given by $\sqrt{h^2 + (v\Delta t)^2}$, so the Δt is:

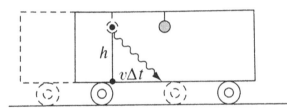

Fig. 2.14 Illustration of a light shining on the train car's floor

$$\Delta t = \frac{\sqrt{h^2 + (v\Delta t)^2}}{c} \tag{2.197}$$

Solving for Δt, we obtain:

$$\Delta t = \frac{h}{c} \frac{1}{\sqrt{1 - (v^2/c^2)}} \tag{2.198}$$

Therefore,

$$\Delta t' = \sqrt{1 - (v^2/c^2)}\,\Delta t \tag{2.199}$$

Evidently, the time elapsed between the *same two events*—(1) light leaves bulb and (2) light strikes center of train floor—is different for the two observers. In fact, the interval recorded on the train clock, $\Delta t'$, is *shorter* by the factor γ as indicated here:

$$\gamma \equiv \frac{1}{\sqrt{1 - (v^2/c^2)}} \tag{2.200}$$

In conclusion, the moving clock on a train runs at a slower speed; thus, such a schema is called *time dilation*. Note that such a scenario has nothing to do with the mechanics of clocks and there is nothing wrong with either the clock on the train nor the ground observer. It is a statement about the nature of time, which applies to *all* properly functioning timepieces.

2.9.3 Time Interval

An extended length of time can be divided into several shorter periods of time, all the same length. These are called *time intervals*. For example, say you want to measure the speed of a car over a journey taking an hour. You could divide the hour into time intervals of 10 min. You record the speed at each 10-min interval. You may find that the car is traveling at a constant speed, or maybe it is slowing down and speeding up during its journey. That is how you can use time intervals.

2.9.4 The Invariant Interval

We can describe the interval as invariant using the displacement 4-vector, by assuming event A occurs at a frame of reference $(x_A^0, x_A^1, x_A^2, x_A^3)$, and event B takes place at $(x_B^0, x_B^1, x_B^2, x_B^3)$. The difference then is:

2.9 The Structure of Spacetime, Interval, and Diagram

$$\Delta x^\mu = (\Delta x)_\mu (\Delta x)^\mu$$
$$= -(\Delta x^0)^2 + (\Delta x^1)^2 + (\Delta x^2)^2 (\Delta x^3)^2 \tag{2.201}$$
$$= -c^2 t^2 + d^2$$

where t is the time interval between the two events and d is their spatial separation. When we transform to a moving system, the *time* between events A and B is altered as $t' \neq t$, and so is *spatial separation* $d' \neq d$, but interval I remains the same [8].

According to Griffiths [8], depending on the two events in question, the interval can be positive, negative, or zero, then the following three circumstances would apply to the *invariant interval*, as defined by interval I as possibilities:

1. If $I < 0$, we call the interval *time-like* because this is the sign we get when the two events, A and B, occur at the *same place*, where $d = 0$, and are separated only temporally.
2. If $I > 0$, we call the interval *space-like* because this is the sign we get when the two occur at the *same time*, $t = 0$, and are separated only spatially.
3. If $I > 0$, we call the interval *light-like* because this is the relation that holds when the two events are connected by a signal traveling at the speed of light.

Thus, if the interval between two events is time-like, there exists an inertial system that is accessible by a Lorentz transformation in which they occur at the same point.

Second, if we get on a train going from A to B at speed $v = (d/t)$, leaving event A when it occurs, while we are in time to pass B when that event takes place. In the train system A and B are taking place at the same point. We cannot do this for a space-like interval, however, because speed V would have to be greater than the speed of light, c, and no observer can exceed the speed of light. If the observer goes faster than the speed of light, the Lorentz transformation factor, $\gamma = 1/\left(\sqrt{1 - v^2/c^2}\right)$, would be imaginary and it would not make any sense at all. On the other hand, if the interval is space-like, then there exists a system in which the two zompp;events occur at the same time (see the following example)

Example 2.1 The coordinates of event A are (x_A, t_A), and the coordinates of event B are (x_B, t_B). Assuming the interval between them is space-like, find the velocity of the system in which they occur at the same time.

Solution Using Eq. 2.199, we can write:

$$\Delta t' = \gamma\left(\Delta t - \frac{v}{c^2}\Delta x\right) = 0 \Rightarrow \Delta t = \frac{v}{c^2}\Delta x$$

$$v = \frac{\Delta t}{\Delta x} c^2 = \boxed{\frac{t_B - t_A}{x_B - x_A}}$$

2.9.5 Lorentz Contraction Length

For this subject we refer to Griffiths book on the introduction to electrodynamics [8], and we explain the *Gedanken* (i.e., German word that means "thought" in English) experiment by imagining that we have set up a lamp at one end of a boxcar and a mirror at the other, as illustrated in Fig. 2.15, so that a light signal can be sent down and back.

Now, we should ask: How long does it take for the signal to complete the round trip to an observer on the train? Logically, the answer in mathematical form is given by:

$$\Delta t' = 2\frac{\Delta x'}{c} \tag{2.202}$$

where $\Delta x'$ is the length of the boxcar and the prime sign, as before, denotes measurements made on the moving train at a speed of v.

To an observer on the ground the process is more complicated because of the motion of the train. If Δt_1 is the time for the light signal to reach the front end and Δt_2 is the return time, as illustrated in Fig. 2.16, then we can write the following mathematical notations:

$$\begin{cases} \Delta t_1 = \dfrac{\Delta x + v\Delta t_1}{c} \\ \Delta t_2 = \dfrac{\Delta x + v\Delta t_2}{c} \end{cases} \tag{2.203}$$

From the set of Eq. 2.203, we solve for Δt_1 and Δt_2, thus we get:

$$\begin{cases} \Delta t_1 = \dfrac{\Delta x_1}{c - v} \\ \Delta t_2 = \dfrac{\Delta x}{c + v} \end{cases} \tag{2.204}$$

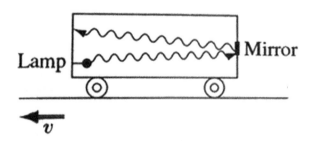

Fig. 2.15 Illustration of train boxcar with lamp and mirror installed

2.9 The Structure of Spacetime, Interval, and Diagram

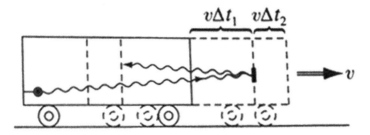

Fig. 2.16 Train boxcar at motion with speed v

Furthermore, from Eq. 2.204, we can conclude that the round-trip time results in the following mathematical form:

$$\Delta t = \Delta t_1 + \Delta t_2 = 2\left(\frac{\Delta x}{c}\right)\left[\frac{1}{(1 - v^2/c^2)}\right] \qquad (2.205)$$

These intervals, however, are related by time dilation (see Eq. 2.191), as was derived previously. Substituting Eq. 2.199 into Eq. 2.202, we achieve the following notation:

$$\Delta x' = \frac{1}{\sqrt{1 - v^2/c^2}} \Delta x \qquad (2.206)$$

The length of the boxcar is not the same when measured by an observer on the ground, as it is when measured by an observer on the train from a ground point of view; it is somewhat *shorter*.

As a conclusion, *"moving objects are shortened,"* and we call this the Lorentz contraction. Note that the same Lorentz factor, γ, appears as before in both time dilation and the Lorentz contraction equation as:

$$\gamma = \frac{1}{\sqrt{1 - v^2/c^2}} \qquad (2.207)$$

As general rule, we can conclude that moving clocks run slower, moving sticks appear shorter, and the factor is always the Lorentz factor, γ, as given in Eq. 2.207 [8].

Nevertheless, the observer on the moving train boxcar does not see everything that way, such as the time on a clock moving slower and a moving stick that is shortened either, so all the measurements are the same as when the train is standing on its track without any movement. In fact, from the point of view of the observer on a train, it is the objects on the *ground* that seem to be getting shorter [10].

References

1. D. Fleisch, *A Student's Guide to Maxwell's Equations*, 1st edn. (Cambridge: Cambridge University Press, New York, 2008)
2. J.R. Reitz, F.J. Milford, R.W. Christy, *Foundations of Electromagnetic Theory*, 3rd edn. (Reading, MA: Addison – Wesley Publishing, New York, 1979)
3. J. Peatross, M. Ware, *Physics of Light and Optics* (Brigham Young University, March 16, 2015 Edition, New York)
4. J.D. Jackson, *Classical Electrodynamics*, 3rd edn. (New York: John Wiley Publisher, 1990), pp. 27–29
5. W.H. Furry, *Examples of Momentum Distributions in the Electromagnetic Field and in Matter*, Am. J. Phys. **37**, 621 (1969)
6. L.D. Landau, E.M. Lifshitz, *The Classical Theory of Fluids*, Third Revised English Edition. (Boston: Pergamon Press, Addison-Wesley Publishing, New York, 1971)
7. A. Einstein et al., *The Principle of Relativity* (Mineola, Dover Publications, 1923), Chapter V
8. D. Griffiths, *Introduction to Electrodynamics*, 3rd edn (Upper Saddle River: Prentice Hall, New York, 1999)
9. R. Fitzpatrick, *Maxwell's Equations and the Principles of Electromagnetism* (Sudbury, MA: Infinity Science Press LLC, New York, 2008)
10. A. Steane, *Relativity Made Relatively Easy*, 1 edn. (Oxford: Oxford University Press, New York, December 1, 2012)

Chapter 3
All About Wave Equations

The fundamental definition of the terminology that is knowns to us as a *wave* consists of a series of examples of various different situations, whichthat we are referring to as waves. However, the one feature that is a common denominator characteristic for of wave types of waves is that they they propagate in one or more other directions, and they create some kind of "disturbance" in their propagation path. of their propagations, for example, in the case of water waves, we can observe the elevation of the water's surface, and in case of sound waves, we experience, pressure variations in its path of traveling, with a velocity characteristic of the medium that the wave goes through. However, for us tTo be able to explain describe a wave, however, we need a more definitive way of to describeing thea wave with usinge of mathematics, and primarily the concept of partial differential equationss type scenarios, which goes beyond the level of most any basic text book.

Thus, in this chapter we first establish what a wave is and, second, describe each form or shape of waves produced both from classic and relativistic mechanics, as well as an electrodynamics point of view. This approach allows us to identify mechanical waves, electromagnetic waves, and quantum mechanical waves as well. Within each of these categories we can establish wave types and classify them—for example, soliton waves, scalar waves, plassma waves, shock waves, and so on.

3.1 Introduction

To put the wave definition in perspective, we say a wave is a *disturbance of a continuity medium that propagates with a fixed shape at constant velocity*. In the presence of absorption, however, a wave diminishes in size as it moves, and the following circumstances apply:

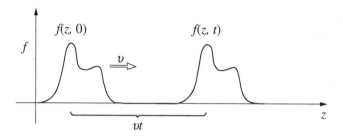

Fig. 3.1 Illustration of a moving wave with a persistent shape and a constant speed

If the medium is dispersive, then dissimilar frequencies travel at various speeds. In two or more dimensions, as the wave spreads out, its amplitude decreases. *Standing waves* do not propagate at all, of course.

Consider a wave with a persistent shape traveling to the right of the page at constant speed v, as illustrated in Fig. 3.1, where the wave is drawn at two different times, once at time $t = 0$ and again at some time later time t; t at each point on the wave is then simply shifting to the right of the page by an amount that is vt.

Let us say that the wave demonstrated in Fig. 3.1 is generated by shaking one end of a rope, and suppose function $f(z, t)$ represents the displacement of the rope at point z at time t. If we assume that the initial shape of the rope is $g(z) \equiv f(z, 0)$, then the displacement at point z at the later time t is the same as the displacement at a distance vt to the left at $(z - vt)$, back at time $t = 0$; therefore, mathematically we can write the following relationship for the wave displacement function:

$$f(z,t) = f(z - vt, 0) = g(z - vt) \tag{3.1}$$

This equation represents the general form of wave equations, and it captures the wave function mathematically for a wave in motion. This equation also tells us that function $f(z, t)$, which might have depended on variables z and t, is only a very special combination, such as $(z - vt)$, where function $f(z, t)$ represents a wave of fixed shape traveling in the z-direction at a constant speed of v. For example, if A and b are constants with the appropriate units, we can write the following relations:

$$\begin{cases} f_1(z,t) = Ae^{-b(z-vt)^2} \\ f_2(z,t) = A \sin\left[b(z - vt)\right] \\ f_3(z,t) = \dfrac{A}{b(z - vt)^2 + 1} \end{cases} \tag{3.2}$$

The equations in the set given in Eq. 3.2 are all representations of the wave function in one form or another; the following functions, however, are not representations of a wave function:

3.1 Introduction

Fig. 3.2 A rope with tension *T*

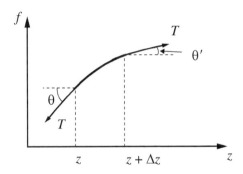

$$\begin{cases} f_4(z,t) = Ae^{-b(bz)^2} \\ f_5(z,t) = A \sin(bz) \cos(bvt)^3 \end{cases} \quad (3.3)$$

Using Fig. 3.2 and Newton's second law in classical mechanics for the rope with tension T, we can write:

$$\Delta F = T \sin \theta' - T \sin \theta \quad (3.4)$$

For small angles of θ and θ' we can reformate Eq. 3.4 by replacing the sine with a tangent of these angles:

$$\begin{aligned} \Delta F &\simeq T(\tan \theta' - \tan \theta) \\ &= T\left(\left.\frac{\partial f}{\partial z}\right|_{z+\Delta z} - \left.\frac{\partial f}{\partial z}\right|_z\right) \\ &\simeq T\left(\frac{\partial^2 f}{\partial z^2}\right)\Delta z \end{aligned} \quad (3.5)$$

If the mass of the rope per unit length is designated by symbol μ, then Newton's second law says that:

$$\Delta F = \mu(\Delta z)\frac{\partial^2 f}{\partial t^2} \quad (3.6)$$

Combining Eq. 3.5 with Eq. 3.6 would produce the following result:

$$\begin{aligned} \mu(\Delta z)\frac{\partial^2 f}{\partial t^2} &= T\left(\frac{\partial^2 f}{\partial z^2}\right)\Delta z \\ \frac{\partial^2 f}{\partial z^2} &= \frac{\mu}{T}\frac{\partial^2 f}{\partial t^2} \end{aligned} \quad (3.7)$$

Evidently, small disturbances on the rope satisfy the following differential equation:

$$\frac{\partial^2 f}{\partial z^2} = \frac{1}{v} \frac{\partial^2 f}{\partial t^2} \tag{3.8}$$

Now, by comparing Eq. 3.8 with Eq. 3.7, we conclude that:

$$\begin{cases} \dfrac{1}{v^2} = \dfrac{\mu}{T} \\ \\ v^2 = \dfrac{T}{\mu} \end{cases} \Rightarrow \boxed{v = \sqrt{\dfrac{T}{v}}} \tag{3.9}$$

where in Eq. 3.9 the symbol v represents the speed of propagation for the wave on the rope; this equation is known as the *classical wave equation*. This type of wave is not the only kind that we know; other types of waves exist, and we describe them in the sections throughout this chapter.

3.2 The Classical Wave Equation and Separation of Variables

In the previous section Eq. 3.9 was developed as a general form of the classical wave equation, and in this section we are in quest of a general solution to be able to solve not only Eq. 3.8, which is in a *one-dimensional space*, but eventually in all the appropriate coordinate systems known. Obviously, the easiest one is the Cartesian coordinate and a one-dimensional problem of a string, such as a piano or violin string, that is stretched tightly and its ends are fastened to supports at $x = 0$ and $x = L$. When the string is vibrating, its vertical displacement u is always very small, and the string's slope, $\partial u/\partial x$, at any point is far away from its stretched equilibrium position. In fact, we do not distinguish between the length of the string and the distance between the supports, although it is clear that the string must stretch a little as it vibrates out of its equilibrium position.

Under these assumptions, displacement u satisfies the one-dimensional classical wave equation and can be written:

$$\frac{\partial^2 u}{\partial x^2} = \frac{1}{v^2} \frac{\partial^2 u}{\partial t^2} \tag{3.10}$$

In this equation the following definitions apply:

u = displacement
v = wave velocity; this depends on the tension and the linear density of the string, as designated by Eq. 3.9

3.2 The Classical Wave Equation and Separation of Variables

Now the question is: How do we solve the second-order, linear, partial differential equation given by Eq. 3.10?

The most important technique we can use is to separate variables by trying an approach like the following form of a general solution as a combination of two separate independent functions, $X(x)$ and $T(t)$:

$$u(x,t) = X(x)T(t) \qquad (3.11)$$

By plugging Eq. 3.11 in to Eq. 3.10, we get the following form of separation of variables:

$$\frac{\partial^2}{\partial x^2}\left[X(x)\underbrace{T(t)}_{\substack{\text{not operated on} \\ \text{by } \frac{\partial}{\partial x}}}\right] = \frac{1}{v^2}\frac{\partial^2}{\partial t^2}\left[\underbrace{X(x)}_{\substack{\text{not operated on} \\ \text{by } \frac{\partial}{\partial t}}}T(t)\right] \qquad (3.12)$$

Multiplying Eq. 3.12 on the left by $\frac{1}{X(x)T(t)}$, we get the following form of equation:

$$\underbrace{\frac{1}{X(x)}\frac{\partial^2 X(x)}{\partial x^2}}_{\text{only } x} = \underbrace{\frac{1}{v^2}\frac{1}{T(t)}\frac{\partial^2 T}{\partial t^2}}_{\text{only } t} \qquad (3.13)$$

In this equation, as can be seen, x and t are independent variables. It only can be valid if both sides are equal to constant K; this is called the *separation constant*.

$$\begin{cases} \dfrac{1}{X}\dfrac{d^2 X}{dx^2} = K \\ \dfrac{1}{v^2 T}\dfrac{d^2 T}{dt^2} = K \end{cases} \qquad (3.14)$$

Note we have *total* not *partial* derivatives: linear, second-order, and ordinary differential equations.

The general solutions of the two sets of Eq. 3.14 have this form of solutions:

$$\begin{cases} K > 0 \\ K < 0 \end{cases} \quad \begin{cases} e^{+kx}, e^{-kx} \\ \sin(kx), \cos(kx) \end{cases} \quad \text{if we let} \quad \begin{cases} K = k^2 \\ K = -k^2 \end{cases} \qquad (3.15)$$

We always have two linearly independent solutions for second-order partial differential equations.

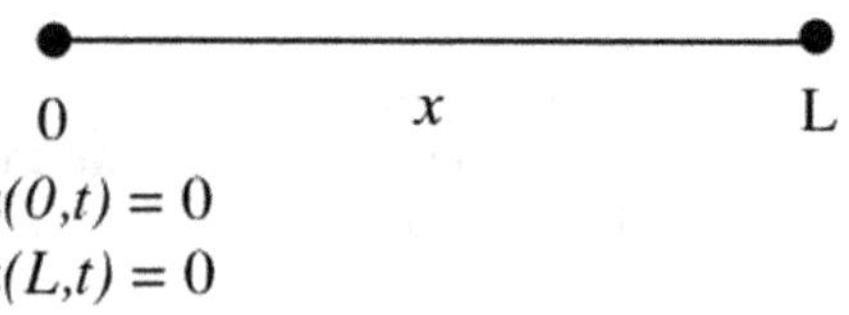

Fig. 3.3 Illustration of boundary conditions

```
●────────────────────────●
0           x            L
```
$u(0,t) = 0$
$u(L,t) = 0$

$$\begin{cases} K > 0 \\ \\ K < 0 \end{cases} \text{general solution is either} \begin{cases} X(x) = Ae^{kx} + Be^{-kx} \\ \\ X(x) = C\sin(kx) + D\cos(kx) \end{cases}$$
(3.16)

Also, for the $T(t)$ part of Eq. 3.14, we have:

$$\frac{d^2T}{dt^2} = v^2KT \qquad (3.17)$$

We now look at the *boundary conditions* that we started with at the beginning of this problem, as depicted in Fig. 3.3. For $K > 0$ try to satisfy the boundary conditions. We can write:

$$X(0) = Ae^0 + Be^{-0} = 0$$
$$A + B = 0 \quad A = -B \qquad (3.18a)$$

and

$$X(L) = 0 = Ae^{kL} + Be^{-kL} = A\left(e^{kL} - e^{-kL}\right)$$
$$e^{kL} - e^{-kL} \quad \text{can never be } 0 \qquad (3.18b)$$
$$A = 0 \qquad u(x,t) = 0$$

Thus the solution of Eq. 3.18 does not look good for the given boundary condition. What about the case when $K < 0$ is the solution?

$$X(0) = C\sin(0) + D\cos(0) = 0$$
$$D = 0 \qquad (3.19a)$$

and

$$X(L) = C\sin(kL) + 0 = 0$$
$$kL = n\pi \quad n = 0, 1, 2, \cdots \qquad (3.19b)$$

where we can say that

$$k_n = \frac{n\pi}{L} \quad n = 0, 1, 2, \cdots \qquad (3.19c)$$

3.2 The Classical Wave Equation and Separation of Variables

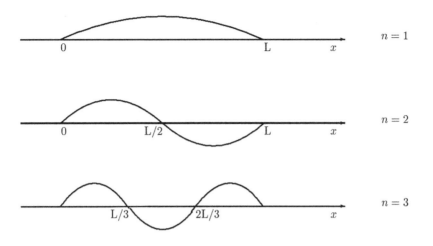

- # nodes is $n-1$
- nodes are equally spaced at $x = L/n$, $\lambda_n = 2(L/n)$.
- all lobes are the same, except for alternating sign

Fig. 3.4 Depiction of quantization solutions

The "quantization" term and pictures in Fig. 3.4 are drawn without looking at the equation or using a computer to plot them.

Now, we look at the $T(t)$ equation for $K < 0$, and we get the following result:

$$\begin{cases} T(t) = E \sin(vk_n t) + F \cos(vk_n t) \\ \omega_n \equiv vk_n \\ T(t) = E \sin(\omega_n t) + F_n \cos(\omega_n t) \end{cases} \quad (3.20)$$

Normal modes are

$$u_n(x,t) = \left[A_n \sin\left(\frac{n\pi}{L}\right)x\right](E_n \sin(\omega_n t) + F_n \cos(\omega_n t)) \quad (3.21)$$

The time-dependent factor of the nth normal mode can be rewritten in "frequency phase" form as

$$E'_n \cos[\omega_n t + \phi_n] \quad (3.22)$$

The next step is to consider the $t=0$ pluck of the system. This pluck is expressed as a linear combination of the normal modes:

$$u(x,t) = \sum_{n=1}^{\infty} (A_n E'_n) \sin\left(\frac{n\pi}{L}x\right) \cos(\omega_n t + \phi_n) \quad (3.23)$$

There is a further simplification based on the trigonometric formula:

$$\sin(A)\cos(B) = \frac{1}{2}[\sin(A+B) + \sin(A-B)] \tag{3.24}$$

that enables us to write Eq. 3.23:

$$u(x,t) = \sum_{n=1}^{\infty} \left[\frac{A_n E'_n}{2}\right] \left\{ \sin\left(\frac{n\pi}{L}x + \omega_n t + \phi_n\right) + \sin\left(\frac{n\pi}{L}x - \omega_n t - \phi_n\right) \right\} \tag{3.25}$$

From this equation we can write the following conclusions:

- A single normal mode is a *standing wave* with no left–right motion and no "breathing."
- A superposition of two or more normal modes with diverse values of n gives a more complicated motion. For two normal modes, where one is even-n and the other is odd-n, the time-evolving wave packet will exhibit left-to-right motion. For two normal modes, where both are odd or both are even, the *wave-packet* motion will be "breathing" rather than showing left-to-right motion.

Here is a crude time-lapse video of a superposition of the $n = 1$ and $n = 2$ (fundamental and first overtone) modes. The period of the fundamental is $T = 2\pi/\omega$. We are going to consider the time-step of $T/8$, as depicted in Fig. 3.5.

As can be seen in Fig. 3.5, the time-lapse video of the sum of two normal modes can be viewed as moving to the left at $t = -T/4$, close to the left-turning point at $t = -T/8$, but at $t = 0$ moving to the right at $t = +T/8$. It will reach the right-turning point but at $t = T/2$.

In *quantum mechanics* you can see wave packets that exhibit motion, breathing, dephasing, and rephasing. The "center of the wave packet" will follow a trajectory that obeys Newton's laws of motion. If we generalize from waves on a string to waves on a rectangular drum head, we get a drum head as shown in Fig. 3.6.

Then, the separation of variable solutions to the wave equation will have the form

$$u(x,y,t) = X(x)Y(y)T(t) \tag{3.26}$$

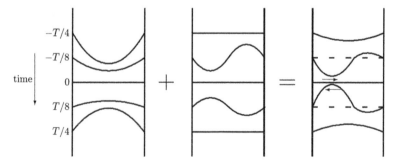

Fig. 3.5 Illustration of time-steps for $T/8$

3.2 The Classical Wave Equation and Separation of Variables

Fig. 3.6 A drum head

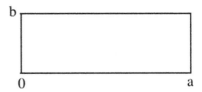

There will be two separation constants, and we will find that the normal mode frequencies are

$$\omega_{nm} = v\pi \left[\frac{n^2}{a^2} + \frac{m^2}{b^2}\right]^{1/2} \tag{3.27}$$

This is a more complicated quantization rule than for waves on a string, and it should be evident to an informed observer that these waves are on a rectangular drum head with edge lengths a and b.

Some important conclusions can be drawn from the preceding discussion. The underlying unity of e^{kx}, e^{-kx} and $\sin(kx)$, $\cos(kx)$ is a solution to the following form of a linear ordinary differential equation:

$$\frac{d^2y}{dx^2} = k^2 y \tag{3.28}$$

To solve Eq. 3.16, take a step back and look at the two simplest second-order ordinary differential equations:

$$\frac{d^2y}{dx^2} = +k^2 y \quad \rightarrow \quad y(x) = Ae^{kx} + Be^{-kx} \tag{3.29a}$$

and

$$\frac{d^2y}{dx^2} = -k^2 y \quad \rightarrow \quad y(x) = C\sin(kx) + D\cos(kx) \tag{3.29b}$$

The solutions to these two equations are more similar than they appear at first glance. *Euler's formula* provides that

$$e^{\pm i\theta} = \cos\theta \pm i\sin\theta \tag{3.30a}$$

or

$$\begin{cases} \frac{1}{2}\left(e^{i\theta} + e^{-i\theta}\right) = \cos\theta \\ \frac{1}{2i}\left(e^{-i\theta} - e^{i\theta}\right) = \sin\theta \end{cases} \tag{3.30b}$$

Therefore we can express the solution of the second differential equation in (complex) exponential form to bring out its similarity to the solution of the first differential equation:

$$y(x) = C \sin(kx) + D \cos(kx)$$
$$= \frac{i}{2}c\left(e^{-ikx} - e^{ikx}\right) + \frac{1}{2}D\left(e^{ikx} + e^{-ikx}\right) \quad (3.31a)$$

Rearranging Eq. 3.31 will yield:

$$y(x) = \frac{1}{2}(D - iC)e^{ikx} + \frac{1}{2}(D + iC)e^{-ikx} \quad (3.31b)$$

The $\sin\theta$, $\cos\theta$ and $e^{i\theta}$, $e^{-i\theta}$ forms are two sides of the same coin: Insight. Convenience. What do we notice? The *general solution* to the second-order differential equation consists of the *sum of two linearly independent functions*, each multiplied by an unknown constant.

3.3 Standing Waves

Standing waves, also called *stationary waves*, are a combination of two waves moving in opposite directions, each of which has the same amplitude and frequency. The phenomenon is the result of interference—that is, when waves are superimposed, their energies are either added together or canceled out. In the case of waves moving in the same direction, interference produces a traveling wave; for waves moving in opposite directions, interference produces an oscillating wave fixed in space. A vibrating rope tied at one end produces a standing wave, as shown in Fig. 3.7; the wave train (*line B*), after arriving at the fixed end of the rope, will be reflected back and superimposed on itself as another train of waves (*line C*) in the same plane. Because of interference between the two waves, the resultant amplitude (R) of the two waves is the sum of their individual amplitudes.

The first part of the figure shows wave trains B and C coinciding so that standing wave R has twice their amplitude. In the second part, 1/8 period later, B and C have each shifted 1/8 wavelength. The third part represents the case, another 1/8 period later, when the amplitudes of component waves B and C are oppositely directed. At all times there are positions (N) along the rope, called nodes, at which no movement occurs at all; the two wave trains are always in opposition. On either side of a node is a vibrating anti-node (A). The anti-nodes alternate in the direction of displacement so that at any instant the rope resembles a graph of the mathematical function called the *sine*, as represented by line R. Both longitudinal (e.g., sound) waves and transverse (e.g., water) waves can form standing waves (Figs. 3.7 and 3.8) [1].

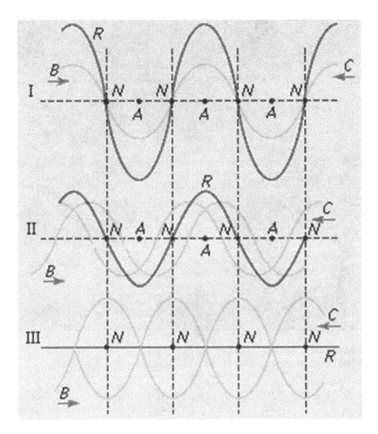

Fig. 3.7 Location of fixed nodes in a standing wave

3.4 Seiches

A *seiche* is a standing wave in an enclosed or partially enclosed body of water. Seiches and seiche-related phenomena have been observed on lakes, reservoirs, swimming pools, bays, harbors, and seas. The key requirement for formation of a seiche is that the body of water be at least partially bounded, allowing the formation of a standing wave [2].

The term was proposed in 1890 by the Swiss hydrologist François-Alphonse Forel, who was the first to make scientific observations of the effect in Lake Geneva (Lac Léman), Switzerland [3]. The word originates from a Swiss-French dialect and means "to sway back and forth"; it apparently had been used for a long time in the region to describe oscillations in alpine lakes. A seiche may last from a few minutes to several hours or for as long as two days.

Seiches may be induced by local changes in atmospheric pressure. They also may be initiated by the motions of earthquakes and, in the case of coastal inlets, by

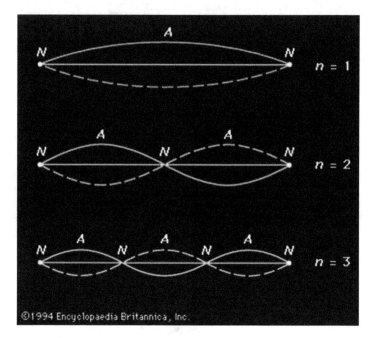

Fig. 3.8 The first three harmonic standing waves in a stretched string. Nodes (N) and anti-nodes (A) are marked. The harmonic number (n) for each standing wave is given on the right (see text).

tsunamis. Seismic surface waves from the Alaska earthquake of 1964, for example, triggered seiche surges in Texas in the southwestern United States when they passed through the area. Studies of seiche behavior have shown that once the surface of the water is disturbed, gravity seeks to restore the horizontal surface and simple vertical harmonic motion ensues. The impulse travels the length of the basin at a velocity dependent on the water depth and is reflected from the basin's end, generating interference. Repeated reflections generate standing waves with one or more nodes, or points, that experience no vertical motion. The length of the lake is an exact multiple of the distance between nodes.

Seiches can disturb shipping by generating strong reversible currents at the entrances to harbors or by causing moored vessels to oscillate against their mooring cables and break free. They also may drown unsuspecting persons on piers and shores. Seiches often are imperceptible to the naked eye, and observers in boats on the surface may not notice that a seiche is occurring because of the extremely long wavelengths.

The effect is caused by resonances in a body of water that has been disturbed by one or more factors, most often meteorological effects (e.g., wind and atmospheric pressure variations), seismic activity, or tsunamis [4]. Gravity always seeks to restore the horizontal surface of a body of water, as this represents the configuration in which the water is in hydrostatic equilibrium. Vertical harmonic motion results, producing an impulse that travels the length of the basin at a velocity that depends on the depth of the water. The impulse is reflected back from the end of the basin, generating interference.

The frequency of the oscillation is determined by the size of the basin, its depth and contours, and water temperature. The longest natural period of a seiche is the period associated with the fundamental resonance for the body of water—corresponding to the longest standing wave. For a surface seiche in an enclosed rectangular body of water, this can be estimated using *Merian's formula* [5, 6]:

$$T = \frac{2L}{\sqrt{gh}} \tag{3.32}$$

where T is the longest natural period, L is the length, h is the average depth of the body of water, and g is the acceleration of gravity. Higher order harmonics also are observed. The period of the second harmonic will be half the natural period, the period of the third harmonic will be one-third of the natural period, and so forth.

3.4.1 Lake Seiches

Low rhythmic seiches almost always are present on large lakes. They are usually unnoticeable among the common wave patterns, except during periods of unusual calm. Harbors, bays, and estuaries often are prone to small seiches with amplitudes of a few centimeters and periods of a few minutes. Among lakes well known for their regular seiches is New Zealand's Lake Wakatipu, which varies its surface height at Queenstown by 20 cm in a 27-min cycle. Seiches also can form in semi-enclosed seas; the North Sea often experiences a lengthwise seiche for a period of about 36 h.

The National Weather Service issues low-water advisories for portions of the Great Lakes when seiches of 2 ft or higher are likely to occur [7]. Lake Erie is particularly prone to seiches caused by wind because of its shallowness and its elongation on a northeast–southwest axis, which frequently matches the direction of prevailing winds and therefore maximizes the reach of those winds. These can lead to extreme seiches of up to 5 m (i.e., 16 ft) between the ends of the lake (Fig. 3.9).

The effect is similar to a storm surge like that caused by hurricanes along ocean coasts, but the seiche effect can cause oscillations back and forth across the lake for some time. In 1954, Hurricane Hazel piled up water along the northwestern shoreline of Lake Ontario near Toronto, causing extensive flooding, and it created a seiche that subsequently caused flooding along the south shore.

Lake seiches can occur very quickly: on July 13, 1995, a large seiche on Lake Superior caused the water level to fall and then rise again by 3 ft (1 m) within 15 min, leaving some boats hanging from the docks on their mooring lines when the water retreated. The same storm system that caused the 1995 seiche on Lake Superior produced a similar effect in Lake Huron; there the water level at Port Huron changed by 6 ft (1.8 m) over 2 h. On Lake Michigan, eight fishermen were swept away from

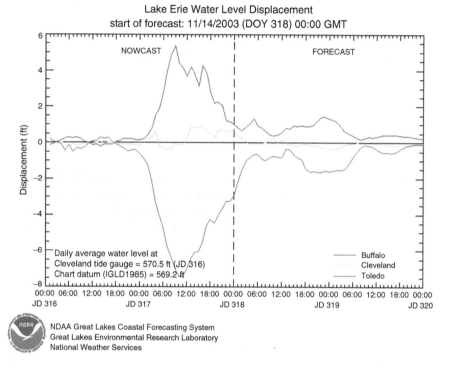

Fig. 3.9 Lake Erie water-level displacement

piers at the Montrose and North Avenue Beaches and drowned when a 10-ft (3.0-m) seiche hit the Chicago waterfront on June 26, 1954.

Lakes in seismically active areas, such as Lake Tahoe in California/Nevada, are significantly at risk from seiches. Geological evidence indicates that the shores of Lake Tahoe may have been hit by seiches and tsunamis as much as 10 m (32.8 ft) high in prehistoric times, and local researchers have called for the risk to be factored into emergency plans for the region.

Earthquake-generated seiches can be observed thousands of miles from the epicenter of a quake. Swimming pools especially are prone to seiches caused by earthquakes because the ground tremors often match the resonant frequencies of small bodies of water. The 1994 Northridge earthquake in California caused swimming pools to overflow across southern California. The massive Good Friday earthquake that hit Alaska in 1964 caused seiches in swimming pools as far away as Puerto Rico. The earthquake that hit Lisbon, Portugal, in 1755 caused seiches 2000 mi (3000 km) away in Loch Lomond, Loch Long, Loch Katrine, and Loch Ness in Scotland and in canals in Sweden. The 2004 Indian Ocean earthquake caused seiches in standing water bodies in many Indian states as well as in Bangladesh, Nepal, and northern Thailand. Seiches were observed in Uttar Pradesh, Tamil Nadu, and West Bengal in India, as well as in many locations in Bangladesh, during the 2005 Kashmir earthquake.

The 1950 Chayu-Upper Assam earthquake is known to have generated seiches as far away as Norway and southern England. Other earthquakes on the Indian subcontinent known to have generated seiches include the 1803 Kumaon-Barahat, the 1819 Allah Bund, the 1842 Central Bengal, the 1905 Kangra, the 1930 Dhubri, the 1934 Nepal-Bihar, the 2001 Bhuj, the 2005 Nias, and the 2005 Teresa Islandones. The February 27, 2010, an earthquake in Chile produced a seiche on Lake Pontchartrain in Louisiana, with a height of around 0.5 ft. The 2010 Sierra El Mayor earthquake produced large seiches that quickly produced an Internet phenomenon. Seiches up to at least 1.8 m (6 ft) were observed in Sognefjorden, Norway, during the 2011 Tōhoku earthquake.

3.4.2 Sea and Bay Seiches

Seiches have been observed in seas such as the Adriatic and the Baltic. These resulted in the flooding of Venice and St. Petersburg, respectively, as both cities are constructed on former marshland. In St. Petersburg seiche-induced flooding is common along the Neva River in the autumn. The seiche is driven by a low-pressure region in the North Atlantic moving onshore, giving rise to cyclonic lows on the Baltic Sea. The low pressure of the cyclone draws larger than normal quantities of water into the virtually land-locked Baltic.

As the cyclone continues inland, long, low-frequency seiche waves with wavelengths up to several hundred kilometers are established in the Baltic Sea. When the waves reach the narrow and shallow Neva Bay, they become much higher, ultimately flooding the Neva embankments [8]. Similar phenomena have been observed at Venice, resulting in the Modulo Sperimentale Elettromeccanico (MOSE) project—a system of 79 mobile barriers designed to protect the three entrances to the Venetian Lagoon [9]. The MOSE *Experimental Electromechanical Module* is intended to prevent the Venetian Lagoon from flooding to protect the city of Venice, Italy.

Seiches also can be induced by a tsunami, a wave train (series of waves) generated in a body of water by a pulsating or abrupt disturbance that vertically displaces the water column. On occasion, tsunamis can produce seiches because of local geographic peculiarities. For instance, the tsunami that hit Hawaii in 1946 had a 15-min interval between wavefronts. The natural resonant period of Hilo Bay is about 30 min. That means that every second wave was in phase with the motion of Hilo Bay, creating a seiche in it. As a result, Hilo suffered worse damage than any other place in Hawaii, with the tsunami/seiche reaching a height of 26 ft along the bayfront, killing 96 people in the city alone. Seiche waves may continue for several days after a tsunami.

Tide-generated internal solitary waves (i.e., *solitons*) can excite coastal seiches at the following locations: Magueyes Island in Puerto Rico, Puerto Princesa in Palawan Island, Trincomalee Bay in Sri Lanka, and in the Bay of Fundy in eastern Canada, where seiches cause some of the highest recorded tidal fluctuations in the world.

A dynamic mechanism exists for the generation of coastal seiches by deep-sea internal waves. These waves can generate a current at the shelf break that is sufficient to excite coastal seiches.

3.5 Underwater or Internal Waves

Although the bulk of the technical literature addresses surface seiches, which can be readily observed, seiches also are observed beneath lake surfaces, acting along the thermocline in constrained bodies of water. The *thermocline* is the boundary between the cold lower layer, the *hypolimnion*), and the warm upper layer, the *epilimnion*), as shown in Fig. 3.10.

Analogous to the Merian formula, the expected period of the internal wave can be expressed as

$$T = \frac{2L}{c} \qquad (3.33)$$

with

$$c^2 = g\left(\frac{\rho_2 - \rho_1}{\rho_2}\right)\left(\frac{h_1 h_2}{h_1 + h_2}\right) \qquad (3.34)$$

where T is the natural period, L is the length of the water body, h_1 and h_2 are the average thicknesses of the two layers separated by stratification (e.g., epilimnion and hypolimnion), ρ_1 and ρ_2 are the densities of these same two layers, and g is the acceleration of the gravity force.

3.6 Maxwell's Equations and Electromagnetic Waves

As discussed in previous chapters, the principles of guidance and propagation of electromagnetic (EM) energy, in its wide variety of applications and regimes, may differ in detail from one application to another; however, they all are governed by one set of equations that we know as Maxwell's equations. They are based on experimental observations and provide the foundations of *all* EM phenomena and their applications. Many of the underlying concepts starting in the early nineteenth century were developed by early scientists, especially Michael Faraday, who was a visual and physical thinker but not enough of a mathematician to express his ideas in a form complete and consistent enough to provide a theoretical framework.

Maxwell, however, started putting Faraday's ideas into the strict perspective of mathematical form and thus established a theory that predicted the existence of EM waves. Based on Maxwell's ideas and his development of his famous equations that

3.6 Maxwell's Equations and Electromagnetic Waves 193

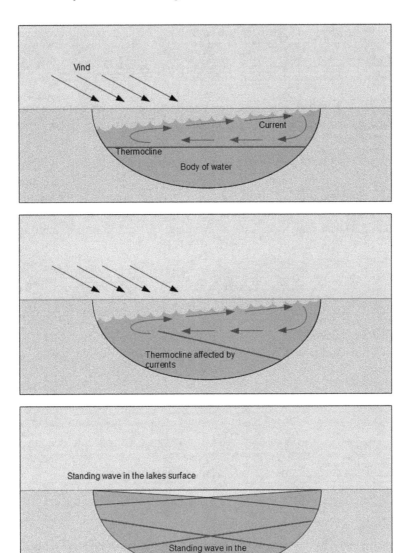

Fig. 3.10 Illustration of the initiation of surface and subsurface thermocline seiches

were described in Chap. 2 and Eqs. 2.31 through 2.34, we now can establish EM waves.

One of the important aspects of Maxwell's equations is the equation for EM wave propagation in a linear medium, analyzed in Chap. 4. Nevertheless, we need to establish the wave equation for an *electromagnetic field* $\vec{H}$ by taking the *curl* of Eq. 2.34 as follows:

$$\begin{cases} \vec{\nabla} \times \vec{H} = \vec{J} + \dfrac{\partial \vec{D}}{\partial t} \\ \vec{\nabla} \times \vec{\nabla} \times \vec{H} = \vec{\nabla} \times \vec{J} + \vec{\nabla} \times \dfrac{\partial \vec{D}}{\partial t} \end{cases} \quad (3.35)$$

Putting $\vec{D} = \varepsilon \vec{E}$ and $\vec{J} = g\vec{E}$ in a linear medium, and assuming that g and ε are constants, we can use the second part of Eq. 3.35 to obtain the following result:

$$\vec{\nabla} \times \vec{\nabla} \times \vec{H} = g \vec{\nabla} \times \vec{E} + \varepsilon \frac{\partial}{\partial t}(\vec{\nabla} \times \vec{E}) \quad (3.36)$$

In this equation, if electric field $\vec{E}$ is a well-behaved function as we assume to be the case, then the order of time and space differentiation can be interchanged. Equation 2.31 can now to be used to eliminate $\vec{\nabla} \times \vec{E}$ from Eq. 3.36, thereby yielding

$$\vec{\nabla} \times \vec{\nabla} \times \vec{H} = g\mu \frac{\partial \vec{H}}{\partial t} + \varepsilon\mu \frac{\partial^2 \vec{H}}{\partial t^2} \quad (3.37)$$

where $\vec{B} = \mu \vec{H}$, with μ a constant, has been used. The following vector identity from Table 1.1 can be used:

$$\vec{\nabla} \times \vec{\nabla} \times \vec{F} = \vec{\nabla}(\vec{\nabla} \cdot \vec{F}) - \nabla^2 \vec{F} \quad (3.38)$$

Therefore, Eq. 3.37 can be reformated as

$$\vec{\nabla}(\vec{\nabla} \cdot \vec{H}) - \nabla^2 \vec{H} = -g\mu \frac{\partial \vec{H}}{\partial t} - \varepsilon\mu \frac{\partial^2 \vec{H}}{\partial t^2} \quad (3.39)$$

Because μ is a constant, the $\vec{B} = \mu \vec{H}$ relation can be written:

$$\vec{\nabla} \cdot \vec{H} = \frac{1}{\mu}(\vec{\nabla} \cdot \vec{B}) = 0 \quad (3.40)$$

Consequently, the first term on the left-hand-side (LHS) of Eq. 3.39 vanishes. The final wave equation then is equal to:

$$\nabla^2 \vec{H} - \varepsilon\mu \frac{\partial^2 \vec{H}}{\partial t^2} - g\mu \frac{\partial \vec{H}}{\partial t} = 0 \quad (3.41)$$

Vector $\vec{E}$ satisfies the same wave equation, as can readily be seen by first taking the curl of Eq. 2.31 as

$$\vec{\nabla} \times \vec{\nabla} \times \vec{E} = -\vec{\nabla} \times \frac{\partial \vec{B}}{\partial t} \quad (3.42)$$

3.6 Maxwell's Equations and Electromagnetic Waves

Using Eq. 2.34 to eliminate the magnetic field and treating g, μ, and $\mathcal{E}$ as constants, Eq. 3.42, yields:

$$\vec{\nabla} \times \vec{\nabla} \times \vec{E} = -g\mu \frac{\partial \vec{E}}{\partial t} - \varepsilon\mu \frac{\partial^2 \vec{E}}{\partial t^2} \tag{3.43}$$

Applying the vector identity of Eq. 3.38 again and restricting the application of the equation to a charge-free medium so that $\vec{\nabla} \cdot \vec{D} = 0$ gives:

$$\nabla^2 \vec{E} - \varepsilon\mu \frac{\partial^2 \vec{E}}{\partial t^2} - g\mu \frac{\partial \vec{E}}{\partial t} = 0 \tag{3.44}$$

The wave equations derived in the preceding equations govern the EM field in a homogeneous, linear medium in which the charge density is zero, whether this medium is conducting or non-conducting. Yet, it is not enough that these equations be satisfied; Maxwell's equations also must be satisfied. It is clear that Eqs. 3.41 and 3.44 are necessary consequences of Maxwell's equations, but the converse is not true [10].

There are many methods available to solve the wave equations, but special care must be taken to obtain a useful solution to Maxwell's equations. One method that works very well for monochromatic waves is to obtain a solution for vector electric file $\vec{E}$. The curl of $\vec{E}$ (i.e., $\vec{\nabla} \times \vec{E}$) then gives the time deviation of magnetic field $\vec{B}$, which for monochromatic waves is sufficiently related to $\vec{B}$ so that $\vec{B}$ can be found easily.

Monochromatic waves may be described as waves that are characterized by a single frequency. The methods of complex variable analysis afford a convenient way of treating such waves and the time dependency of vector field $\vec{E}$ is taken to be $e^{-i\omega t}$ so that:

$$\vec{E}(\vec{r}, t) = \vec{E}(\vec{r}) e^{-i\omega t} \tag{3.45}$$

Note that the physical electric field is obtained by taking the real part of Eq. 3.45, which is the convenient mathematical description in terms of complex variables to physical quantities by taking either the real or imaginary part of the complex quantity. Furthermore, $\vec{E}(\vec{r})$ is complex in general so that the actual electric field is proportional to $\cos(\omega t + \phi)$, where ϕ is the phase of $\vec{E}(\vec{r})$. Using Eq. 3.45 in Eq. 3.44 provides the following result:

$$e^{-i\omega t} \left(\nabla^2 \vec{E} + \omega^2 \varepsilon\mu \vec{E} + i\omega g\mu \vec{E} \right) = 0 \tag{3.46}$$

In this equation the common factor $e^{-i\omega t}$ for governing the spatial variation of the electric field can be dropped out. Now, we put our effort behind solving Eq. 3.46 in various spatial cases of interest to determine the spatial variation of the EM field. This is treated in Chap. 4 and, in a simple form of possibility, we assume that the "medium" is empty space so that $g = 0$, $\varepsilon = \varepsilon_0$, and $\mu = \mu_0$ (see Chaps. 1 and 2). Furthermore, we suppose that $\vec{E}(\vec{r})$ varies in only one-dimensional form in the z-direction, and it is independent of x and y coordinates. Then Eq. 3.46 becomes:

$$\frac{d^2 \vec{E}(z)}{dz^2} + \left(\frac{\omega}{c}\right)^2 \vec{E}(z) = 0 \tag{3.47}$$

where we have the following relation:

$$\varepsilon_0 \mu_0 = \frac{1}{c^2} \tag{3.48}$$

So the solution to Eq. 3.47 is given by

$$\vec{E}(z) = \vec{E}_0 e^{\pm ikz} \tag{3.49}$$

where $\vec{E}_0$ is a constant vector, providing:

$$k = \frac{\omega}{c} \tag{3.50}$$

Substituting $\vec{E}(\vec{r})$ into Eq. 3.45, we get the full solution:

$$\vec{E}(\vec{r},t) = \vec{E}_0 e^{-i(\omega t \mp kz)} \tag{3.51}$$

Or taking the real part of Eq. 3.51, we get

$$\vec{E}(\vec{r},t) = \vec{E}_0 \cos(\omega t \mp kz) \tag{3.52a}$$

With Eq. 3.50, an equivalent form is:

$$\vec{E}(\vec{r},t) = \vec{E}_0 \cos(\omega t \mp \kappa z) \tag{3.52b}$$

The final form of Eq. 3.52 is a representation of a sinusoidal wave traveling to the right or left in the z-direction, depending on whether the minus or plus sign is applied. The velocity of propagation of the wave is the speed of light c and if light is a form of EM radiation, then Maxwell's equations *predict* that $c = 1/\sqrt{\varepsilon_0 \mu_0} = 2.9979 \times 10^8$ m/s is the velocity of light in a vacuum. The format in Eq. 3.52a shows that the wave frequency is $f = \omega/2\pi$ and the wavelength is $\lambda = 2\pi/k$. Thus, Eq. 3.50 is the familiar result for a wave:

$$\lambda f = c \tag{3.53}$$

In a non-conducting, non-magnetic dielectric, we still maintain that $g = 0$ and $\mu = \mu_0$, but now $\varepsilon = \varepsilon_0 \kappa$. The preceding derivation will carry through just the same, except that now Eq. 3.50 becomes:

$$k = \sqrt{K}(\omega/c) \tag{3.54}$$

Now, if we define $n = \sqrt{K}$, we can see that the results are the same in vacuum, except that the velocity of wave propagation is now c/n instead of just c, and quantity n is the *index of refraction* of the dielectric medium and for vacuum $n = 1$. The value

3.6 Maxwell's Equations and Electromagnetic Waves

of 1 for n accounts for refractive effects in transparent materials, as can be seen in the discussion that follows.

If the medium is conducting, then $g > 0$; thus, the third term in Eq. 3.46 must be retained. When g is small, the result will be merely that the wave is damped [10]. By small g we mean that the third term of Eq. 3.46 is small compared with the second term, which gets the wave solution, or

$$\omega g \mu \ll \omega^2 \varepsilon \mu$$
$$g \ll \omega \varepsilon \tag{3.55}$$

In the other extreme, when $g \gg \omega\varepsilon$, we may neglect the second term of Eq. 3.46. Again, restricting attention to the one-dimensional case, we get the following wave equation:

$$\frac{d^2 \vec{E}(z)}{d^2 z} + i\omega g \mu \vec{E}(z) = 0 \tag{3.56}$$

We can make the coefficient of $\vec{E}(z)$ real if we assume that $\alpha = i\omega$ is real or, in other words, that the frequency is imaginary. Then, if

$$k = \sqrt{\alpha g \mu} \tag{3.57}$$

the spatial dependence $\vec{E}(\vec{r})$ of the solution is just the same as before. The difference is that the time dependence Eq. 2.54, becomes:

$$\vec{E}(\vec{r},t) = \vec{E}(\vec{r}) e^{-\alpha t} \tag{3.58}$$

That is, the field simply decays exponentially with time instead of oscillating in a wave-like manner. The transition between the decaying and the wave behavior occurs when:

$$|\omega| = |\alpha| \tilde{=} \left|\frac{g}{\varepsilon}\right| = \frac{1}{t_c} \tag{3.59}$$

where t_c is the relaxation time of the medium's material. Note that, we repeat, caution is needed when this condition is applied to a metal because g/ε is itself solidly dependent on ω.

Now tracing the derivation of Eq. 3.46 back to Maxwell's equations in sets of Eqs. 2.31 through 2.34, we notice that the second term, or $\partial^2 \vec{E}(\vec{r},t)/\partial t^2$ in Eq. 3.44, is derived from displacement current $\partial \vec{D}/\partial t$ in Eq. 2.34, whereas the third term, or $\partial \vec{E}(\vec{r},t)/\partial t$ in Eq. 3.44, is derived from transport current density $\vec{J}$. Thus, the very existence of EM wave propagation depends on Maxwell's introduction of the displacement current. Without it, only exponential decay of the fields could occur.

Based on our discussion so far, we can claim that the EM plane waves are transverse and in vacuum; we can write the general form of the wave in the z-direction as:

$$\begin{cases} \vec{\nabla} \cdot \vec{E} = 4\pi\rho \Rightarrow \rho = 0 \quad \text{in vacuum} \\ \vec{\nabla} \cdot \vec{E} = 0 \\ \vec{\nabla} \cdot \vec{E} = \dfrac{\partial E_z(z,t)}{\partial z} = 0 \end{cases} \qquad (3.60)$$

This equation says that E_z is independent of z. That E_z is also independent of time t can be seen by considering Maxwell's "displacement current," Eq. 2.32. Thus,

$$\frac{\partial \vec{E}}{\partial t} = c^2 \vec{\nabla} \times \vec{B} \qquad (3.61)$$

Take the z-component of Eq. 3.61. The right-hand side (RHS) involves $\partial B_y/\partial x$ and $\partial B_x/\partial y$, both of which are zero; thus, $\partial E_z/\partial t$ is zero. We conclude that E_z is a constant. For simplicity, we take the constant to be zero; similarly, the fact that we have, from Eq. 2.33, $\vec{\nabla} \cdot \vec{B} = 0$ tells us that $B_z(z,t)$ has no z-dependence. Also no time-dependence is seen by considering the z-component of Faraday's Law, Eq. 2.31, which gives $\partial B_z/\partial t$ to be zero. Although there may be some steady magnetic fields because of large steady currents somewhere, they have no space or time-dependence and are not of interest to us presently. We therefore take B_z to be zero by using the superposition principle.

A traveling harmonic wave was established by Eq. 3.52b and assumes the x-component of electric E_x is given by the following relation:

$$E_x = A \cos(\omega t - kz) \qquad (3.62)$$

Then, by coupling E_x and B_y, we have the following sets of equations:

$$\frac{\partial E_x}{\partial t} = -c^2 \frac{\partial B_y}{\partial z} \qquad \frac{\partial B_y}{\partial t} = -\frac{\partial E_x}{\partial z} \qquad (3.63)$$

Similarly,

$$\frac{\partial E_y}{\partial t} = c^2 \frac{\partial B_x}{\partial z} \qquad \frac{\partial B_x}{\partial t} = \frac{\partial E_y}{\partial z} \qquad (3.64)$$

Then, by using Eq. 3.50, Eq. 3.63 becomes:

$$\begin{aligned} \frac{\partial B_y}{\partial z} &= -\frac{1}{c^2} \frac{\partial E_x}{\partial t} \\ &= \frac{\omega}{c^2} \frac{\partial E_x}{\partial t} = \frac{1}{c} \frac{\partial E_x}{\partial t} \end{aligned} \qquad (3.65)$$

$$\frac{\partial B_y}{\partial t} = -\frac{\partial E_x}{\partial z}$$
$$= -kA \sin(\omega t - kz) \quad (3.66)$$
$$= -\frac{1}{c}\frac{\partial E_x}{\partial t}$$

According to Eqs. 3.65 and 3.66, the variation of B_y with respect to z and t is the same as that of E_x. Thus, we can see that in a traveling harmonic plane wave propagation in the +z-direction, B_y and E_x are equal aside from uninteresting additive constant, which we "superpose to zero."

If we consider a harmonic traveling wave propagating in the $-z$-direction, we find that B_y is the negative of E_x, as you can see easily by replacing k with $-k$ in the previous equations. Both directions of propagation are included in these summarizing equations as:

$$\text{Traveling Waves:} \begin{cases} |\vec{E}(z,t)| = |\vec{B}(z,t)| \\ \vec{E} \cdot \vec{B} = 0 \\ \widehat{E} \times \widehat{B} = \widehat{v} \end{cases} \quad (3.67)$$

Standing harmonic waves then come into play, assuming that E_x is given by:

$$E_x(z,t) = A \cos \omega t \cos kz \quad (3.68)$$

Then we can show that:

$$B_y(z,t) = A \sin \omega t \sin kz = E_x\left(z - \frac{1}{4}\lambda, t - \frac{1}{4}T\right) \quad (3.69)$$

From Eqs. 3.68 and 3.69 we can see that in an EM standing plane wave in vacuum $\vec{E}$ and $\vec{B}$ are perpendicular to one another and to unit vector $\hat{z}$, have the same amplitude, and are 90° out of phase both in space and in time. This behavior is similar to that of the pressure and velocity in a standing sound wave or that of the transverse tension and velocity in a standing wave on a string or rope, as described in Sect. 3.3.

3.7 Scalar and Vector Potentials

As we have learned so far, *Maxwell's equations* consist of a set of coupled *first-order partial differential equations* relating the various components of electric and magnetic fields as follows:

$$\begin{cases} \vec{\nabla} \times \vec{H} = \dfrac{4\pi}{c}\vec{J} + \dfrac{1}{c}\dfrac{\partial \vec{D}}{\partial t} & \vec{\nabla} \cdot \vec{D} = 4\pi\rho \\ \vec{\nabla} \times \vec{E} + \dfrac{1}{c}\dfrac{\partial \vec{B}}{\partial t} = 0 & \vec{\nabla} \cdot \vec{B} = 0 \end{cases} \quad (3.70)$$

Given that in linear media $\vec{D} = \varepsilon \vec{E}$ and $\vec{B} = \mu_0 \vec{H}$, a different format of the set of Eq. 3.70 can be written:

$$\begin{cases} (i) \;\; \vec{\nabla} \cdot \vec{E} = \dfrac{1}{\varepsilon_0}\rho & (iii) \;\; \vec{\nabla} \times \vec{E} = -\dfrac{\partial \vec{B}}{\partial t} \\ (ii) \;\; \vec{\nabla} \cdot \vec{B} = 0 & (iv) \;\; \vec{\nabla} \times \vec{B} = \mu_0 \vec{J} + \mu_0 \varepsilon_0 \dfrac{\partial \vec{E}}{\partial t} \end{cases} \quad (3.71)$$

Given that $\rho(\vec{r},t)$ and $\vec{J}(\vec{r},t)$, what are the fields $\vec{E}(\vec{r},t)$ and $\vec{B}(\vec{r},t)$? In the static case, in a situation of time-independent Maxwell's equations configuration, as we saw in Chap. 1, *Coulomb's Law* and the *Biot-Savart Law* provide the answer. What we are looking for then, however, is the generalization of those laws to *time-dependent Maxwell's equations configuration*, as discussed in Chap. 1 of this book.

These sets of equations can be solved as they stand in simple situations, but it often is convenient to introduce potentials, obtaining a small number of second-order equations, while satisfying some of Maxwell's equations identically. We already are familiar with the concept in electrostatics and magnetostatics, where we used the *scalar potential* ϕ and *vector potential* $\vec{A}$.

In electrostatics $\vec{\nabla} \times \vec{E} = 0$ allows us to write $\vec{E}$ as a gradient of a scalar potential as $\vec{E} = -\vec{\nabla}\phi$; however, in electrodynamics this is no longer possible because the curl of $\vec{E}$ is non-zero (i.e., $\vec{\nabla} \times \vec{E} \neq 0$). But $\vec{B}$ remains divergence and $\vec{\nabla} \cdot \vec{B} = 0$ still holds, so we can define $\vec{B}$ in terms of a vector potential as in magnetostatics as follows:

$$\vec{B} = \vec{\nabla} \times \vec{A} \quad (3.72)$$

Putting Eq. 3.72 into Faraday's Law of (*iii*) in Eq. 3.71 will yield the following result:

$$\vec{\nabla} \times \vec{E} = -\dfrac{\partial}{\partial t}(\vec{\nabla} \times \vec{A}) \quad (3.73a)$$

or

$$\vec{\nabla} \times \left(\vec{E} + \dfrac{1}{c}\dfrac{\partial \vec{A}}{\partial t} \right) = 0 \quad (3.73b)$$

This means that the quantity with vanishing curl in Eq. 3.73b can be written as the gradient of some scalar function—namely, a scalar potential ϕ—as:

3.7 Scalar and Vector Potentials

$$\begin{cases} \vec{E} + \dfrac{1}{c}\dfrac{\partial \vec{A}}{\partial t} = -\vec{\nabla}\phi \\ \vec{E} = -\vec{\nabla}\phi - \dfrac{1}{c}\dfrac{\partial \vec{A}}{\partial t} \end{cases} \qquad (3.74)$$

The definition of $\vec{B}$ and $\vec{E}$ in terms of the potentials $\vec{A}$ and ϕ, according to Eqs. 3.72 and 3.74, identically satisfies the two homogeneous Maxwell's equations. The dynamic behavior of $\vec{A}$ and ϕ will be determined by the two homogeneous equations in Eq. 3.71.

At this point it is convenient to restrict our considerations to the microscopic form of Maxwell's equations. Then the inhomogeneous Eq. 3.71 can be written in terms of the both vector and scalar potentials $\vec{A}$ and ϕ accordingly:

$$\nabla^2 \phi + \dfrac{1}{c}\dfrac{\partial}{\partial t}(\vec{\nabla}\cdot\vec{A}) = -4\pi\rho \qquad (3.75)$$

and

$$\nabla^2 \vec{A} - \dfrac{1}{c^2}\dfrac{\partial^2 A}{\partial t^2} - \vec{\nabla}\left(\vec{\nabla}\cdot\vec{A} + \dfrac{1}{c}\dfrac{\partial\phi}{\partial t}\right) = -\dfrac{4\pi}{c}\vec{j} \qquad (3.76)$$

We have now reduced the set of four Maxwell's equations to two equations, but they still are coupled equations. The uncoupling can be accomplished by exploiting the arbitrariness involved in the definition of the potentials $\vec{A}$ and ϕ. Because $\vec{B}$ is defined through Eq. 3.72 in terms of $\vec{A}$, the vector potential is arbitrary to the extent that the gradient of some scalar function Λ can be added. Thus, magnetic intensity $\vec{B}$ is left unchanged by a transformation such as [11]:

$$\vec{A} \rightarrow \vec{A}' = \vec{A} + \vec{\nabla}\Lambda \qquad (3.77)$$

For electric field $\vec{E}$ in Eq. 3.74 to stay unchanged as well, scalar potential ϕ must be simultaneously transformed as follows [11]:

$$\phi \rightarrow \phi' = \phi - \dfrac{1}{c}\dfrac{\partial \Lambda}{\partial t} \qquad (3.78)$$

The freedom implied by Eqs. 3.77 and 3.78 means that we can choose a set of potentials $\vec{A}$ and ϕ such that:

$$\vec{\nabla}\cdot\vec{A} + \dfrac{1}{c}\dfrac{\partial \phi}{\partial t} = 0 \qquad (3.79)$$

This equation will uncouple the pairs of Eqs. 3.75 and 3.76 and leave two inhomogeneous wave equations, one for scalar potential ϕ and one for vector potential $\vec{A}$, as follows:

$$\nabla^2 \phi - \frac{1}{c^2} \frac{\partial^2 \phi}{\partial t^2} = -4\pi\rho \tag{3.80}$$

and

$$\nabla^2 \vec{A} - \frac{1}{c^2} \frac{\partial^2 \vec{A}}{\partial t^2} = -\frac{4\pi}{c} \vec{j} \tag{3.81}$$

Equations 3.80 and 3.81, plus Eq. 3.79, form a set of equations equivalent in all respects to Maxwell's equations.

3.8 Gauge Transformation: Lorentz and Coulomb Gauges

In this section we refer to Jackson [11] and describe each of these topics in conjunction with scalar and vector potentials of in the chapter's previous section.

The transformation of Eqs. 3.77 and 3.78 is called a *gauge transformation*, and the invariance of the fields under such transformations is called *gauge invariance*. The relation given by Eq. 3.79 between $\vec{A}$ and ϕ is called the *Lorentz condition*. To see that potentials can always be found to satisfy the Lorentz condition, suppose that the potentials $\vec{A}$ and ϕ that satisfy Eqs. 3.75 and 3.76 do not satisfy Eq. 3.79. Then let us make a gauge transformation to potentials $\vec{A}'$ and ϕ', and demand that $\vec{A}'$ and ϕ' satisfy the Lorentz condition:

$$\vec{\nabla} \cdot \vec{A}' + \frac{1}{c} \frac{\partial \phi'}{\partial t} = 0 = \vec{\nabla} \cdot \vec{A} + \frac{1}{c} \frac{\partial \phi}{\partial t} + \nabla^2 \Lambda - \frac{1}{c^2} \frac{\partial^2 \Lambda}{\partial t^2} \tag{3.82}$$

Thus, provided a gauge function Λ can be found to satisfy the following:

$$\nabla^2 \Lambda - \frac{1}{c^2} \frac{\partial^2 \Lambda}{\partial t^2} = -\left(\vec{\nabla} \cdot \vec{A} + \frac{1}{c} \frac{\partial \phi}{\partial t}\right) \tag{3.83}$$

Potentials $\vec{A}'$ and ϕ' will satisfy the Lorentz condition and wave Eqs. 3.80 and 3.91.

Even for potentials, which satisfy the Lorentz condition in Eq. 3.79, there is arbitrariness. Evidently, the *restricted gauge transformation* as:

$$\begin{cases} \vec{A} \to \vec{A} + \vec{\nabla}\Lambda \\ \phi \to \phi + \frac{1}{c} \frac{\partial \Lambda}{\partial t} \end{cases} \tag{3.84}$$

where

$$\nabla^2 \Lambda - \frac{1}{c^2} \frac{\partial^2 \Lambda}{\partial t^2} = 0 \tag{3.85}$$

3.8 Gauge Transformation: Lorentz and Coulomb Gauges

preserve the Lorentz condition, provided $\vec{A}$ and ϕ satisfy it initially. All potentials in this restricted class are said to belong to the *Lorentz gauge*. It is commonly used, first because it leads to wave Eqs. 3.80 and 3.81, which are independent of the coordinate system chosen and fits naturally into the considerations of special relativity. See Chaps. 2 and 6 of this book.

Another useful gauge for the potentials is the so-called *Coulomb radiation* or *transverse gauge*. This is the gauge in which we have:

$$\vec{\nabla} \cdot \vec{A} = 0 \tag{3.86}$$

From Eq. 3.80 we can see that the *scalar potential* satisfies Poisson's equation (see Eq. 1.83b) as:

$$\nabla^2 \phi = -4\pi\rho \tag{3.87}$$

with a solution:

$$\phi(\vec{x}, t) = \int \frac{\rho(\vec{x}', t)}{|\vec{x} - \vec{x}'|} d^3\vec{x}' \tag{3.88}$$

The scalar potential is just *instantaneous Coulomb potential* because of charge density $\rho(\vec{x}, t)$ using the Cartesian coordinate notation. This the origin of the name *Coulomb gauge*. The *vector potential* satisfies the inhomogeneous wave equation as follows:

$$\nabla^2 \vec{A} - \frac{1}{c^2} \frac{\partial^2 \vec{A}}{\partial t^2} = -\frac{4\pi}{c} \vec{J} + \frac{1}{c} \vec{\nabla} \left(\frac{\partial \phi}{\partial t} \right) \tag{3.89}$$

The term involving the scalar potential, in principle, can be calculated from Eq. 3.88. Because it involves the gradient operator it is a term that is *irrotational*—that is, it has a vanishing *curl*. This suggests that it may cancel a corresponding piece of the current density. The current density or any vector field can be written as the sum of two terms:

$$\vec{J} = \vec{J}_l + \vec{J}_t \tag{3.90}$$

where $\vec{J}_l$ is called the *longitudinal current* or *irritational current* and has $\vec{\nabla} \times \vec{J}_l = 0$, while $\vec{J}_t$ is called the *transverse current* or *solenoidal current* and has $\vec{\nabla} \cdot \vec{J}_t = 0$. Substituting from the vector identity of Table 1.1, we get the following:

$$\vec{\nabla} \times (\vec{\nabla} \times \vec{J}) = \vec{\nabla}(\vec{\nabla} \cdot \vec{J}) - \nabla^2 \vec{J} \tag{3.91}$$

Together with $\nabla^2(1/|\vec{x} - \vec{x}'|) = -4\pi\delta(\vec{x} - \vec{x}')$ as per the definition of the Dirac delta function in Chap. 1, Sect. 1.8.3, it can be shown that $\vec{J}_l$ and $\vec{J}_t$ can be constructed explicitly from $\vec{J}$ as follows:

$$\vec{J}_l = -\frac{1}{4\pi} \int \frac{\vec{\nabla}' \cdot \vec{J}}{|\vec{x} - \vec{x}'|} d^3 \vec{x}' \tag{3.92}$$

and

$$\vec{J}_t = \frac{1}{4\pi} \vec{\nabla} \times \vec{\nabla} \times \int \frac{\vec{J}}{|\vec{x} - \vec{x}'|} d^3 \vec{x}' \tag{3.93}$$

With the help of the continuity equation and Eq. 3.93, it can be seen that:

$$\vec{\nabla}\left(\frac{\partial \phi}{\partial t}\right) = 4\pi \vec{J}_l \tag{3.94}$$

Therefore, the source for the wave equation for $\vec{A}$ can be expressed entirely in terms of the transverse current, Eq. 3.93, as:

$$\nabla^2 \vec{A} - \frac{1}{c^2}\frac{\partial^2 \vec{A}}{\partial t^2} = -\frac{4\pi}{c} \vec{J}_t \tag{3.95}$$

This, of course, is the origin of the name "transverse gauge." The name "radiation gauge" stems from the fact that transverse radiation fields are given by the vector potential alone, the instantaneous Coulomb potential contributing only to the near fields. This gauge is useful particularly in quantum electrodynamics. A quantum–mechanical description of photons necessitates quantization of only the vector potential. The Coulomb, or transverse, gauge is often used when *no sources* are present. Then $\phi = 0$ and $\vec{A}$ satisfy the homogeneous wave equation. The fields are given by

$$\begin{cases} \vec{E} = -\frac{1}{c}\frac{\partial \vec{A}}{\partial t} \\ \vec{B} = \vec{\nabla} \times \vec{A} \end{cases} \tag{3.96}$$

In passing we note a peculiarity of the Coulomb gauge. It is well known that EM disturbances propagate with finite speed. Yet, Eq. 3.93 indicates that the scalar potential "propagates" instantaneously everywhere in space. The vector potential, on the other hand, satisfies wave Eq. 3.95, with its implied finite speed of propagation c. At first glance it is puzzling to see how this obviously unphysical behavior is avoided. A preliminary remark is that it is the fields, not the potentials, that are of concern. A further observation is that the transverse current of Eq. 3.93 extends over all space, even if $\vec{J}$ is localized [12].

3.9 Infrastructure, Characteristics, Derivation, and Scalar Wave Properties

More details about this chapter's interesting subject are in this section and are also provided in Chap. 6 of this book. The *scalar wave* (SW), however, is a member of the wave family that we are discussing in this chapter; thus, we need to describe it here.

Starting from Faraday's discovery—instead of the formulation of the law of induction according to Maxwell—an extended field theory is derived, which goes beyond Maxwell's theory with the description of potential vortices (noise vortices) and their propagation as a SW, that contains the Maxwell theory as a special case. The new field theory with that does not collide with the textbook opinion but extends it in an essential way with the discovery and addition of potential vortices. Likewise, the theory of objectivity, which follows from the discovery, is compared in the form of a summary with the subjective and the relativistic point of view, and the consequences for variable velocity of propagation of SWs, formed from potential vortices, are discussed.

From Maxwell's field equations only, the well-known transverse or Hertzian can be derived, whereas the calculation of *longitudinal scalar waves (LSW)* give zero as a result. This is a flaw of the field theory because SWs exist for all particle waves (e.g., plasma waves, as photons or neutrino). Starting from Faraday's discovery, instead of the formulation of the law of induction according to Maxwell, an extended field theory is derived that goes beyond Maxwell's theory with the description of potential vortices (e.g., noise vortices) and their propagation as a SW. With that the extension is allowed and does not contradict textbook physics.

It was a transverse wave (TW) for which the electric and the magnetic field pointers oscillate perpendicular to the direction of propagation. This can be seen as the reason that the velocity of propagation is showing itself to be field-independent and constant. It is the speed of light c. With that Hertz had experimentally proved the properties of this wave, previously calculated in a theoretical way by Maxwell, and at the same time proved the correctness of the Maxwellian field theory. The scientists in Europe were just saying to each other "well done!" as completely different words came from a private research laboratory in New York: "Heinrich Hertz is mistaken, it by no means is a transverse wave but a longitudinal wave!"

Besides the mathematical calculation of SWs this section contains a voluminous material collection concerning information about the technical use of SWs, infrastructure, derivation, and properties of such waves. If the useful signal, usually the interfering noise signal, changes places, a separate modulation of frequency and wavelength makes a parallel image transmission possible. It may concern questions of environmental compatibility for the sake of humankind, such as bioresonance among others; harm to humanity (e.g., electro-smog); or even high-energy weapon applications, like in Star Wars—also known as the Strategic Defense Initiative (SDI) [13].

With regard to the environmental compatibility, a decentralized electrical energy technology should be required, which manages without overhead power lines, without combustion, and without radioactive waste. The liberalization of the energy markets will not in any way solve the energy problem; it will only accelerate the way into a dead end. A useful energy source could be represented by space quanta that hits the Earth from the Sun or space. They, however, only are revealed to a measurement technician if they interact. It will be shown that the particles oscillate and an interaction or collection, with the goal of the energy's technical use, is possible only in the case of resonance.

Because these space quanta as oscillating particles have almost no charge and mass averaged over time, they have the ability of penetration, as proved for neutrinos. In the case of the particle radiation discovered 100 years ago by Tesla, it obviously concerns neutrinos. We proceed from the assumption that in the future decentralized neutrino converters will solve the current energy problem. Numerous concepts from nature and engineering—such as on the one hand lightning or photosynthesis and, on the other hand, the railgun or the Tesla converter—are illustrated and can be discussed.

Given all the preceding scenarios, we start the discussion of SWs in this section by asking: What is a *scalar wave* exactly? A SW is just another name for a "longitudinal" wave (LW). The term "scalar" is sometimes used instead because the hypothetical source of these waves is thought to be a "scalar field" of some kind, similar to the Higgs field (i.e., boson), for example.

There is nothing particularly controversial about *longitudinal waves* in general as illustrated in Fig. 3.11. They are a ubiquitous and well-acknowledged phenomenon in nature. Sound waves traveling through the atmosphere (or underwater) are longitudinal, as are plasma waves propagating through space—also known as *Birkeland currents*. Longitudinal waves moving through the Earth's interior are known as *Telluric currents*. They can all be thought of as pressure waves of sorts.

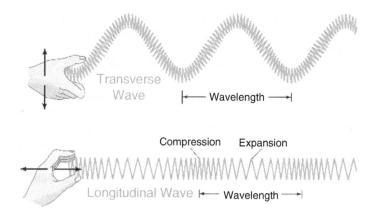

Fig. 3.11 Illustration of transverse versus longitudinal waves

3.9 Infrastructure, Characteristics, Derivation, and Scalar Wave Properties

In modern-day electrodynamics (both classical and quantum), *electromagnetic waves(EMW)* traveling in "free space" (e.g., photons in the "vacuum") are generally considered to be a TW. But, this was not always the case. When the preeminent mathematician James Clerk Maxwell first modeled and formalized his unified theory of electromagnetism in the late nineteenth century neither the EM SW or LW nor the EM TW had been experimentally proved, but he had postulated and calculated the existence of both.

After *Heinrich Hertz* demonstrated experimentally the existence of transverse radio waves in 1887, theoreticians (e.g., Heaviside, Gibbs, and others) went about revising Maxwell's original equations (who was now deceased and could not object). They wrote out the SW/LW component from the original equations because they felt the mathematical framework and theory should be made to agree only with experiments. Obviously, the simplified equations worked—they helped make the AC/DC electrical age engineerable. But at what cost?

Soon after Hertz's claim of discovering Maxwell's transverse EMWs, Tesla visited him and personally demonstrated his experimental error. Hertz agreed with Tesla and had planned to withdraw his claim, but varying agendas intervened and set the stage for a major rift in the "accepted" theories that soon became transformed into the fundamental "laws" of the electric science that have held sway in industry and halted academia to the present day (Fig. 3.12).

Then in the 1889 Nikola Tesla—a prolific experimental physicist and originator of the alternating current (AC)—threw a proverbial wrench in the works when he discovered via experimental proof for the elusive electric SW. This seemed to

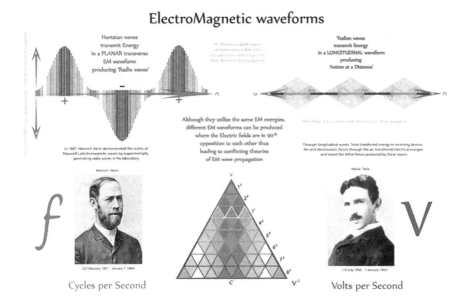

Fig. 3.12 Illustration of an electromagnetic wave

The different waveforms

1. *H. Hertz:* **electromagnetic wave** (transverse)

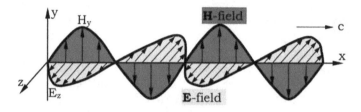

2. *Nikola Tesla:* **electric wave** (longitudinal)

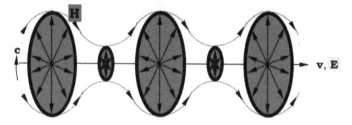

3. **magnetic wave** (longitudinal)

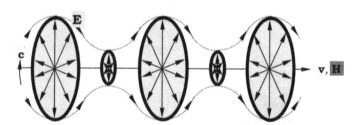

The three basic types according to the wave equation
(electric, magnetic and electromagnetic wave)

Fig. 3.13 Illustration of various waveforms

suggest that SWs/LWs, as opposed to the *transverse wave*, could propagate as pure electric waves or as pure magnetic waves. Tesla also believed these waves carried a hitherto-unknown form of excess energy he referred to as "radiant." This intriguing and unexpected result was said to have been verified by Lord Kelvin and others soon after (Figs. 3.13 and 3.14).

Instead of merging their experimental results into a unified proof for Maxwell's original equations, however, Tesla, Hertz, and others decided to bicker and squabble over who was more correct. In actuality, they both derived correct results.

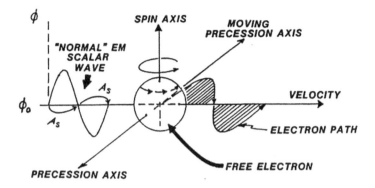

Fig. 3.14 Illustration of electron path and normal EM scalar wave

Nonetheless, because humans (even "rational" scientists) are fallible and prone to fits of vanity and self-aggrandizement, both sides insisted dogmatically that they were right and the others were wrong.

The issue was allegedly settled after the dawn of the twentieth century when:

A) The concept of the mechanical (passive/viscous) Ether was purportedly disproved by Michelson-Morley and replaced by Einstein's Relativistic Space–Time Manifold
B) Detection of SW/LW's proved much more difficult than initially thought (mostly because of the wave's subtle densities, fluctuating frequencies, and orthogonal directional flow). As a result, the truncation of Maxwell's equations was upheld.

Scalar and longitudinal waves in free space, however, are quite real. Besides Tesla, empirical work carried out by electrical engineers (e.g., Eric Dollard, Konstantin Meyl, Thomas Imlauer, and Jean-Louis Naudin, to name only some) clearly have demonstrated their existence experimentally. These waves seem able to exceed the speed of light, pass through EM shielding, also known as *Faraday cages*, and produce overunity effects—that is, more energy out than in. They seem to propagate in a yet unacknowledged counterspatial dimension, also known as hyperspace, pre-space, false-vacuum, aether, implicit order, and so on (Fig. 3.15).

Because the concept of an all-pervasive material ether was discarded by most scientists, the thought of vortex-like electric and/or magnetic waves existing in free space, without the support of a viscous medium, was thought to be impossible. Later experiments carried out by Dayton Miller, Paul Sagnac, E. W. Silvertooth, and others, however, have contradicted the findings of Michelson and Morley. More recently Italian mathematician-physicist Daniele Funaro, American physicist-systems theorist Paul LaViolette, and British physicist Harold Aspden have all conceived of (and mathematically formulated) models for a free space ether that is dynamic, fluctuating, self-organizing, and allows for the formation and propagation of SWs/LWs.

A harmonic set of bidirectional longitudinal EMW pairs in 3-space is depicted in Fig. 3.16. Unseen here is the time-polarized EMW in the time domain, which reacts with the source charge to produce the 2-space bi-wave potential. The potential, as

Fig. 3.15 Imaginary hyperspace

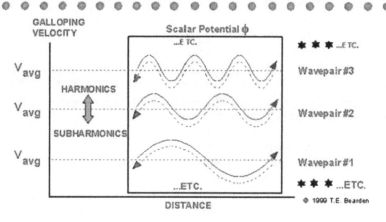

Fig. 3.16 A harmonic set of bidirectional longitudinal EMW pairs in 3-space

Fig. 3.17 Shape of a Möbius supercoils

observed or detected, is a harmonic set of bidirectional LWs in 3-space. That is, this potential is the "effect" of transduction of an incoming time-polarized EMW interacting with the source charge.

Therefore, defining the characteristics of SWs, we can state that SWs are produced when two EMWs of the same frequency are exactly out of phase (opposite to each other), and the amplitudes subtract and cancel or destroy each other. The result is not exactly an annihilation of magnetic fields but a transformation of energy back into a SW. This scalar field has reverted to a vacuum state of potentiality. Scalar waves can be created by wrapping electrical wires around a figure eight in the shape of a *Möbius coil*, as illustrated in Fig. 3.17. When electrical current flows through the wires in opposite directions, the opposing EM fields from the two wires cancel each other and create a SW. As an example of what we just stated, within our day-to-day life there is a DNA antenna in our body and cells.

The DNA antenna in our cells' energy production centers (*mitochondria*) assumes the shape of what is called a supercoil. Supercoil DNA look like a series of Möbius coils. These Möbius supercoil DNA are hypothetically able to generate scalar waves. Most cells in the body contain thousands of these Möbius supercoils, which are generating scalar waves throughout the cell and throughout the body.

The standard definition of SWs is that they are created by a pair of identical waves (usually called the wave and its anti-wave) that are in phase spatially (space), but out of phase temporally (time). That is, the two waves are physically identical, but 180° out of phase in terms of time. They even look different—like an infinitely projected Möbius pattern on an axis. The DNA antenna in our cells' energy production centers (mitochondria) assumes the shape of what is called a supercoil. Scalar energy can regenerate and repair itself indefinitely. This also has important implications for the body's DNA synthesis (Fig. 3.18).

Mitochondrial DNA is only a small portion of the DNA in a cell; most of it can be found in the cell nucleus. In most species on Earth, including human beings, mitochondrial DNA is inherited solely from the mother. Mitochondria have their own genetic material, and the mechanism to manufacture their own RNAs and new proteins. This process is called *protein biosynthesis*, which refers to the process whereby biological cells generate new sets of proteins (Fig. 3.19).

A SW also is called a *standing wave* (see Sect. 3.3); it is a pattern of moving energy that stays in one place. We generally think of waves as moving through space as well as vibrating "up and down" but a SW is stationary or standing. Scalar waves

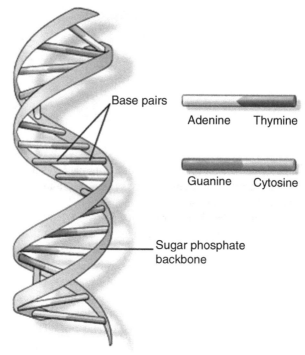

Fig. 3.18 Illustration of DNA's structure

Fig. 3.19 Image of mitochondrial DNA

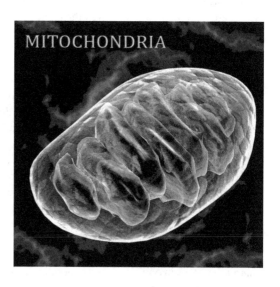

3.9 Infrastructure, Characteristics, Derivation, and Scalar Wave Properties

are used by controllers to generate interference or feedback systems or to stimulate the nervous system of bodies to repeatedly loop in a certain manner; if you can visualize the kind of waveform described, you can probably imagine it.

In water resonance the DNA is sending a LW that propagates in the direction of the magnetic field vector. The computed frequencies from the structure of the DNA agree with those of biophoton radiation as predicted. The optimization of efficiency by minimizing the conduction losses leads to the double-helix structure of DNA. The vortex model of the magnetic SW not only covers many observed structures within the nucleus from perfect, but also introduces the hyperboloid channels in the matrix—if two cells communicate with each other.

Physical results revealed in 1990 form the basis of the essential component of a potential vortex SW. The need for an extended field theory approach has been known since 2009 with the discovery of magnetic monopoles. For the first time this provides the opportunity to explain the physical basis of life not only from the biological discipline of science's understanding. Nature covers the whole spectrum of known scientific fields of research; for the first time this interdisciplinary understanding is explaining such complex relationships. Decisive are the characteristics of the potential vortex. With its concentration effect, it provides a miniaturization down to a few nanometers, which for the first time allows the outrageously high information density in the nucleus.

Here the magnetic SW theory explains dual-based pair-stored information of the genetic code and a process of converting it into an electrical modulation—say, "piggyback" information transfer from the cell nucleus to another cell. At the receiving end the reverse process takes place when writing a chemical structure physically. The energy required to power the chemical process comes from the SW itself. As an example of such a piggyback scenario, we can observe the carrier wave piggybacking in *Scientific Consciousness Interface Operation (SCIO)* technology.

Figure 3.20 is an illustration of such technology; when we set our radio or TV to a wavelength, such as 103.5, we get the wave of the station. The music and voice are superimposed onto the master wavelength in a piggyback manner known as a *carrier wave*. Therefore, when we make a master signal from the SCIO/Eductor we can superimposes a piggyback signal on the carrier wave.

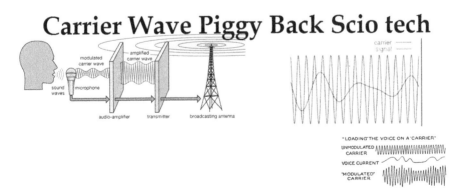

Fig. 3.20 Demonstration of carrier wave piggyback SCIO technology

3.9.1 Derivation of the Scalar Waves

Imagine a three-dimensional space that is continuous; it is governed by a uniform and constant flow of time for every point in the space. This space contains no matter and no particles of any kind. It is at absolute rest, and furthermore, it defines *absolute rest*, where no concept of velocity for a body exists. This space allows for the existence of three fields. There are two vector fields—the electric field $\vec{E}$ and the magnetic field $\vec{B}$—as they are viewed classically through sets of *Maxwell's equations*. There is also a third field, the scalar field, which we call the *scalar magnetic field* or just *scalar field* for short and is designated by the symbol β. These fields are related by the following equations:

$$\vec{\nabla} \cdot \vec{E} = \frac{\partial \beta}{\partial t} \tag{3.97}$$

$$\vec{\nabla} \times \vec{E} = -\frac{\partial \vec{B}}{\partial t} \tag{3.98}$$

$$\vec{\nabla} \cdot \vec{B} = 0 \tag{3.99}$$

$$\vec{\nabla} \times \vec{B} + \nabla \beta = \frac{1}{c^2} \frac{\partial \vec{E}}{\partial t} \tag{3.100}$$

where c is the speed of light; we call these four equations the *field equations*. Note that if $\beta = 0$, then the preceding equations postulate Maxwell's equations in free space along with their energy content. See the next sets of equations for further details.

Note that the difference between a scalar field and a scalar vector can be expressed simply, in general; any scalar is a quantity (number) without direction (e.g., speed and temperature) although the vector has quantity and direction (e.g., velocity, acceleration, force). The field is a physical entity that occupies a certain domain where its effect appears—for example, the gravitational field is a vector field, the EM field is a vector field. Even though the Higgs field is scalar, it is everywhere around us and has no specific direction; also a gauge field is a scalar field. Any vector has direction in addition to quantity, whereas any scalar is a quantity without specific direction.

Mathematically, a scalar field on a region U is a real or complex-valued function or distribution on U. The region U may be a set in some *Euclidean space*, *Minkowski space*, or more generally a subset of a manifold. It is typical in mathematics to impose further conditions on the field, such that it is continuous or often continuously differentiable to some order. A scalar field is a tensor field of order zero, and the term "scalar field" may be used to distinguish a function of this kind with a more general tensor field, density, or differential form.

Physically, a scalar field is distinguished additionally by having units of measurement associated with it. In this context a scalar field also should be independent of the coordinate system used to describe the physical system—that is, any two

3.9 Infrastructure, Characteristics, Derivation, and Scalar Wave Properties

observers using the same units must agree on the numerical value of a scalar field at any given point in physical space. Scalar fields can be compared with other physical quantities, such as vector fields, which associate a vector to every point of a region, as well as tensor fields and spinor fields. More subtly, scalar fields often are contrasted with pseudo-scalar fields. So, to put it in simple form, we express that:

1. *Scalar field:* If at every point in a region, a scalar function has a defined value, the region is called a scalar field.
    Example: Temperature distribution in a rod
2. *Vector field:* If at every point in a region, a vector function has a defined value, the region is called a vector field.
    Example: Velocity field of a flowing fluid

Now, back to our discussion about the field equations, which do not tell us what the fields are but only how they interact. These three fields could exist alone in a region of space or simultaneously in any combination of magnitudes and directions, if they do not violate Eqs. 3.97 through 3.100. Besides these relations, there is a dynamic content that can be conveniently expressed by the energy of each of the fields, given by

$$\xi_E = \frac{\varepsilon}{2} \int E^2 d\tau \tag{3.101}$$

$$\xi_B = \frac{1}{2\mu} \int B^2 d\tau \tag{3.102}$$

$$\xi_\beta = \frac{1}{2\mu} \int \beta^2 d\tau \tag{3.103}$$

where τ is the volume element, ε is the permittivity, and μ is the permeability of free space. We will call these three equations the *energy equations*.

The seven, Eqs. 3.97 through 3.103, and the static particle-less space constitute the postulates of the *stationary field theory*;that is the reason for saying if $\beta = 0$, these postulates are Maxwell's equations in free space and their energy content ca any textbook on the subject (e.g., Jackson [11] or Griffiths [18]). In addition, the nature of the space that these fields occupy is consistent with the classical view of electromagnetics.

Now the question is: Why do we need to postulate a stationary field in our theory of static particle-less space, called stationary field theory? To answer this question, we start by decoupling the four field equations, Eqs. 3.97 through 3.100, to obtain ones related independently to each field so that we easily can see that each field is a wave equation. This means that, typically, given the presence of any of the fields, they essentially would work in tandem to propagate themselves through space at the speed of light [19].

At this stage we are interested in the properties of the space itself, where we derived our four sets of equations that we called *field equations* and have not yet added anything to the space to produce the fields. Imagine that you have a large pan filled with water and the water is just standing motionless in the pan; it seems undisturbed. We know that if we touch the water, say in the center of the water in the pan, it would produce waves. This is because water has certain properties that support the production of waves.

We know that a wave in space travels at the speed of light as experimentally proved and also verified theoretically by Maxwell and others. The question here, particularly from a modern physics point of view, is: "Velocity relative to what?" There are no particles or matter in this space, which we assumed at the beginning, by which any wave velocity would be measured. In other words, any velocity in the space is absolute—that is, just like any wave velocity of water in our pan—based on the frame of reference of the pan wall holding the water and not the motion of the finger touching it. This is provided that there is a constraint—the water is not flowing around in the pan. Thus, there is no reason to consider that different regions of space would have some form of associated velocity, like water flowing around the pan, and that Maxwell's equations contain nothing that would suggest the idea. Therefore, another property of the space is that it is stationary, thus the name: *stationary field theory.*

Now, if we take into consideration that waves are traveling at various velocities in dissimilar media, then intrinsically each velocity is absolute because of the nature of wave propagation. Nevertheless, this scenario totally contradicts what we know from a modern physics point of view, where absolute velocity is very much non-existant and only relative velocity has any significance.

Many scientists take such a contradiction as an acknowledged fact because there is no way to measure any sort of absolute velocity experimentally; that is, most are designed to measure absolute velocity and experiments fail to do so. Furthermore, the strict and pervasive employment of absolute velocity here, which we are describing, would conflict with experiments if we perform them.

So, although the field equations described by sets of Eqs. 3.97 through 3.100 show the relationship between the three fields, and one needs some initial conditions to set up an applicable problem; in other words, we should imagine that some fields exist and then from these conditions the equations will tell us what will happen. To get a better understanding of this idea and the situation we are in and what would happen, it will be useful to decouple the equations. The decoupling of these field equations will take place in the following manner:

First, we take the curl of Eq. 3.98, as follows:

$$\vec{\nabla} \times (\vec{\nabla} \times \vec{E}) = -\nabla \times \frac{\partial \vec{B}}{\partial t}$$
$$\nabla \times \frac{\partial \vec{B}}{\partial t} = -\vec{\nabla} \times (\vec{\nabla} \times \vec{E})$$
(3.104)

Next, we take the gradient of Eq. 3.97 as:

$$\vec{\nabla}(\vec{\nabla} \cdot \vec{E}) = \vec{\nabla}\left(\frac{\partial \beta}{\partial t}\right)$$
$$\vec{\nabla}\left(\frac{\partial \beta}{\partial t}\right) = \vec{\nabla}(\vec{\nabla} \cdot \vec{E})$$
(3.105)

3.9 Infrastructure, Characteristics, Derivation, and Scalar Wave Properties

Then, we take the *time derivative* of Eq. 3.100 as:

$$\frac{\partial}{\partial t}(\vec{\nabla} \times \vec{B} + \nabla \beta) = \frac{\partial}{\partial t}\left(\frac{1}{c^2}\frac{\partial \vec{E}}{\partial t}\right)$$

$$\vec{\nabla} \times \frac{\partial \vec{B}}{\partial t} + \vec{\nabla}\left(\frac{\partial \beta}{\partial t}\right) = \frac{1}{c^2}\frac{\partial^2 \vec{E}}{\partial t^2} \quad (3.106)$$

$$\vec{\nabla}\left(\frac{\partial \beta}{\partial t}\right) + \vec{\nabla} \times \frac{\partial \vec{B}}{\partial t} = \frac{1}{c^2}\frac{\partial^2 \vec{E}}{\partial t^2}$$

Now, we can take the results in Eqs. 3.104 and 3.105 and substitute them into the final form in Eq. 3.106, which then yields the following:

$$\vec{\nabla}(\vec{\nabla} \cdot \vec{E}) - \vec{\nabla} \times (\vec{\nabla} \times \vec{E}) = \frac{1}{c^2}\frac{\partial^2 \vec{E}}{\partial t^2} \quad (3.107)$$

By using the vector identity $\vec{\nabla} \times (\vec{\nabla} \times \vec{F}) = \vec{\nabla}(\vec{\nabla} \cdot \vec{F}) - \nabla^2 \vec{F}$ in Table 1.1, Eq. 3.107 ends up in a final form as:

$$\nabla^2 \vec{E} = \frac{1}{c^2}\frac{\partial^2 \vec{E}}{\partial t^2} \quad (3.108)$$

Equation 3.108 indicates that an electric field can propagate as a wave through the space of interest, although it appears identical to a classic electromagnetism-type wave, not a solenoidal-type.

This wave equation also has a longitudinal component made possible by the scalar field. Therefore, if the initial conditions are that some electric vector field exists, then the electric field will propagate at the speed of light. From a localized initial field, the wave will spread out in all directions. For a somewhat artificial example, we consider a localized electric field where the field vector is always in the same direction, as illustrated in Fig. 3.21. As shown in the figure, the electric field at the center will propagate outward along with the magnetic field.

Second, we take the curl of Eq. 3.100, thus we have:

$$\vec{\nabla} \times (\vec{\nabla} \times \vec{B}) + \vec{\nabla} \times (\vec{\nabla}\beta) = \frac{1}{c^2}\vec{\nabla} \times \frac{\partial \vec{E}}{\partial t} \quad (3.109)$$

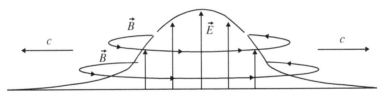

Fig. 3.21 Depiction of electromagnetic wave propagation

Then we take the time derivative of Eq. 3.98, so we have:

$$\vec{\nabla} \times \frac{\partial \vec{E}}{\partial t} = -\frac{\partial^2 \vec{B}}{\partial t^2} \tag{3.110}$$

We can then substitute Eq. 3.110 into Eq. 3.109, which yields:

$$\vec{\nabla} \times (\vec{\nabla} \times \vec{B}) = \frac{1}{c^2}\frac{\partial^2 \vec{B}}{\partial t^2} \tag{3.111}$$

Using vector identity $\vec{\nabla} \times (\vec{\nabla} \times \vec{F}) = \vec{\nabla}(\vec{\nabla} \cdot \vec{F}) - \nabla^2 \vec{F}$ from Table 1.1 again,

$$\vec{\nabla}(\vec{\nabla} \cdot \vec{B}) - \nabla^2 \vec{B} = \frac{1}{c^2}\frac{\partial^2 \vec{B}}{\partial t^2} \tag{3.112}$$

In addition, with Ddivergence of the magnetic field being zero, we have:

$$\nabla^2 \vec{B} = \frac{1}{c^2}\frac{\partial^2 \vec{B}}{\partial t^2} \tag{3.113}$$

Equation 3.113 shows that the magnetic field propagates along the electric field. The magnetic propagation is only solenoidal and tracks the solenoidal propagation of the electric field.

Third, we take the divergence of Eq. 3.100, so we have:

$$\vec{\nabla} \cdot (\vec{\nabla} \times \vec{B}) + \vec{\nabla} \cdot (\vec{\nabla}\beta) = \frac{1}{c^2}\vec{\nabla} \cdot \frac{\partial \vec{E}}{\partial t} \tag{3.114}$$

Using vector identity $\vec{\nabla} \cdot (\vec{\nabla} \times \vec{F}) = 0$ from Table 1.1, we can see the first term on the LHS of Eq. 3.114 is zero; thus, we are left with the following result:

$$\nabla^2 \beta = \frac{1}{c^2}\vec{\nabla} \cdot \frac{\partial \vec{E}}{\partial t} \tag{3.115}$$

Then, taking the time derivative of Eq. 3.97, we obtain:

$$\vec{\nabla} \cdot \frac{\partial \vec{E}}{\partial t} = \frac{\partial^2 \beta}{\partial t^2} \tag{3.116}$$

We can then substitute Eq. 3.116 into Eq. 3.115 to yield the following result:

$$\nabla^2 \beta = \frac{1}{c^2}\frac{\partial^2 \beta}{\partial t^2} \tag{3.117}$$

This equation indicates that the *scalar field* propagates along with the electric field and tracks the *longitudinal propagation* of the electric field, as shown in Eqs. 3.108, 3.113, and 3.115. These will be called the *wave equations*.

3.9 Infrastructure, Characteristics, Derivation, and Scalar Wave Properties

Note that the scalar field only exists when the electric field has divergence. If there is no divergence, the electric field has only solenoidal propagation. These three equations, Eqs. 3.108, 3.113, and 3.115, then revert to typical EMW propagation in free space because they can be derived from Maxwell's equations; see Chap. 1 of this book.

The scalar field propagates as a wave along the other fields as well. In other words, the electric field propagates through space, whereas the magnetic field propagates along the supporting propagation of the electric field that is perpendicular to the direction of propagation in solenoidal wave conditions; the scalar field propagates along with the electric field that is longitudinal to the direction of propagation, which we know by now as a *longitudinal wave*.

This longitudinal propagation might seem to defy observations immediately, but it actually does not occur EM radiation for complicated reasons that need to be explained (see Ensle [19]); they are discussed in the next section as well. Nonetheless, scalar propagation does exist in a limited form and shape, or context.

Further analysis of Eq. 3.108 will provide some mathematical reasoning and shed some light on the proof of a SW. Taking apart Eq. 3.108 of the Laplace operator, according to the rules of vector identity analysis in Table 1.1, it becomes $\nabla^2 \vec{F} = \vec{\nabla}(\vec{\nabla} \cdot \vec{F}) - \vec{\nabla} \times \vec{\nabla} \times \vec{F}$; the result is what we show in the following form with a more detailed breakdown:

$$\underbrace{\nabla^2 \vec{E}}_{\text{Wave}} = \underbrace{\vec{\nabla}(\vec{\nabla} \cdot \vec{E})}_{\text{Longitudinal}} - \underbrace{\vec{\nabla} \times (\vec{\nabla} \times \vec{E})}_{\text{Transverse}} \qquad (3.118)$$

We can compare Eq. 3.118 to the solution of Maxwell's field equations with the following interpretation:

$$\text{Hertzian Wave} = \text{Transverse Wave}$$

1. *Maxwell's field equations*: The solution of them with no source means Eq. 3.118 reduces to:

$$\vec{\nabla} \cdot \vec{E} = 0 \quad \text{and} \quad -\vec{\nabla} \times \vec{\nabla} \times \vec{E} = \frac{1}{c^2} \frac{\partial^2 \vec{E}}{\partial t^2} \qquad (3.119)$$

2. *Transverse wave*: Field pointers oscillate crosswise to the direction of propagation, which occurs with the speed of light c. The claim we have had so that:

$$\text{Tesla Radiation} = \text{Longitudinal Wave}$$

3. *Special case*: Irrotationally,

$$\vec{\nabla} \times \vec{E} = 0 \quad \text{and} \quad \vec{\nabla}(\vec{\nabla} \cdot \vec{E}) = \frac{1}{c^2} \left(\frac{\partial^2 \vec{E}}{\partial t^2} \right) \qquad (3.120)$$

4. *Longitudinalwave, shock wave, and standing wave*: File pointer oscillates in the direction of propagation. Velocity of propagation is variable.

By taking apart the wave equation, one plunges into the adventure of an entirely new field theory; it first is that all should be traced and analyzed. What the latest textbooks say about SWs is that they are non-existant so far except for a few notes here; they are by different researchers and scientists.

We did introduce scalar and vector potentials earlier. There the constant of dielectricity κ is written in a complex variable, although it physically concerns a material constant, only to be able to calculate it with this trick artificially—a loss angle, which should indicate the losses occurring in a dielectric, where, in reality, it concerns vortex losses. Of course, one can explain the dielectric losses of a capacitor or the heating in a microwave oven entirely without vortex physics with such a label "fraud," but it should be clear to anyone that some complex constant lies buried in an inner contradiction, which is incompatible with physical concepts.

We are used to such auxiliary descriptions so much so that the majority of today's physicists tend to attribute physical reality to this mathematical nonsense. As pragmatists, they put themselves on this stance: If experimental results can be described, then such an auxiliary description cannot be so wrong after all. By doing so the circumstance is forgotten, so here the ground of pure science is abandoned and is replaced by dogmas.

We find everything we need so far in the wave equation, as can be found in all classic textbooks, including wave Eq. 3.108. Behind this formulation two completely different kinds of waves are hiding because the usage of the Laplace operator consists of two parts according to the rules of vector analysis given by Table 1.1, and we demonstrated them in the preceding; thus, again we can write:

$$\underbrace{\vec{\nabla}(\vec{\nabla} \cdot \vec{E})}_{\text{Longitudinal}} - \underbrace{\vec{\nabla} \times (\vec{\nabla} \times \vec{E})}_{\text{Transverse}} = \underbrace{\nabla^2 \vec{E}}_{\text{Wave}} \qquad (3.121)$$

We should now discuss two special cases.

If we put the left part in Eq. 3.121 to zero ($\vec{\nabla} \cdot \vec{E} = 0$), which is tantamount to no sources of the field, then the well-known radio wave remains; italso is called the *Hertzian wave*, after Heinrich Hertz. As noted, it had been experimentally detected by Karlsruhe in 1888; it is written in the form of Eq. 3.119 as:

$$\vec{\nabla} \cdot \vec{E} = 0 \quad \text{and} \quad -\vec{\nabla} \times \vec{\nabla} \times \vec{E} = \frac{1}{c^2} \frac{\partial^2 \vec{E}}{\partial t^2} \quad \text{(Special Case)} \qquad (3.122)$$

This equation concerns the TW, described by Maxwell, for which the field pointers oscillate crosswise to the direction of propagation, which again occurs with the speed of light c. So much for the state of the art concerning technology.

Nonetheless, as we can see the mathematical formulation of the wave equation is hiding, yes, even more than only the generally known EMW. The no sources approach is an abandonment, which only is valid under certain prerequisites! The

3.9 Infrastructure, Characteristics, Derivation, and Scalar Wave Properties

other thing we mentioned is the Tesla claim and mathematical reasoning behind the SW—that is,

$$\text{Tesla Radiation} = \text{Longitudinal Wave}$$

$$\vec{\nabla} \times \vec{E} = 0 \quad \text{and} \quad \vec{\nabla}(\vec{\nabla} \cdot \vec{E}) = \frac{1}{c^2}\left(\frac{\partial^2 \vec{E}}{\partial t^2}\right) \quad (3.123)$$

Then, in the condition of a special case, we can write the source field because $\vec{\nabla} \cdot \vec{E} \neq 0$, then:

$$\text{Source} = \text{Charge Carriers} \quad (\text{Plasma Waves})$$
$$\text{Source} = \text{Vortex Structure}$$

Then, we take the approach where $\vec{\nabla} \cdot \vec{E} \neq 0$ is a SW, which gives us the indication that the $\vec{E}$-field vector can be derived from scalar potential ϕ via its gradient, as we have said all along—that is:

$$\vec{E} = -\vec{\nabla}\phi \quad (3.124)$$

and

$$\vec{\nabla} \cdot \vec{E} = -\vec{\nabla} \cdot \vec{\nabla}\phi = \nabla^2 \phi \quad (3.125)$$

Insert Eq. 3.125 into Eq. 3.123, then we obtain *homogeneous scalar wave equation* as:

$$\nabla^2 \phi = \frac{1}{c^2}\left(\frac{\partial^2 \phi}{\partial t^2}\right) \quad (3.126)$$

which is analogous to Eq. 3.117 in the preceding.

For the case of an additional space charge density, where we need to consider ρ_{electric} in the matter, then we can write:

$$\text{div}\,\vec{D} = \rho_{\text{electric}} \quad (3.127)$$

where $\vec{D}$ is called *electric displacement*; see Eq. 1.115 as well. In this case we get a *inhomogeneous scalar wave equation*, which is nothing more than a plasma wave and that:

$$\nabla^2 \phi = \frac{1}{c^2}\left(\frac{\partial^2 \phi}{\partial t^2}\right) - \frac{\rho_{\text{electric}}}{\varepsilon} \quad (3.128)$$

where ε is permittivity of materials. One solution to the plasma wave of Eq. 3.128 is *Langmuir waves*, where $\omega^2 = c^2 k^2 + \omega_{\text{plasma}}^2$. In this relationship ω frequency of oscillation (i.e., *Longmuir oscillation*) that is propagating and ω_{plasma} is the frequency of plasma, while k is a wave number and c is the speed of light (see Zohuri [20]).

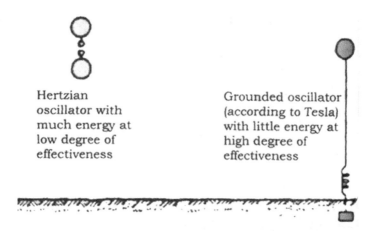

Fig. 3.22 Comparison between Hertzian and scalar wave per Tesla

Fig. 3.23 Hertzian wave versus scalar wave

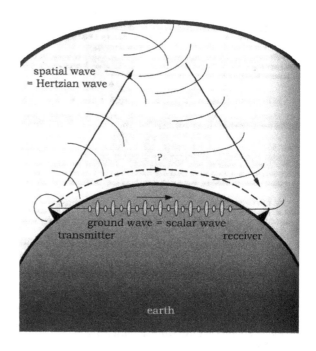

Equation 3.128 is a derivation of the plasma wave as an example of the existence of a SW in the wave equation. The solution to Eq. 3.128 describes the dispersion relation of plasma wave 20, which is LW movements plus Langmuir oscillation of the electron density. Nikola Tesla explains the difference between his SW radiation and the Hertzian wave; it is depicted in Figs. 3.22 and 3.23.

In Fig. 3.23 we see the ground waves that follow the curvature of the Earth and radio waves reflected at the ionosphere. This figure also indicates that the

3.9 Infrastructure, Characteristics, Derivation, and Scalar Wave Properties

interference and fading with which the radio amateur is fighting are a result of the various fast-arriving wave parts and by doing so the SW part tunnels as a straight line right through the Earth, as shown in Fig. 3.23. Driving the plasma wave can be done without formation of the gradient in order to obtain a homogeneous equation that is tantamount to an integration of the equation. Thus, under certain conditions, we must expect the occurrence of an integration constant.

This is the case if in addition a space charge density occurs as a source of the field, which according to Maxwell's Eq. 1.115 can be considered as the divergence of dielectric displacement $\vec{D}$ in respect to the material relation such as:

$$\vec{D} = \varepsilon \vec{E}$$
$$\vec{\nabla} \cdot \vec{D} = \rho_{electric} \qquad (3.129)$$
$$\vec{\nabla} \cdot \vec{E} = \frac{\rho_{electric}}{\varepsilon} = -\nabla^2 \phi$$

If we complete these contributions with a possible present field source, then the inhomogeneous SW equation yields, as illustrated in Eq. 3.128, a solution for this wave that has been established by many plasma physicists. They have the same form as the well-known dispersion relations of Langmuir waves—that is, electron plasma waves; thus, LW movements associated with Langmuir oscillations of the electron density.

With that, it has been proved that SWs and longitudinally propagating standing waves are described by the wave equation and are contained in it. This, in any case, is valid in general just as in the special case of a plasma wave, as could be derived here mathematically. From our previous discussion and derivation of plasma waves in Eq. 3.128, we easily can see that the SWs by all means are nothing new and their existence has been known to us both theoretically and experimentally.

One important scenario that can be taken away from all this is that wave absorption means nothing; however, TWs in the case of a disturbance rolling up to vortices in the measurement of a localized wave and vortices for a standing wave, localized vortex, and broadband antenna for *electromagnetic compatibility (EMC)* measurements as shown in Fig. 3.24. As far as decoupling of the wave parts is concerned, the set of difficulties of ground waves makes clear the coupling of LWs and TWs as two aspects or parts of a wave. As the corresponding Eq. 3.108, mathematically taken apart into Eq. 3.121, dictates every transmitter emits both parts.

In this case from other areas of application of these waves—for instance, from flow dynamics point of view, or for body sound—it is generally known that both wave parts exist and in addition occur jointly. In the case of a propagation through the Earth, like for an earthquake, both parts are received and used. Because their propagation is inversely fast, the faster oscillations arrive first and are the longitudinal ones. From the time delay with which the TWs arrive at the measurement station, the distance to the epicenter of the quake is determined by means of the diverse velocity of propagation. For geophysicists this tool is part of everyday knowledge (Fig. 3.25).

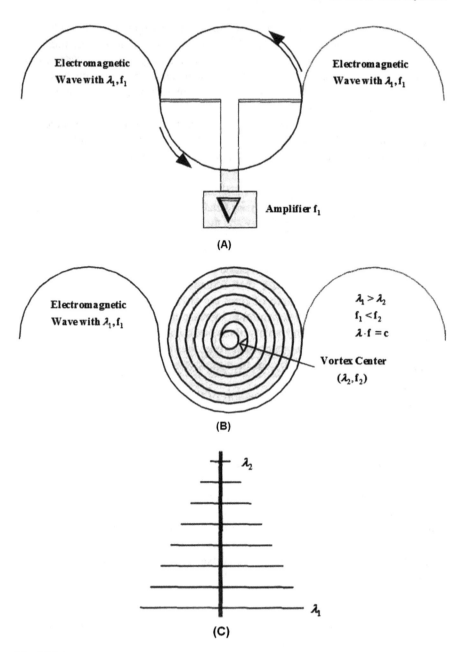

Fig. 3.24 Measurement of localized waves and vortices: (**a**) Standing wave; (**b**) Localized vortex; (**c**) Broadband antenna for EMC measurements

3.9 Infrastructure, Characteristics, Derivation, and Scalar Wave Properties 225

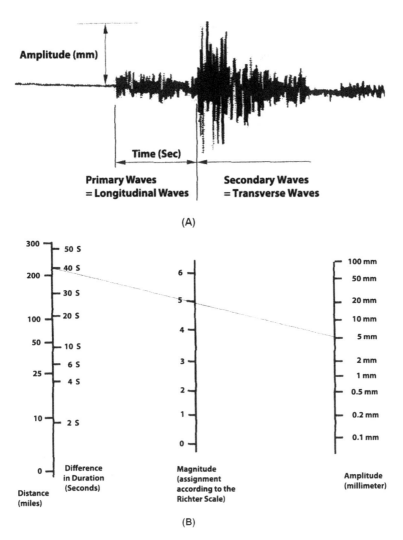

Fig. 3.25 Illustration of earthquake waves and magnitude. (**a**) Longitudinal and transverse earthquake waves. (**b**) Analysis according to Richter scale (e.g., 40 S duration between *S*-wave and *P*-wave for 5 mm amplitude means an earthquake of strength 5 at a distance of 220 miles)

As we can see, it is obvious that the EMW is not just purely transverse and that the sound wave is not purely longitudinal either. It is true that a transverse sound wave does not get too far in air because that sound as a rule is considered as a purely LW by neglecting this part; however, such neglect may not be sustainable in general. It must be checked as to whether it is legitimate from case to case and an error factor should be considered..

Further examples for the coupling of the wave parts are furnished by the latest tunnel experiments. Here so-called pure TWs are sent into a tunnel through which they do not fit at all. The Maxwell theory then dictates that behind the tunnel no signal should be measurable. Nonetheless, a signal is being measured, which in the tunnel was faster than allowed.

So, this should tell us that thoughts of the phase velocities of an EMW, which is not present at all, before instantaneous tunneling, during which the clocks should stop. The wave equation, however, supplies the only possible answer; that is, the tunnel filters out the SW parts and lets it to pass from them only if they are sufficiently small and correspondingly fast [13].

1. **Near-Field Difficulties**

Sets of difficulties that we face in the near-field also are discussed, as follows, in high-frequency technology that distinguishes between the near-field and the far-field. Both fundamentally have other properties [13].

Heinrich Hertz did experiments in the short-wave range at wavelengths of some meters. From today's viewpoint his work would be assigned the far-field. As a professor in Karlsruhe, he had shown that his EMW propagates like a light wave and can be refracted and reflected in the same way. It is a TW for which the field pointers of the electric and the magnetic fields oscillate perpendicular to each other and both are perpendicular to the direction of propagation. Thus, it should be obvious, if in the case of the Hertzian wave, it would concern the far-field. Besides propagation with the speed of light is a characteristic that no phase shift between the $\vec{E}$-field and the $\vec{H}$-field [13] occurs.

For the approach of vortex and closed-loop field structures, derivations for the near-field are known. Doing so it must be emphasized that the structures do not follow from the field equations according to Maxwell, but that the calculations are based on assumed rotation of symmetrical structures. The Maxwell theory by no means is capable of such structure shaping in principle. The calculation provides an important result that, in the proximity of the emitting antenna, a phase shift exists between the pointers of the $\vec{E}$-field and the $\vec{H}$-field. The antenna current and the $\vec{H}$-field coupled with it lag the $\vec{E}$-field of the oscillating dipole charges for 90° (Fig. 3.26). These charges form a longitudinal standing wave and the antenna rod or antenna dipole. For this reason, at first the fields produced by high-frequency currents have the properties of a LW in the proximity of the antenna.

These two fields—namely, the $\vec{E}$-field and the $\vec{H}$-field—are completely different in their proximity. The proximity concerns distance to the transmitter of less than the wavelength divided by 2π. Nikola Tesla broadcast in the range of long waves, around 100 kHz, in which case the wavelength already is several meters. For the experiments concerning the resonance of the Earth he operated his transmitter in Colorado Springs at frequencies down to 6 Hz. By doing so the whole Earth moves into the proximity of his transmitter. We probably have to proceed from the assumption that the Tesla radiation primarily concerns the proximity, which also is called the *radiant range* of the transmitting antenna.

The near-field already is used in practice in anti-theft devices because they are installed in the entrance area of stores. The customer walks through the SW

3.9 Infrastructure, Characteristics, Derivation, and Scalar Wave Properties

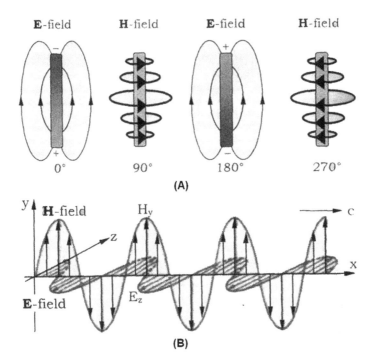

Fig. 3.26 The antenna current of oscillating dipole charges. (a) Fields of the oscillating dipole antenna. (b) Planar electromagnetic wave in the proximity

transmitters. If the coupling coil has not been removed at the cash point, then a signal from the alarm system sounds. The coils work entirely passively; that is, they are supplied with electric energy per SW and stimulated to oscillate for their part. Then the effect back on the transmitter is being used. Even if the principle is functioning, people still should be warned not to use a technology that is not understood completely; if they do unexplained catastrophes are inevitable.

2. Far-Field Transition

We talked about transition in the near-field in the preceding, and now we discuss the transition in the far-field as well. At sufficient distance to the transmitting antenna is the far-field where the transverse EMW occurs (Fig. 3.27b). It is distinguished by not occurring anymore at a phase shift between the $\vec{E}$- and $\vec{H}$-fields. Every change of the electric alternating field is followed immediately, and at the same time, by a change of the magnetic alternating field and vice versa [13]. In proximity, however, the phase shift amounts to 90°. Somewhere and somehow between the causing antenna current and the far-field, a conversion from a LW into a TW occurs. How should one imagine the transition?

The coming off of a wave from a dipole is represented in Fig. 3.27a. The fields come off the antenna, the explanation reads. If we consider the structure of the fields

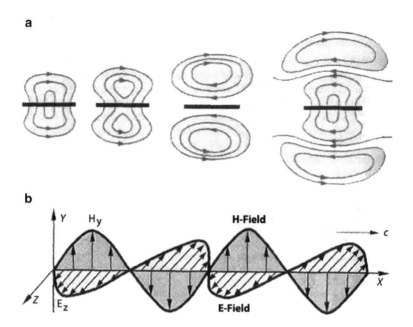

Fig. 3.27 (a) Forming vortex structures found in a longitudinal electric wave carrying an impulse. (b) Electromagnetic wave (transverse)

coming off, then we see field vortices, which run around a point, that we can call the vortex center. Such field structures naturally are capable of forming standing waves and carrying an impulse. The SW field, in general, and the near-field, in special cases, we will understand only with suitable vortex physics and with a field theory extended for corresponding vortices. We also will be able to calculate it. Postulates *cannot* replace field physics [13].

Be that as it may, the vortex, after having left the antenna, for getting greater distances at some time seems to unroll to propagate further as an EMW. A transition from longitudinal to transverse takes place, or spoken figuratively, from vortex to wave. How completely this conversion takes place and how large the respective wave parts are afterwards, on the one hand, depends on the structure and the dimensions of the antenna. Information is given by the measurable degree of the antenna's effectiveness...

The vortex structures, on the other hand, are stable the smaller and faster they are. If they are as fast as the speed of light or even faster, then they become stable elementary particles (e.g., neutrinos). Slower vortex structures, however, predominantly are unstable. They preferably unwind to waves. Vortices and waves prove to be two possibilities and under certain conditions are even stable field configurations.

Let us emphasize: A Hertzian dipole does not emit Hertzian waves! An antenna a as near-field without exception emits vortices, which only at the transition to the far-field unwind to EMWs. A Hertzian wave just as small can be received with a dipole antenna. At the receiver the conditions are reversed. Here, the wave is rolling

3.9 Infrastructure, Characteristics, Derivation, and Scalar Wave Properties 229

Fig. 3.28 Diversion of light by a strong gravitation field

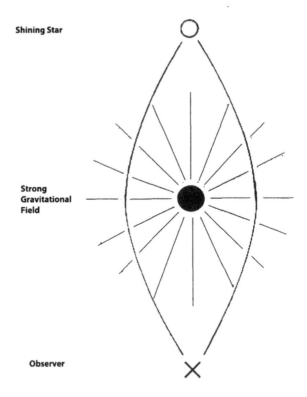

up to a vortex, which usually is called and conceived of as a "standing wave." Only this field vortex causes an antenna current in the rod, which afterward the receiver amplifies and utilizes.

The mostly unknown or not understood near-field properties prove to be the key to the understanding of the wave equation and of the method for the functioning of the transmitting and receiving antenna. Then, the question is asked: How should one imagine the rolling up of waves to vortices and vice versa, the unrolling? How would a useful vortex model look?

3. A Scalar Wave Model

Light, as an EMW, in the presence of a heavy mass or of strong fields, is bent toward the field source (Fig. 3.28). The wave normally propagating in a straight line, thus, can be diverted.

The square of the speed of light further is inversely proportional to the permeability and dielectricity short; in the presence of matter it is more or less significantly slowed down. If this slowing down of the wave occurs one-sidedly, then a bending of the path can be expected as well. At the end of the antenna a reflection and a going back of the wave can occur, which at the other end again hits itself. Now the wave has found a closed-loop structure that can be called a vortex. Figures 3.29b and 3.30a show the two possible structures.

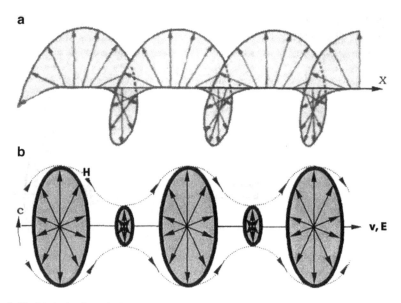

Fig. 3.29 (a) A circular polarized wave (transverse). (b) An electric wave (logitudinal)

In most technical textbooks the vortex with the properties of a *standing wave* is described; a *near-field* and standing waves are two examples that can be seen. Their description mathematically does not go through much of the details of SW properties, however. It is important to have a good awareness of vortex physics and to have a better understanding of pure SW transmission according to Nikola Tesla and the properties of this wave type. With the vortex concept of an extended field in physics a new horizon can be opened. Note the following details with respect to Fig. 3.29:

<u>Vortex and Wave = Two Stable Field Configuration</u>

<u>Electromagnetic Wave = Transverse Wave Propagation in a Straight Line</u>

----Ring Like Vortex = Transverse Wave Running in Circles

----Vortex Velocity = Speed of Light c

Change of Structure = If the Field is Disturbed without Expense of Energy

If we pay further attention to Figs. 3.29b and 3.30a, they reveal that, in both cases, EMWs are represented, where they propagate with the speed of light; the exception isthe wave that does not go forward in a straight line, but instead runs around in circles. Furthermore, the wave is transverse simply because the field pointers of the $\vec{E}$-field and the $\vec{H}$-field are oscillating perpendicular to the direction of the speed of light c. So by virtue of the orbit, the speed of light c now does become the vortex velocity, as well as the wave; thus, the vortices turn out to be two possible and stable field configurations. For the transition from one into the other no energy is used; it

3.9 Infrastructure, Characteristics, Derivation, and Scalar Wave Properties

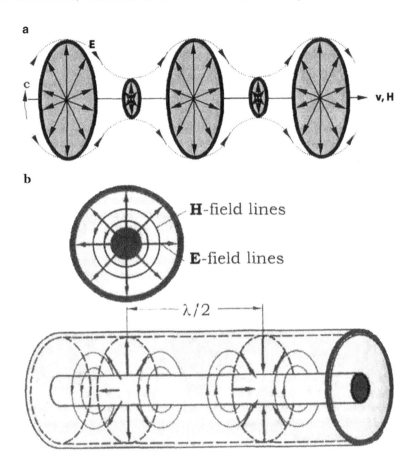

Fig. 3.30 (a) A magnetic wave (longitudinal). (b) Wave propagation in a coaxial cable

only is a question of structure. The vortex structure therefore stabilizes itself by means of the field dependency of the speed of light.

Under the circumstance that the vortex direction of the ring-like vortex is determined, and further that the field pointers are standing perpendicular to it, as well as perpendicular to each other; this results in two theoretical formations for the SW. In the first case (see Fig. 3.29b), the vector of the $\vec{H}$-field points in the direction of the vortex center and that of the $\vec{E}$-field points axially to the outside. The vortex, however, will propagate in this direction in space and appear as a SW, so the propagation of the wave takes place in the direction of the electric field. We call this an *electric wave*.

In the second case the field vectors switch their places. The direction of propagation this time coincides with the oscillating magnetic field pointer (see Fig. 3.30a), for which reason we speak of a *magnetic wave*. The vortex picture of the rolled-up wave already fits very well because the propagation of a wave direction of its field pointer characterizes a LW because all measurement results are perfectly covered by the vortex model. It even is clear that no energy has to be used for the conversion because merely the structure has changed. If it becomes a vortex, the wave just does not run in a straight line anymore but in circles to either wrap around the magnetic field vector (see Fig. 3.29b) or the electric-field vector (see Fig. 3.30a).

4. Double-Frequent Oscillation of Size

Because of the fact that a LW propagates in the direction of the field, the field pointer also will oscillate with the velocity of propagation v. This therefore is not constant at all; it can differ significantly from that of the light and can take arbitrary values. According to the theory of objectivity, the field oscillating with it determines its momentary size:

$$E, H \sim \frac{1}{v} \qquad (3.130)$$

The velocity of propagation v of the SW thus oscillates double-frequently and with an opposite phase to the corresponding field. A detailed description would be: if the field strives for its maximum value, velocity v of the wave reaches its lowest value.

In the field minimum the SW, and vice versa, accelerates to its maximum value. For LWs therefore, only an averaged velocity of propagation is given and measured because, for instance, this is usual for the sound wave and can vary significantly as is well known (i.e., body sound compared to air sound, etc.).

The two dual-field vectors of $\vec{E}$ and $\vec{H}$, one in the direction of propagation and one standing perpendicular to it, occur jointly. Both oscillate with the same frequency and both form the ring-like vortex in the respective direction. As a result, the ring-like vortex also oscillates in its diameter double-frequently and with the opposite phase to the corresponding field (see Figs. 3.29b and 3.30a).

This circumstance owes the ring-like vortex its property to tunnel. No Faraday cage is able to stop it, as can be demonstrated in experiments. Therefore, only the ground wave runs through the Earth and not along the curvature of it. A further example is the coaxial cable (see Fig. 3.30b). Also, this acts as a long tunnel and so it is not further astonishing that the electric field lines have the same orientation as for magnetic scalar wave. As a practical consequence in this place, there should be a warning about open cable ends, wave guides, or horn radiators with regard to uncontrolled emitted SWs!

Currently, we see a lot of discussion about this; if the cable network runs, there is a possibly of distribution and the impact of it on, for example, airline radio traffic. The original opening for cable frequencies that actually are reserved for airline radio traffic, based on the erroneous assumption that conflicts are unthinkable. Nonetheless, planes were disturbed in their communication. As the cause, TV cables were singled out because they had not been closed with a resistor, according to the rules; by all means, this can occur on building sites and during renovation work.

On the other hand, this is being argued about because of the small current that flows through the coaxial cable, and the great distance to the planes is cited too. According to that a Hertzian wave actually cannot be of concern. It presumably is a

3.9 Infrastructure, Characteristics, Derivation, and Scalar Wave Properties

SW that escapes from the open cable ends and that is collected by a receiver in the plane. There indeed is very little field energy, but this is of concern because it is being collected and bundled; the SW can exceed the part of the radio wave by great distances and can cause problems. For such examples from practice, the SW theory is taking effect entirely.

5. Electric and Magnetic Scalar Waves

Per our discussion of wave propagation so far, there are three possible and stable states as illustrated in Fig. 3.31 as follows:

The transverse EMW according to Heinrich Hertz (Fig. 3.31a).
The longitudinal electric wave according to Nikola Tesla (Fig. 3.31b).
A longitudinal magnetic wave, which is not connected yet with a discoverer's name (Fig. 3.31c),.

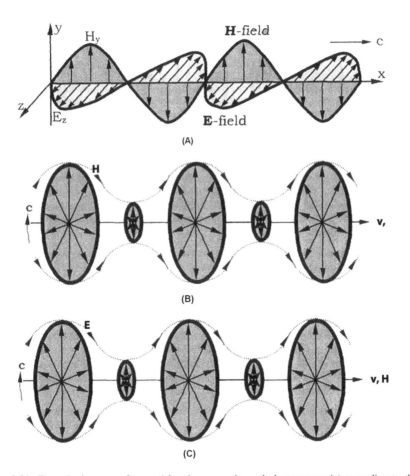

Fig. 3.31 Three basic types of wave (electric, magnetic, and electromagnetic) according to the wave equations, (**a**) H. Hertz: Electromagnetic wave (transverse). (**b**) Nikola Tesla: Electric wave (longitudinal). (**c**) Magnetic wave (longitudinal).

The last one is a pure product of this author's theoretical derivation. The question is asked: What practical meaning could the magnetic wave have?

The wave is formed by individual electric-field vortices, which was discovered in 1990 by Meyl [13] and are called *potential vortices*. He did proceed from the assumption that the youngest of the three waves will play by far the biggest role in the future because its properties are unattainable, both regarding technical energy and the information's technical use. One example for each should support this valuable research at the level of a doctorate thesis.

The experiments concerning the electric wave according to Nikola Tesla, where he was working with electrically charged spheres that do not show a particularly high power. Magnetic converters, per the experiences of this author's laboratory activities, are by far superior to an electrostatic converter as a collector for free energy. That even can be expected because a magnetic engine is much smaller than an electrostatic engine of the same power, as is well known.

At a medical congress a talk was given on the basic regulation of the cells and on the communication of the cells with each other. Professor Heine in his decades of research work has found that the cells, for the purpose of communication, build up channels (e.g., in the connective tissue); after having conducted research on the information again, it was found that they collapse. Interestingly, the channels have a hyperboloid structure for which no conclusive explanation exists.

The structure of the data channels, however, is identical to the one of a magnetic SW, as shown earlier in Fig. 3.31c. Through a channel formed, which functions like a tunnel or a dissimilarly formed waveguide, only one very particular SW can run through it. Waves with different frequencies or wavelengths do not fit through the hyperboloid-formed tunnel at all in the first place. Through that, the information transmission obtains an extremely high degree of safety from interference.

To the biologist here, a completely new view at the function of a cell and the basic regulation of the whole organism is beginning . The information tunnel temporarily forms, more or less a vacuum through which only potential vortices can behandled without any losses—simply perfect! From this example it is becoming clear that nature is working with SWs—namely, with magnetic waves.

One other point should be noted: The mentioned tunnel experiments, in which speeds faster than light are being measured with various devices, impressively confirm the presence of SWs. But if they exist at faster than light and othersare slower, then it is almost obvious that such SWs exist that will propagate exactly with the speed of light. These then will have all the properties of the light and will not differ from the corresponding EMW in the observable result. A SW is, however, formed by vortex configurations that unambiguously have a particle nature. Nothing would be more obvious than to equate these quantum structures with the photons.

6. Scalar-Wave Properties

Famous physicists and scientists of high caliber came together, and they were concerned about the question of whether the light quanta hypothesis is a wave, or a particle, or even has both properties simultaneously. For both variants experimental proof was presented, so the discussion became agitated and things boiled over.

3.9 Infrastructure, Characteristics, Derivation, and Scalar Wave Properties 235

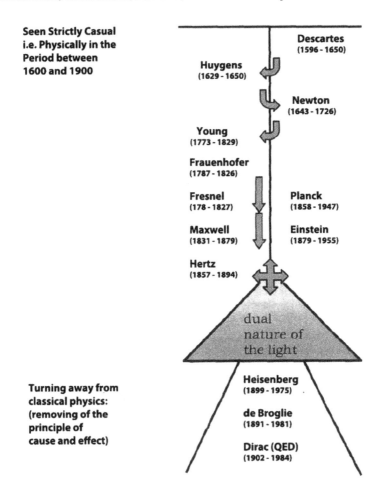

Fig. 3.32 The view of some physicists concerning the nature of the light as a wave or as a particle

Finally, they were as smart as before when *Werner Heisenberg* presented his ideas concerning the uncertainty principle.

This compromise, on which they eventually came to an agreement with good cause, may come to be recognized as the worst in the history of physics. It dictates to this author what I shall see, and exactly how we may look at waves. With it the contradiction should be overcome—that is, the light contrary to every connection should be a wave and a particle at the same time (Fig. 3.32). Such fixings not only have a funny but also a tragic side. Because it was authorities who approved the compromise and the whole community of science had confidence in the statements of them, they immediately and unfiltered were incorporated into all textbooks.

At the meeting, which simply and solely concerned the wave equation, they could have supplied the correct and only possible answer: It falls into two parts and this explains why light at one time appears as an EMW and the next time as a vortex particle, which is called a *photon*. The conversion can take place at any time spontaneously and without putting on energy, so it depends on the measuring technique used whether the particle appears as a wave or as a particle; however, of course, never as both at the same time!

Looking back, one can say that the funny thing about the situation was that all discussed the wave and its properties known at that time, and that all should have known the wave equation. An equation, as is well known, contains more than a thousand words, and one look would have sufficed entirely to answer the controversial question once and for all. It would have saved us a lot of further effort.

Concerning the measurement of light, the uncertainty principle with the interpretation of Heisenberg—light is a wave and a particle at the same time—is incompatible with the wave equation. Heisenberg puts an equal sign where in the wave equation, in reality,an addition sign is present for both wave parts. Fortunately, in mathematics there is no need of speculating; there a derivation is either right or wrong. Nothing has changed even if all physicists of the world should go in the wrong direction following the prevailing opinion. The wave equation exerts influence on the interpretation of the light experiments, on the one hand—that is, the ones concerning the interference and refraction of the light, where EMWs are becoming visible (Fig. 3.33) and on the other hand the photo-electric effect, as proof of light quanta (Fig. 3.34).

Already the wave theory of Huygens requires interference patterns of light rays because, for instance, they are observed behind a slit and demonstrate the wave nature with that. If on that occasion the particle nature is lost, the photons present before the slit cannot be detected behind the slit anymore, then plain and simple the measuring method; thus, the slit is to blame for that. The vortices have unrolled themselves to waves at the slit.

Corresponding experiments also have been carried out with matter. At the Massachusetts Institute of Technology whole sodium atoms were converted into waves. At the detector pure interference patterns were observed, which function as evidence for successful dematerialization. But, to the vortex physicist, they show

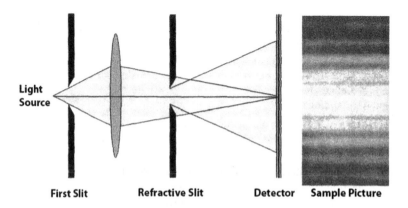

Fig. 3.33 Depiction of light forms' interreference at the slit

3.9 Infrastructure, Characteristics, Derivation, and Scalar Wave Properties 237

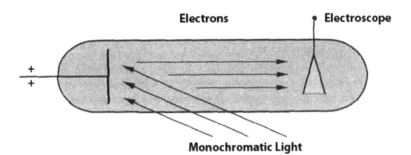

Fig. 3.34 The photo-electric effect

still more; they reveal that atoms are merely waves rolled up to spherical vortices that, at any time and spontaneously, can again unroll to waves at a lattice.

In Fig. 3.33 light strips are formed, where the waves oscillate in phases (i.e., dark stripes) and out of phase (see Fig. 3.34). The common interpretation of the wave's detectable nature behind a slit must have been present in the same form before the slit; this is untenable and in the end wrong, as the experiment with the sodium atoms makes clear.

The photoelectric effect, which on the other hand shows the quantum nature of the light, was discovered by *Heinrich Hertz*, further investigated by Lenard, and finally rendered more precisely by *Robert Millikan* in 1916 (see Fig. 3.34b). It is based on the circumstance that light of higher frequency (i.e., blue light) has more energy than red light of lower frequency. Nonetheless, if electrons are knocked out of a metal plate by light, then what occurs is that the waves roll up into vortices. Now, indeed, photons are at work and can be detected indirectly with an electroscope.

In the same way a photon ray in a bubble chamber can be photographed. But also, here the measuring method is responsible for what is being observed. A good example is the human eye, the rods and cones of which merely can pick up potential vortices and pass them on to the nerves as so-called *reaction potentials*. Incident waves can be detected only if they first have rolled up into vortices in the corpus vitreum of the eye, which is a transparent jelly-like substance filling the interior of the eyeball behind the lens; it is composed of a delicate network (vitreous stroma) enclosing its meshes, a watery fluid (vitreous humor).. For us to see it does not play a role in whatpercent of vortices and waves the light contains.

Behind a sheet of glass, for instance, a larger vortex part can be expected and still the light has the same brightness as without the sheet; the glass is perceived as transparent. We nevertheless must assume that light with a large wave part has another quality, then such light behind glass or artificial light has a large share of photons.

7. Comparison of the Parts of Tesla and Hertz

A rule that has been established for light from a quantum physics perspective is that it always is formed as a photon, even at the surface of the Sun. If at the end the Sun's radiation waves arrive on Earth, then the vortices sometimes on the way to us from such wave radiation must have unrolled to waves.

Photon radiation after all is SW radiation that generally is predominant in the near-field of the source of radiation. There is no reason why the light should act in an alternate way than the wave radiated by a radio transmitter, which also form vortices in the near-field area, as we already have discussed. For different interpretations of wave properties of one and the same physical phenomenon there is no place in a unified theory.

If we want to take under consideration the comparison of the parts of Tesla and Hertz, as indicated in Table 3.1, then it is not an individual case that an experimental setup is responsible for what is being measured and observed. A parallel case to the experiments concerning the nature of the light is the one that concerns wave propagation. Hertz received and used the transverse part, and Tesla the longitudinal part; they both claimed only he is right. There does not exist another equation, which has been and is being ignored and misunderstood so thoroughly, as the wave equation.

Table 3.1 shows a survey of the two parts of the wave equation in the assignment of terms and forms, where the RHS is the EMW description according to Heinrich Hertz, and the LHS is the SW described by Nikola Tesla. The terms, on one hand, are TWs and, on the other hand, LWs related to the kind of wave propagation.

If the field pointers oscillate crossways to the direction of propagation, as a consequence the velocity of propagation is decoupled from the oscillating fields. The result in all cases is the speed of light, and from what we know is a constant value. It is usual to make a list for increasing frequency, starting at the longest waves—that is, *extremely low frequency* (ELF) and *very low frequency (VLF)*)— over the radio waves—that is,*long frequency (LW), medium wave (MW), short wave (SW), ultra high frequency (UHF)*, TV channels (VHF, UHF), microwaves, IR radiation, light, and X-rays, up to cosmic radiation.

Table 3.1 The Two Parts of the Wave Equation

$$\nabla^2 \vec{E} = \nabla(\vec{\nabla} \cdot \vec{E}) - \vec{\nabla} \times \vec{\nabla} \times \vec{E} = \frac{1}{c^2}\left(\frac{\partial^2 \vec{E}}{\partial t^2}\right) \quad \text{Eq. 3.121}$$

Nikola Tesla	Heinrich Hertz
Scalar wave = (electric or magnetic)	Electromagnetic wave
Longitudinal wave	**Transverse wave**
Form (each time for velocity of propagation v):	Form (each time for frequency):
($v > c$): Neutrino radiation, morphogenetic fields, ...	Cosmic radiation
($v = c$): Photons	X-rays
($v < c$): Plasma wave, thermal	Light
Vortices, biophotons, Earth's radiation	UV radiation
($v = 0$): Noise	Microwave
	Radio waves
	VLF, ULF, ...

It really is interesting that it concerns one and the same phenomenon despite the different forms! As long as Maxwell only had published a theory for light, in the world of science for 24 years nothing at all happened. Only Heinrich Hertz with his short-wave experiments opened everyone's eyes. Now, all suddenly started at the same time to research various phenomena on the frequency scale, from Madame Curie to Konrad Rontgen up to Nikola Tesla, who primarily researched the area of long waves.

With regard to the SW so far there are not many corresponding and collective documents or textbooks one can identify, except a reference, such as Meyl [13], or a few papers here and there that are published by various scientists and physicists. The immense area is new ground scientifically that is awaiting to be explored systematically. Thus, what we need to try is to make a contribution by rebuilding a SW transmission line according to the plans that were proposed by Tesla.

8. Noise, a Scalar-Wave Phenomenon

Longitudinal waves can take arbitrary velocities between zero and infinity because they propagate in the direction of an oscillating field pointer and as a consequence their velocity of propagation oscillates as well and by no means is constant. It does make sense to list the forms of SWs according to their respective velocity of propagation (see Table 3.1, *left column*).

If we start with a localized vortex, a wave rolling up is contracting further. Doing so the wavelength gets smaller and smaller, whereas the frequency increases. An even frequency mixture distributed over a broad frequency band is observed. This phenomenon is called *noise* (Fig. 3.35). But besides localized noise, noise vortices can be on the way with a certain velocity as a SW (e.g., for radio noise). In this case they show the typical properties of a *standing wave* with nodes and anti-nodes.

Also, the Earth's radiation is said to have a standing wave nature, which can be interpreted as slowed down neutrino radiation. If it is slowed down on the way through the Earth, then the neutrino properties are changing; this was measured in the Kamiokande detector in Japan recently. Unfortunately, the proof occurs only indirectly because there still does not exist measuring devices for SWs. We will discuss this problem area later and are content with the clue that already is within living memory—the standing wave property has been used to find water and deposits of ore and still is used today (Fig. 3.36).

Let us continue our considerations concerning the forms of SWs, as they are listed in Table 3.1. Scalar waves, which are slower on the way than light, are joined by plasma waves. This is confirmed by measurements and calculations. For thermal vortices, as they have been investigated by Max Planck, and for biophotons, as they have been detected in living cells by colleague Popp, the velocity of propagation is unknown however. It still has not been measured at all and doing so is needed now more than ever. The research scientists have confidence in the assumption that all waves go with the speed of light, but that is a big mistake.

For all wave types there is at least one vortex variant (e.g., for radio waves); it is radio noise that propagates with a velocity different from c. The velocity is the product of frequency and wavelength:

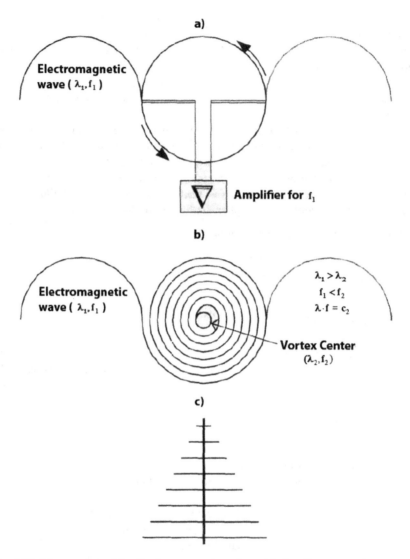

Fig. 3.35 Measurement of localized waves and vortices. (**a**) Standing wave. (**b**) Localized vortex = noise. (**c**) Broadband antenna for EMC-mesurements.

$$v = f \cdot \lambda \qquad (3.131)$$

From the three variables v, f, and λ at least two must be measured (Eq. 3.131), if one has a suspicion that it could concern SWs. At this point most errors are made in the laboratories.

Countless experiments concerning biological compatibility, concerning medical therapy methods, and similar experiments must be repeated because as a rule only the frequency is being measured; it has been omitted so as to at least check the

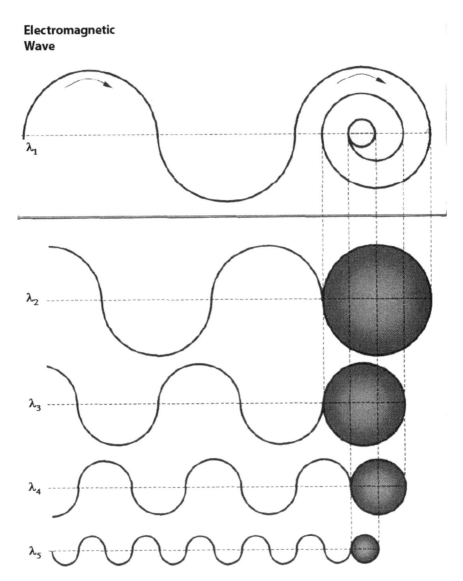

Fig. 3.36 The wave rolling up to a vortex

wavelength or the velocity of the wave. Countless research scientists must put up with this accusation. Much too blind scientists, who now again may start from the very beginning with their work, have had confidence in the predominance of the speed of light.

9. Neutrino Radiation

From a historical point of view, Tesla was the first person to discover *neutrino radiation* and in an issue of the *New York Times* (Feb. 6, 1932, p. 16, Col. 8), he wrote that he discovered and investigated the phenomenon of the cosmic radiation long before others started their research. He claims and says:

According to my theory a radioactive body is only a target, which constantly is being bombarded by infinitely small balls (neutrinos), which are projected from all parts of the Universe. If this, at present unknown, cosmic radiation could be interrupted completely, then no radioactivity would exist any longer. I made some progress regarding the solution of the mystery, until I in the year 1898 attained mathematical and experimental evidence, that the Sun and similar celestial bodies emit energy-rich radiation, which consist[s] of inconceivable small particles and have velocities, which are considerabl[y] faster than the speed of light. The ability of penetration of this radiation is so large, that it penetrates thousands of kilometers of solid matter, without their velocity being reduced noticeably.

It must be respected how Tesla, guided by experimental observations and a reliable instinct, came to the correct result. With the conclusion, because of the missing interaction the neutrinos have, to be inconceivably small; he was not quite right. Their size rather depends on the velocity because the high-speed neutrinos are being length-contracted significantly. Tesla, however, hit the nail exactly on the head if he, on the occasion of the press conference for his 81st birthday, declares that the radioactivity is clear proof of the existence of an outer radiation of cosmic origin. "If Radium could be shielded against this radiation in an effective way.", Tesla wrote in an essay of 1934: "...then it would not be radioactive anymore." At this occasion he contradicted Albert Einstein, without speaking his name and was indignant about the incorrect working methods of the scientists. As part of neutrino radiation, however, the neutrino scientists made the error of proceeding from the assumption that their particles were on the way with a speed somewhat less than the speed of light c.

This contradicts the observation according to which black holes should represent strong sources of neutrinos, which are black only for the reason that no particle radiation is able to escape them, which is on the way with c or even slower. If a black hole does hurl neutrino radiation into space, then that must be considerably faster than c, as normal neutrino scientists still by no means can imagine today [13].

Nevertheless, neutrino radiation only can be detected after it has been slowed down to a value that is smaller than c. If the slowing down occurs slightly asymmetrical, then a mean of the mass different from zero appears. The "measurement" of such a rest mass, as it is presently propagated and celebrated, is a classical measurement error! As long as a neutrino is on the way to us it still is faster than the light; the mean of its mass is generally zero. The effective value of the mass of a neutrino, however, is considerable. It only is able to give an account for the sought-for dark matter, as far as it must exist in today's supposed form anyway (Fig. 3.37) [13].

The radiation that the discoverer, Nikola Tesla, already had found in his own experiments , is faster than the speed of light as it was described at the beginning of this section. According to the description because this Tesla radiation is identical with neutrino radiation, it forms a subset. We will call the neutrino radiation in this form *scalar waves*, which are faster than the speed of light c. This extends from the weak radiation at low frequencies up to the hard neutrino radiation of cosmic origin.

3.9 Infrastructure, Characteristics, Derivation, and Scalar Wave Properties   243

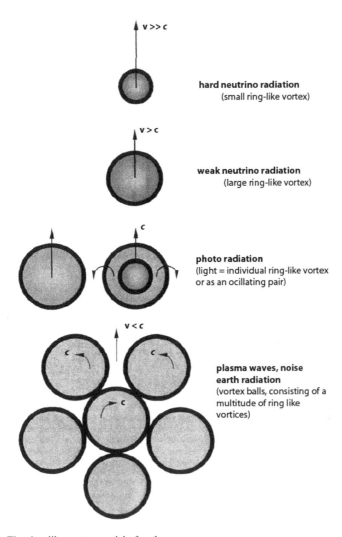

Fig. 3.37 The ring-like vortex model of scalar waves

Nonetheless, the hardness of the radiation does not only increase with the frequency; in particular, it increases with the velocity.

The neutrino radiation first of all is carrying energy. On top of this basic wave radiation information can be modulated. Doing so is an extremely complex modulation of variants that are offered. Of this kind we must envision them as being complex modulated vortices that can propagate as SWs in space. Rupert Sheldrake calls this vortex field a *morphogenetic field*. At this point he is merely indicating his very interesting research results.

Thoughts can be standing in space in the form of localized noise, but they also can move with speeds faster than light. According to that, a communication with

Fig. 3.38 The electron–neutrino as a ring-like vortex

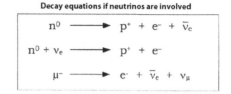

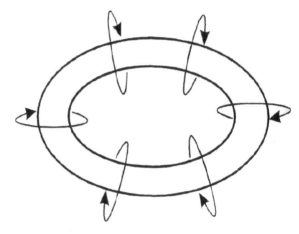

Fig. 3.39 The photon as an oscillating electron–positron pair

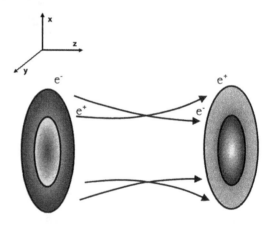

intelligent beings from other star systems, by all means, would not be a Utopia anymore. Every fast neutrino forms an individual ring-like vortex (Fig. 3.38).

The slower the SW is, the more dependent the vortices become. The photon already can consist of two ring-like vortices (Fig. 3.39), whereas plasma waves and other slow SWs can form from a multitude of individual vortices that are rotating around each other to form vortex balls and vortex streets. From this circumstance there already are very different SW behavior results in various areas of the velocity of propagation. This trend for small velocities can be observed toward lower

3.9 Infrastructure, Characteristics, Derivation, and Scalar Wave Properties 245

frequencies as well. For a certain wavelength the frequency, after all (according to Eq. 3.131), is proportional to the velocity of propagation v.

Keep in mind that the vortex principle is self-similar [21]. This means that the properties of an individual vortex for the collection of numerous vortices can appear again and can be observed in a similar manner too. That is why a vortex ball behaves entirely similar to an individual isolated vortex. The same concentration effect, which keeps the vortex together, shows its effect for the vortex ball and keeps it together as well.

Something consistent holds for a basic property of potential vortices being of a completely different nature. It is the property to bind matter in the vortex and carry it away with the vortex. The vortex rings that skillful cigarette smokers can blow in the air are well known. Of course, non-smokers also can produce these air eddies with their mouth, but they remain invisible. Solely by the property of the vortex ring to bind the smoke it becomes visible to the human eye.

If the potential vortex transports something out, then it should be a dielectric material such as water preferably. Therefore, if in the environmental air we are surrounded by potential vortices that we can detect (e.g., as noise), then they are capable, with their "phenomenon of transport," to pick up water and to keep it in the vortex. In this way the atmospheric humidity is explicable as the ability of the air particles to bind comparatively heavy water molecules. If the vortex falls apart, then it inevitably releases the water particles and it rains. This is merely a charming alternative for the classical representation without a claim to completeness [13].

10. **Parallel Instead of Serial Image Transmission**

We continue with our considerations concerning the special properties of SWs, represented in Table 3.1's left column, and compare these with the well-known behavior of EMWs in the right column; this is also found in Table 3.1 and later in Fig. 3.41. If we take up the possibilities for modulation and the transmission of information again, it becomes very clear from the comparison that we work with a technology, which we, more or less, can master but that is everything else but optimum.

For the Hertzian wave the velocity of propagation is constant and, meanwhile, the frequency of the wavelength is being modulated at the same time; however that limits information transmission significantly. An image, for instance, must be transmitted serially from one point to another, as well as line after line. Serial image transmission takes place very slowly, for this reason the velocity of personal computers (PCs) must be increased permanently so that the quantity of data can be managed accordingly.

With clock frequency, on the other hand, losses increase also so that in the end the CPU-cooler limits the efficiency of modern PCs. Something engineers obviously do incorrectly compared to the human brain. The brain works without a fan. For it, a clock frequency of 10 Hz is sufficient. It needs neither megahertz (MHz) nor gigahertz (GHz) frequencies and despite that is considerably more efficient.

Nature works only with the best technology. The second-best technology, as has been noted, to use in our machines in the evolution would not have had the slightest chance of surviving. The strategies to optimize nature are merciless. In a free economy that is completely different. There the "bunglers" are joining together in companies to dominate the market, buying up the innovative ideas without further ado, to let them disappear in the drawer so that they can bungle along in the way they have done until now. It after all has been the lousy products that have made them into the companies they are today. The ego of power is incompatible with the interests of nature [13].

Nature works with SWs and their velocity of propagation is arbitrary. Wavelength and frequency now can be modulated, and information can be recorded separately. In this manner a whole dimension is gained to modulate; image transmission can take place in parallel, which means it is considerably faster, safer, and more reliable. As anyone knows from our own experience, assembling the image takes place all at once; the memory of past images takes place ad hoc. Nature is indescribably more efficient than technology with the SW technique (Fig. 3.40).

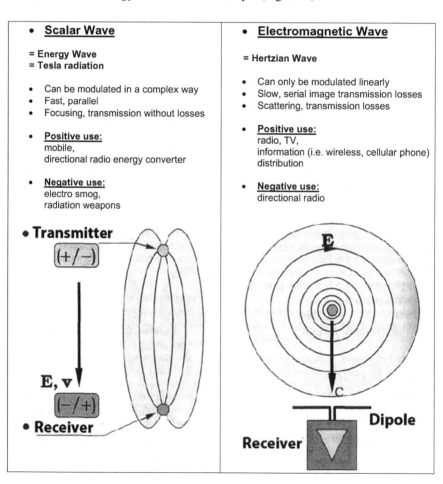

Fig. 3.40 Comparison of radio waves according to Hertz and electric scalar waves according to Tesla

3.9 Infrastructure, Characteristics, Derivation, and Scalar Wave Properties   247

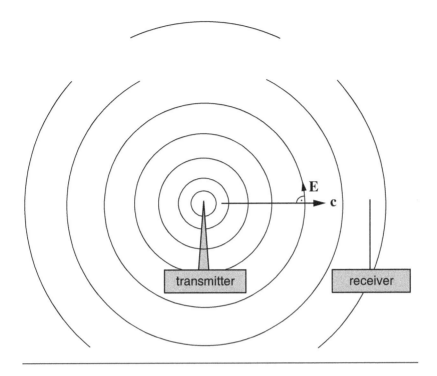

Tesla radiation (radiations) = scalar wave, longitudinal wave propagation:

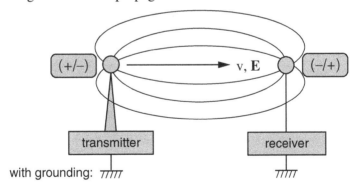

Fig. 3.41 Comparison of scalar waves and radio waves properties

We again take the RHS of Fig. 3.41, with the properties of the Hertzian wave. In the opinion of Nikola Tesla, it is a gigantic waste of energy. The broadcasting power is scattered in all directions and the transmission losses are enormous. At the receiver virtually no power arrives anymore. To receive a radio signal the antenna power has to be amplified significantly. It is a wave thatcan be used only as a radio wave, thus as a wave with which arbitrarily many participants should be reached [13].

11. Research on Scalar Waves

Scalar waves are still an unexplored area, scientific new ground as it were. Individual research scientists already have selectively ventured forward in this area and have described properties of the SW investigated by them in their special research area, mostly in measurement of technical matters. As a rule of thumb, however, they lack the physical relations, as derived in this book for the first time. If we do not proceed from individual measurements, but instead from the wave equations and the mathematical physical derivation of SWs, we have a great chance to understand something as belonging together. On the one hand, is noise, photons, neutrinos. and lots of other known phenomena as well as; on the other hand, there are still unknown phenomena, which are called para-scientific.

We should remember that we, without the theory of Maxwell and the representation in a frequency band, today we still would not know that radio waves (i.e., LW, MW, KW, UHF), microwaves (μW), IR thermal radiation, light, and X-rays relate to the same phenomenon. The graphic representation of both waves in one diagram at this point is extremely helpful.

3.9.2 Wave Energy

In this section we describe the fields or, actually, the space itself that contains energy, and as the fields propagate through the space, energy also propagates through space. The space remains stationary, just like the water in the pan, as we described the situation before; however, energy can travel through space. This is the same as waves in any material where the material itself does not move with waves, but the energy is transmitted to a new location. is Thisvery similar to the situation where we tap the stationary water in the pan at the center of its surface and see waves created that travel in all directions through the interior of the pan. The derivation that follows gives us the exact quantities involved in this process.

So because the fields contain energy, this energy is transported when the waves propagate. It will be useful to derive an expression for this process. First, multiply Eq. 3.98 by $-\vec{B}$ to obtain:

$$-\vec{B} \cdot \vec{\nabla} \times \vec{E} = \vec{B} \cdot \left(\frac{\partial \vec{B}}{\partial t}\right) \tag{3.132}$$

Then, multiply Eq. 3.132 by $1/\mu$ and bring the magnetic field into a time derivative using a product rule; we obtain the following result:

$$-\frac{1}{\mu}\left(\vec{B} \cdot \vec{\nabla} \times \vec{E}\right) = \frac{1}{2\mu}\left(\frac{\partial}{\partial t}B^2\right) \tag{3.133}$$

3.9 Infrastructure, Characteristics, Derivation, and Scalar Wave Properties

Notice that the RHS of Eq. 3.133 is now representing the time derivative of the energy density of the magnetic field.

Now we multiply Eq. 3.100 by $\vec{E}$ so that we obtain the following result:

$$\vec{E} \cdot \vec{\nabla} \times \vec{B} + \vec{E} \cdot \vec{\nabla}\beta = \frac{1}{c^2} \vec{E} \cdot \frac{\partial \vec{E}}{\partial t} \qquad (3.134)$$

Multiply Eq. 3.134 by $1/\mu$ and bring the electric field into the time derivative using a product rule; it yields the following:

$$\frac{1}{\mu}\vec{E} \cdot \vec{\nabla} \times \vec{B} + \frac{1}{\mu}\vec{E} \cdot \vec{\nabla}\beta = \frac{\varepsilon}{2}\left(\frac{\partial}{\partial t}E^2\right) \qquad (3.135)$$

This equation gives the time change in the electric field energy density as well.

We then multiply Eq. 3.97 by β to get:

$$\beta(\vec{\nabla} \cdot \vec{E}) = \beta\left(\frac{\partial}{\partial t}\beta\right) \qquad (3.136)$$

Now multiply Eq. 3.136 by $1/\mu$ to obtain the following form:

$$\frac{1}{\mu}\beta(\vec{\nabla} \cdot \vec{E}) = \frac{1}{2\mu}\left(\frac{\partial}{\partial t}\beta^2\right) \qquad (3.137)$$

Now, we can add all the energy densities together to get this conclusion:

$$\frac{\partial}{\partial t}\left(\frac{\varepsilon}{2}E^2 + \frac{1}{2\mu}B^2 + \frac{1}{2\mu}\beta^2\right) \\ = \frac{1}{\mu}\left\{(\vec{E} \cdot \vec{\nabla} \times \vec{B} - \vec{B} \cdot \vec{\nabla} \times \vec{E}) + \vec{E} \cdot \vec{\nabla}\beta + \beta\vec{\nabla} \cdot \vec{E}\right\} \qquad (3.138)$$

Using vector quantity $\vec{\nabla} \cdot (\varphi\vec{F}) = (\vec{\nabla}\varphi) \cdot \vec{F} + \varphi\vec{\nabla} \cdot \vec{F}$ from Table 1.1 on the third term on the LHS of Eq. 3.138 yields:

$$\frac{\partial}{\partial t}\left(\frac{\varepsilon}{2}E^2 + \frac{1}{2\mu}B^2 + \frac{1}{2\mu}\beta^2\right) \\ = \frac{1}{\mu}\left\{(\vec{E} \cdot \vec{\nabla} \times \vec{B} - \vec{B} \cdot \vec{\nabla} \times \vec{E}) + \vec{\nabla} \cdot (\beta\vec{E})\right\} \qquad (3.139)$$

Using vector quantity $\vec{\nabla} \cdot (\vec{F} \times \vec{G}) = \vec{G} \cdot (\vec{\nabla} \times \vec{F}) - \vec{F} \cdot (\vec{\nabla} \times \vec{G})$ from Table 1.1 on the first term and second term on the LHS of Eq. 3.139 yields the following result:

$$\frac{\partial}{\partial t}\left(\frac{\varepsilon}{2}E^2 + \frac{1}{2\mu}B^2 + \frac{1}{2\mu}\beta^2\right) = \frac{1}{\mu}\vec{\nabla} \cdot (\vec{B} \times \vec{E} + \beta\vec{E}) \qquad (3.140)$$

Fig. 3.42 Energy flow between two regions

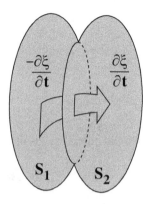

Then, integrating both sides over a volume, we have:

$$\frac{\partial}{\partial t}\left(\frac{\varepsilon}{2}\int E^2 d\tau + \frac{1}{2\mu}\int B^2 d\tau + \frac{1}{2\mu}\int \beta^2 d\tau\right) \\ = \frac{1}{\mu}\int \vec{\nabla}\cdot(\vec{B}\times\vec{E} + \beta\vec{E})d\tau \quad (3.141)$$

Finally, using the divergence theorem on the RHS and using Eqs. 3.101 through 3.103, results in:

$$\frac{\partial}{\partial t}(\xi_E + \xi_B + \xi_\beta) = \frac{1}{\mu}\oint_S (\vec{B}\times\vec{E} + \beta\vec{E})\cdot d\vec{s} \quad (3.142)$$

This equation represents what we call the *energy continuity equation*. It shows that the fields as they are on a specific closed surface indicate how the energy in the surrounding volume is changing. This is a statement that energy is conserved as the fields propagate because, for example, as the energy decreases in the enclosed region, the energy "flows" through the surface to the exterior region with energy that will increase by the same amount as the decrease of energy in the enclosed region.

Imagine two regions, as illustrated in Fig. 3.42, with two surfaces S_1 and S_2, where the two surfaces intersect giving them a mutual area. If the energy flow is confined to this area, the RHS of Eq. 3.142 is the same magnitude for both regions, but the sign is reversed. Therefore, the change in energy is reversed and the sum of the energy in both regions remains constant. Notice that if $\beta = 0$, the energy of the scalar field will be zero and Eq. 3.142 reverts to a *Poynting's vector* or theorem as we presented in Eq. 2.86.

3.9.3 Particles or Charge Field Expressions

If we start with an electric field of a specific form as an initial condition, it will not propagate at all and will just sit there at absolute rest. This is analogous to the

3.9 Infrastructure, Characteristics, Derivation, and Scalar Wave Properties

example of water inside the pan, where one could be touching the surface, and instead of generating a wave, an indentation appears and just sits there without moving even after taking the finger away. In reality, because of the wave properties of water, this is not actually possible; however, the wave properties of space are different because of the vector nature of the fields, which allows such a possibility. This is not just an imaginary suggestion; it is a proof that comes directly from the equations that define the wave behavior of empty space.

Therefore, it is possible for an electric field to exist that, being governed by these equations, would not propagate. The answer to such a postulate is "yes." All we need to do is set the time derivative of the wave in Eq. 3.108 to zero. Thus, the equation reduces to the Laplacian form of the electric field:

$$\nabla^2 \vec{E} = 0 \tag{3.143}$$

The electric field that satisfies Eq. 3.143 will not propagate as one can see, but remains at absolute rest.

To begin we will not solve it in complete generality. We will find just a solution as we did in Chap. 1 of this book. Again, we can try with this approach if there exists a solution when only a radial electric field occurs. Using the spherical coordinate from Appendix E and setting $E_\theta = 0$ and $E_\phi = 0$ yields:

$$\frac{\partial^2 E_r}{\partial r^2} + \frac{2}{r}\frac{\partial E_r}{\partial r} + \frac{1}{r^2}\frac{\partial^2 E_r}{\partial \theta^2} + \frac{\cot\theta}{r^2}\frac{\partial E_r}{\partial \theta} + \frac{1}{r^2 \sin^2\theta}\frac{\partial^2 E_r}{\partial \phi^2} - \frac{2}{r^2}E_r = 0 \tag{3.144}$$

$$\frac{2}{r^2}\frac{\partial E_r}{\partial \theta} = 0 \tag{3.145}$$

$$\frac{2}{r^2 \sin\theta}\frac{\partial E_r}{\partial \phi} = 0 \tag{3.146}$$

Substituting Eqs. 3.146 and 3.145 into Eq. 3.144 will yield:

$$\frac{\partial^2 E_r}{\partial r^2} + \frac{2}{r}\frac{\partial E_r}{\partial r} - \frac{2}{r^2}E_r = 0 \tag{3.147}$$

The result in this equation is a second-order differential equation that is dependent only on variable r; thus, if we multiply both sides of this equation by variable r^2, we obtain a new form of Eq. 3.147:

$$r^2\frac{\partial^2 E_r}{\partial r^2} + 2r\frac{\partial E_r}{\partial r} - 2E_r = 0 \tag{3.148}$$

This equation is an *Euler equation* that can be solved by assuming a solution that is some power of r; thus, we can write:

$$\begin{cases} E_r = r^a \\ \dfrac{\partial E_r}{\partial r} = ar^{a-1} \\ \dfrac{\partial^2 E_r}{\partial r^2} = a(a-1)r^{a-2} \end{cases} \quad (3.149)$$

Substituting Eq. 3.149 into Eq. 3.148 for each appropriate term, we obtain the following result that is the auxiliary form of Eq. 3.148:

$$a(a-1)r^a + 2ar^a - 2r^a = 0 \quad (3.150)$$

We can divide Eq. 3.150 by r^a, which then provides us with the characteristic form of the equation as a quadric one:

$$a(a-1) + 2a - 2 = 0 \quad (3.151)$$

Solving for a will provide two different answers as should be expected; thus, we have:

$$\begin{cases} a = 1 \\ a = -2 \end{cases} \quad (3.152)$$

Equation 3.150 will give us two different power-type solutions to Eq. 3.148 to form the solution; thus, it will be:

$$E_r = k_1 r + \frac{k_2}{r^2} \quad (3.153)$$

As we can observe, Eq. 3.153, has two terms; the first is invalid at infinity, and the second is invalid at the center of the radial field. This suggests that we should use the two different solutions in the dissimilar region of space that, although being piecewise continuous, still covers the whole space. We will designate an *internal field* (i.e., E_i) and an *external field* (i.e., E_e) such that:

$$\vec{E}_e = \frac{k\hat{r}}{r^2} \quad (3.154)$$

and

$$E_i = \frac{kr\hat{r}}{r_0^3} \quad (3.155)$$

Thus, which solution is used is dependent on whether we are considering the region within r_0 or outside of r_0. We have set it up so that at the boundary the fields match, and they then both relate to the same constant of integration k. The field along a line bisecting the center then would have an intensity as shown in Fig. 3.43.

3.9 Infrastructure, Characteristics, Derivation, and Scalar Wave Properties

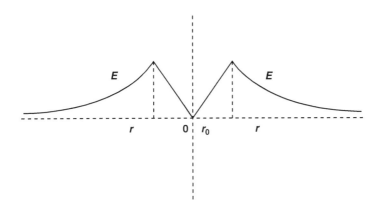

Fig. 3.43 Magnitude of a charge field

From now on we will call this stationary field a *charge field*; as it is clearly described in the following two sections. It is interesting that, although typically electric fields will propagate at the speed of light and spread out in all directions, when the electric field is of a certain form it does not behave this way. In this case, it does not propagate at all.

As we continue, we will find out other distinct and interesting electric fields that propagate in special ways. To designate these cases in general we will use the term *constrained propagation*.

3.9.4 Particle Energy

Thus far we have seen that by Maxwell's equations that there is energy in the space itself. If this is not the scenario, then the waves could not transmit through the so-called void of space. As modern physics indicates, there is mass involved in energy. One can use this postulate to determine what is sometimes called the *classical radius* of the electron. By calculating the energy in the charge field as we discussed, which would be similar to the field created by an electron, one can calculate a mass.

Because the mass of the electron is known, as we sum up energy by stating at an infinite distance and approaching its center, we come to a distance where the energy in the electric field matches the mass of the electron. It is known, as you sum up the energy, stating that at an infinite distance and approaching its center, you come to a distance where the energy in the electric field matches the mass of the electron; then we are done at this stage because all the energy is accounted for. If there were something else inside this radius, the energy would no longer balance, which actually is an argument against the existence of a material body in space. We also, in the previous section, demonstrated that a physical particle cannot exist at all, and it is important to point out that the idea that all particles must be wave pulses is *not* a postulate of this theory. We know that these postulates are simply classical electromagnetism points of view.

Thus, the charge field has energy as describe by Eq. 3.101, assuming the external field as:

$$E_e = \frac{q_e}{4\pi\varepsilon r^2} \qquad (3.156)$$

By substituting Eq. 3.156 into Eq. 3.101, we have:

$$\xi_e = \frac{\varepsilon}{2}\int \left(\frac{q_e}{4\pi\varepsilon r^2}\right)^2 d\tau \qquad (3.157)$$

Performing the integration over Eq. 3.157, we obtain:

$$\begin{aligned}\xi_e &= \frac{\varepsilon}{2}\int_{r_0}^{\infty}\left(\frac{q_e}{4\pi\varepsilon r^2}\right)^2 4\pi r^2 dr \\ \xi_e &= \frac{q_e^2}{8\pi\varepsilon}\int_{r_0}^{\infty}\frac{dr}{r^2} \\ \xi_e &= \frac{q_e^2 - 1}{8\pi\varepsilon\ r}\bigg|_{r=r_0}^{\infty} \\ \xi_e &= \frac{q_e^2}{8\pi\varepsilon r_0}\end{aligned} \qquad (3.158)$$

The final result of Eq. 3.158 is the energy of the external field of the particle. As mentioned, we easily can find that this energy represents a mass. If we have the mass of the particle, we can determine the radius of this particle in the final step of Eq. 3.158. This imposes an interesting problem, however, for if we consider some particle (e.g., an electron), all the mass would be accounted for by the time the summation reaches this radius, as shown in Fig. 3.44.

This situation might tell us there could be nothing inside the radius or the energy could no longer balance. This typically is called the *classical radius* of the electron and is well known because it appears in textbooks such as Jackson [22]; on pages 681 and 790 he refers to it as classical *diameter* versus radius. This supports the

Fig. 3.44 Energy in the electric field of a particle

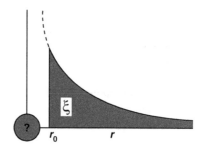

3.9 Infrastructure, Characteristics, Derivation, and Scalar Wave Properties

concept of *stationary field theory*, where everything is based on the field with no separate object existing in the space. More details on this discussion can be found in the book by Ensle [19]; readers should refer to it if there is any interest on their part.

3.9.5 Velocity

So, for what we have done one may notice that our charge field is at absolute rest; of course, charges are not all stationary and sitting around at absolute rest. They are moving at various velocities and none of these velocities is the speed of light, being the speed that the wave properties of space tell us they should be going. Furthermore, typically EMWs spread out. They do not move like a wave pulse in a single direction. Thus, we come to perhaps the most significant derivation that Ensle's book [19] shows by driving the *particle equation*, which when solved tells us what form of electricity is needed. Details of such a derivation is beyond the discussion here; we just write the final answers and refer readers to Ensle [19] for further information.

The final form of the particle equation is written as follows:

$$\nabla^2 \vec{E} - \frac{1}{c^2}(\vec{v} \cdot \vec{\nabla})(\vec{v} \cdot \vec{\nabla})\vec{E} = 0 \tag{3.159}$$

The reason that this equation is called a particle equation is because the electric field that satisfies this equation will propagate as a coherent pulse in a single direction, behaving kinematically as a particle in a vacuum. As it turns out this is not an easy equation to deal with; however, if we consider the space external to the charge, where the divergence is zero, we can do some vector calculus manipulation in such a way as to find a set of solutions. They are not in a general form though, but in one working solution (see Ensle [19] for details). Using a spherical coordinate system and assuming the following relation for a particle of spherical shape, we can find a separable partial differential equation as the result; thus, we have:

$$\vec{E} = E\hat{r}$$
$$\vec{v} = v\cos\theta\hat{r} - v\sin\theta\hat{\theta} \tag{3.160}$$

Therefore, we find:

$$\vec{v} \cdot \vec{E} = v\cos\theta E \tag{3.161}$$

and

$$\vec{v}(\vec{v} \cdot \vec{E}) = v^2\cos^2\theta E\hat{r} - v^2\sin\theta\cos\theta E\hat{\theta} \tag{3.162}$$

The final form of Eq. 3.159 reduces to a separable partial differential equation with two variables, r and θ:

$$\frac{v^2}{c^2}E - \frac{v^2}{c^2}r\frac{\partial E}{\partial r} - \frac{\left(1 - \frac{v^2}{c^2}\sin^2\theta\right)}{\sin\theta\cos\theta}\frac{\partial E}{\partial \theta} = 0 \qquad (3.163)$$

Assuming $E = R\Theta$, where R is dependent only on r and Θ is dependent only on θ and using the separation method by substituting $E = R\Theta$ into Eq. 3.163, as well as doing the proper algebra manipulation, we obtain the following:

$$\frac{v^2}{c^2} - \frac{v^2}{c^2}\frac{r}{R}\frac{\partial R}{\partial r} - \frac{\left(1 - \frac{v^2}{c^2}\sin^2\theta\right)}{\sin\theta\cos\theta}\frac{1}{\Theta}\frac{\partial \Theta}{\partial \theta} = 0 \qquad (3.164)$$

The form of Eq. 3.164 suggests two possible forms by the separation of variables methodology of using λ as a separation constant that could be anything and will be as follows:

$$\frac{v^2}{c^2} - \frac{v^2}{c^2}\frac{r}{R}\frac{\partial R}{\partial r} = \lambda \qquad (3.165)$$

and

$$\frac{\left(1 - \frac{v^2}{c^2}\sin^2\theta\right)}{\sin\theta\cos\theta}\frac{1}{\Theta}\frac{\partial \Theta}{\partial \theta} = -\lambda \qquad (3.166)$$

The two differential equations, Eqs. 3.165 and 3.166, could be rearranged to the new form as follows:

$$\frac{v^2}{c^2}r\frac{\partial R}{\partial r} + \left(\frac{v^2}{c^2} - \lambda\right)R = 0 \qquad (3.167)$$

$$-\frac{\left(1 - \frac{v^2\sin^2\theta}{c^2}\right)}{\sin\theta\cos\theta}\frac{\partial \Theta}{\partial \theta} + \lambda\Theta = 0 \qquad (3.168)$$

The solution to both Eqs. 3.167 and 3.168 for their characteristic parts are as follows [19]:

$$R = kr^{\left(1 - \lambda\frac{c^2}{v^2}\right)} \qquad (3.169)$$

More analysis of Eq. 3.169, however, tells us that the radial field has to fall off by the power of -2 (i.e., $1/r^2$), as demonstrated at the end of Sect. 1.9. This is because the divergence of the electric field must be zero. Therefore, the radial solution becomes:

$$R = \frac{k}{r^2} \qquad (3.170)$$

3.9 Infrastructure, Characteristics, Derivation, and Scalar Wave Properties

and the separation constant is given by

$$\lambda = 3\frac{v^2}{c^2} \tag{3.171}$$

Now if we also consider the characteristic solution of Eq. 3.168; with some mathematical manipulation, we get the following result:

$$\Theta = \frac{k}{\left(1 - \frac{v^2}{c^2}\sin^2\theta\right)^{3/2}} \tag{3.172}$$

where k is the constant of integration in the solution process of both Eqs. 3.167 and 3.168. Now if we take both results in Eqs. 3.170 and 3.172 and plug them into relationship $E = R\Theta$, we obtain the final solution for Eq. 3.163:

$$\vec{E} = \frac{k\hat{r}}{r^2\left(1 - \frac{v^2}{c^2}\sin^2\theta\right)^{3/2}} \tag{3.173}$$

We can easily show that Eq. 3.173 is the final solution to Eq. 3.163 by plugging it back into the original wave Eq. 3.163. If this is done, the wave equation will show a coherent propagation in one direction at velocity $\vec{v}$. This means that the particle equation was the correct method to determine the field and that a correct solution was found to the particle equation.

Notice that the electric field retains its cylindrical symmetry but becomes asymmetrical on the longitudinal axis as in the diagram in Fig. 3.45 and the case of the SW will create the vortex shape depicted earlier in Fig. 3.31. From Fig. 3.45 we see that the image on the LHS is the representation of the electric filed at rest, while the image on the RHS of it shows the field when the charge field is propagating at velocity $\vec{v}$.

Note that the transverse field is increased and the longitudinal field actually is decreased because the constant of integration k itself is dependent on velocity. To determine the value of the constant of integration is not an easy task; however, Ensle [19] shows that this constant with its dependency on velocity $\vec{v}$ is calculated as follows:

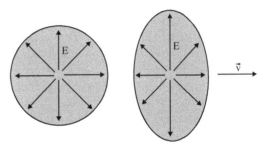

Fig. 3.45 Electric field of a charge field at rest and moving with velocity $\vec{v}$

$$k = \frac{q_e}{4\pi\varepsilon}\sqrt{1-\frac{v^2}{c^2}} \qquad (3.174)$$

Therefore, the electric field, Eq. 3.173, for a moving charge will yield as a final form:

$$\vec{E} = \frac{\frac{q_e}{4\pi\varepsilon}\sqrt{1-\frac{v^2}{c^2}}}{r^2\left(1-\frac{v^2}{c^2}\sin^2\theta\right)^{3/2}}\hat{r} \qquad (3.175)$$

Note that the electric field is transverse to its velocity ($\theta = \pi/2$), thus it is given by:

$$E_T = \frac{q_e}{4\pi\varepsilon r^2}\frac{1}{\left(1-\frac{v^2}{c^2}\right)} \qquad (3.176)$$

and it increases with velocity, while the field longitudinal to its velocity ($\theta = 0$) is:

$$E_T = \frac{q_e}{4\pi\varepsilon r^2}\left(1-\frac{v^2}{c^2}\right) \qquad (3.177)$$

So, it decreases with velocity. The velocity being discussed here is absolute. It is a wave velocity in a continuum that itself defines absolute rest. Equation 3.173 of the field is representing a type of constrained propagation that emulates a particle.

3.9.6 The Magnetic Field

So far, we have not stated as a postulate that a moving charge produces a magnetic field as is done in Maxwell's equations. In fact, we began without any charge at all, stationary or moving. Thus, at this time all we know is that a changing electric field causes a magnetic field. Imagine that a charge field is approaching a point in space. As it approaches, the electric field at that point increases; therefore, the charging electric field produces a magnetic field. It then becomes a simple matter of how much the electric field changes at a stationary point in relation to the velocity that the charge approaches the point.

We then get a relation of the velocity of the charge field, the electric field at a certain point, and the resulting magnetic field at the same point. This gives the expected quantity. Therefore, the result is not really anything new, but it is interesting because of the way in which we have found it. From the postulates, the magnetic field is created when the electric field changes over time. This time change, however, also can be caused by the coherent propagation of the charge. Thus, given that the charge fields are ready-made and can move at some velocity, it is appropriate to know how this velocity produces a magnetic field.

3.9 Infrastructure, Characteristics, Derivation, and Scalar Wave Properties 259

Because we are interested in the field outside of the charge we can say that:

$$\begin{cases} \vec{\nabla} \cdot \vec{E} = 0 \\ \beta = 0 \end{cases} \tag{3.178}$$

We can replace the time derivative of Eq. 3.100 with the motion operator; thus, we have the following relationship:

$$\vec{\nabla} \times \vec{B} = -\frac{1}{c^2} (\vec{v} \cdot \vec{\nabla}) \vec{E} \tag{3.179}$$

Using vector quantity $\vec{\nabla} \times (\vec{F} \times \vec{G}) = (\vec{\nabla} \cdot \vec{G}) \vec{F} - (\vec{\nabla} \cdot \vec{F}) \vec{G} + (\vec{G} \cdot \vec{\nabla}) \vec{F} - (\vec{F} \cdot \vec{\nabla}) \vec{G}$ from Table 1.1 will result in:

$$\vec{\nabla} \times \vec{B} = -\frac{1}{c^2} (\nabla \times (\vec{v} \times \vec{E})) \tag{3.180}$$

Considering the case where the electric field is based on a signal charge field for an elementary charge, we have:

$$\vec{E} = \frac{q_e \hat{r}}{4\pi \varepsilon r^2} \tag{3.181}$$

Substituting Eq. 3.181 into Eq. 3.180 yields:

$$\begin{cases} \vec{\nabla} \times \vec{B} = -\frac{1}{c^2} \left(\left(\vec{v} \times \frac{q_e \hat{r}}{4\pi r^2} \right) \right) \\ \vec{\nabla} \times \vec{B} = -\frac{1}{c^2} \vec{\nabla} \times \left(\left(\vec{v} \times \frac{q_e \hat{r}}{4\pi \varepsilon r^2} \right) \right) \end{cases} \tag{3.182}$$

There is not enough information to solve for the magnetic field unless we also know the divergence of the field, but we do so that is not a problem; however, we still do not know the complete vector field on the RHS of Eq. 3.182 unless we know its divergence. Now in the special condition where both sides have a divergence of zero, the equivalence of the curl of the two vector fields leads to the vector field being equivalent, plus a constant of integration. So, first we must verify that the vector field of the RHS of Eq. 3.182 has zero divergence that is:

$$\vec{\nabla} \cdot \left(\vec{v} \times \frac{q_e \hat{r}}{4\pi \varepsilon r^2} \right) = 0 \tag{3.183}$$

Using vector quantity $\vec{\nabla} \cdot (\vec{F} \times \vec{G}) = \vec{F} \cdot (\vec{\nabla} \times \vec{G})$ from Table 1.1, we obtain the following:

$$-\vec{v} \cdot \left(\vec{\nabla} \times \frac{q_e \hat{r}}{4\pi \varepsilon r^2} \right) = 0 \tag{3.184}$$

Thus, from a relation in $\vec{\nabla} \cdot (\vec{\nabla} \times \vec{F}) = 0$, we can see that the divergence is zero. Therefore, Eq. 3.182 leads to the following result:

$$\vec{B}_e = \frac{1}{c^2}\left(\vec{v} \times \frac{q_e \hat{r}}{4\pi\varepsilon r^2}\right) + k \qquad (3.185)$$

where k is a constant. Note that when $v = 0$, the magnetic field is zero, so the constant is zero. Therefore, we can write:

$$\vec{B}_e = \frac{1}{c^2}\left(\vec{v} \times \frac{q_e \hat{r}}{4\pi\varepsilon r^2}\right) \qquad (3.186)$$

There is a problem with Eq. 3.186 however. We assumed that the electric field remained the same regardless of its velocity. This is not true as we saw in Chap. 1 of this book.

This relationship in Eq. 3.186 is useful for velocities much less than the velocity of light but becomes more inaccurate as velocity increases. If we simply replace the charge field with the electric field, however, the magnetic field now relates directly to the electric field, and as it changes with velocity, the magnetic field tracks the change. Thus, we have simply the following equation:

$$\vec{B} = \frac{1}{c^2}\left(\vec{v} \times \vec{E}\right) \qquad (3.187)$$

This equation is valid at all velocities. Notice that it corresponds directly to the electric field as it exists at the point in space where the magnetic field is created. The magnetic field, however, is in some way independent of the electric field, so if the magnetic field is created by an electric field somewhere in space and in one more region, the electric field is canceled out by the other electric field. It would be possible for the magnetic field, which is not being canceled out, to extend into a space where the electric field is zero [19].

We now know how a moving elementary charge produces a magnetic field. Then by superposition it can be applied to any collection of charges in motion. This again gives the expected result, which was defined as the *Biot-Savart Law*, as we saw in Chap. 1 of this book. Given the magnetic field from a moving charge as we found it in Eq. 3.186, we obtain the following new form:

$$\vec{B}_e = \frac{\mu}{4\pi}\left(q_e \vec{v} \times \frac{\hat{r}}{r^2}\right) \qquad (3.188)$$

Equation 3.166 would be the magnetic field of moving change, as shown in Fig. 3.46.

Because of superposition we can sum any number of charge fields that could be treated as a charge density spread out over a volume; thus:

3.9 Infrastructure, Characteristics, Derivation, and Scalar Wave Properties

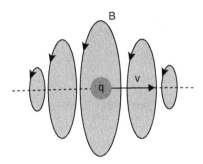

Fig. 3.46 Magnetic field of a charge in motion

$$\vec{B} = \frac{\mu}{4\pi} \int \rho \vec{v} \times \frac{\vec{r} - \vec{r}_0}{|r - r_0|^3} d\tau \quad (3.189)$$

where r_0 is the position vector of the charge element. We will define *current density* such that:

$$\vec{J} \equiv \rho \vec{v} \quad (3.190)$$

So, we can write:

$$\vec{B} = \frac{\mu}{4\pi} \int \vec{J} \times \frac{\vec{r} - \vec{r}_0}{|r - r_0|^3} d\tau \quad (3.191)$$

which again is just the Biot-Savart Law as before. If one imagines an infinitely thin line of charge density, Eq. 3.169 becomes:

$$\vec{B} = \frac{\mu}{4\pi} \int \vec{I} \times \frac{\hat{r}}{r^2} dl = \frac{\mu I}{4\pi} \int d\vec{l} \times \frac{\hat{r}}{r^2} \quad (3.192)$$

Finally, notice that if we take the time derivative of the charge density in Eq. 3.190, we can obtain the following relationship:

$$\frac{\partial \rho}{\partial t} = -\vec{v} \cdot \vec{\nabla} \rho \quad (3.193)$$

Or with the vector identity given in Table 1.1, we convert Eq. 3.193 into the following form:

$$\frac{\partial \rho}{\partial t} = -\vec{\nabla} \cdot \vec{v} \rho \quad (3.194)$$

Using Eq. 3.168 equality, Eq. 3.172, provides a different form:

$$\frac{\partial \rho}{\partial t} = -\vec{\nabla} \cdot \vec{J} \quad (3.195)$$

This equation is typically called the *continuity equation*. Furthermore, if we consider the charge, we have:

$$\frac{\partial q}{\partial t} = -\int \vec{\nabla} \cdot \vec{J} \, d\tau \tag{3.196}$$

In addition, by the divergence theorem, Eq. 3.196 creates a different format:

$$\frac{\partial q}{\partial t} = -\oint_S \vec{J} \cdot d\vec{s} \tag{3.197}$$

This equation is called the *charge continuity equation*, which states that charge is conserved as it moves from one location to another. It is analogous to the energy continuity in Eq. 3.142.

3.9.7 The Scalar Field

Now to get back to our electrodynamic study of a charge field in an electric field, we look at the case when a charge field approaches a point; the change in the electric field at the point also causes a *scalar field*. This situation, however, only occurs when the electric field has a divergence. In this case we treat the charge field as a point charge to simplify calculations, but it is still useful because by superposition it can be generally applied. Even though the non-divergent electric field does not contribute to the situation, by the continuity rule, the scalar field extends into the non-divergent region. The way the scalar field extends into the non-divergent electric field makes it possible to relate the scalar field directly to the non-divergent electric field as we could do for a magnetic field, but they are not directly linked [19].

It is just aresemblance , although it could be useful in practical calculations. The field is the most positive in front of the moving charge and the most negative behind the charge motion. On the plane bisecting the change perpendicular to its motion, the scalar field is zero. This is opposite in a sense to the magnetic field because the magnetic field is strongest on the perpendicular plane; however, directly in front or behind, the magnetic field is zero.

To continue with this matter, we now look at the magnetic field, and determine that the scalar magnetic field in relation to the velocity of a charge field will be derived by taking the divergence of Eq. 3.100 to obtain the following:

$$\begin{cases} \vec{\nabla} \times \vec{B} + \vec{\nabla}\beta = \dfrac{1}{c^2}\left(\dfrac{\partial \vec{E}}{\partial t}\right) \\ \vec{\nabla} \cdot (\vec{\nabla} \times \vec{B}) + \vec{\nabla} \cdot (\vec{\nabla}\beta) = \dfrac{1}{c^2}\left(\vec{\nabla} \cdot \dfrac{\partial \vec{E}}{\partial t}\right) \end{cases} \tag{3.198}$$

3.9 Infrastructure, Characteristics, Derivation, and Scalar Wave Properties

Using Table 1.1, we can see that the divergence of a curl is zero; thus, Eq. 3.198 reduces to the following form:

$$\nabla^2 \beta = \frac{1}{c^2}\left(\vec{\nabla} \cdot \frac{\partial \vec{E}}{\partial t}\right) \qquad (3.199)$$

or

$$\nabla^2 \beta = \frac{1}{c^2}\frac{\partial}{\partial t}(\vec{\nabla} \cdot \vec{E}) \qquad (3.200)$$

Now let us do this for a charge field using the interior field, where the divergence is not zero, and use the relation of ρ/ε to signify it. It then becomes:

$$\nabla^2 \beta_e = -\mu \vec{v} \cdot \vec{\nabla} \rho_e \qquad (3.201)$$

Using vector identity $\vec{\nabla} \cdot (\varphi \vec{F}) = (\vec{\nabla}\varphi) \cdot \vec{F} + \varphi \vec{\nabla} \cdot \vec{F}$ from Table 1.1, Eq. 3.201 yields:

$$\nabla^2 \beta_e = -\mu \vec{\nabla} \cdot \vec{v} \rho_e \qquad (3.202)$$

This equation looks like Poisson's equation and thus has a solution such as:

$$\beta_e = \frac{\mu}{4\pi} \int \left[\frac{\vec{\nabla} \cdot \vec{v} \rho_e}{|\vec{r} - \vec{r}_0|}\right] d\tau \qquad (3.203)$$

If we use the same vector quantity again, as in the preceding, we can break Eq. 3.203 into two parts:

$$\beta_e = \frac{\mu}{4\pi} \int \vec{\nabla} \cdot \frac{\vec{v}\rho_e}{|\vec{r} - \vec{r}_0|} d\tau - \frac{\mu}{4\pi} \int \vec{v}\rho_e \cdot \vec{\nabla}\left(\frac{1}{|\vec{r} - \vec{r}_0|}\right) d\tau \qquad (3.204)$$

Applying divergence theorem in Eq. 1.61 to the first term integral in Eq. 3.204, we obtain the following:

$$\beta_e = \frac{\mu}{4\pi} \oint_S \frac{\vec{v}\rho_e}{|\vec{r} - \vec{r}_0|} d\vec{s} - \frac{\mu}{4\pi} \int \vec{v}\rho_e \cdot \vec{\nabla}\left(\frac{1}{|\vec{r} - \vec{r}_0|}\right) d\tau \qquad (3.205)$$

Notice that the velocity's direction is the direction of the it and the velocity into the closed surface is the same as the velocity out of the closed surface; thus, we have:

$$\frac{\mu}{4\pi} \oint_S \frac{\vec{v}\rho_e}{|\vec{r} - \vec{r}_0|} \vec{v} \cdot d\vec{s} = 0 \qquad (3.206)$$

Therefore, Eq. 3.206 reduces to the following form:

$$\beta_e = \frac{\mu}{4\pi} \int \vec{v}\rho_e \cdot \vec{\nabla}\left(\frac{1}{|\vec{r}-\vec{r_0}|}\right) d\tau \qquad (3.207)$$

Resolving the gradient within the integral of Eq. 3.207, yields the following:

$$\beta_e = \frac{\mu}{4\pi} \int \frac{\vec{v}\rho_e|\vec{r}-\vec{r_0}|}{|r-r_0|^3} d\tau \qquad (3.208)$$

Then finishing the integration of Eq. 3.208 for a point charge, we obtain:

$$\beta_e = \frac{\mu q_e}{4\pi} \frac{\vec{v} \cdot \hat{r}}{r^2} \qquad (3.209)$$

This equation gives us the scalar field of a moving charge; however, it suffers from the same defect as the magnetic field does [19].

This relation is useful for velocities much less than the velocity of light but becomes more inaccurate as the velocity increases. Yet, if we simply replace the charge field with the electric field, the magnetic field now relates directly to the electric field, and as it changes with velocity, the scalar field tracks the charge. Thus, we simply can have the following relationship:

$$\beta_e = \frac{1}{c^2}\left(\vec{v} \cdot \vec{E}\right) \qquad (3.210)$$

This equation is valid at all velocities and note that we can relate the scalar field to the electric field like we have done for magnetic field. Unlike the magnetic field, however, the scalar field has no direct connection to the electric field, where the divergence is zero. Although the scalar field extends into the non-divergent region, it does not exist unless the electric field has divergence somewhere. Nonetheless, Eq. 3.210 in relation to the electric field can be employed provided the electric field in question is related to a charge field and, as we said in the case of the magnetic field, the scalar field has a certain independence. If the electric field is canceled out somewhere, the scalar field can exist even where the electric field is zero [19].

Like with the magnetic field we use superposition to relate it to current density where it is then analogous to the *Biot-Savart Law*. Nevertheless, there are significant differences in this relationship. If there is a closed loop or current, no scalar field is generated. The current must be open and at the points where the charge accumulates so that a scalar field is generated.

Given the scalar field from a moving charge, as shown in Eq. 3.209, we have:

$$\beta_e = \frac{\mu}{4\pi}\left(q_e \vec{v} \cdot \frac{\hat{r}}{r^2}\right) \qquad (3.211)$$

3.9 Infrastructure, Characteristics, Derivation, and Scalar Wave Properties

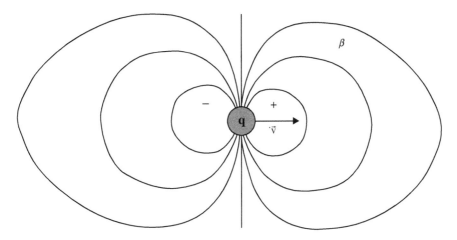

Fig. 3.47 Scalar magnetic field of a charge in motion

This equation is the indication of a scalar magnetic field of a moving charge, as depicted in Fig. 3.47. Because of superposition, we can sum any number of charge fields that could be treated as a charge density spread out over a volume; thus,

$$\beta = \frac{\mu}{4\pi} \int \rho \vec{v} \cdot \left(\frac{\vec{r} - \vec{r}_0}{|r - r_0|^3} \right) d\tau \qquad (3.212)$$

Using our definition for current density as before (i.e., $\vec{J} = \rho \vec{v}$), Eq. 3.146, becomes:

$$\beta = \frac{\mu}{4\pi} \int \vec{J} \cdot \left(\frac{\vec{r} - \vec{r}_0}{|r - r_0|^3} \right) d\tau \qquad (3.213)$$

if one imagines an infinitely thin line of charge density. This equation results in:

$$\beta = \frac{\mu}{4\pi} \int \vec{I} \cdot \frac{\hat{r}}{r^2} dl = \frac{\mu I}{4\pi} \int \frac{\hat{r}}{r^2} \cdot d\vec{l} \qquad (3.214)$$

where I is the current. Notice that the integrand can be viewed as a potential, and we can then apply the gradient theorem; thus:

$$\beta = -\frac{\mu I}{4\pi} \int \nabla \left(\frac{1}{r} \right) \cdot d\vec{l} = -\frac{\mu I}{4\pi} \left(\frac{1}{r_2} - \frac{1}{r_1} \right) \qquad (3.215)$$

The scalar field exists when the current is discontinuous. If the current is in a loop, no scalar field is created because $r_1 = r_2$. It should be noted though that microscopically the scalar field exists for each charge field that is propagating, but this is not observed at any point external to the current. Note that for a $1/r$ field the energy is infinite, but the scalar field is always created as two opposite poles, that it is a dipole, and that the distant field falls off as $1/r^2$.

3.9.8 Scalar Fields: From Classical Electromagnetism to Quantum Mechanics

As we have learned so far, in classical electrodynamics the vector and scalar potentials conveniently were introduced by using Maxwell's equations as an aid for calculating these fields and from there we managed to analyze them in quantum electrodynamics via Lagrangian and Hamiltonian aspects of gauge theory and its invariance postulation. See Chap. 5 of this book for more details. It is a known fact, however, that in order to obtain a classical canonical formalism, the potentials are needed. Nevertheless, the fundamental equation of motion always can be described directly in terms of the fields alone.

In quantum mechanics, however, the canonical formalism is necessary, and as a result, the potentials cannot be eliminated from the basic equations. Still, these equations, as well as the physical quantities, are all gauge-invariant so that it may seem that even in quantum mechanics, the potentials themselves have no independent significance [23].

To be able to take these fields, in particular, the scalar field and its prediction from *classical electromagnetism* and interaction from *quantum mechanics*, we assume that readers have encountered the Principle of Least Action (see Chap. 5 as well) in *classical mechanics*, as well as the related concepts of the *Lagrangian*, the *Hamiltonian*, and the *Euler-Lagrange* equations. This section briefly examines them from a special relativistic point of view and leaves the details for Chap. 5; ; so we begin with a summary of the classical results both as a reminder and to introduce the notation we need for the subject in hand.

Prudently, in classical mechanics classical Lagrangian mechanics is branching off from Newton's law of force and acceleration, as well as momentum as:

$$\vec{f} = m\vec{a} = m\frac{d\vec{v}}{dt} = \frac{dm\vec{v}}{dt} = \frac{d\vec{p}}{dt} \quad (3.216)$$

which we can write in a new form as:

$$-\nabla \vec{V} = m\frac{d\dot{\vec{x}}}{dt} \quad (3.217)$$

The basic idea of Lagrangian mechanics is to replace this vector treatment by an action based on a scalar quantity called the Lagrangian that allows vector equations to be extracted by taking derivatives just as $\nabla \vec{V}$ is a vector extracted from potential energy V. This approach proves to be more flexible, and it simplifies many problems in classical mechanics.

At any given instant in time the state of a physical system is described by a set of n variables q_i called *coordinates* and their time derivatives $\dot{q}_i$ called *velocities*. For example, these could be the positions and velocities of a set of particles making up the system. Define function $\mathcal{L}$, called the *Lagrangian*, given by

3.9 Infrastructure, Characteristics, Derivation, and Scalar Wave Properties

$$\mathcal{L} = T - V \tag{3.218}$$

In this equation T and V are the kinetic energy and potential energies of the system, respectively. The Lagrangian therefore is a function of the positions and velocities, and it can be a function of time. This is indicated by the notation $\mathcal{L} = \mathcal{L}(\{q_i\}, \{\dot{q}_i\}, t)$, which we abbreviate to $\mathcal{L} = \mathcal{L}(q, \dot{q}, t)$. For particle motion with no external time-dependent fields, the Lagrangian has no explicit dependence on time.

The phrase "no explicit dependence on time" means it has no dependence on time over and above what is already implied by the fact that q and $\dot{q}$ may depend on time. For example, a single particle undergoing simple harmonic motion has the Lagrangian $\mathcal{L} = (1/2)m(\dot{x}^2 - \omega^2 x^2)$. An example motion of the particle is $x = x_0 \sin(\omega t)$, and for this motion the Lagrangian also can be written as $(m\omega^2 x_0^2/2)\cos 2\omega t$, which is a function of time. The latter form, however, hides dependence on x and $\dot{x}$, which is what we mainly are interested in; furthermore, in general the Lagrangian cannot be deduced from the motion, but the motion can be deduced from the Lagrangian when the latter is written as a function of coordinates and velocities.

For this reason the variables $\{q, \dot{q}_i\}$ are said to be the "natural" or "proper" variables of $\mathcal{L}$. A similar issue arises in the treatment of functions of state in thermodynamics as well. The time integral of the Lagrangian along a path $q(t)$ is called *action* S and is written:

$$S[q(t)] = \int_{q_1, t_1}^{q_2, t_2} \mathcal{L}(q, \dot{q}, t)\, dt \tag{3.219}$$

The *Principle of Least Action* states that the path followed by the system is the one that gives an extreme value either a maximum or a minimum of S with respect to small changes in the path. The word "least" comes from the fact that in practice a minimum is more usual than a maximum. The path is to be taken between given starting and finishing "properties" of q_1, q_2 at time t_1, t_2 [24]. To find the extremum of S we need to ask for a zero derivative with respect to changes in all the variables describing the path. The calculus of variation can be used to show that the conclusion is that action S reaches an extremum for the path satisfying it.

We are now at the point that we can establish and define the *Euler-Lagrange equations* as:

$$\frac{d}{dt}\left(\frac{\partial \mathcal{L}}{\partial \dot{q}_i}\right) = \frac{\partial \mathcal{L}}{\partial q_i} \tag{3.220}$$

The physical interpretation of Eq. 3.320 is found by discovering its implications. The end result of such a study is summarized as follows:

$$\underbrace{\frac{d}{dt}}_{\text{(Rate of change of)}} \underbrace{\left(\frac{\partial \mathcal{L}}{\partial \dot{q}_i}\right)}_{\text{('Momentum')}} = \underbrace{\left(\frac{\partial \mathcal{L}}{\partial q_i}\right)}_{\text{('Force')}} \tag{3.221}$$

The "force" here is called a *generalized force*, and the "momentum" is called *canonical momentum*, defined by:

$$\tilde{p}_i \equiv \left(\frac{\partial \mathcal{L}}{\partial \dot{q}_i}\right) \qquad (3.222)$$

In the simplest cases, such as motion of a free particle or a particle subject to conservative forces, the canonical momentum may be equal to a familiar momentum (e.g., linear or angular), but this does not have to happen. An opposing exampleoccurs for the motion of a particle in a magnitude field.

The Hamiltonian of a system is defined as:

$$\mathcal{H}(q,\tilde{p},t) \equiv \left(\sum_i^n \tilde{p}_i \dot{q}_i\right) - \mathcal{L}(q,\dot{q}_i,t) \qquad (3.223)$$

In this equation $\dot{q}_i$ is to be written as a function of q_i and $\tilde{p}_i$ so that the result is a function of coordinates and canonical momentum as the natural variables of the Hamiltonian. For conservative forces one finds that the sum in Eq. 3.223 calculates to twice the kinetic energy, and then the final form of Hamiltonian can be written:

$$\mathcal{H} = T + V \qquad (3.224)$$

This equation is clearly the total energy of the system.

The Euler-Lagrange equation will yield, in the *Hamilton's canonical equation*:

$$\begin{cases} \dfrac{dq_i}{dt} = \dfrac{\partial \mathcal{H}}{\partial \tilde{p}_i} \\ \dfrac{d\tilde{p}_i}{dt} = -\dfrac{\partial \mathcal{H}}{\partial q_i} \end{cases} \qquad (3.225)$$

Thus, the Hamiltonian with the canonical equations offers an alternative to the Lagrangian with Euler-Lagrange equations.

Now that we have developed and refreshed our knowledge of classical mechanics and the motion of particles based on Lagrangian and Hamiltonian definitions, we can go back to the stream of the original goal for this section on the topic of the significance of EM potentials in quantum theory and scalar fields, as well as their prediction from classical electromagnetism and interpretation from quantum mechanics.

If we begin the discussion with a charged particle inside a "Faraday cage" connected to an external generator, as was suggested by Aharonov and Bohm in 1959 [23], that causes the potential on the cage to alternate in time. Such a condition adds to the Hamiltonian of the particle term $V(x, t)$, which is for the region inside the cage, a function of time only. In the non-relativistic limit we assume this situation almost everywhere and have for the region inside the cage:

3.9 Infrastructure, Characteristics, Derivation, and Scalar Wave Properties

$$\mathcal{H} = \mathcal{H}_0 + V(t) \tag{3.226}$$

In this equation the quantity $\mathcal{H}_0$ is the Hamiltonian when the generator is not functioning and $V(t) = \exp \phi(t)$. If $\psi_0(x, t)$ is a solution of the Hamiltonian, $\mathcal{H}_0$, then the solution for $\mathcal{H}$ is:

$$\psi = \psi_0 e^{-iS/\hbar}$$
$$S = \int V(t) dt \tag{3.227}$$

which follows from

$$i\hbar \frac{\partial \psi}{\partial t} = \left(i\hbar \frac{\partial \psi_0}{\partial t} + \psi_0 \frac{\partial S}{\partial t} \right) e^{-iS/\hbar} = [\mathcal{H}_0 + V(t)]\psi = H\psi \tag{3.228}$$

This new solution differs from the old one just by a phase factor, and it corresponds to no change in physical results.

Building the analysis based on Eq. 3.228, Aharonov and Bohm [23] have shown that, from relativistic considerations, it can be seen that the covariance of their analysis includes a demand for a similar result involving vector potential $\vec{A}$. As we have learned so far, a scalar can be defined non-rigorously as a point with magnitude but no direction in contrast to a vector, which is defined as a point with both magnitude and direction.

As we said previously, Hertzian waves, which consist of oscillating transverse electric and magnetic fields, are vector waves and therefore cannot be the SWs that Tesla used. The other types of fields, which show promise in describing scalar fields, are the potential fields and the paper written by Aharonov and Bohm [23]; these authors described a subtle effect arising from potential fields when both the electric and the magnetic fields were absent. A short account of these two authors' published paper presents a phenomenon that today we know it as the *Aharonov-Bohm effect*; a summary follows.

1. The Aharonov-Bohm Effect

In a region with electric and magnetic fields $\vec{E}$ and $\vec{B}$, classical physics predicts that a particle of charge q will encounter a force $\vec{F}$, which was introduced as the *Lorentz force* in Chap. 1 of this book and is written as:

$$\vec{F} = q\vec{E} + \frac{q}{c}(\vec{v} \times \vec{B}) \tag{3.229}$$

where c is the speed of light and $\vec{v}$ is particle velocity. If the $\vec{E}$ and $\vec{B}$ fields are radiation fields with time varying, then the energy flow of Poynting's vector is written:

$$\vec{S} = \frac{1}{2} \frac{c}{4\pi} (\vec{E} \times \vec{B}) \tag{3.230}$$

If both the fields—namely, electric and magnetic fields—are zero, it can be said that all forces are zero and all energy flows are zero accordingly. Thus, no physical consequences can be observed if the $\vec{E}$ and $\vec{B}$ fields are zero.

Quantum mechanics requires the use of potential fields rather than $\vec{E}$ and $\vec{B}$ force fields. *Vector potential* $\vec{A}$ and *scalar potential* ϕ are related to $\vec{E}$ and $\vec{B}$ via Eqs. 3.72 and 3.74 as before; they are repeated here:

$$\vec{E} = -\nabla\phi - \frac{1}{c}\frac{\partial \vec{A}}{\partial t} \qquad (3.231)$$

and

$$\vec{B} = \vec{\nabla} \times \vec{A} \qquad (3.232)$$

We assume again that the use of ϕ and $\vec{A}$ are equivalent to the use of the $\vec{E}$ and $\vec{B}$ fields and that Eq. 3.229 was created based on the kinematic state of the charged particle of interest with velocity $\vec{v}$ depending on Lorentz force $\vec{F}$, where the EM continuum is described in classical physics by the local values of both electric field $\vec{E}$ and magnetic field $\vec{B}$ strengths, which are given by Eqs. 3.231 (3.72) and 3.232 (3.74).

Bear in mind that the distribution of EM potentials is not uniquely determined by the distribution of field strengths because of a change in the gauge of the potentials such as:

$$\phi' = \phi - \frac{1}{c}\frac{\partial f}{\partial t} \qquad (3.233a)$$

and

$$\vec{A}' = \vec{A} + \vec{\nabla}f \qquad (3.233b)$$

which leaves field strengths unaffected for an arbitrary gauge function f of space and time. Therefore, in classical electromagnetism, the potentials are considered as mathematical entities without physical significance.

Aharonov and Bohm (1959) [23] in their paper, however, showed that potential fields ϕ and $\vec{A}$ can result in physical consequences when the $\vec{E}$ and $\vec{B}$ fields are zero. The quantum mechanical *Schrödinger's equation* for an electron in an EM field is given by the following form:

$$\frac{1}{2m}\left[\frac{\hbar}{c}\nabla - \frac{q}{c}\vec{A}\right]\Psi(x,t) + q\phi\,\Psi(x,t) = \vec{E}\Psi(x,t) \qquad (3.234)$$

In this equation $\Psi(x,t)$ is the electron wave function. Aharonov and Bohm [23] suggested an experiment where an electron beam was split into two beams enclosed in tightly wound long solenoid under a special case, where only a path in space was considered (i.e., $t =$ constant). The experiment is shown in Fig. 3.48.

3.9 Infrastructure, Characteristics, Derivation, and Scalar Wave Properties

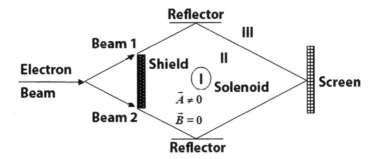

Fig. 3.48 Setup for Aharonov-Bohm interface with time-independent experiment

The beams are recombined at a screen and their interference pattern is observed. If neither magnetic field $\vec{B}$ nor vector potential field $\vec{A}$ are present, as shown in the figure, and the path lengths are the same, then the electrons will interface constructively because their phases were the same in the initial beam. Now the solenoid is turned on and a magnetic flux exists inside the solenoid. Yet, because it is a long and highly wounded solenoid, no magnetic flux exists in *Region* II. At the same time a vector potential field will be in the form because the following equation exists in *Region* II:

$$\vec{A} = (A_r, A_\theta, A_z) = \left(0, 2Ni\pi a^2/cr, 0\right) \qquad (3.235)$$

In this equation,

i = the current
N = the number of turns per unit length
a = the radius of the solenoid
r = the distance from solenoid axis

It can be seen from Eq. 3.232 that indeed magnetic field $\vec{B}$ is zero (i.e., $\vec{B} = 0$), thus we can write:

$$\vec{B} = \vec{\nabla} \times \vec{A} = 0 \qquad (3.236)$$

Considering this equation, we encounter a situation where $\vec{B} = 0$ but, on the other hand, $\vec{A} \neq 0$. So the question is: Does this have any physical consequences? The solution to Schrödinger's equation for an electron traversing a path of *Beam 1* in Fig. 3.48, however, is:

$$\Psi = \Psi_0 \exp\left[\frac{iq}{\hbar c}\int_{X_1} \vec{A} \cdot d\vec{l}\right] \qquad (3.237)$$

where Ψ_0 is the free particle solution and the integral measure the summation of $\vec{A}$ along the path of travel is X_1. The combined solution for electron path 1 and another electron path 2 designed as X_2 will provide this new form of solution for Schrödinger's equation [23]:

$$\Psi = \Psi_1 + \Psi_2 = \Psi_0 \exp\left[\frac{iq}{\hbar c}\int_{X_1} \vec{A}\cdot d\vec{l}\right] \\ + \Psi_0 \exp\left[\frac{iq}{\hbar c}\int_{X_2} \vec{A}\cdot d\vec{l}\right] \quad (3.238)$$

What can be observed is the probability density as shown here:

$$\Psi^*\Psi = 2|\Psi_0| + 2|\Psi_0|^2 \exp\left[\frac{iq}{\hbar c}\oint \vec{A}\cdot d\vec{l}\right] \quad (3.239)$$

where the integral denotes the summation of $\vec{A}$ along the closed loop, path 1 plus path 2, and Ψ^* is the conjugate of wave function Ψ. The remarkable effect is that the observable $\Psi^*\Psi$ varies with $\vec{A}$ even through no $\vec{B}$ field interacts with the electrons! Thus, Aharonov and Bohm [23] theoretically predicted electron phase shifts when no force fields are present. Many experiments have been performed that confirm the existence of this effect as can be seen from Eqs. 3.232, 3.234, and 3.235.

2. The Existence of Potential and Scalar Fields and Waves

The preceding described the experimental setup used for locally generated potential fields and the electron beams enclosed a magnetic flux. In fact, the phase integral is equal to the enclosed magnetic flux as follows:

$$\oint \vec{A}\cdot d\vec{l} = \phi_B \quad (3.240)$$

These are severe and serious limitations. If potential fields have any correspondence to scalar fields, they should exist independently of enclosed fluxes and can originate from non-local sources.

In fact, EM theory does predict the existence of potential waves traveling at the speed of light. If we substitute Eqs. 3.231 and 3.232 into all four of Maxwell's equations (see Eq. 3.71) and use charge e for electron in place of charge q, this yields the following form of sets of equations:

$$\vec{\nabla}\cdot\vec{E} = 4\pi e \\ \vec{\nabla}\cdot\vec{B} = 0 \\ \vec{\nabla}\times\vec{B} = \frac{4\pi}{c}\vec{j} + \frac{1}{c}\frac{\partial\vec{E}}{\partial t} \\ \vec{\nabla}\times\vec{E} = -\frac{1}{c}\frac{\partial\vec{B}}{\partial t} \quad (3.241)$$

3.9 Infrastructure, Characteristics, Derivation, and Scalar Wave Properties

and using the Lorentz gauge condition, we obtain (see Sect. 3.8 as well as Eq. 3.80):

$$\vec{\nabla} \cdot \vec{A} + \frac{1}{c}\frac{\partial \phi}{\partial t} = 0 \tag{3.242}$$

Assuming we are far away from the sources, we arrive with two potential wave equations, as before (see Sect. 3.8 as well as Eq. 3.81); they are as follows:

$$\nabla^2 \phi - \frac{1}{c^2}\frac{\partial^2 \phi}{\partial t^2} = 0 \tag{3.243}$$

$$\nabla^2 \vec{A} - \frac{1}{c^2}\frac{\partial^2 \vec{A}}{\partial t^2} = 0 \tag{3.244}$$

The important question now is whether the $\vec{A}$ and ϕ waves always need to be in association with the $\vec{B}$ and $\vec{E}$ waves. For the static or quasi-static case (i.e., the *Aharonov-Bohm effect*) [35] we see that $\vec{A}$ can decouple from $\vec{B}$. For the time-varying case it is more complicated because if we initially have:

$$\vec{B} = \vec{\nabla} \times \vec{A}(t) = 0 \tag{3.245}$$

then

$$\vec{E} = -\vec{\nabla}\phi - \frac{1}{c}\frac{\partial \vec{A}(t)}{\partial t} \tag{3.246}$$

Therefore, $\vec{E}$ is not necessarily zero. An $\vec{E}$ not equal to zero automatically generates a $\vec{B}$; thus, $\vec{A}$ will be forced to couple with $\vec{B}$.

If generating source force $\vec{B}$ is to be zero at all times, however, then we automatically ensure that $\vec{E} = 0$ for all times because:

$$E_0 = cB_0 \tag{3.247}$$

where B_0 and E_0 are the magnitude of the $\vec{B}$ and $\vec{E}$ fields. $\vec{E} = 0$ means that, from Eq. 3.246, we can obtain the following result:

$$-\vec{\nabla}\phi - \frac{1}{c}\frac{\partial \vec{A}}{\partial t} = 0 \tag{3.248}$$

Equation 3.249 can always be satisfied with scalar super-potential field χ in Weber units as:

$$\vec{A} = \vec{\nabla}\chi \tag{3.249}$$

and

$$\phi = -\frac{1}{c}\frac{\partial \chi}{\partial t} \tag{3.250}$$

If Eqs. 3.249 and 3.250 are substituted into Eq. 3.242, the yield would be:

$$\nabla^2 \chi - \frac{1}{c^2} \frac{\partial^2 \chi}{\partial t^2} = 0 \qquad (3.250)$$

which is a wave equation for χ. Thus, we have predicted the existence of *scalar waves*, and this is what we were looking for.

> **Scalar Super-Potential**
> The scalar super-potential is the substrate of physicality, the Ether permeating and underlying the Universe from which all matter and force fields derive [1].
>
> Super-potential → Potential → Force field
>
> It is a scalar field, meaning each point in that field has one value associated with it. This value is the degree of magnetic flux at that point, with a unit that is the Weber. This is not the magnetic force field we all know, composed of vectors with Wb/m² units, but a magnetic *flux* field of scalar values with a unit of measure that is simply Wb.
>
> The scalar super-potential symbol is defined by the Greek letter Chi χ. The scalar super-potential may be written as $\chi = \chi(x, y, z, t)$, an equation assigning a flux value to each coordinate in spacetime. By itself the absolute flux value has no direct physical significance in terms of measurable forces; however, it is closely associated with quantum phase θ of a wave function as follows:
>
> $$\chi = \frac{h}{q} \theta \qquad (I)$$
>
> where q = electric charge and h = Plank's constant. So, its effects are limited to the quantum domain and determine the degree of intersection and interaction between different probable realities. Nevertheless, certain distortions in its distribution do give rise to measurable forces.
>
> If we expand on *magnetic vector potential* $\vec{A}$, then we can express that $\vec{A}$ is the gradient of the scalar super-potential, meaning the flux must change over some distance to comprise a vector potential; it can be written as:
>
> $$\vec{A}(x, y, z, t) = \nabla \chi \qquad (II)$$
>
> The absolute value of flux or super-potential does not figure into this, just as altitude above sea level does not figure into the measurement of the "inclination" of a hillside:
>
> (continued)

$$\nabla \chi = \nabla(\chi + \chi_0) \tag{III}$$

Although certain perturbations of the vector potential give rise to certain force fields, by itself $\vec{A}_0$ has no physical significance in terms of measurable forces; however, because it is made of super-potential, it alters quantum phase θ of charged particles per the Aharonov-Bohm effect:

$$\chi = \int \vec{A} \cdot d\vec{l} \tag{IV}$$

$$\theta = \frac{q}{h} \int \vec{A} \cdot d\vec{l} \tag{V}$$

James Maxwell also considered $\vec{A}$ the primary field in electrodynamics and likened it to EM momentum.

Super-Potential of the Magnetic Field

To define this matter, we ask the question: What is the underlying scalar super-potential of $\vec{A} = \hat{\phi}/s$, where s is the element of the surface area? The answer is the gradient of the super-potential, which gives rise to vector potential $\vec{A}$ as defined by Eq. VI:

$$\nabla \chi = \frac{1}{s}\hat{\phi} \tag{VI}$$

By comparing this to the definition of a gradient in cylindrical coordinates, we have the following relation:

$$\begin{cases} \dfrac{1}{s} = \dfrac{1}{s}\dfrac{\partial \chi}{\partial \phi} \\ \chi = \dfrac{s}{s}\phi \end{cases} \tag{VII}$$

It is outside the origin surrounded by the element surface, where $s \neq 0$; the results in Eq. VII simplify to $\chi = \phi$, while at the origin the flux can be analyzed as being equal to 2π. Thus, we can write:

$$\chi = \begin{cases} 2\pi & s = 0 \\ \phi & s \neq 0 \end{cases} \tag{VIII}$$

So, this is the fundamental super-potential field of a non-rotational vector potential, which has a singularity at the central axis of rotation that produces a non-zero $\vec{B}$ at the origin. Because $\vec{B}$ is zero everywhere else, χ is allowed to have a gradient everywhere besides at the origin.

(continued)

But then again what does this mean? The χ field is a corkscrew of infinite width that winds around the z-axis. The infinite width is not a problem; it simply means that the phenomena that depend on the path around the flux do not depend on the distance from it.

One example is the Aharonov-Bohm effect, where an electron traveling around a long thin solenoid picks up a phase factor that depends on the magnetic flux inside the solenoid but not the distance from it. If this solenoid were bent into a closed toroid so that all flux was absolutely confined inside, the effect would still exist.

Another example is a loop of wire wound around a ferromagnetic rod in which there is a changing magnetic field. The electromotive force induced by the changing magnetic flux is independent of the diameter of the loop. If the flux were completely confined inside a toroidal core, it would still produce the same electromotive force. This is because the electron is not actually experiencing the flux itself, but rather the corkscrew super-potential surrounding the flux lines does.

A changing flux creates a changing gradient in the super-potential, and an electron in that path will be pumped along the gradient. Stated another way, a changing gradient generates an electric field that places a force on the electron as expected.

1. **Website[1]: A Brief Introduction to Scalar Fields by Thomas Minderle, Version 0.2**

Here the SWs replace the *potential waves* when the physical $\vec{E}$ and $\vec{B}$ fields are zero.

The scalar fields are more primitive than the potential fields in that the latter are derived from the former. If we assume that χ varies harmonically with time, then we can state that:

$$\vec{A} = i\vec{K}\chi \quad (3.252)$$

In this equation $\vec{K}$ is the representation of the wave vector pointing in the direction of travel. It can be seen from the preceding equation that the waves of $\vec{A}$, in the absence of electric and magnetic fields, are purely longitudinal in that the $\vec{A}$ vector points in the direction of travel. It generally has been assumed by physicists that longitudinal modes of motion can be supported only in a plasma; however, now it can be seen that pure potential fields also can support LWs.

When we substitute Eq. 3.249 into Eq. 3.237 the result is:

$$\Psi = \Psi_0 \exp\left[\frac{iq}{\hbar c}\chi\right] \quad (3.253)$$

---

[1] See scalarphysics.com, May 23, 2014.

3.9 Infrastructure, Characteristics, Derivation, and Scalar Wave Properties

But, for two interfering electrons it is:

$$\Psi^*\Psi = |\Psi_1|^2 + |\Psi_2|^2 + 2|\Psi_1||\Psi_2|\exp\left[\frac{iq}{\hbar c}(\chi_1 - \chi_2)\right] \quad (3.254)$$

and the exponential phase factor may be directly observed.

Consider the radiation from an oscillating dipole. The vector potential field is given by:

$$A \sim \frac{\sin[\omega(t - r/c)]}{r} \quad (3.255)$$

In the radiation zone, where $r \gg \lambda$, we can write:

$$E \sim \frac{\cos[\omega(t - r/c)]}{r} \quad (3.256)$$

and

$$B \sim \frac{\cos[\omega(t - r/c)]}{r} \quad (3.257)$$

The energy received at the receiving antenna will be:

$$S \sim E \times B \propto \frac{1}{r^2} \quad (3.258)$$

The scalar phases parameter, χ, also appears to have a $1/r$ dependence and is written:

$$\chi \sim \frac{\cos[\omega(t - r/c)]}{r} \quad (3.259)$$

Thus, we can see that in detecting a scalar field, such as χ fields, a $1/r$ drop in intensity will be observed, while in detecting radio waves a $1/r^2$ drop in intensity will be observed. It also is interesting that $\vec{A}$ fields, when decoupled from electric field $\vec{E}$ and magnetic field $\vec{B}$ of Maxwell's equations, become χ scalar fields that penetrate all objects because there will be no energy transfer to objects when $\vec{E}$ and $\vec{B}$ are zero. However, if $\vec{A}$ interacts slightly with a highly non-linear media, $\vec{A}$ has a weak coupling with $\vec{B}$ and $\vec{E}$; then it is possible that $\vec{A}$ can be rotated enough such that:

$$\vec{B} = \vec{\nabla} \times \vec{A} \neq 0 \quad (3.260)$$

In such situations, scalar fields may be detected using highly non-linear or metastable systems.

2. Scalar Field Generators

A simple $\vec{A}$ field generator consists of a toroid coil such as the one demonstrated in Fig. 3.49, the $\vec{B}$-field flux is enclosed inside the coil and only $\vec{A}$ fields exist outside. At *extreme low frequency* (ELF) this setup can generate strong $\vec{A}$ fields [35].

At radio frequencies, however, the $\vec{E}$ and $\vec{B}$ fields will be generated outside as well. A true scalar generator is shown in Fig. 3.50. A finished Möbius coil should look something like the image that is shown in Fig. 3.52. This coil is composed of a series quadrifocal cables with a 45° helical twist; the cable is then wound with a toroidal winding pattern (see Fig. 3.51). The first wrap of the cable serves as the core around which to wind the toroid. Realistically, it will seldom be a perfect 45° if wound by hand. Angles between say 38° and 45° seem to work well enough, but the closer to 45° the better.

A Möbius coil merely consists of a loop configuration, as illustrated earlier in Fig. 3.17, with a double twist at the neck and folds over itself. This configuration ensures that nearly all EM fields and potential fields are canceled. Nevertheless, the scalar field need not be zero. The scalar field measures the phase of the current and an electron at the beginning of the loop will have a different phase than at the end of the loop because of lattice scattering. Thus, there should be an average phase difference between the top loop and the bottom loop. The scalar fields will reflect this average phase difference.

An Internet search shows that such a device has been developed and certain properties of scalar fields have been qualitatively confirmed (see Eq. 3.236). The

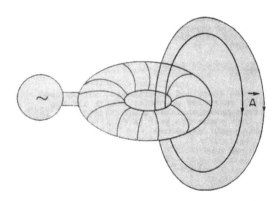

Fig. 3.49 Toroid $\vec{A}$-field generator

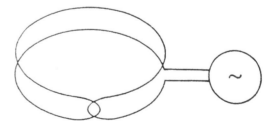

Fig. 3.50 Möbius scalar generator [35]

3.9 Infrastructure, Characteristics, Derivation, and Scalar Wave Properties

Fig. 3.51 The toroidal Möbius coils configuration

generated fields appear to penetrate everything and seem to have a $1/r$ dependence as we discussed here and at the end of Sect. 1.9. Another feature of a device operating near the Earth's Schumann resonance frequencies (about 7–10 Hz) was that it has a calming effect on humans and animals. See slso Sects. 1.11, 3.13, and 3.14 in this chapter.

3. **Detection of Scalar Fields**

Biological organisms and gas tubes serve as highly non-linear media, and for this reason they can serve as scalar field detectors. However, the efficiency may be rather low. For high-efficiency detection, an electron interference device should be used. The setup can be as shown earlier in Fig. 3.48. It is the understanding of this book's author that this setup can measure fields from Region III as well as from Region II, as depicted in Fig. 3.48. For a local source closer to beam 1, this beam will be phase shifted more than beam 2. Thus, interference effects will be observed on the screen if the source intensity changes.

4. **Power Transmission and the Extension to Nuclear Fields**

Tesla's claims for wireless power transmission [25] at his time and Bearden's qualitative theory [26] for this process can now be examined in light of the previously described scalar fields.

The wireless transference of power ultimately requires the conversion of SWs into the usual EMWs. Bearden claimed that this conversion may be accomplished through the interference of two SWs at the target location [26]. Maxwell's equations and gauge conditions are linear equations. The superposition of two scalar fields, which depend on linear equations at least to the first order, can yield only more scalar fields and not the usual EM fields. Therefore, it can be concluded that the predicted SWs cannot be used for power transmission.

The EM field is just one of four fundamental fields known to science. The others are the nucleonic (weak), strong, and gravitational fields. The nucleonic field represents the next level of mathematical complexity in comparison with the EM field. The nucleonic field transforms under rotation as a tensor, represented by a

2 × 2 matrix—the *special unitary* (SU) (2) group. This field is inherently non-linear. It can be speculated that two forceless nucleonic fields can interact to produce energy fields at a distance. The forceless nucleonic fields can be found through a procedure analogous to the procedure for finding the scalar fields.

Note: In mathematics the special unitary group of degree n, denoted SU(n), is the *Lie* group of $n \times n$ unitary matrices with determinant 1. The group operation is that of matrix multiplication. The special unitary group is a subgroup of the unitary group, U(n), consisting of all $n \times n$ unitary matrices, which is itself a subgroup of the *general linear group*, GL(n, C). More general unitary matrices may have complex determinants with absolute value 1, rather than real 1 in the special case.

First, the potential fields are generalized using spacetime notations and can be written:

$$A^\mu = (\phi, \vec{A}) \quad \mu = 0, 1, 2, 3 \qquad (3.261)$$

Now Eqs. 3.245 and 3.246 are combined into the field strength tensor and can be written:

$$f^{\mu\nu} = \partial^\mu A^\nu - \partial^\nu A^\mu \qquad (3.262)$$

where ∂^μ and ∂^ν partials are shorthand for partial derivatives. Equation 3.262 describes the forceless $\vec{A}$ field from which the scalar field was derived. Similarly, the field strength tensor for the nucleonic field may be obtained via Eq. 3.237 and set to zero. Thus, we can write:

$$F^{\mu\nu} = \partial^\mu W^\nu - \partial^\nu W^\mu + gW^\mu \times W^\nu = 0 \qquad (3.263)$$

where W^μ is the nucleonic field and g is a coupling strength parameter. The solution for Eq. 3.263 is being studied by the author, but it is recommended that other interested investigators pursue and share interesting solutions for this equation.

The original description of the nucleonic field by Yang and Mill [27] predicted massless exchange particles (i.e., vector bosons) between two nucleons such as a proton and a neutron. These massless particles, in analogy with photons, have a long range; thus, they can be used for power transmission. The standard unified field theory, U(l) × SO (2) × SO (3), however, uses the Higgs mechanism to introduce mass to the vector bosons and their rang becomes too short for power transmission. Consequently , the original Yang-Mill fields are the best candidates for the Tesla-Bearden scalar fields for wireless power transmission.

5. Conclusion

In this section certain claims for the existence of non-Hertzian waves were examined in light of classical electromagnetism and quantum mechanics. In the process the existence of scalar fields and waves were predicted. These are fields with a long range and great penetrating power. The associated potential waves are LWs in contrast to the familiar transverse EMWs. The SWs cannot be detected directly because they do not impart energy and momentum to matter. On the other hand, they impart phase shifts to matter, and they may be detected through interference means. Because of their elusive nature they also may be called *scalar vacuum waves*. The underlying scalar field already is known to physicists in the context of quantum field theory and is known as the *scalar gauge field*. It is gratifying that in this period of study other researchers reported the observation of fields that behaved qualitatively similar to the predicted scalar fields. The extension of the forceless field concept to the nucleonic field should yield higher order fields with even more interesting properties than the scalar fields. This matter is under investigation.

3.9.8.1 Scalar Interactions

Up to now we have considered only the coupling of particles to the EM field. This interaction distinguishes particles from anti-particles. There is another type of interaction that plays an important role in elementary particle physics—that is, a scalar interaction that treats particles and anti-particles alike. Suppose that non-relativistically we have potential energy $S(\vec{r},t)$ that is the same for particles and anti-particles. The Schrödinger equation (see also Appendix B of this book) with this potential is given as:

$$i\hbar \frac{\partial \Psi(\vec{r},t)}{\partial t} = \left[-\frac{\hbar^2 \nabla^2}{2m} + S(\vec{r},t)\right] \Psi(\vec{r},t) \tag{3.264}$$

We would like to be able to include such a potential in the relativistic wave equation.

One possible way, which as we will see is a relativistic generalization of Eq. 3.265, is the following. Suppose that $S(\vec{r},t)$ is a real scalar function—that is, its value at a given spacetime point is the same in all Lorentz frames. Then if we add $2mS(\vec{r},t)$ onto m^2c^2 in the *Klein-Gordon equation*—Eq. B.70 in Appendix B—we obtain the following relation:

$$\left(\frac{i\hbar}{c^2} \frac{\partial}{\partial t}\right)^2 \Psi(\vec{r},t) = \left(\frac{\hbar}{i} \nabla\right)^2 \Psi(\vec{r},t) + m^2 c^2 \Psi(\vec{r},t) \tag{3.265}$$

This equation is the Klein-Gordon equation that is derived in Appendix B; thus, adding the term $2mS(\vec{r},t)$ onto m^2c^2 would yield:

$$-\frac{\hbar^2}{c^2} \frac{\partial^2}{\partial t^2} \Psi(\vec{r},t) = \left[-\hbar^2 \nabla^2 + m^2 c^2 + 2mS(\vec{r},t)\right] \Psi(\vec{r},t) \tag{3.266}$$

We find that $\Psi(\vec{r},t)$ is still a scalar under a *Lorentz transformation*.

Equation 3.266 is the wave equation corresponding to a modified energy momentum relation; thus, we can write the following mathematical relationship using a similar approach to Einstein's Theory of General Relativity of energy (i.e., $E = mc^2$) and momentum (i.e., $p = mc$). Or, a final version of relativistic energy using Lorentz factor $\gamma = [1 + (p/mc)]^{1/2}$; then the energy equation from a relativistic point of view will be (see Appendix B as well):

$$E = mc^2 \sqrt{1 + \left(\frac{p}{mc}\right)^2} \qquad (3.267)$$

By adding the term $2mS(\vec{r}, t)$ for a relativistic condition, we obtain:

$$E^2 = c^2 p^2 + m^2 c^4 + 2mS(\vec{r}, t) \qquad (3.268)$$

In the non-relativistic limit Eq. 3.268 reduces to:

$$E = mc^2 + S(\vec{r}, t) + \frac{p^2}{2m} \qquad (3.269)$$

Because $S(\vec{r}, t)$ is real, both wave function $\Psi(\vec{r}, t)$ and its conjugate $\Psi^*(\vec{r}, t)$ are solutions for Eq. 3.267; therefore, anti-particles behave the same in potential energy $S(\vec{r}, t)$ as what takes place for the particles.

If $S(\vec{r}, t)$ is independent of time and is just a function of position $\vec{r}$ (i.e., $S = S(\vec{r})$), then Klein-Gordon Eq. 3.264 is no longer in transient mode and for stationary states becomes:

$$\frac{E^2 - m^2 c^4}{2mc^2} \Psi(\vec{r}) = \left[-\frac{\hbar^2 \nabla^2}{2m} + S(\vec{r})\right] \Psi(\vec{r}) \qquad (3.270)$$

This equation is the relativistic generalization of the Schrödinger equation in potential energy $S(\vec{r})$ because in the non-relativistic limit $E \approx mc^2$; so we can write the following equation:

$$\frac{E^2 - m^2 c^4}{2mc^2} = \frac{(E - mc^2)(E + mc^2)}{2mc^2} \approx E - mc^2 \qquad (3.271)$$

The fact that Eq. 3.270 depends only on E^2 means that if E is an eigenvalue so is $-E$. There is a symmetry between positive and negative energy solutions because scalar potential S does not distinguish particles and anti-particles.

Solving Eq. 3.270 is equivalent to solving the non-relativistic equation as:

$$E' \Psi(\vec{r}) = \left[-\frac{\hbar^2 \nabla^2}{2m} + S(\vec{r})\right] \Psi(\vec{r}) \qquad (3.272)$$

The eigenvalues of this equation obviously are related to those of Eq. 3.270 by the following relation:

$$E^2 = c^2 p^2 + m^2 c^4 E' \qquad (3.273)$$

3.9 Infrastructure, Characteristics, Derivation, and Scalar Wave Properties   283

For E^2 to be positive, only the eigenvalues of Eq. 3.272 will satisfy:

$$E' > -\frac{1}{2}(mc^2) \qquad (3.274)$$

This corresponds to the physical solution of relativistic Eq. 3.270.

3.9.8.2 Quantum Gauge Invariance

We begin this section with a brief review of gauge invariance in quantum mechanics. The classical Hamiltonian, $\mathcal{H}$, describing the motion of a charged particle in the presence of vector potential ϕ and scalar potential $\vec{A}$ is given by:

$$\mathcal{H} = \frac{1}{2m}\left[\vec{p} - \frac{q}{c}\vec{A}\right]^2 + qc \qquad (3.275)$$

where:

$m =$ the particle's mass
$\vec{p} =$ the dynamical momentum of the particle
$q =$ the particle's charge
$c =$ the speed of light

From Eq. 3.275 we obtain the Schrödinger equation (i.e., non-relativistic and neglecting spin) in the usual way by the substitutions of the following terms; thus, we have:

$$\mathcal{H} \rightarrow i\hbar\frac{\partial}{\partial t} \qquad (3.276)$$

$$\vec{p} \rightarrow i\hbar\vec{\nabla}$$

Therefore, we can write Eq. 3.275 in the following form:

$$\frac{\hbar}{i}\frac{\partial \Psi(\vec{r},t)}{\partial t} = \frac{1}{2m}\left[\frac{\hbar}{i}\vec{\nabla} - \frac{q}{c}\vec{A}\right]\left[\frac{\hbar}{i}\vec{\nabla} - \frac{q}{c}\vec{A}\right]\Psi(\vec{r},t) + q\phi\Psi(\vec{r},t) \qquad (3.277)$$

where:

$h =$ Planck's constant/2π (i.e., $\hbar = h/2\pi$)
$\Psi(\vec{r},t) =$ the wave's function
$\vec{\nabla} =$ the Del or Nabla operator (i.e., $\partial/\partial x, \partial/\partial y, \partial/\partial z$)

By convention the Schrödinger equation is expressed in the *Coulomb gauge* with $\vec{\nabla} \cdot \vec{A} = 0$.

The quantum mechanical transformation that preserves the gauge invariance of the Schrodinger equation is a simple extension of classical gauge invariance. Thus, if

$$\vec{E} = -\nabla\phi - \frac{1}{c}\frac{\partial \vec{A}}{\partial t} \quad (3.278)$$

and

$$\vec{B} = \vec{\nabla} \times \vec{A} \quad (3.279)$$

where again $\vec{E}$ is the electric field and $\vec{B}$ is the magnetic field; and if χ is any scalar, then the substitutions of the following terms would leave $\vec{E}$ and $\vec{B}$ invariant:

$$\vec{A}' = \vec{A} - c\nabla x \quad (3.280a)$$

$$\phi' = \phi + \frac{\partial x}{\partial t} \quad (3.280b)$$

In quantum mechanics the additional transformation of the wave function by minimal substitution (see Gates [30]) is required. This is given by the following equation:

$$\Psi' = e^{-ix}\Psi \quad (3.281)$$

The invariance of the Schrödinger equation is shown as follows because:

$$\frac{\partial}{\partial x}\Psi' = \frac{\partial}{\partial x}e^{-ix}\Psi = e^{-ix}\frac{\partial \Psi}{\partial x} - i\frac{\partial x}{\partial x}\Psi e^{-ix} \quad (3.282)$$

or

$$i\hbar \nabla \left(e^{-ix}\Psi\right) = e^{-ix}(-i\hbar\nabla - \hbar\nabla x)\Psi \quad (3.283)$$

and

$$\left(i\hbar\nabla - \frac{q}{c}\vec{A}\right)e^{-ix}\Psi = e^{-ix}\left(-i\hbar\nabla - \hbar\nabla x - \frac{q}{c}\vec{A}\right)\Psi \quad (3.284)$$

Presuming partial differential with respect to the variable time over Eq. 3.284, we have:

$$\frac{\partial}{\partial t}\Psi' = \frac{\partial}{\partial t}e^{-ix}\Psi = e^{ix}\frac{\partial \Psi}{\partial t} - i\frac{\partial x}{\partial t}\Psi e^{-ix} \quad (3.285)$$

The term $i(\partial x/\partial t)\Psi\, e^{-ix}$ can be collected with the $\phi\Psi e^{-ix}$ term. Thus, the substitution into Eq. 3.285 will yield the following form:

$$\Psi' = e^{-ix}\Psi \quad (3.286a)$$

3.9 Infrastructure, Characteristics, Derivation, and Scalar Wave Properties

$$\vec{A}' = \vec{A} - \frac{\hbar c}{q}\nabla x \qquad (3.286b)$$

$$\phi' = \phi + \frac{\hbar}{q}\frac{\partial x}{\partial t} \qquad (3.286c)$$

Thus, leaving the Schrödinger equation unchanged.[2] From this we can see that a gauge transformation changes the waveform by an arbitrary phase constant (i.e., e^{-ix}). Because the phase of a wave function is not observable, it does not refer to any physically observable effect.

In what follows we will confine the discussion to the vector potential and drop the terms that include the scalar potential. It is important to emphasize that vector potential $\vec{A}$ appearing in quantum equations is formally identical with the classical vector potential and that quantum effects associated with it do not imply that the vector potential appearing in quantum equations differs in any way from the classical vector potential.[3] Therefore, for static conditions the vector potential used in the Schrodinger equation may be calculated in the standard classical way from normal source currents[4] by solving for a volume integral with suitable boundary conditions.[5] Thus, we have the following equation:

$$\vec{A} = \frac{1}{4\pi}\int \frac{\vec{J}}{r}dV + \text{(gauge terms)} \qquad (3.287)$$

where

$\vec{J}$ = the current density of the source
r = the distance from the source current element to the position in space where the vector potential is to be determined
dV = an element of volume

---

[2] Some authors write these substitutions with an inverted sign—that is, $\Psi' = e^{ix}\Psi$, $\vec{A}' = \vec{A} + (\hbar/q)(\partial x/\partial t)\nabla x$; $\phi' = \phi - (\hbar/q)(\partial x/\partial t)$. However, the choice of sign convention does not alter the gauge-invariant properties of the Schrödinger equation.

[3] In quantum gauge transformation, the gauge terms differ from the classical terms by the factor $\hbar/q$ for the vector potential and $\hbar/q$ for the scalar potential. However, this offers no difficulty because the gauge terms may be completely arbitrary functions $\nabla F(x, y, z, t)$ and still leave observables unchanged. For proof see Landau and Lifschitz, Section 124 [31].

[4] These arguments may not apply to a vector potential directly associated with a supercurrent. Normal currents are classically gauge-invariant while in London's theory supercurrents do not receive a gauge-invariant expression. However, superconductivity is a special instance of broken gauge symmetry and therefore does not yield results that contradict the general case. For a treatment of superconductivity as an instance of spontaneously broken symmetry of the electromagnetic gauge, see Forster [32].

[5] For a derivation of this equation from the law of Biot and Savart, see book by Panofsky and Phillips, Section 7-8 [33].

As is well known, Eq. 3.287 can be defined only up to an arbitrary gradient of a scalar (i.e., pseudo-potential surfaces) or gauge terms. Such gauge terms can be interpreted as gradients of pseudo-potential surfaces that arise from the source configurations of pseudo-charges. By convention, the pseudo-potential surfaces can be expressed in terms of the divergence of the vector potential by solving Poisson's equation:

$$\nabla^2 \mathbb{S} = \vec{\nabla} \cdot \vec{A} \qquad (3.288)$$

Equation 3.288 enables definition of a specific gauge in terms of the divergence of the vector potential. For the special condition, where $\vec{\nabla} \cdot \vec{A} = 0$ (i.e., *Coulomb gauge*), all pseudo-source terms are zero; thus, all pseudo-potential surfaces, $\mathbb{S}$, are zero. Under these conditions a *curl-free vector potential (CFVP)* effect is solenoidal [34].

3.9.8.3 Gauge-Invariant Phase Difference

A difference in the phase of a wave function at two points in space is, in principle, an observable. Following Gates [30], we define phase δ of a wave function by

$$\delta = -i \left(\frac{1}{2} \right) \ln \left(\frac{\Psi}{\Psi^*} \right) \qquad (3.289)$$

where Ψ^* is the complex conjugate of wave function Ψ. The phase difference, $\Delta\delta$, can be expressed by:

$$\Delta\delta = i \left(\frac{1}{2} \right) \left[\ln \left(\frac{\Psi(2)}{\Psi^*(2)} \right) - \ln \left(\frac{\Psi(1)}{\Psi(2)} \right) \right] = \delta_2 - \delta_1 \qquad (3.290)$$

with

$$\begin{cases} \Psi(1) = u e^{i\delta_1} \\ \Psi(2) = u e^{i\delta_2} \end{cases} \qquad (3.291)$$

Clearly, $\Delta\delta$ is a gauge-dependent quantity. Yet, note that the expression for $\Delta\delta^*$ is given by

$$\Delta\delta^* = \delta_2 - \delta_1 - \frac{q}{\hbar c} \int_1^2 \vec{A} \cdot d\vec{x} \qquad (3.292)$$

Equation 3.292 is gauge invariant after substitution of the gauge transformation given by Eqs. 3.286a and 3.286b. Thus, we can write the following expressions:

3.9 Infrastructure, Characteristics, Derivation, and Scalar Wave Properties

$$\Delta\delta'^* = (\delta_2 - x_2) - (\delta_1 - x_1) - \frac{q}{\hbar c}\int_1^2 \left(\vec{A} - \frac{\hbar c}{q}\nabla x\right) \cdot d\vec{x}$$

$$= (\delta_2 - \delta_1) - \frac{q}{\hbar c}\int_1^2 \left[\vec{A}\cdot d\vec{x} - \left((x_2 - x_1) - \int_1^2 \nabla x \cdot d\vec{x}\right)\right] = \Delta\delta^*$$

(3.293)

We therefore conclude that quantum gauge invariance requires that any measurable quantity that depends on such a difference in phase angles also must depend on the integral of the vector potential, as in Eq. 3.292.

We now derive the gauge-invariant phase difference in a manner that ensures that P, appearing in Hamiltonian Eq. 3.275, is consistently defined as dynamic momentum. To do this we first form the probability current density as:

$$\vec{J}_p = \frac{1}{2}\left[\left(\frac{(\hbar/i)\nabla - (q/c)\vec{A}}{m}\Psi\right)^* \Psi - \Psi^* \left(\frac{(\hbar/i)\nabla - (q/c)\vec{A}}{m}\Psi\right)\right] \quad (3.294)$$

where we continue to neglect terms involving the scalar potential. Now, again using $\Psi = ue^{i\delta}$ and after some algebraic manipulation, we obtain the "generalized momentum" in terms of phase gradient and vector potential:

$$m\frac{\vec{J}_p}{u^2} = \hbar\nabla\delta - \frac{q}{c}\vec{A} \quad (3.295)$$

Gauge-invariant phase difference[6] $\Delta\delta^*$ is readily obtained by forming the line integral:

$$\Delta\delta^* = \int_1^2 \left(\nabla\delta - \frac{q}{\hbar c}\vec{A}\right) \cdot d\vec{x} = \delta_2 - \delta_1 - \frac{q}{\hbar c}\int_1^2 \vec{A}\cdot d\vec{x} \quad (3.296)$$

Many physicists are surprised to learn that gauge invariance of a phase difference between two points of a wave function involves an expression, including a line integral of the vector potential, that is not closed in a loop. Although the line integral segment of the vector potential in Eq. 3.296 is not gauge invariant, the complete argument, including the initial phase terms, is gauge invariant. Thus, in quantum mechanics gauge-invariant expressions of the phase difference between spatially separated points in a wave function of a charged particle do not require a closed line integral of the vector potential [34].

In classical EM theory the equations for the fields provide gauge invariance by prescriptions that make both the gauge terms and any CFVPs vanish by the rules of

---

[6]Note that quantity $\hbar\nabla\delta$ appears as the generalized dynamic momentum and has correspondence with the dynamic momentum, $\vec{p}$, appearing in Hamiltonian Eq. 3.275.

elementary vector calculus. In quantum mechanics, however, minimal substitution gives a prescription for gauge invariance in which the gauge terms cancel but the CFVP terms remain. Minimal substitution requires careful interpretation of the gauge terms in order that they may be inserted appropriately into arguments that contain line integral segments of the vector potential. Thus, a calculation of the vector potential in a gauge that includes pseudo-potential terms (i.e., $\vec{\nabla} \cdot \vec{A} \neq 0$) must carry a definition of those terms according to Eq. 3.288; and they must be inserted into the argument defining the phase difference, as demonstrated by Eq. 3.293, in order to preserve its integrity as a gauge-invariant expression. Such a procedure is immediately seen to be equivalent to an initial calculation of a vector potential in the Coulomb gauge followed by a gauge transformation, as given by the substitutions shown in Eqs. 3.286a and 3.286b [34].

It is interesting to note that such a gauge transformation in quantum mechanics has the effect of reducing the expression for the gauge-invariant phase difference to an equivalent expression with the vector potential in the Coulomb gauge. Also note that the gauge transformation given by Eqs. 3.86a and 3.286b does not allow a gauge expression of $\Delta \delta^*$ to be formed such that both the gauge and the vector potential terms will cancel simultaneously. Thus, setting a gauge such that the vector potential is reduced to zero—that is, $\vec{A}' = \vec{A} - (\hbar c/q)\nabla x = 0$—the gauge pseudo-potentials remain, thereby ensuring that the phase difference does not change after a gauge transformation. This result holds whether the vector potential has curl or is curl-free.

Finally, note that the Aharonov-Bohm effect may be modeled by using the difference between two gauge-invariant phase differences of de Broglie waves calculated separately for two paths,[7] Γ_1 and Γ_2, that enclose a magnetic flux bundle, ϕ. Thus, for a CFVP $\vec{A}_{\text{eff}}$ arising from ϕ we have:

$$\Delta \delta^*_{\Gamma_1} = \delta_2 - \delta_1 - \frac{q}{\hbar c} \int_1^2 \vec{A}_{\text{eff}_{\Gamma_1}} \cdot d\vec{x} \qquad (3.297)$$

and

$$\Delta \delta^*_{\Gamma_2} = \delta_2 - \delta_1 - \frac{q}{\hbar c} \int_1^2 \vec{A}_{\text{eff}_{\Gamma_2}} \cdot d\vec{x} \qquad (3.298)$$

In addition, because we have:

$$\int_1^2 \vec{A}_{\text{eff}_{\Gamma_2}} \cdot d\vec{x} = -\int_1^2 \vec{A}_{\text{eff}_{\Gamma_1}} \cdot d\vec{x} = -\int_1^2 \vec{A}_{\text{eff}_{\Gamma_2}} \cdot d\vec{x} \qquad (3.299)$$

then we have:

---

[7]The more conventional procedure is to form the wavefunction: $\Psi_{\Gamma_1}(x) + \Psi_{\Gamma_2}(x)$ superposition of the two wavefunctions that arrive at the interference region after transversing the separate paths, Γ_1 and Γ_2 where,

3.9 Infrastructure, Characteristics, Derivation, and Scalar Wave Properties

$$\Delta\delta^*_{\Gamma_1} - \Delta\delta^*_{\Gamma_2} = \frac{q}{\hbar c} \oint_{1\to 2} \vec{A}_{\text{eff}} \cdot d\vec{x} = \frac{q}{\hbar c}\phi \qquad (3.300)$$

The enclosed flux bundle can be expressed in terms of the closed line integral of a CFVP that is gauge invariant by a remedy in classical EM theory (i.e., Stokes's theorem). Nevertheless, the phase difference of the wave function in each leg, $\Delta\delta^*_{\Gamma_1}$ and $\Delta\delta^*_{\Gamma_2}$, taken separately are each gauge invariant by the quantum prescription. Although the closed line integral of $\vec{A}_{\text{eff}}$ provides a convenient description of the *Aharonov-Bohm effect* in terms of the enclosed flux, it is not required for a gauge-invariant description of the this effect. This analysis shows that the Aharonov-Bohm effect is a consequence of quantum gauge invariance rather than of classical EM gauge invariance [34].

We conclude that the gauge invariance of the Schrödinger equation determines the gauge invariance of a phase difference given by

$$\Delta\theta^* = \delta_2 - \delta_1 - \frac{q}{\hbar}\int_1^2 \vec{A}\cdot d\vec{x} \qquad (3.301)$$

such that the substitutions of the following terms in gauge framresults in:

$$\delta'_1 = \delta_1 - x_1, \quad \delta'_2 = \delta_2 - x_2, \quad \vec{A}' = \vec{A} - \frac{\hbar c}{q}\nabla x \qquad (3.302)$$

Therefore, the phase difference is unchanged.

Note: There are a couple of experiments proposed by Raymond C. Gelinas [3] in respect to demonstration of the CFVP in a simply connected space; they involve a toroidal coil-carrying current that provides a source of CFVP that extends over a large region of space. Thus, readers who are interested in further investigation of this subject should follow his approach and suggestions; this is beyond the scope of the book.

3.9.8.4 The Matrix of Spacetime

The two basic requirements for a fundamental theory of matter are that the equations obey wave mechanics and that they also must be covariant (i.e., obey special relativity). The Dirac delta-function equation for the electron fits these two requirements. The equation, however, gives not only the energy states of electrons but also predicts the existence of electrons with negative energy. These negative energy states are not physical and therefore are unobservable. Nevertheless, by stimulating the negative energy states with sufficient energy (i.e., gamma rays) electrons may be kicked into positive energy states and become real. The holes left behind are the positrons. Thus, we can imagine that we live within a sea of virtual (unobservable) electrons and other particles in the Dirac sea.

Consider a single point that comprises all of pre-existence. There is no spacetime because there is only a point, which can contain within itself an infinity of states (e.g., the electron can be in many states although it is a point-like particle). Some of these are Fermion states (i.e., anti-symmetrical). They obey the *Pauli exclusion principle*, which says that each state must differ from another state in at least one way. Suppose fluctuations create a pair of electrons with exactly the same state. This pair must "separate" from each other in order to satisfy the Pauli exclusion principle. Because there are infinite numbers of states, the principle "squeezes" an infinity of spacetime out of a single point. In this view, then, spacetime is a concretization of the variance of virtual states. The Dirac sea, thus, is seen to be the matrix of spacetime [29].

In special relativity we are allowed to use only inertial frames to assign coordinates to events. There are many different types of inertial frames. Still, it is convenient to adhere to those with *standard coordinates*. That is, spatial coordinates that are right-handed rectilinear Cartesian based on a standard unit of length, and time scales based on a standard unit of time. We will continue to assume that we are employing standard coordinates. From now on, however, we intend to make no assumptions about the relative configuration of the two sets of spatial axes, and the origins of time, when dealing with two inertial frames. Thus, the most general transformation between two inertial frames consists of a Lorentz transformation in the standard configuration plus a translation (including a translation in time) and a rotation of the coordinate axes. The resulting transformation is called a *general Lorentz transformation*, as opposed to a Lorentz transformation in the standard configuration, which hereafter will be termed a *standard Lorentz transformation*.

As part of Lorentz transformation analysis, we proved quite generally that corresponding differentials in two inertial frames, S and S', satisfy the relation:

$$dx^2 + dy^2 + dz^2 - c^2 dt^2 = dx'^2 + dy'^2 + dz'^2 - c^2 dt'^2 \qquad (3.303)$$

Thus, we expect this relation to remain invariant under a general Lorentz transformation. Because such a transformation is *linear*, it follows that

$$\begin{aligned}(x_2 - x_1)^2 + (y_2 - y_1)^2 + (z_2 - z_1)^2 - c^2(t_2 - t_1)^2 \\ = (x'_2 - x'_1)^2 + (y'_2 - y'_1)^2 + (z'_2 - z'_1)^2 - c^2(t'_2 - t'_1)^2\end{aligned} \qquad (3.304)$$

where (x_1, y_1, z_1, t_1) and (x_2, y_2, z_2, t_2) are the coordinates of any two events in S, and the primed symbols denote the corresponding coordinates in S'. It is convenient to write:

$$-dx^2 - dy^2 - dz^2 + c^2 dt^2 = ds^2 \qquad (3.306)$$

and

3.9 Infrastructure, Characteristics, Derivation, and Scalar Wave Properties

$$-(x_2 - x_1)^2 - (y_2 - y_1)^2 - (z_2 - z_1)^2 + c^2(t_2 - t_1)^2 = s^2 \quad (3.311)$$

The differential ds, or the finite number s, defined by these equations is called the *interval* between the corresponding events. Equations 3.305 and 3.306 express the fact that the interval *between two events* is invariant, in the sense that it has the same value in all inertial frames. In other words, the interval between two events is invariant under a general Lorentz transformation.

Let us consider entities defined in terms of four variables:

$$x^1 = x, \quad x^2 = y, \quad x^3 = z, \quad x^4 = ct \quad (3.307)$$

and that *transform as tensors* under a general Lorentz transformation. From now on such entities will be referred to as *4-tensors*.

Tensor analysis cannot proceed very far without the introduction of a non-singular tensor gg_{ij}, the so-called *fundamental tensor* that is used to define the operations of raising and lowering suffixes. The fundamental tensor is usually introduced using metric $ds^2 = g_{ij} dx^i dx^j$, where ds^2 is a differential invariant. We have already come across such an invariant—namely,

$$\begin{aligned} ds^2 &= -dx^2 - dy^2 - dz^2 + c^2 dt^2 \\ &= -(dx^1)^2 - (dx^2)^2 - (dx^3)^2 + (dx^4)^2 \\ &= g_{\mu\nu} dx^\mu dx^\nu \end{aligned} \quad (3.308)$$

where μ, ν run from 1 to 4. Note that the use of Greek suffixes is conventional in 4-tensor theory. Roman suffixes are reserved for tensors in three-dimensional Euclidian space, so-called *3-tensors*. The 4-tensor $g_{\mu\nu}$ has the components $g_{11} = g_{22} = g_{33} = -1, g_{44} = 1$, and $g_{\mu\nu} = 0$ when $\mu \neq \nu$ is in all permissible coordinate frames.

From now on $g_{\mu\nu}$, as defined before, is adopted as the fundamental tensor for 4-tensors. $g_{\mu\nu}$ can be thought of as the *metric tensor* of the space with points that are the events (x^1, x^2, x^3, x^4). This space usually is referred to as *spacetime* for obvious reasons. Note that spacetime cannot be regarded as a straightforward generalization of Euclidian 3-space to four dimensions, with time as the fourth dimension. The distribution of signs in the metric ensures that time coordinate x^4 is not on the same footing as the three-space coordinates. Thus, spacetime has a non-isotropic nature that is quite unlike Euclidian space, with its positive definite metric. According to the relativity principle, all physical laws are expressible as interrelationships between 4-tensors in spacetime.

A tensor of rank one is called a *4-vector*. We also will have occasion to use ordinary vectors in three-dimensional Euclidian space. Such vectors are called *3-vectors* and are conventionally represented by boldface symbols. We will use the Latin suffixes, i, j, k, and so on, to denote the components of a 3-vector; these suffixes are understood to range from 1 to 3. Thus, $\vec{u} = u^i = dx^i/dt$ denotes a velocity vector. For 3-vectors, we will use the notation $u^i = u_i$ interchangeably—that is, the level of the suffix has no physical significance.

When tensor transformations from one frame to another actually have to be computed, we usually will find it possible to choose coordinates in the standard configuration, so the standard Lorentz transform applies. Under such a transformation, any contravariant 4-vector, T^μ, transforms according to the same scheme as the difference in coordinates $x_2^\mu - x_1^\mu$ between two points in spacetime. It follows that

$$T^{1'} = \gamma(T^1 - \beta T^4) \tag{3.309}$$

$$T^{2'} = T^2 \tag{3.310}$$

$$T^{3'} = T^3 \tag{3.311}$$

$$T^{4'} = \gamma(T^4 - \beta T^1) \tag{3.312}$$

where $\beta = v/c$. Higher ranked 4-tensors transform according to the rules of the formal definition of a tensor, as follows:

1. An entity component, $A_{ij\cdots k}$ in the x^i system and $A_{i'j'\cdots k'}$ in the $x^{i'}$ system, is said to behave as a *covariant tensor* under the transformation $x^i \to x^{i'}$ if:

$$A_{i'j'\cdots k'} = A_{ij\cdots k} p_{i'}^i p_{j'}^j \cdots p_{k'}^k \tag{3.313}$$

2. Similarly, $A^{ij\cdots k}$ is said to behave as a *contravariant tensor* under $x^i \to x^{i'}$ if:

$$A^{i'j'\cdots k'} = A^{ij\cdots k} p_i^{i'} p_j^{j'} \cdots p_k^{k'} \tag{3.314}$$

3. Finally, $A_{k\cdots l}^{i\cdots j}$ is said to behave as a *mixed tensor* (contravariant in $i\cdots j$ and covariant in $k\cdots l$) under $x^i \to x^{i'}$ if:

$$A_{k'\cdots l'}^{i'\cdots j'} = A_{k\cdots l}^{i\cdots j} p_i^{i'} \cdots p_j^{j'} \cdots p_{k'}^k \tag{3.315}$$

When an entity is described as a tensor, it generally is understood that it behaves as a tensor *under all* non-singular differentiable transformations of the relevant coordinates. An entity that only behaves as a tensor under a certain subgroup of non-singular differentiable coordinate transformation is called a *qualified tensor* because its name conventionally is qualified by an adjective recalling the subgroup in question. For instance, an entity that only exhibits tensor behavior under Lorentz transformations is called a *Lorentz tensor* or, more commonly, a *4-tensor*.

Given the conditions defined in Eqs. 3.313 through 3.315, the transformation coefficients take the following form:

3.9 Infrastructure, Characteristics, Derivation, and Scalar Wave Properties

$$p_\mu^{\mu'} = \begin{bmatrix} +\gamma & 0 & 0 & -\gamma\beta \\ 0 & 1 & 0 & 0 \\ 0 & 0 & 1 & 0 \\ -\gamma\beta & 0 & 0 & +\gamma \end{bmatrix} \quad (3.316)$$

$$p_\mu^{\mu'} = \begin{bmatrix} +\gamma & 0 & 0 & +\gamma\beta \\ 0 & 1 & 0 & 0 \\ 0 & 0 & 1 & 0 \\ +\gamma\beta & 0 & 0 & +\gamma \end{bmatrix} \quad (3.317)$$

Often the first three components of a 4-vector coincide with the components of a 3-vector. For example, the x^1, x^2, and x^3 in $R^\mu = (x^1, x^2, x^3, x^4)$ are the components of $\vec{r}$, the position 3-vector of the point at which the event occurs. In such cases we adopt the notation exemplified by $R^\mu = (\vec{r}, ct)$. The covariant form of such a vector is simply $R_\mu = (-\vec{r}, ct)$. The squared magnitude of the vector is $(R)^2 = R_\mu R^\mu = -r^2 + c^2 t^2$. The inner product $g_{\mu\nu} R^\mu Q^\nu = R_\mu Q^\mu$ of R^μ with a similar vector $Q^\mu = (q, k)$ is given by $R_\mu Q^\mu = -\vec{r} \cdot \vec{q} + ctk$. The vectors R^μ and Q^μ are said to be *orthogonal* if $R_\mu Q^\mu = 0$.

Because a general Lorentz transformation is a *linear* transformation, the partial derivative of a 4-tensor is also a 4-tensor:

$$\frac{\partial A^{\nu\sigma}}{\partial x^\mu} = A^{\nu\sigma}_{,\mu} \quad (3.318)$$

Clearly, a general 4-tensor acquires an extra covariant index after partial differentiation with respect to contravariant coordinate x^μ. It is helpful to define a covariant derivative operator:

$$\partial_\mu \equiv \frac{\partial}{\partial x^\mu} = \left(\vec{\nabla}, \frac{1}{c}\frac{\partial}{\partial t}\right) \quad (3.319)$$

where

$$\partial_\mu A^{\nu\sigma} \equiv A^{\nu\sigma}_{,\mu} \quad (3.320)$$

There is a corresponding contravariant derivative operator:

$$\partial^\mu \equiv \frac{\partial}{\partial x_\mu} = \left(-\vec{\nabla}, \frac{1}{c}\frac{\partial}{\partial t}\right) \quad (3.321)$$

where

$$\partial^\mu A^{\nu\sigma} \equiv g^{\mu\tau} A^{\nu\sigma}_{,\tau} \quad (3.322)$$

The four-divergence of a 4-vector, $A^\mu = (\vec{A}, A_0)$, is the invariant:

$$\partial^\mu A_\mu = \partial_\mu A^\mu = \vec{\nabla} \cdot \vec{A} + \frac{1}{c}\frac{\partial A^0}{\partial t} \tag{3.323}$$

The four-dimensional Laplacian operator, or *d'Alembertian*, is equivalent to the invariant contraction, as follows:

$$\Box \equiv \partial_\mu \partial^\mu = -\nabla^2 + \frac{1}{c^2}\frac{\partial^2}{\partial t^2} \tag{3.324}$$

Recall that we still need to prove from the Lorentz transformation, the invariance of the differential metric:

$$ds^2 = dx'^2 + dy'^2 + dz'^2 - c^2 dt'^2 = dx^2 + dy^2 + dz^2 - c^2 dt^2 \tag{3.325}$$

Between two general inertial frames implies that the coordinate transformation between such frames is necessarily linear. To put it another way, we need to demonstrate that a transformation that transforms metric $g_{\mu\nu} dx^\mu dx^\nu$ with constant coefficients into metric $g_{\mu'\nu'} dx^{\mu'} dx^{\nu'}$ without constant coefficients must be linear. Now,

$$g_{\mu\nu} = g_{\mu'\nu'} p^{\mu'}_\mu p^{\nu'}_\nu \tag{3.326}$$

Differentiating with respect to x^σ, we get:

$$g_{\mu'\nu'} p^{\mu'}_{\mu\sigma} p^{\nu'}_\nu + g_{\mu'\nu'} p^{\mu'}_\mu p^{\nu'}_{\nu\sigma} = 0 \tag{3.327}$$

where

$$p^{\mu'}_{\mu\sigma} = \frac{\partial p^{\mu'}_\mu}{\partial x^\sigma} = \frac{\partial^2 x^{\mu'}}{\partial x^\mu \partial x^\sigma} = p^{\mu'}_{\sigma\mu} \tag{3.328}$$

and so on. Interchanging the indices μ and σ yields:

$$g_{\mu'\nu'} p^{\mu'}_{\mu\sigma} p^{\nu'}_\nu + g_{\mu'\nu'} p^{\mu'}_\sigma p^{\nu'}_{\nu\mu} = 0 \tag{3.329}$$

Interchanging the indices ν and σ gives:

$$g_{\mu'\nu'} p^{\mu'}_\sigma p^{\nu'}_{\nu\mu} + g_{\mu'\nu'} p^{\mu'}_\mu p^{\nu'}_{\nu\sigma} = 0 \tag{3.330}$$

where the indices μ' and ν' have been interchanged in the first term. It follows from Eqs. 3.327, 3.329, and 3.330 that

$$g_{\mu'\nu'} p^{\mu'}_{\mu\sigma} p^{\nu'}_\nu = 0 \tag{3.331}$$

Multiplication by $p^\nu_{\sigma'}$ yields:

$$g_{\mu'\nu'}p^{\mu'}_{\mu\sigma}p^{\nu'}_{\nu}p^{\nu}_{\sigma'} = g_{\mu'\sigma'}p^{\mu'}_{\mu\sigma} = 0 \qquad (3.332)$$

Finally, multiplication by $g^{\nu'\sigma'}$ gives:

$$g_{\mu'\sigma'}g^{\nu'\sigma'}p^{\mu'}_{\mu\sigma} = p^{\nu'}_{\mu\sigma} = 0 \qquad (3.333)$$

This proves that the coefficients $p^{\nu'}_\mu$ are constants and, thus, that the transformation is linear.

3.9.9 A Human's Body Works with Scalar Waves

We have stated so far that a SW is a non-linear, non-Hertzian (does not diminish with distance) standing wave capable of supporting significant effects including carrying information and inducing higher levels of cellular energy, which greatly enhances the performance and effectiveness of the body and immune system. Additionally, it helps to clear cellular memory by shifting polarity, similar to erasing the memory of a cassette tape with a magnet. Scalar waves travel faster than the speed of light and do not decay over time or distance.

Doctors and biologists are not able to tell us what life is. They cannot explain in a proper way how cells communicate with each other. Modern science focuses on chemical reactions, however, that is not the whole picture. The most important communication tool in the body has not been detected by mainstream medicine yet because of the huge conspiracy against SWs.

A human's body works with SWs (an aspect of neutrinos). The body also constantly is generating these universal SWs. Scalar waves are produced when two EMWs of the same frequency are exactly out of phase (opposite to each other) and the amplitudes subtract and cancel or destroy each other. The result is not exactly an annihilation of magnetic fields but a transformation of energy back into a SW. This scalar field has reverted to a vacuum state of potentiality. Scalar waves can be created by wrapping electrical wires around a figure eight in the shape of a Möbius coil. When electric current flows through the wires in opposite directions, the opposing EM fields from the two wires cancel each other and create a SW as we stated in the previous section.

There is another Möbius coil configuration found within the vascular system. The continuous flow of blood through the arterial system that runs next to the venous system but in opposite directions contains Möbius coil properties. The circulation of blood throughout the body resembles the figure-eight shape (see Fig. 3.17 as well as Fig. 3.52) of the Möbius coil. Excerpts from the book, *"The Heart of Health: The Principles of Physical Health and Vitality"* by Stephen Linsteadt, discuss these issues thoroughly.

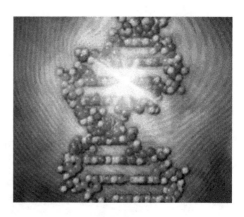

Fig. 3.52 Illustration of how DNA emits and absorbs light energy

The whole body is able to send information and free energy (e.g., acupuncture works with that) back and forth to every cell. This goes faster than light because SWs are extremely fast. Scalar waves can send pictures with parallel technology and holographic images. That is why our memory pictures always come quick, three-dimensional, and clear without any "pixels" lacking. In a computer you can see pixels missing in a picture if the connection is bad because computers use more primitive digital technology compared to humans' brains.

Even the brain works mostly with SWs. It can change them into EMWs and back again. The left and right part of the brain can send different SWs and let them interfere, establishing a scalar-wave interferometer (i.e., thoughts are SWs). That is why they can be extremely quick and expend little energy to penetrate the Earth and reach remote places; we call this *telepathy*.

Because the body uses SWs as its most important tool, *mind control* technology is forced to use the same tool in order to be successful. Otherwise, it is difficult to induce pain in a part of the body at distance; read or change thoughts; control, manipulate, or erase memory; make the body a remote-controlled camera; or force someone to act as a honeytrap. We know that this torture technology works perfectly after more than 60 years of secret research. Every Targeted Individuals (TIs) can give testimony on that. Therefore, the conclusion is close to a mind control fight with its own technology: scalar waves. There are already a lot of devices on the market using SWs to divert psychotronic (Russian wording) attacks. Some of them absorb SWs and transform them into something useful.

Scalar waves increase the energy covalent level of every single hydrogen atom in the body, as verified by spectrographs. This is significant in that hydrogen bonds are what hold deoxyribonucleic acid (DNA) together. Dr. Glen Rein, a quantum biologist, points out from his experiments with SWs that they positively influence the immune and nervous systems independent of the belief systems of the individual. Every cell in the human body, when functioning at its maximum health potential, ranges between 70–90 mV. Diseases and aging occur when the cellular energy declines to levels below this range.

The human body has crystalline structures in every cell wall that can hold a charge. The shape of SWs is reminiscent of the multiple helical structure of DNA as it folds in on itself. Quantum mechanical models describe subatomic particles that can store and carry biological information along helical macromolecules such as DNA. This indicates that scalar energy is capable of imprinting itself in the DNA.

As part of the health implications of SWs, it is a known scientific fact that everything is energy and that everything vibrates at different frequencies. Scalar energy is a unique energy form that can be harnessed and directed into solid objects or bodies placed in its field. You can use scalar energy to raise the vibratory frequency of supplements, food, and water and other liquids. When used on living cells, they become more vibrant, naturally energized, and more capable of absorbing nutrients and eliminating wastes and toxins. When embedded in nutritional supplements, food, and beverages, SWs will make these substances more absorbable and bio-available, thus further raising the energy levels in cells.

The higher the level of energy in cells, the greater their ability to absorb nutrients, eliminate toxins and wastes; and build healthy tissues, bones, organs, glands, and nerves. High cellular energy greatly enhances the performance and effectiveness of the immune system and the body's ability to heal itself. Everything is vibrations and all life is energy. [[repeat]] Degenerative diseases begin developing when overall cellular mV is reduced significantly. Cancer cells measure 15–20 mV. Whole foods and fresh foods measure at a higher mV; cooked, processed, and refined foods measure at a lower mV; canned food has 0 mV.

Scalar waves are cumulative and tend to build up in the body. By frequently exposing yourself to SW energy, you build stronger, healthier cells; enhance your immune system; and support your body's ability to neutralize manufactured waves (i.e., e.m.f.s) that surround you. Increased exposure to SWs means a greater accumulation of them in cells that helps raise their vibratory rate to the ideal 70–90 mV level.

Scalar waves are non-linear and support the neutralization of all artificial 50–60 cycle waves in the body from cell phones, cordless phones, computers, microwaves, Wi-Fi routers, and numerous other sources. Scalar-wave fields always have existed, and now we have the knowledge and the technology to increase them in the body and people's lives. They are well known in astrophysics, geology, and hydrodynamics. A scalar-wave field is known as a fifth-dimension non-linear field; thus, the SW is not bound by three-dimension laws. Scalar-wave fields function in a self-referral and self-generating manner. They are unbounded and capable of passing through solid matter.

According to research done over several decades, the key benefits of scalar-wave technology are as follows:

Eliminates and nullifies the effects of artificial frequencies (50–60 cycle) in the human body.
Increases overall body energy levels because of increasing cellular energy for trillions of cells.
Cleanses the blood, improving chylomicron levels (i.e., protein/fat particles floating in the blood) and triglyceride profiles and fibrin patterns.

Increases the energy level of every single cell in the body to the ideal 70–90 mV range.

Improves immune function by as much as 149% as proved in laboratory studies.

Increases the energy covalent level of every single hydrogen atom in the body as verified by spectrographs. This is significant because covalent hydrogen bonds are what hold DNA together.

Improves mental focus, as demonstrated by the increased amplitude of EEG frequencies.

Improves cell wall permeability, thus facilitating the intake of nutrients into every cell and the elimination of waste from all cells.

Balances out the two hemispheres of the brain, as measured by EEG tests.

Decreases the surface tension of substances such as food, water, and supplements, thus enhancing the body's ability to assimilate and hydrate.

Inhibits the uptake of noradrenaline by PC12 nerve cells to support a better mood.

Catalyzes heightened states of awareness and creativity while advancing the process of manifestation in regard to health, wealth, and happiness.

As we have stated, scalar energy is non-Hertzian, non-linear, and possesses magnitude only, not direction. Additionally, scalar energy never degrades nor experiences entropy; therefore, the information of scalar energy remains unchanged regardless of the environment. Scalar energy is capable of propagating across time and space without having its information impinged on by the background radiation in the vastness of space.

Regarding the aforementioned, scalar energy is eternal, thus, serves as the archival, information system of the Universe. All events, past, present, and future, are recorded and archived by scalar energy that is the quintessential informational system of all prayers, thoughts, words, and actions. Scalar energy programs all DNA. Furthermore, scalar energy transcends space, thus, is not subject to time and space. The carrier wave for time is scalar energy; it is responsible for time. That is, scalar energy is present in all time frames because it transmits information without transmitting energy.

As an information system, scalar energy programs all DNA of every living species, includinghumans, animals, and plants. Thus, scalar energy is the informational input that is responsible for the genetic code. Scalar energy assumes the shape of a double-helix spiral and this shape subsequently is conveyed to all DNA. Correspondingly, the DNA of each species alive has a double-helix spiral structure (Fig. 3.53).

From a double-helix spiral point of view of scalar energy, each rotation of a SW detects a mathematical value of Phi 1.618..., an irrational number. That is, for each rotation of the *major groove* of a SW, the length of it is 1.618 times greater than the width. This identical, structural motif is imparted to all DNA whereby each rotation of the DNA double helix likewise detects the same mathematical value of Phi. That is, the length of a DNA double helix's major groove is 1.618 times greater than the width of it for each rotation of the DNA. Thus, scalar energy is a perfect Phi-spiral and is responsible for the formation of a DNA double helix. In short, scalar energy

3.9 Infrastructure, Characteristics, Derivation, and Scalar Wave Properties 299

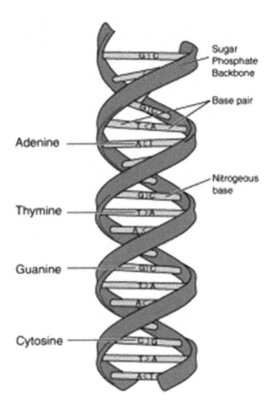

Fig. 3.53 DNA double-helix structure

Fig. 3.54 Double-helix DNA

provides the light information responsible for assembling and maintaining all DNA (Fig. 3.54).

Scalar energy as a carrier wave contains an infinite number of harmonics that serve as the instructions to assemble physical forms. Succinctly, one scalar energy harmonic contains the instructions for the formation of one physical form. Thus, each physical form in the Universe is assembled from one (1) specific scalar energy harmonic. In specific, one scalar energy harmonic serves to assemble the DNA code of one species of life. The DNA of a species is unique because only one scalar energy harmonic is responsible for assembling its physical form. Thus, the DNA of each species is unique and divinely chosen in their bodies.

As we have mentioned, there are some people out there who are using scalar energy in healing sessions for their patients; this book's author, however, has no personal experience with this, thus it is up to readers to do their own investigation on this matter. Such healers are claiming that their sessions may consist of a *pathogen cleanse, nutritional therapy*, and a *Chakra balance*.

3.9.10 The Scalar-Wave Superweapon Conspiracy Theory

According to Tom Bearden, the Scalar Potential Interferometer is a powerful superweapon that the Soviet Union used for years to modify weather in the rest of the world [14]. It taps quantum vacuum energy, using a method discovered by T. Henry Moray in the 1920s [15]. It may have brought down the *Columbia* spacecraft [16–17]. Nevertheless, some conspiracy theorists believe Bearden is an agent of disinformation on this topic, thus we leave this matter to readers to make their own conclusions and follow up on their own findings. This author does not make claims that any of these matters are false or true.

In the 1930s, however, Tesla announced other bizarre and terrible weapons: a death ray, a weapon to destroy hundreds or even thousands of aircraft at a range of hundreds of miles; his ultimate weapon to end all war was purported to be the Tesla shield that nothing could penetrate. But, by this time no one paid any real attention to the forgotten great genius. He died in 1943 without ever revealing the secret of these great weapons and inventions. Tesla called his superweapon a scalar potential howitzer, or death ray, as artistically depicted in Fig. 3.55. Later the Soviets demonstrated a weaponized Telsa howitzer at their Saryshagan missile range during the peak of the Strategic Defense Intuitive (SDI), and it was mentioned during the SALT treaty negotiations.

According to Bearden, in 1981 the Soviet Union discovered and weaponized the Tesla SW effects. Here we detail only the most powerful of these frightening Tesla weapons, which Brezhnev undoubtedly was referring to in 1975 when at the SALT talks the Soviet suddenly suggested limiting development of new weapons "more frightening than the mind of man had imagined." A high-level U.S. official considered Telsa's superweapon to be either a high-energy laser or a particle beam weapon.[8] As Fig. 3.56 illustrates, the Saryshagan howitzer had four modes of operation (Fig. 3.57). He also claims:

[T]he Saryshagan howitzer actually is a huge Tesla scalar interferometer with four modes of operation. One continuous mode is the Tesla shield, that places a thin, impenetrable hemispherical shell of energy over a large defended area. The three-

---

[8]Excerpted from *Aviation Week & Space Technology*, July 28, 1980.

3.9 Infrastructure, Characteristics, Derivation, and Scalar Wave Properties 301

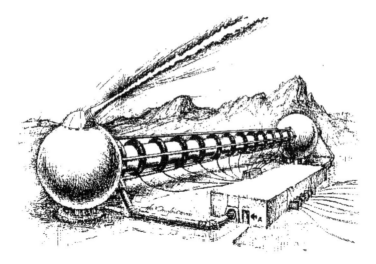

Fig. 3.55 Scalar potential interferometer (multimode Tesla weapon)

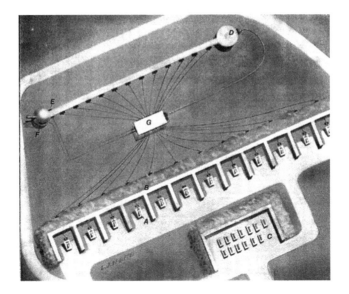

Fig. 3.56 The photo is taken by U.S. High-Resolution Reconnaissance Satellite, KH-11

dimensional shell is created by interfering two Fourier-expansion, three-dimensional scalar hemispherical patterns in space so they pair-couple into a dome-like shell of intense, ordinary electromagnetic energy. The air molecules and atoms in the shell are totally ionized and thus highly excited, giving off intense, glowing light. Anything physical that hits the shell receives an enormous discharge of electrical energy and is instantly vaporized—it goes pfft! like a bug hitting one of the electrical bug killers now so much in vogue.

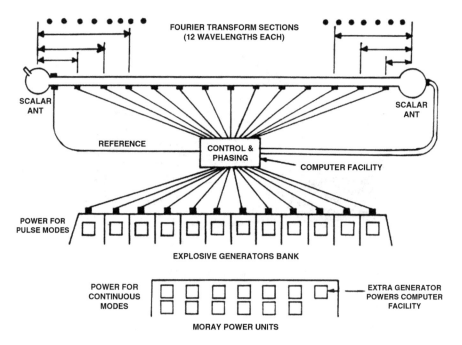

Fig. 3.57 Tesla weapons at Saryshagan

Bearden goes on to say that if several of these hemispherical shells are concentrically stacked, even the gamma radiation and EMP from a high-altitude nuclear explosion above the stack cannot penetrate all the shells because of repetitive absorption, reradiation, and scattering in the layered plasmas. In the continuous shield mode, the Tesla interferometer is fed by a bank of Moray free energy generators so that enormous energy is available in the shield. A diagram of the Saryshagan-type Tesla howitzer is shown earlier in Fig. 3.57. Hal Crawford's fine drawing of the interferometer end of the Tesla howitzer (see Fig. 3.54. Hal's exceptional rendition of the Tesla shield produced by the howitzer is shown in Figs. 3.58 and 3.59.

In the pulse mode, a single intense three-dimensional scalar phi-field pulse form is fired using two truncated Fourier transforms, each involving several frequencies, to provide the proper three-dimensional shape (Fig. 3.60). This is why two scalar antennas separated by a baseline are required. After a time, delay calculated for the particular target, a second and faster pulse form of the same shape is fired from the interferometer antennas. The second pulse overtakes the first, catching it over the target zone and pair-coupling with it to instantly form a violent EMP of ordinary vector (Hertzian) electromagnetic energy. There is thus no vector transmission loss between the howitzer and the burst. Further, the coupling time is extremely short, and the energy will appear sharply in an *electromagnetic pulse (EMP)* strikingly similar to the two-pulsed EMP of a nuclear weapon.

3.9 Infrastructure, Characteristics, Derivation, and Scalar Wave Properties 303

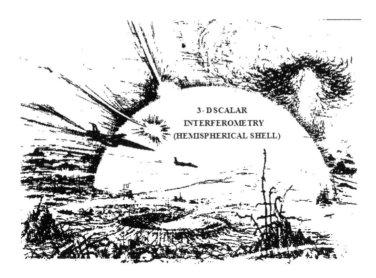

Fig. 3.58 The Tesla shield

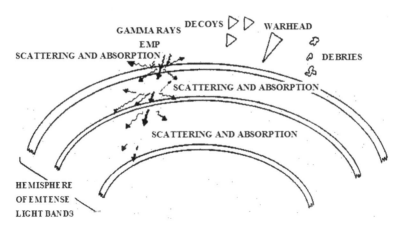

Fig. 3.59 Tesla Terminal Area Defense System

This type weapon is what actually caused the mysterious flashes off the southwest coast of Africa, picked up in 1979 and 1980 by Vela satellites. The second flash was in the IR only, with no visible spectrum. Nuclear flashes do not do that, and neither do super-lightning, meteorite strikes, meteors, and so on. In addition, one of the scientists at the Arecibo Ionospheric Observatory observed a gravitational wave disturbance—signature of the truncated Fourier pattern and the time-squeezing effect of the Tesla potential wave—traveling toward the vicinity of the explosion.

The pulse mode may be fed from either Moray generators or, if the Moray generators have suffered their anomalous "all fail" malfunction, ordinary explosive generators. Thus, the Tesla howitzer can always function in the pulse mode, but it

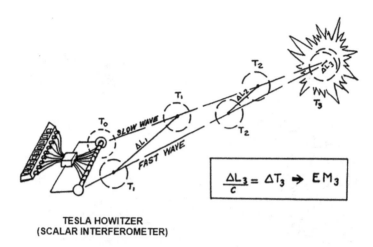

**TESLA HOWITZER
(SCALAR INTERFEROMETER)**

Fig. 3.60 Conceptual nuclear flash theory

will be limited in power if the Moray generators fail. In the continuous mode, two continuous scalar waves are emitted—one faster than the other—and they pair-couple into vector energy at the region where they approach an in-phase condition. In this mode, the energy in the distant "ball" or geometric region would appear continuously and be sustained—and this is Tesla's secret of wireless transmission of energy at a distance without any losses. It is also the secret of a "continuous fireball" weapon capable of destroying hundreds of aircraft or missiles at a distance. This mode of operation is shown in Fig. 3.61. The volume of the Tesla fireball can be vastly expanded to yield a globe that will not vaporize physical vehicles but will deliver an EMP to them to make their electronics useless. An artistic test of this mode is shown in Fig. 3.62.

If the Moray generators fail anomalously, then a continuous mode limited in power and range could conceivably be sustained by powering the interferometer from more conventional power-sources such as advanced magnetohydrodynamic generators. Typical strategic ABM uses of Tesla weapons are shown in Fig. 3.63. In addition, of course, smaller Tesla howitzer systems for anti-tactical ballistic missile defense of tactical troops and installations could be created from more conventional field missile systems using paired or triplet radars, of conventional external appearance, in a scalar interferometer mode.

Bearden also suggests that, with Moray generators [28] as power sources and multiply deployed reentry vehicles with scalar antennas and transmitters, ICBM reentry systems now can become long range "blasters" of the target areas from thousands of km distances (Fig. 3.64). Literally, "Star Wars" is liberated by the Tesla technology. In addition, in an air attack jammers and ECM aircraft become "Tesla blasters." With the Tesla technology, emitters become primary fighting components of stunning power.

3.9 Infrastructure, Characteristics, Derivation, and Scalar Wave Properties 305

Fig. 3.61 Continuous Tesla fireball

Fig. 3.62 Artistic illustration of Tesla EMP globe

The potential peaceful implications of Tesla waves are enormous too. By using the "time squeeze" effect, one can get anti-gravity, materialization and dematerialization, transmutation, and mind-boggling medical benefits. One also can get subluminal and superluminal communications, see through the Earth and through the ocean, and so on. The new view of phi-field also provides a unified field theory, higher orders of reality, and a new super-relativity; these need to be investigated and tested as well.

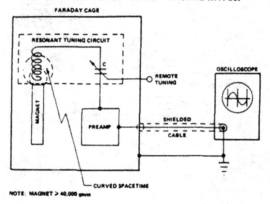

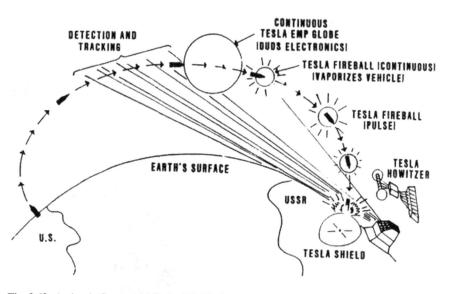

Fig. 3.63 Antistatic Conceptual Tesla ABM Defenses

3.9.11 Deployment of Superweapon Scalar Wave Drive by Interferometer Paradigm

Considering a SW, sometimes called a *longitudinal scalar wave (LSW)*, as one of the superweapons of directed energy, and to overcome the controversy of such phenomena (i.e., SW or LSW) to be used as a high-energy weapon is the possible

3.9 Infrastructure, Characteristics, Derivation, and Scalar Wave Properties

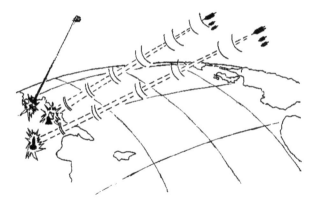

Fig. 3.64 Moray/Tesla technology

deployment of the interferometer concept that has been known to physicists and scientists for a while. Before taking this approach, however, we need to have some understanding of what interference is and as a result interferometers as well. How do interferometers work?

To understand interferometry, you need to understand *interference*. In everyday life interference simply means getting in the way or meddling, but in physics it has a much more specific meaning. Interference is what happens when two waves carrying energy meet up and overlap. The energy they carry gets mixed together so, instead of two waves, you get a third wave with a shape and size that depends on the patterns of the original two waves. When waves combine like this, the process is called *superposition.*

If you have ever sat making waves in a bathtub, you will have seen interference and superposition in action. If you push your hand back and forth, you can send waves of energy out from the center of the water to the walls of the tub. When the waves get to the walls, they bounce back off the hard surface more or less unchanged in size but with their velocity reversed. Each wave reflects off the tub just as if you had kicked a rubber ball at the wall. Once the waves come back to where your hand is, you can make them much *bigger* by moving your hand in step with them. In effect, you create new waves that add themselves to the original ones and increase the size of their peaks (amplitude).

There are two types of interference waves as described here and depicted in Fig. 3.65: (1) *constructive interference*, which means combining two or more waves to get a third wave that is bigger. The new wave has the same wavelength and frequency but more amplitude (higher peaks). (2) *Destructive interference*, which means waves subtracting and canceling out. The peaks in one wave are canceled by the troughs in the other.

In summary, when waves add together to make bigger waves, scientists call it constructive interference. If you move your hands a different way, you can create waves that are out of step with your original waves. When these new waves add to the originals, they subtract energy from them and make them smaller. This is what scientists call destructive interference.

Fig. 3.65 Illustration of two types of interference

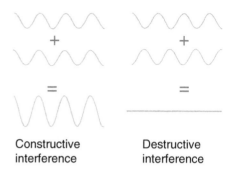

Analysis of Fig. 3.65 indicates that an interferometer is nothing more than being just a phase. The extent to which one wave is in step with another is known as its *phase*. If two identical waves are "in phase," it means their peaks align so, if we add them together, we get a new wave that is twice as big but otherwise exactly the same as the original waves. Similarly, if two waves are completely out of phase (in what we call anti-phase), the peaks of one exactly coincide with the troughs of the other, so adding the waves together gives you nothing at all.

In between these two extremes are all sorts of other possibilities where one wave is partly in phase with the other. Adding two waves like this creates a third wave that has an unusual rising and falling pattern of peaks and troughs. Display a wave like this onto a screen and you get a characteristic pattern of light and dark areas called interference fringes. This pattern is what you study and measure with an interferometer.

Thus, the interferometer is an instrument that uses the interference patterns formed by waves (usually light, radio, or sound waves) to measure certain characteristics of the waves themselves or of materials that reflect, refract, or transmit the waves. Interferometers also can be used to make precise measurements of distance. Interference patterns are produced when two identical series of waves are brought together.

Optical interferometers can be used as spectrometers for determining wavelengths of light and for studying fine details in the lines of a spectrum. Optical interferometers also are used in measuring lengths of objects in terms of wavelengths of light, providing great precision, and in checking the surfaces of lenses and mirrors for imperfections. In astronomy, optical interferometers make it possible to determine the diameter of large, relatively nearby stars and the separation of very close double stars. Radio interferometers are used in astronomy for mapping celestial sources of radio waves. Acoustic, or sound, interferometers are used for measuring the speed and absorption of sound waves in liquids and gases.

In Fig. 3.66, the two light rays with a common source combine at the half-silvered mirror to reach the detector. They may either interfere constructively (i.e., strengthening in intensity) if their light waves arrive in phase or interfere destructively (i.e., weakening in intensity) if they arrive out of phase, depending on the exact distances between the three mirrors.

3.9 Infrastructure, Characteristics, Derivation, and Scalar Wave Properties

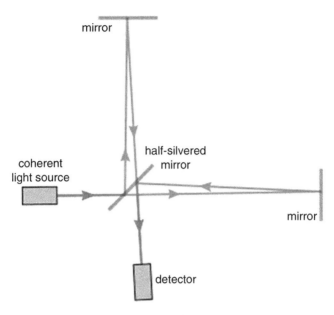

Fig. 3.66 The light path through a Michelson interferometer

Interferometry is a family of techniques in which waves, usually electromagnetic waves, are superimposed causing the phenomenon of interference in order to extract information. Interferometry is an important investigative technique in the fields of astronomy, fiber optics, engineering metrology, optical metrology, oceanography, seismology, spectroscopy (and its applications to chemistry), quantum mechanics, nuclear and particle physics, plasma physics, remote sensing, biomolecular interactions, surface profiling, microfluidics, mechanical stress–strain measurement, velocimetry, and optometry.

Interferometers are widely used in science and industry for the measurement of small displacements, refractive index changes, and surface irregularities. In an interferometer light from a single source is split into two beams that travel different optical paths, then combine again to produce interference. The resulting interference fringes give information about the difference in optical path length. In analytical science, interferometers are used to measure lengths and the shape of optical components with nanometer precision; they are the highest precision length-measuring instruments that exist. In *Fourier transform spectroscopy* they are used to analyze light-containing features of absorption or emission associated with a substance or mixture. An astronomical interferometer consists of two or more separate telescopes that combine their signals, offering a resolution equivalent to that of a telescope of a diameter equal to the largest separation between its individual elements. Furthermore, the way interferometers work is as follows.

An interferometer is really a precise scientific instrument designed to measure things with extraordinary accuracy. The basic idea of interferometry involves taking a beam of light (or another type of electromagnetic radiation) and splitting it into two

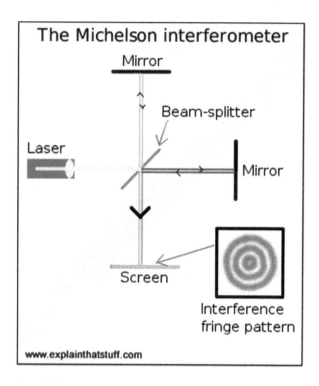

Fig. 3.67 Artwork of a basic Michelson interferometer

equal halves using what is called a beam-splitter (also called a half-transparent mirror or half-mirror). This is simply a piece of glass with a surface that is very thinly coated with silver. If you shine light at it, half the light passes straight through and half of it reflects—that is, the beam-splitter is like a cross between an ordinary piece of glass and a mirror (Fig. 3.67). One of the beams (known as the reference beam) shines onto a mirror and from there to a screen, camera, or other detector. The other beam shines at or through something you want to measure onto a second mirror, back through the beam splitter, and onto the same screen. This second beam travels an extra distance (or in some other slightly different way) to the first beam, so it gets slightly out of step (out of phase).[9]

[[moved to footnote]]When the two light beams meet up at the screen, they overlap and interfere, and the phase difference between them creates a pattern of light and dark areas (i.e., a set of *interference fringes*). The light areas are places where the two beams have added together (constructively) and become brighter; the dark areas are places where the beams have subtracted from one another (destructively). The exact pattern of interference depends on the different ways or the extra distances that one of the beams has traveled. By inspecting and measuring the

---

[9]If we take the green beam to be the reference beam, we would subject the blue beam to some sort of change we wanted to measure. The interferometer combines the two beams and the interference fringes that appear on the screen are a visual representation of the difference between them.

fringes, you can calculate this with great accuracy—and that gives you an exact measurement of whatever it is you are trying to find.

Instead of the interference fringes falling on a simple screen, often they are directed into a camera to produce a permanent image called an *interferogram*. In another arrangement the interferogram is made by a detector (e.g., the *charge couple device, CCD* image sensor used in older digital cameras) that converts the pattern of fluctuating optical interference fringes into an electrical signal that can be very easily analyzed with a computer. As part of enhancing our knowledge further about interferometer is that what are the various types of interferometers and which one is more appropriate for deployment of an application and which one drives SWs or in essence LSWs?

Interferometers became popular toward the end of the nineteenth century, and there are several different kinds, each based roughly on the principle we have outlined previously and named for the scientist who perfected it. Six common types are the Michelson, Fabry-Perot, Fizeau, Mach-Zehnder, Sagnac, and Twyman-Green interferometers; the following list describes them:

The *Michelson interferometer* (named for Albert Michelson, 1853–1931) is probably best known for the part it played in the famous Michelson-Morley experiment in 1881. That was when Michelson and his colleague Edward Morley (1838–1923) disproved the existence of a mysterious invisible fluid called "the ether" that physicists had believed filled empty space. The Michelson-Morley experiment was an important stepping-stone toward Albert Einstein's theory of relativity.

The *Fabry-Perot interferometer* (invented in 1897 by Charles Fabry, 1867–1945, and Alfred Perot, 1863–1925), also known as an *etalon*, evolved from the Michelson interferometer. It makes clearer and sharper fringes that are easier to see and measure.

The *Fizeau interferometer* (named for French physicist Hippolyte Fizeau, 1819–1896) is another variation and is generally easier to use than a Fabry-Perot. It is widely used for making optical and engineering measurements.

The *Mach-Zehnder interferometer* (invented by German Ludwig Mach and Swissman Ludwig Zehnder) uses two beam splitters instead of one and produces two output beams that can be analyzed separately. It is widely used in fluid dynamics and aerodynamics—the fields for which it was originally developed.

The *Sagnac interferometer* (named for Georges Sagnac, a French physicist, 1869–1928) splits light into two beams that travel in opposite directions around a closed loop or ring (hence its alternative name, the ring interferometer). It's widely used in navigational equipment, such as ring-laser gyroscopes (optical versions of gyroscopes that use laser beams instead of spinning wheels).

The *Twyman-Green interferometer* (developed in 1916 by Frank Twyman, 1876–1959, and Arthur Green) is a modified Michelson mainly used for testing optical devices.

Most modern interferometers use laser light because it is more regular and precise than ordinary light and produces *coherent* beams (in which all the light waves travel in phase), as shown schematically in Fig. 3.68. The pioneers of interferometry did

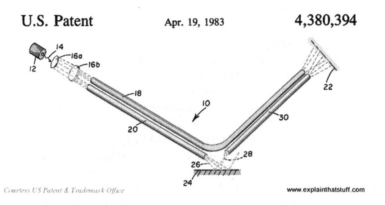

Fig. 3.68 *Fiber-optic interferometry*

not have access to lasers (i.e., were not developed until the mid-twentieth century) so they had to use beams of light passed through slits and lenses instead. As the figure indicates:

[M]ost interferometers pass their beams through the open air, but local temperature and pressure variations can sometimes be a source of error. If that matters, one option is to use a fiber-optic interferometer like [the one in Fig. 3.68]. A laser (red, 12) shoots its beam through lenses (gray, 16a/b) into a pair of fiber-optic cables. One of them (blue, 18) becomes the reference beam, bouncing its light straight onto a screen (orange, 22). The other (green, 20) allows its beam to reflect off something that is being measured (e.g., a vibrating surface) into a third cable (green, 30). The reference and reflected beams meet up and interfere on the screen in the usual way.[10] Another concern we should have in order to deploy the concept of interferometers is: How accurate can they be?

A state-of-the-art interferometer can measure distances to within 1 nanometer (nm), but like any other kind of measurement, it is subject to errors. The main source of error is likely to come from changes in the wavelength of the laser light, which depends on the refractive index of the material through which it is traveling. The temperature, pressure, humidity, and concentration of dissimilar gases in the air all change its refractive index, altering the wavelength of the laser light passing through it and potentially introducing measurement errors. Fortunately, good interferometers can compensate for this. Some have separate lasers that measure the air's refractive index, while others measure air temperature, pressure, and humidity and calculate the effect on the refractive index indirectly; either way, measurements can be corrected, and the overall error is reduced to perhaps one or two parts per million.

The common applications of interferometers are depicted in Fig. 3.69, and they are widely used in all kinds of scientific and engineering applications for making

---

[10] Artwork from US Patent 4,380,394: *Fiber optic interferometer* by David Stowe, Gould Inc., April 19, 1983, courtesy of US Patent and Trademark Office.

3.9 Infrastructure, Characteristics, Derivation, and Scalar Wave Properties 313

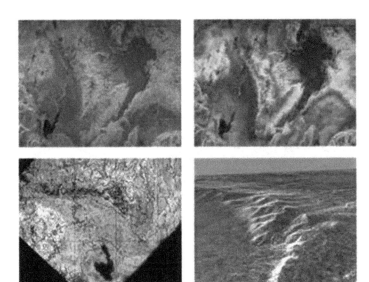

Fig. 3.69 Interferometry in action. *Photo courtesy of NASA Jet Propulsion Laboratory (NASA-JPL).*

accurate measurements. By scanning interferometers over objects, you also can make very precise maps of surfaces. The 3D topographical maps of Long Valley, California, that are shown in Fig. 3.69, were made from the Space Shuttle using a technique called radar interferometry in which beams of microwaves are reflected off the Earth's contours and then recombined.

By "accurate" and "precise," this author really does mean *accurate* and *precise*! The interference fringes that an optical (light-based) interferometer produces are made by light waves traveling fractionally out of step. Because the wavelength of visible light is in the hundreds of nm, interferometers can theoretically measure lengths a couple of hundred times smaller than a human hair! In practice, everyday laboratory constraints sometimes make that kind of accuracy difficult to achieve. Albert Michelson, for example, found his ether-detecting apparatus was affected by traffic movements about one-third of a km away!

Astronomers also use interferometers to combine signals from telescopes; they work in the same way as larger and much more powerful instruments that can penetrate deeper into space. Some of these interferometers work with light waves; others use radio waves similar to light waves, but with much longer wavelengths and lower frequencies (Fig. 3.70). Astronomers have linked the two 10-m (33-ft) optical telescopes in domes on Mauna Kea, Hawaii, to make what is effectively a single, much more powerful telescope.

An astronomical interferometer achieves high-resolution observations using the technique of aperture synthesis—that is, mixing signals from a cluster of comparatively small telescopes rather than a single very expensive monolithic telescope. Note that the use of white light will result in a pattern of colored fringes as illustrated

Fig. 3.70 The Keck interferometer. *Photo courtesy of NASA Jet Propulsion Laboratory (NASA-JPL)*

in Fig. 3.71 The central fringe representing equal path length may be light or dark depending on the number of phase inversions experienced by the two beams as they traverse the optical system. (See *Michelson interferometer* for a discussion of this in an optics textbook or on the Internet.)

Early radio telescope interferometers used a single baseline for measurement. Later astronomical interferometers, such as the *very large array (VLA)*, illustrated in Fig. 3.72, used arrays of telescopes arranged in a pattern on the ground. A limited number of baselines result in insufficient coverage. This was alleviated by using the rotation of the Earth to turn the array relative to the sky. Thus, a single baseline could measure information in multiple orientations by taking repeated measurements, a technique called *Earth-rotation synthesis*. Baselines thousands of km long were achieved using very long baseline interferometry.

Astronomical optical interferometry has had to overcome several technical issues not shared by radio telescope interferometry. The short wavelengths of light necessitate extreme precision and stability of construction. For example, a spatial resolution of 1 milliarcsecond (mas) requires 0.5 μm stability in a 100 m baseline. Optical interferometric measurements require high sensitivity, low noise detectors that did not become available until the late 1990s. Astronomical "seeing," of the turbulence that causes stars to twinkle, introduces rapid, random phase changes in the incoming light, requiring kHz data-collection rates to be faster than the rate of turbulence.

Fig. 3.71 Colored and monochromatic fringes in a Michelson interferometer. (a) White light fringes where the two beams differ in the number of phase inversions. (b) White light fringes where the two beams have experienced the same number of phase inversions. (c) Fringe pattern using monochromatic light (sodium D lines).

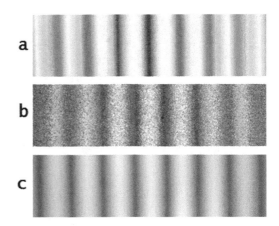

Fig. 3.72 Very large array interferometer

Figure 3.73 is a picture of ALMA, which is an astronomical interferometer located on the Chajnantor Plateau at Caltech.

Despite these technical difficulties, roughly a dozen astronomical optical interferometers are now in operation, offering resolutions down to the fractional mas range. There is a movie, assembled from aperture synthesis images of the Beta Lyrae system, that shows a binary star system approximately 960 light-years (290 parsecs) away in the constellation Lyra; it was observed by the CHARA array with the MIRC instrument. The brighter component is the primary star, or the *mass donor*. The fainter component is the thick disk surrounding the secondary star, or the *mass*

Fig. 3.73 ALMA interferometer

gainer. The two components are separated by 1 mas. Tidal distortions of the mass donor and the mass gainer are both clearly visible.

Microwave technology plays a more important role in modern industrial sensing applications. Pushed by the significant progress in monolithic microwave integrated-circuit technology over the past decades, complex sensing systems operating in the microwave and even millimeter-wave range are available for reasonable costs combined with exquisite performance. In the context of industrial sensing, this stimulates new approaches for metrology based on microwave technology. An old measurement principle nearly forgotten over the years has recently gained more and more attention in both academia and industry. Both industry and manufacturing technology are experiencing a revolution these days. The demand for products with a high degree of individuality, leading to a huge variety of configurations, is increasing more and more.

As illustrated in Figure 3.74, as another application of interferometry, a VHF and microwave interferometric and phase amplitude noise measurements technique allows close-to-the-carrier measurements of both phase and amplitude noise, improving the instrument noise floor by 10–25 dB compared to the traditional method based on a saturated mixer. As part of EMW development and its types, we should consider an extremely interesting property of EMWs that propagate in homogeneous waveguides. This will lead to the concept of "modes" and their classification, as follows:

Transverse electric and magnetic (TEM): The transverse EMW cannot be propagated within a waveguide but is included for completeness. It is the mode that is commonly used within coaxial and open wire feeders. The TEM wave is characterized by the fact that both electric vector $\vec{E}$ and magnetic vector $\vec{H}$ are perpendicular to the direction of propagation.

3.9 Infrastructure, Characteristics, Derivation, and Scalar Wave Properties 317

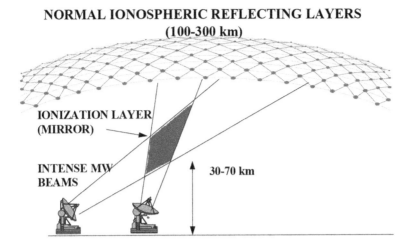

Fig. 3.74 Crossed-beam approach for generating an artificial ionospheric mirror

Transverse electric (TE): This waveguide mode is dependent on the transverse electric waves, also sometimes called H waves, characterized by the fact that electric vector $\vec{E}$ is always perpendicular to the direction of propagation.

Transverse magnetic (TM): Transverse magnetic waves, also called E waves, are characterized by the fact that magnetic vector $\vec{H}$ is always perpendicular to the direction of propagation.

Figure 3.74 is artist's schematic of Earth Ionosphere of *transvers electric* waves in a *localized wave (LW)* scenario, where the Earth and ionosphere can be used to launch TE waveguide modes, thus providing a way to send LWs beyond the line of sight around the globe.

An array of antennas in the MW range can ionize particles in the atmosphere and heat up the ionosphere causing it to change shape, which affects weather. This was shown by the *High Frequency Active Auroral Research Program (HAARP)* technology. The HAARP was initiated as an ionospheric research program jointly funded by the U.S. Air Force, the U.S. Navy, the University of Alaska–Fairbanks, and the Defense Advanced Research Projects Agency (DARPA). It was designed and built by BAE Advanced Technologies (BAEAT). Its original purpose was to analyze the ionosphere and investigate the potential for developing ionospheric enhancement technology for radio communications and surveillance. As a university-owned facility, HAARP is a high-power, high-frequency transmitter used for study of the ionosphere.

The most prominent instrument at HAARP is the *Ionospheric Research Instrument (IRI)*, a high-power radio frequency transmitter facility operating in the HF band. The IRI is used to temporarily excite a limited area of the ionosphere. Other instruments (e.g., VHF and UHF radar, a fluxgate magnetometer, a *digisonde* (i.e., an ionospheric sounding device), and an induction magnetometer, are used to study the physical processes that occur in the excited region. HAARP is a target of conspiracy theorists, who claim that it is capable of "weaponizing" weather. Commentators and scientists say that advocates of this theory are uninformed because claims made fall well outside the abilities of the facility, if not the scope of natural science.

Creation of an artificial uniform ionosphere was first proposed by Soviet researcher A. V. Gurevich in the mid-1970s. An *artificial ionospheric mirror (AIM)* would serve as a precise mirror for EM radiation of a selected frequency or a range of frequencies. It would thereby be useful for both pinpoint control of friendly communications and interception of enemy transmissions. By causing a variation in the ionosphere's particles, the image on the ground causes Earth currents and high enough magnetic variations to harm or destroy grid systems. This concept has been described in detail by Paul A. Kossey et al. in a paper entitled "Artificial. Ionospheric Mirrors" [44].

These authors describe how one could precisely control the location and height of the region of artificially produced ionization using crossed MW beams, which produce atmospheric breakdown (ionization) of neutral species. The implications of such control are enormous; one would no longer be subject to the vagaries of the natural ionosphere but would instead have direct control of the propagation environment. Ideally, the AIM could be created rapidly and then would be maintained only for a brief operational period. A schematic depicting the crossed-beam approach for generation of an AIM was shown earlier in Fig. 3.74.

An AIM theoretically could reflect radio waves with frequencies up to 2 GHz, which is nearly two orders of magnitude higher than those waves reflected by the natural ionosphere. The MW radiator power requirements for such a system are roughly an order of magnitude greater than 1992 state-of-the-art systems; however, by 2025 such a power capability is expected to be easily achievable. However, "the capability of influencing the weather even on a small scale could change it from a force degrader to a force multiplier."

In 1977, the UN General Assembly adopted a resolution prohibiting the hostile use of environmental modification techniques. Besides providing pinpoint communication control and potential interception capability, this technology also would provide communication capability at specified frequencies, as desired. Figure 4.2 shows how a ground-based radiator might generate a series of AIMs, each of which would be tailored to reflect a selected transmission frequency. Such an arrangement would greatly expand the available bandwidth for communications and also eliminate the problem of interference and crosstalk by allowing one to use the requisite power level.

The first huge scalar potential interferometers [36–38] of strategic range and power (Fig. 3.75) were deployed by the Soviets in 1963, and one was used to destroy the *U.S.S. Thresher* nuclear attack submarine (Fig. 3.76),[11] leaving clearly recognizable signatures [39]. With her controls jammed, the hull of the doomed

---

[11]This is a claim by T. E. Bearden [39] at the end of this chapter.

3.9 Infrastructure, Characteristics, Derivation, and Scalar Wave Properties 319

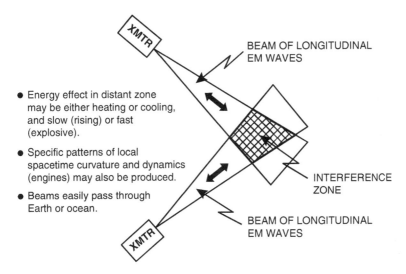

Fig. 3.75 Artist's schematic of a scalar (longitudinal EMW) interferometer

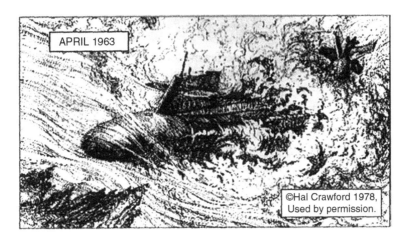

Fig. 3.76 Artistic schematic of the "kill" of the *U.S.S. Thresher*

submarine implodes when she reaches crush depth as a part of deployment and immediate use of scalar interferometers.

Since 1963, the Russians have had the equivalent of more than seven additional Manhattan Projects (using the Russian five-year program instead of the four-year Manhattan Project), back-to-back, in development of energetics weaponry. The energetics weapons have never been given to the regular Russian forces. Instead, all research, production, siting, manning, and employment are by the KGB and still under its ruthless control by die-hard communist factions.

As part of *Aharonov-Bohm effect* one can conclude that, even in the absence of electrical field $\vec{E}$ and magnetic field $\vec{B}$, the potentials cause real effects to occur in the field-free regions. Using this principle, beams of pure potential without vector force fields (i.e., without electrical field $\vec{E}$ and magnetic field $\vec{B}$) may be deliberately produced and intersected at a distance to cause effects in the interference zone, incongruous to classical mechanics. These effects are in fact required by quantum mechanics. Essentially, energy may be produced directly at the distant interference site or extracted from it, without energy transmission through space.

Implications for weapons built on these concepts were given by U.S. intelligent community scientists during the period of SDI (Star Wars) activities within the United States and Soviet Union of the time, and several types of such scalar EM weapons were discussed. One of the claims in the open literature by some individuals, particularly among U.S. scientists, was that there had been evidence of massive Soviet weaponization of these effects for nearly three decades, including scalar EM weapons' testing on a global scale. Evidence of this exists in the literature.

With the advent of Maxwell's equations and the more completeness of the Ampère-Maxwell's equation, as well as decoupling such equations that was established in the previous sections of this chapter and a previous one in this book, electricity and magnetism were combined into an elegant EM theory; these equations then served as the basis for the development of the modern theory of *quantum-electro-dynamics (QED)*. Gradually, potentials were relegated to a position of inferior importance, and they even came to be regarded as purely mathematical convenience by most scientists.

With the release of the Aharonov and Bohm seminal paper [23], however, it became crystal clear that potentials are in fact real entities, and they directly can affect and control charged particle systems even in a region where all the fields and thus the forces on the particles have vanished. Although this, of course, is completely contrary to the conclusions of classical mechanics, it follows inescapably from quantum mechanics.

With Chambers's direct experimental proof [40] of the predicted Aharonov-Bohm effects in 1960 [23], this new viewpoint was firmly established in general for *quantum mechanics (QM)* and *quantum electromagnetics (QE)*. The same effect even was demonstrated toaffect gauge theories [41–42], requiring introduction of the concept of non-integrable phase factors and global formulation of gauge fields. Thus, increasingly, it is the potentials that are primary physical entities, and the fields that are of secondary, resulting in importance in modern QED.

Nevertheless, the full weapon implications of the Aharonov-Bohm discovery have not yet penetrated the minds and consciousnesses of Western physicists, academia, weapons engineers, and SDI scientists. Indeed, an extended treatment of such implications has not even been addressed at least not in the open literature. To some degree this may be understandable because it required almost 30 years for physicists to realize the primary actuality of the potentials in the first place, as the Nobel Prize winner Richard Feynman [41] stated it succinctly: "It is interesting that something like this can be around for thirty years but, because of certain prejudices of what is and is not significant, continues to be ignored."

Slowly, the overwhelming importance of the scalar electromagnetics indicated by Aharonov and Bohm [23] has been recognized and noted by Western world weapons analysts, and work to investigate and apply this rich new region of QED is now most certainly warranted in order to show that the LSW indeed can be used as a superweapon of *directed energy* (DE). Such a possibly may be more effective than other directed energy weapons (DEWs). Thus, it is imperative that a high- priority effort be mounted as soon as possible, for our very own survivability and to be considered as part of our National Security Policy. Because we are threatened by such tremendous energy from scalar electromagnetics weapons that evidently already are in the hands of the Russia today, as well as other adversaries (e.g., China and North Korea). Given this matter as a fact of its existence in the Russian new age weapons arsenal, we absolutely have no counter-measure and no defense system whatsoever against such a threat.

3.9.11.1 Wireless Transmission of Energy Driven by Interferometry at a Distance

To illustrate one remarkable though typical implication of this new breakthrough area, we point out that, by changing the potentials while keeping the force-field zero [23], one can directly produce energy at a distance as if it were transported through space in the normal fashion and traditional way it occurs. Indeed, it may even be possible to use pure potential waves to "*transport*" the energy at any velocity (i.e., not limited by the speed of light) that will not dissipate at constraint limit of $1/r^2$, yet may at the rate of $1/r$. Because in some cases a potential, for example, *electrostatic scalar potential (ESP)* can be regarded as having infinite velocity, simply appearing "everywhere at once" [43].

Furthermore, the ESP may be regarded as a sort of "locked-in stress energy" of vacuum, as can any other vacuum potential. Changing the potential in a region or at a point changes the amount of "locked-in" or "enfolded" vacuum energy available or stored in that region. Yet, simply changing the potential at that point or in that region need not involve any local expenditure of work there; the work may be expended elsewhere, and the results realized at a distant region by a change in that regions potential, according to the Aharonov-Bohm effect [23].

In the remote region, charged particles are imbedded in vacuum potential by their virtual particle charge flux, and in the induced potential gradients, the imbedded particles move, producing electrical and magnetic forces and fields and performing work. This is somewhat analogous to "putting energy in here at point A" and "extracting it out there at point B" without any travel or losses in between—Nikola Tesla's old "wireless transmission of energy at a distance without losses" idea. Note that, quantum mechanically, we may take the view that this is a very special class of macroscopic "energy tunneling" phenomena, as illustrated in the Aharonov and Bohm original paper [23].

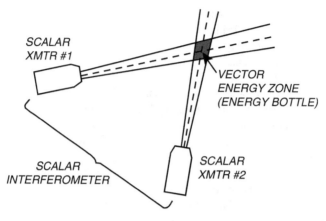

Fig. 3.77 Illustration of a vector energy zone produced by two scalar transmitters driven by interferometry

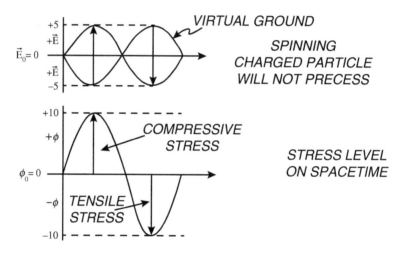

Fig. 3.78 Zero-field or scalar-wave beam

Essentially energy is put into the system at the location where the scalar potentials are produced and is recovered at the distant interference zone where particle effects are produced (Fig. 3.77); see also Fig. 3.75 from interferometry point of view, and Figs. 3.79 and 3.80.

To pursue the preceding single example, depicted in Fig. 3.77, and further show its implications, we point out that in theory one may deliberately make beam-containing zero electric and magnetic fields simply by properly phase-locking together two or more beams of oscillating ordinary $\vec{E}$ and $\vec{B}$ electrical energy—all at the same frequency. In the perfect hypothetical case, for example, two single-frequency beams are phase-locked 180° apart to create such a zero-field or scalar beam, as shown in Fig. 3.78.

3.9 Infrastructure, Characteristics, Derivation, and Scalar Wave Properties   323

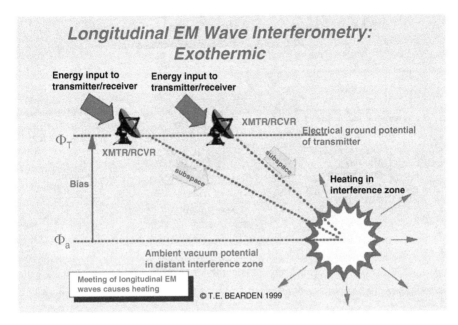

Fig. 3.79 Exothermic LSW interferometry-1 (using HAARP to create explosive steam pockets underground)

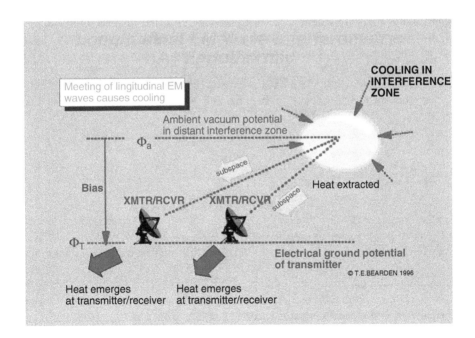

Fig. 3.80 Exothermic LSW interferometry-1

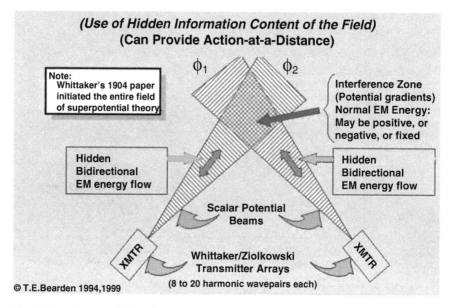

Fig. 3.81 Scalar potential interferometry

In the real bandwidths, how much zero-field beam is obtained at the center of the band depends on the sharpness "Q" of each beam. To purify the beam, it could first be transmitted through a ground Faraday cage or shield, which would remove most of the orthodox $\vec{E}$- and $\vec{B}$-field components not properly zeroed out. By successive Faraday "stripping" of the beam, a scalar beam as pure as desired can be obtained (see Fig. 3.55 earlier). Figure 3.81 is an artist's drawing of scalar potential interferometry between the two sets of bidirectional longitudinal EMW-pair function, produced by all EM force fields and waves.

3.10 Quantum Waves

So far, we have defined three principle types of waves: (1) mechanical waves, (2) electromagnetic waves, and (3) quantum waves. Although mechanical waves may be the most obvious in everyday life, and EMW may be the most useful to the technological society; a case can be made that quantum waves are the most fundamental within a paradigm of all things in the Universe. They generate patterns of energy resulting from their motion, thus every bit of matter in the Universe behaves like a wave under certain circumstances.

Quantum wave theory (QWT) is a model of nature that grew in response to several questions:

3.10 Quantum Waves

What, exactly, is gravity?
How are charge and gravity related?
What gives rise to the fundamental unit of energy?
What, especially, *is* space?

Our attempt to answer these questions evolved into conversations that continued for more than a decade. *Quantum wave* theory is an artwork, a prose poem, that is the result of that collaboration. The theory attempts to unify energy, mass, and force as manifestations of a single entity. We refer to that entity as space, as argued by Amy Robinson and John Holland [45].

According to Einstein, space becomes curved by large bodies (e.g., stars and planets). The results of that curvature are that the Earth and planets orbiting the Sun, the moon orbiting the Earth, and things "fall" toward the center of the Earth. But gravity here is not acting on us from under our feet; rather, it surrounds us. We are not so much "pulled" down by Earth's gravity as "pushed" down by curved space (Fig. 3.82).

What happens between the Sun and planets, and the Earth and moon, tells us something not only about the relationships among those large bodies, but also about space. Part of Einstein's genius was his ability to see the Sun, the planets, and *space* as a unified system, rather than separate, independent phenomena. Space has been described in various ways throughout the history of physics and cosmology including particulate aether, empty vacuum, the Higgs field or Higgs ocean, and space "atoms."

Quantum wave theory proposes a new model of space. The theory describes space as a continuous, quantized, flexible "field"; nowhere divided or divisible, but capable of discrete motions of compression and rebound. Space comprises tiny regions able to act independently of the space that surrounds them, although there are no actual

Fig. 3.82 The Universe as we see it

boundaries between them. Quantum wave theory refers to these fundamental units of space as "space quanta."

Space quanta are tiny, flexible regions of space with dynamics that include compression and rebound. When energized, a quantum of space compresses and deforms along the path of least resistance, which is determined by the origin of a force, its magnitude, and the state of surrounding space. One quantum of space is the amount of space that will compress in response to the fundamental unit of energy, which is described by Planck's constant.

Schrödinger's equation is generalized to a spacetime four-manifold, using standard concepts from differential geometry and operator replacement. This fourth-order equation, which reduces and specializes to the Klein–Gordon equation in the flat space limit, can also be obtained from a variational principle; it must be solved in tandem with the Einstein field equations with suitable stress energy. The propagator, for large momenta, varies like $1/p^4$. A further attractive feature is that no external currents or stress energies need be imposed—these arise naturally. A generalization to fields with arbitrary spin is proposed. Solving the equation would lead to a determination of the mass, just as energies are found in solving Schrödinger's equation. Flat-space plane wave solutions consist of the superposition of two independent waves, which can be interpreted as propagating strings [46].

Wave and particle characteristics have very different behaviors: the way in which they occupy space, the way they travel through openings, and the way they interact with other particles or waves. All particles spin, or rotate, and oscillate (tip back and forth about a point of equilibrium). Imagine a ball under water, spinning as it tips back and forth. The ball's spin disturbs the surrounding water in a circular path, while its oscillation generates alternating wave densities and rarities as it tips back and forth. These combined motions create a disturbance in the surrounding water that translates outward as a distinct, wavy pattern of crests and troughs. That pattern is determined by the size of the ball, its speed of rotation, and the rate and extent of oscillation.

Both the particles themselves, and the traveling wave patterns they generate, are the result of the traveling waves from which they formed. Traveling waves generated by complementary particles form new particles when they intersect. Each wave carries a portion of the particle's "instructions" for forming a new particle. Particle motion is the "genetic code" of matter (Fig. 3.83).

Every particle is a record of those waves from which it formed, the places from which they came, and the particles from which they were generated. Superposition, wavefunction collapse, and the uncertainty principle in quantum physics shows real and imaginary components of quantum wavefunctions for free particles and confined particles.

As can be seen from Fig. 3.82, according to quantum mechanics, all particles in the Universe are described by what we refer to as wavefunction Ψ. Properties, such as position and momentum, do not have defined values until they are observed. A particle exists in a well-defined amount of space, however, if you could pause time at some instant during a particle motion in space. This property means a particle is

3.10 Quantum Waves

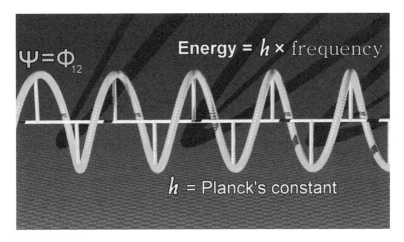

Fig. 3.83 Wave function depiction

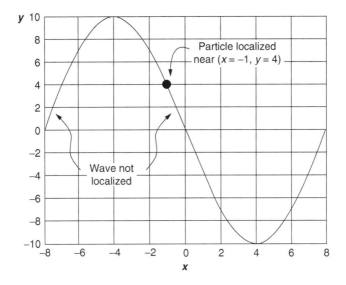

Fig. 3.84 A non-localized wave and a localized particle

"localized" because the particle is in a particular location at a given time. Waves, on the other hand, exist over an extended region of space.

The probability of the particle being at a particular position is given by the square of the amplitude of the wavefunction at that location. As described before, a harmonic function, such as $\sin(kx - \omega t)$, exists over all values of x, from $-\infty$ to $+\infty$, with no way to distinguish one cycle from another. Thus, a single-frequency wave is inherently non-localized; it exists everywhere. The difference is illustrated in Fig. 3.84, which shows a particle represented in time, where a "snapshot" of the particle and wave are depicted. At the instant of the snapshot, all observers can agree that the particle exists at positions $x = -1$ and $y = 4$. But, where is the wave at that instant?

The wave exists at $x = -8$, $x = 0$, $x = +8$, and all other values of x, including locations not shown to the left and right of the graph and in between the x values listed in Fig. 3.84. Any given peak, trough, or zero-crossing of the wave exists at a specific location at a certain time; however, every cycle looks exactly like another cycle, and the wave itself consists of all its positions. Thus, the wave is non-localized.

We have touched up Schrödinger's equation and general relativity briefly and more details can be found in Appendix B of this book. Now we need to briefly show results of derivation of the field equation [46]. The free space in Schrödinger's equation reads:

$$i\hbar \frac{\partial \Psi}{\partial t} = -\frac{\hbar^2}{2m} \nabla^2 \Psi \qquad (3.334)$$

This equation should also hold in a frame that is co-moving or nearly co-moving with the particle. The idea, then, is to transform this equation into coordinates that are adapted to the particle's world line. Vuille [46] shows that the first step in the derivation is to reinsert mass energy m that has essentially been scaled out of the traditional equation as follows:

$$i\hbar \frac{\partial \Psi}{\partial t} = -\frac{\hbar^2}{2m} \nabla^2 \Psi + m^2 \Psi \qquad (3.335)$$

From Eq. 3.335 Vuille shows that Schrödinger's equation can be generalized to the equivalent equation on a four-manifold by using standard differential geometry and operator replacements and it is developed; the equations describe a scalar field on a curved spacetime [46]. The derivation of Eq. 3.335, however, made no demands on the mathematical nature of wave Ψ. This being the case, the general *spinor* equation might be written as:

$$\nabla^{AA'} \nabla^{BB'} \nabla_{AA'} \nabla_{BB'} \Psi^{D_1 \cdots D_n} + \frac{m^2 c^2}{\hbar^2} \nabla^{AA'} \nabla_{AA'} \Psi^{D_1 \cdots D_n} = 0 \qquad (3.336)$$

We encourage readers to refer to Vuille's paper for more details on Eq. 3.31 and its step-by-step derivation [46].

Related to this section it is worth showing some overall calculations and analysis of Edmond T. Whittaker [47] as part of "Relativity and Electromagnetic Theory," where he shows that when any number of electrons are moving in any manner, the functions that define the resulting electrodynamics field—namely, the three components of dielectric displacement in the aether and the three components of the magnetic force at every point of the field—can be expressed in terms of the derivates of two scalar potential functions.

There, some express it in terms of a scalar potential function and a vector potential function, which are equivalent to *four* scalar potential functions (see also Chap. 5 of

3.10 Quantum Waves

this book). Whittaker [47], however, suggests that these two scalar potential functions are explicitly evaluated in terms of the charges and coordinates of the electrons. From these results it has been shown that the general functional form of an electrodynamic disturbance because of electrons can be derived.

Whittaker also goes on to extend his work with other activities related to presentation of the EM entity in vacuo that ends up with the result of his calculations extended to the production of transverse fields and energy by scalar interferometry as well. In this, he shows that a transverse field and energy can be produced in vacuo by the interferometry of two beams prepared in such a way that the only component present in each beam is the physical time-like potential, proportional directly to magnetic flux $(\partial F/\partial t) = i(\partial G/\partial t)$ under conditions of circular polarization in both beams. This indicates, when interference occurs, that the condition no longer holds, and transverse field and energy appear through the interference of the two beams. It also can be shown that energy is conserved during this process.

Whittaker's first paper on theoretical physics excluding dynamics, published in 1904 [47], stated that any electric field can be specified in terms of the derivatives of *two* real SW functions, F and G, by means of the formula as part of the six components of the dielectric displacement and the magnetic force; this can be expressed in terms of the derivatives of these two scalar potentials defined by the equations:

$$F(x,y,z,t) = \sum \frac{e}{4\pi} \sinh^{-1}\left\{\frac{z'-z}{\left[(x'-x)^2 + (y'-y)^2\right]}\right\} \quad (3.337)$$

$$G(x,y,z,t) = \sum \frac{e}{4\pi} \tan^{-1}\left\{\frac{y'-y}{x'-x}\right\} \quad (3.338)$$

where in these two equations, the summation is taken over all the electrons in the field and the new form of Maxwell's equations in terms of $\vec{E}$ and $\vec{H}$ fields, as well as force can be written:

$$\vec{E} = \vec{\nabla} \times \vec{\nabla} \times \vec{F} + \vec{\nabla} \times \left(\frac{\vec{G}}{c}\right)$$

$$\vec{H} = \vec{\nabla} \times \left(\frac{\vec{F}}{c}\right) - \vec{\nabla} \times \vec{\nabla} \times \vec{G} \quad (3.339)$$

$$\vec{F} = (0,0,F) \quad , \quad \vec{G} = (0,0,G)$$

Thus, the usual specification in terms of the Maxwellian *scalar potential* ϕ and *vector potential* $\vec{A}$ is equivalent to *four* real functions connected by the relation $\frac{\partial \phi}{\partial t} + c\vec{\nabla} \cdot \vec{A} = 0$. These two SW functions were evaluated in terms of the charges and coordinates of the electrons generating the field.

It also can be shown without any difficulty that if any number of electrons with a total charge that is zero are moving in any manner so as always to remain in the vicinity of a given point (i.e., to be *stationary*), then the EM field generated as a result is the type given by:

$$F = \frac{1}{r} f\left(t - \frac{r}{c}\right) \qquad G = 0 \qquad (3.340)$$

where r is the distance from the point and f is an arbitrary function; or, more generally, of a field of this type superimposed on fields of the same type, but related to axes y and x the same asthis is related to the axis of z. This is perhaps of some interest in connection with the view advocated by some physicists that the atoms of the chemical elements consist of sets of electrons, with a total charge that is zero in stationary motion.

3.11 The X-Waves

To get a better description of the X-waves, we need to have some basic understanding of the behavior ofwaves or beams , which among them is a *superluminal phenomenon*.

A superluminal phenomenon is a frame of reference traveling with a speed greater than the speed of light c. There is a putative class of particles dubbed *tachyons* that are able to travel faster than light. Faster-than-light phenomena violate the usual understanding of the "flow" of time, a state of affairs that is known as the *causality problem*, also called the *Shalimar Treaty*.

It should be noted that although Einstein's theory of special relativity prevents (real) mass, energy, or information from traveling faster than the speed of light c (see, e.g., Lorentz et al., 1952 [48]; Brillouin and Sommerfeld, 1960; [49]; Born and Wolf, 1999 [50]; Landau and Lifschitz, 1997 [51]), there is nothing preventing "apparent" motion faster than c (or in fact with negative speeds, implying arrival at a destination before leaving the origin). For example, the phase velocity and group velocity of a wave may exceed the speed of light, but in such cases, no energy or information actually travels faster than c. Experiments showing group velocities greater than c include that of Wang et al. (2000) [52], who produced a laser pulse in atomic cesium gas with a group velocity of $(-310 \pm 5)c$. In each case, the observation is superluminal.

It turns out that all relativistic wave equations possess infinity families of formal solutions with arbitrary speeds ranging from zero to infinity, called *undistorted progressive waves (UPWs)* by Rodrigues and Lu (1997) [53]. Like the arbitrary speed plane wave solutions, however, UPWs have infinite energy and therefore cannot be produced in the physical world. Still, approximations to these waves with finite energy, called *finite aperture approximations (FAA)*, can be produced and observed experimentally (Maurino and Rodrigues, 1999 [54]). Among the infinite family of exact superluminal solutions of the homogeneous wave equation and Maxwell's equations are waves known as X-waves. They do not violate special relativity because all superluminal X-waves have wavefronts that travel with the speed parameter c that appears in the corresponding wave equation. The superluminal motion of the peak is therefore a transitory phenomenon similar to the reshaping phenomenon that occurs (under very special conditions) for waves in dispersive media with absorption or gain that is responsible for superluminal (or even negative) group velocities (Maiorino and Rodrigues [54]).

Several authors have published theories claiming that the speed-of-light barrier imposed by relativity is illusionary. Although these "theories" continue to be rejected by the physics community as ill-informed speculation, their proponents continue to promulgate them in rather obscure journals. An example of this kind is the Smarandache hypothesis, which states that there is no such thing as a speed limit in the Universe (Smarandache, 1998) [55]. Similarly, Shan (1999a,b) has concluded that superluminal communication must exist in the Universe and that it does not result in the casual loop paradox [70–71].

The existence of non-diffractive waves, also known as localized pulses, were predicted by many scientists [56–59], as well as more recent articles [60, 61]. Note that the so-called *localized waves (LW)*, also known as non-diffracting waves, are indeed able to resist diffraction for a long distance in free space. Such solutions to the wave equations and, in particular, to Maxwell's equations, under weak hypotheses were theoretically predicted a long time ago [56–58, 68] (cf. also Recami et al., 2004 [59]). In summary, LWs are the type of waves that focus energy at a point in space that can be generated with acoustic, microwave, particle, and light energy.

The non-diffracting solutions to the wave equations (i.e., scalar, vectorial, spinorial, etc.) have been in fashion, both in theory and in experiments, for a couple of decades. Rather well known are the ones with luminal or superluminal peak velocity. Like the so-called X-shaped waves, which are supersonic in acoustics [67], and superluminal in electromagnetism (see [58]; see also [66, 69]).

In 1983 Brittingham [62] set forth a luminal ($v = c$) solution to the wave equation (more particularly, to Maxwell's equations), which travels rigidly (i.e., without diffraction). The solution proposed possessed infinite energy, however, once more the problem of overcoming it arose [62]. A way out was first found by Sezginer [63], who showed how to construct finite-energy luminal pulses that do not propagate without distortion for an infinite distance. As expected, however, travel with constant speed and approximately without deformation for a certain (long) depth of field—much longer, in this case, than that of ordinary pulses such as *Gaussian pulses* [64].

At the beginning of the 1990s, however, Lu et al. [66, 67] constructed, both mathematically and experimentally, new solutions to the wave equation in free space—namely, an X-shaped localized pulse with the form predicted by so-called *extended special relativity* [58, 65]; for the connection between what Lu et al. [66, 67] called *X-waves* and extended relativity, see Recami, Zamboni-Rached, and Dartora, 2004[59].

Fig. 3.85 An X-shaped wave

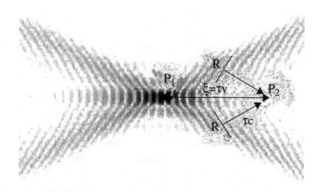

An X-shaped wave that is a localized superluminal pulse is illustrated in Fig. 3.85 [64]. It is an X-wave possessing velocity $v > c$; it shows the fact that if its vertex or central spot is located at point P_1 and at time t_0, it will reach position P_2 at time $t + \tau$, where $\tau = (|P_2 - P_1|/v) < (|P_2 - P_1|/c)$. This is different from the illusory "scissor effect," even if the *feeding energy* coming from region R has traveled with the ordinary speed c (i.e., the speed of light in the EM case, the sound speed in acoustics, etc.).

Such X-shaped waves result in interesting and flexible localized solutions, even if their velocity V is supersonic or superluminal ($v > c$) and have been reported in several papers. Actually, when the phase velocity does not depend on the frequency, it is known that such a phase velocity becomes the group velocity! Remembering how a superposition of Bessel beams is generated (i.e.., by a discrete or continuous set of annular slits or transducers), clearly the energy forming the X-waves coming from those rings travels at ordinary speed c of plane waves in the medium considered [72–75]. (Here c, representing the velocity of the plane waves in the medium, is the speed of sound in the acoustic case, the speed of light in the EM case, and so on.) Nevertheless, the peak of the X-shaped waves is faster than c. Figure 3.86 is a representation of various configurations of all dissimilar waves, with demonstration of their energy locally.

It is possible to generate (in addition to the "classic" X-wave produced by Lu et al., 1992 [66]) infinite sets of new X-shaped waves, with their energy concentrated more and more in a spot corresponding to the vertex region [76]. It therefore may appear rather intriguing that such a spot (even if no violation of special relativity is obviously implied. All the results come from Maxwell's equations or from the wave equations [77, 78]) that travel superluminally when they are EMWs. We shall call all the X-shaped waves *superluminal* even when, for example, the waves are acoustic.

Furthermore, the word *beam* refers to a monochromatic solution to the wave equations considered, with a transverse localization of its field. Yet, our reflections, of course, hold for any wave equation (e.g., vectorial, spinorial, scalar, etc.); this isparticularly the case for acoustic waves . Additionally, in optical physics, the most common type of optical beam is the Gaussian one with transverse behavior that is described by a Gaussian function. But all the common beams suffer a diffraction that

3.11 The X-Waves 333

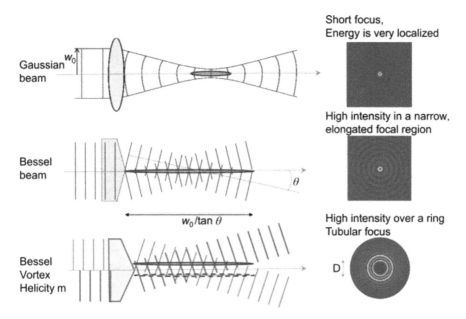

Fig. 3.86 Propagation and configuration of various waves

spoils the transverse shape of their field, widening it gradually during propagation. As an example, the transverse width of a Gaussian beam doubles when it travels a distance of $z_{\text{dif}} = \sqrt{3}\pi\Delta\rho_0^2/\lambda_0$, where $\Delta\rho_0$ is the beam's initial width and λ_0 is its wavelength. One can verify that a Gaussian beam with an initial transverse aperture of the order of its wavelength already will double its width after having traveled along a few wavelengths.

Generally, it was believed that the only wave devoid of diffraction was the plane wave, which does not suffer any transverse changes. Some authors had shown that it is not actually the only one. For instance, in 1941 Stratton [68] obtained a monochromatic solution to the wave equation with a transverse shape that was concentrated in the vicinity of its propagation axis and represented by a Bessel function. Such a solution, now called a *Bessel beam*, was not subject to diffraction because no change in its transverse shape took place with time. Figure 3.87 illustrates the shape and difference between, for example, Gaussian and Bessel beams.

Mathematically, each of these beams can be defined as:

$$S(k_\rho, \omega) = 2a^2 e^{-a^2 k_\rho^2} \delta(\omega - \omega_0) \quad \text{Gaussian Beam} \quad (3.341)$$

and

$$S(k_\rho, \omega) = \frac{\delta[k_\rho - (\omega/c)\sin\theta]}{k_\rho} \delta(\omega - \omega_0) \quad \text{Bessel Beam} \quad (3.342)$$

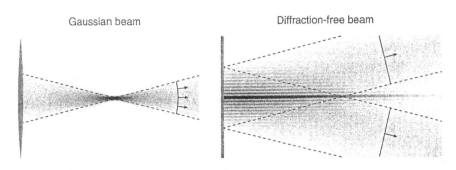

Fig. 3.87 Gaussian versus Bessel beams

Table 3.2 Comparison of Diffraction and Dispersion

Diffraction	Dispersion
Diffraction is a gradual spatial broadening	This effect creates temporal boardening
A limiting factor in that affects transverse localization	A variation in the refraction index with frequency

Table 3.2 compares these two beams from the point of view of diffraction and dispersion, whichhave been long known to be phenomena limiting the application of beams (e.g., optical) or pulses.

Diffraction always is present, affecting any waves that propagate in two- or three-dimensional unbounded media, even when homogeneous. Pulses and beams are constituted by waves traveling in different directions, which produces a gradual spatial broadening. This effect is really a limiting factor whenever a pulse is needed to maintain its transverse localization (e.g., in free space communications, image forming, optical lithography, EM tweezers,etc.).

Dispersion, on the other hand, acts on pulses propagating in material media, causing mainly a temporal broadening: An effect is known to be because of the variation of the refraction index with the frequency so that each spectral component of the pulse possesses a different phase velocity. This entails a gradual temporal widening, which constitutes a limiting factor when a pulse is needed to maintain its time width (e.g., in communication systems).

As we have emphasized, it is important to develop techniques able to reduce those phenomena. The so-called LWs, known also as non-diffracting waves, are indeed able to resist diffraction for a long distance in free space. Such solutions to the wave equations (in particular, Maxwell's equations under a weak hypotheses) were theoretically predicted a long time ago based on the different behaviors between diffraction and dispersion in Table 3.2, Bessel beams are able to resist the effects that make them useful! For more details and definitions of the Bessel beam, see Sect. 3.13 of this chapter.

In summary, among the infinite family of exact superluminal solutions to Maxwell's equations are waves known as X-waves (Rodrigues and Lu, 1997 [79]). Scalar X-waves have been measured experimentally by Lu and Greenleaf (1992 [66]), and subsequently they showed that the peak of a finite aperture approximation to an acoustical X-wave can travel with a speed greater than the sound speed parameter appearing in the homogeneous wave equation [66, 79].

Rodrigues and Lu (1997) also performed several simulations for the propagation of X-waves, showing that their peaks can move with superluminal speed, an effect subsequently verified by Saari and Reivelt (1997) [80]. These results do not violate special relativity because all the produced superluminal X-waves have wavefronts that travel with the speed parameter c (i.e., the speed of light) that appears in the corresponding wave equation. The superluminal motion of the peak is therefore a transitory phenomenon similar to the reshaping phenomenon that occurs (under very special conditions) for waves in dispersive media with absorption or gain that is, in this case, responsible for superluminal (or even negative) group velocities (Maiorino and Rodrigues, 1999 [54]).

In physics X-waves are localized solutions of the wave equation that says that it travels at a constant velocity in a given direction. X-waves can be sound, electromagnetic, or gravitationalones. They are built as a non-monochromatic superposition of Bessel beams. Ideal X-waves carry infinite energy; however, finite-energy realizations have been observed in various frameworks. Electromagnetic X-waves travel faster than the speed of light, and X-wave pulses can have a superluminal phase and group velocity [81]. In optics X-wave solutions have been reported within a quantum mechanical formulation [82].

As final a summary for this section, we can say that one novel family of generalized non-diffracting waves are X-waves, and they are exact non-diffracting solutions of the isotropic/homogenous SW equation and are a generalization of *some of* the previously known non-diffracting waves—for example, the plane wave, *Durnin's beams*, and the non-diffracting portion *of* the Axicon beam equation, in addition to an infinity of new beams. One subset of the new non-diffracting waves has X-like shapes that are termed "*X* waves."

These non-diffracting *X* waves can be almost exactly realized over a finite depth of field with finite apertures and by either broadband or band-limited radiators. With a 25 mm diameter planar radiator, a zeroth-order broadband *X* wave will have about 2.5 mm lateral and 0.17 mm axial -6-dB beam widths with a -6-dB depth of field of about 171 mm. The phase of the *X* waves changes smoothly with time across the aperture of the radiator; therefore, *X* waves can be realized with physical devices. A zeroth-order band-limited *X* wave was produced and measured in water by 10 elements, 50 mm diameter, 2.5 MHz PZT ceramic/polymer composite J_0 Bessel non-diffracting annular array transducer with -6-dB lateral and axial beam widths of about 4.7 mm and 0.65 mm, respectively, over a -6-dB depth of field of about 358 mm.

3.12 Non-linear X-Waves

As we stated before, in physics a *non-linear X-wave (NLX)* is a multi-dimensional wave that can travel without distortion. At variance with X-waves, a NLX does exist in the presence of non-linearity, and in many cases, it self-generates (in any

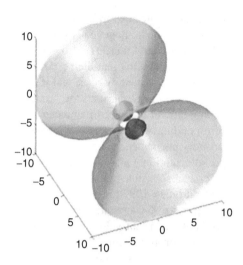

Fig. 3.88 Non-linear X-wave schematic view

direction) from a Gaussian wave packet. The distinctive feature of an NLX is its "biconical" shape that appears as an "X" in any section plane containing the wave peak and the direction of propagation. So far, NLX have been only observed in non-linear optics experiments and have been predicted to occur in a variety of non-linear media, including Bose-Einstein condensates. Figure 3.87 is a schematic view of a NLX.

Nevertheless, note that what we mean by a localized solution for these non-diffraction waves is that anything local is a part of the entire space. If the function goes through a portion of the space, and you solve it on some smaller domain, the solution is local. This is seen classically in optimization problems that are not global. Rather, they give local solutions, which potentially are not globally optimum. A global optimum covers the entire domain (Fig. 3.88).

There are other types, such as ocean waves, traveling water waves, traveling harmonic waves, as well as energy density and flux in traveling waves that are discussed extensity in other textbooks. One reference that this author recommends is the book by Crawford; we encourage readers to refer to it because the details there are beyond the scope of this book [83].

3.13 Bessel's Waves

As we mentioned previously, a *non-diffracting wave (NDW)*, known also as localized waves, indeed, are able to resist diffraction for a long distance (see Table 3.2). Today, NDWs are analyzed and well-established, both theoretically and experimentally, and have very important and innovative applications not only in vacuum, but also in both linear and non-linear material. Recently, in meta-materials media, resistance to dispersion has been shown, thus materials can travel a long distance.

3.13 Bessel's Waves

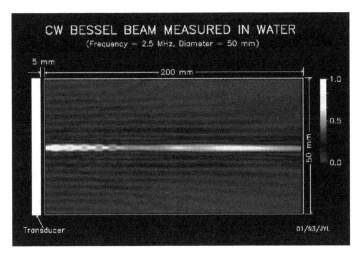

Fig. 3.89 Illustration of continuous wave Bessel beam in water

Their potential applications are being expanded and explored as well as implemented extensively in military and medical industries. The results are very surprising and mesmerizing in fields such as acoustics, microwaves, and optics; they also are very promising in mechanical engineering and geophysics [64], and even in elementary particle physics [84], as well gravitational waves.

One interesting acoustic application already has been obtained in high-resolution ultrasound scanning of moving organs in the human body because of the unique behavior of NDWs that are suitable superpositions of Bessel beams. Furthermore, worth noticing is that peculiar superposition of Bessel beams can be used to obtain a "static" NDW field, with high transverse localization, and with a longitudinal intensity pattern that can assume a desired shape within a chosen interval of the propagation axis (i.e., $0 \leq z \leq L$); thus, such a wave with a *static* envelope is called a *frozen wave* (FW) in terms of continuous Bessel beam superpositions. These FWs promise to have very important applications, even in the field of medicine (e.g., curing of tumors) [85].

Figure 3.89 is a presentation of a *continuous wave* (CW) Bessel beam as part of its acoustic application in water, where it was measured with a frequency of 2.5 MHz. It is worth mentioning that, in consideration of Bessel beam applications, a Bessel beam can travel, without deformation, approximately a distance 28 times greater than a Gaussian beam.

Lu et al. actually managed to introduce classic X-waves for acoustics [66, 67], then later after having mathematically and experimentally constructed their "classic" acoustic X-wave, they started applying waves to ultrasonic scanning, directly obtaining very high-quality 3D images. Thus, they produced images such as Fig. 3.90; it depict the real part of an ordinary X-wave with velocity $v = (1.1) c$ and with $a = 3 m$, where a is a positive constant as part of a *Heaviside* step function for the wave equation solution of X-waves described in Sect. 3.14 of this chapter.

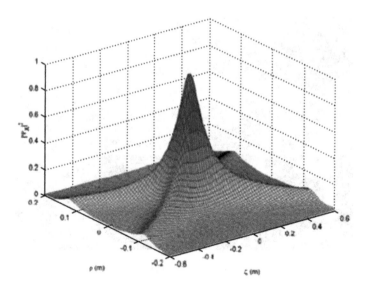

Fig. 3.90 A 3D plot of the real part of the ordinary X-wave

Fig. 3.91 All the X-waves

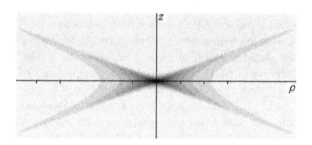

In this figure ρ is the distance from the axis, while ξ is equal to $z - vt$ (i.e., $\xi = z - vt$). Note that X-waves are the superposition of Bessel beams and they are plane waves, where the transverse components of the wave are no longer planar, but rather make up the surface of two cones with an apex that touches, as shown in Fig. 3.91.

In this figure all the X-waves (truncated or not) must have a leading cone in addition to the rear cone; such a leading cone has a role even for peak stability. Long ago, this also was predicted, in a sense, by non-restricted special relativity. One in fact should not forget that all wave equations, not only Maxwell's, have an intrinsic relativistic structure. *Note:* The localization in space of the wave can be changed along z, which means a soliton-like wave of energy can be accelerated toward a target. If there are particles within this soliton, they could act as an electromagnetic/kinetic.

A general solution of the Bessel beam function in cylindrical coordinates (see Sect. 3.14 for more details) for a localized solution of the wave can be as follows:

$$\Psi(\rho, z, t) = J_0\left(\frac{\omega_0}{c}\rho \sin(\theta)\right) \exp\left[i\frac{\omega_0}{c}\cos(\theta)\left(z - \frac{c}{\cos(\theta)}t\right)\right] \quad (3.343)$$

As the equation indicates, this beam possesses phase velocity $v_{ph} = c/\cos(\theta)$ and a field transverse shape represented by Bessel function $J_0(\cdots)$ so that its field is concentrated in the surroundings of propagation axis z. This is also the reason this beam is called a Bessel beam because the solution form in Eq. 3.343 involves a Bessel function of zero-kind such as $J_0(\cdots)$.

3.14 Generalized Solution to Wave Equations

To confine ourselves to electromagnetism, we can consider the present-day studies on EM tweezers, optical or acoustic scalpels, optical guiding of atoms or charged or neutral corpuscles, optical lithography, optical or acoustic images, communications in free space, remote optical alignment, optical acceleration of charged corpuscles, and so on. One of the things that would be interesting to touch on here in a little more detail is the introduction to the general solution of a differential equation known as a homogeneous wave equation; this is a simple form of the wave equations so important in acoustics, electromagnetism (i.e., microwave, optics, etc.), geophysics, and even gravitational waves and elementary particle physics, as we noted before.

Stratton [68] obtained a monochromatic solution to the wave equation with a transverse shape that was concentrated near its propagation axis and represented by a Bessel function. This can be justified partially because that Bessel beam was associated with an infinite power flux (as much as the plane waves, incidentally); it is not square-integrable in the transverse direction. An interesting problem, therefore, was that of investigating what would happen to the ideal Bessel beam solution when truncated by a finite transverse aperture.

To continue with our solution to wave equations in a generalized manner, we take under consideration a general form of solution for wave function $\psi(x, y, z, t)$ that can be written in as a Cartesian coordinate as follows:

$$\left(\frac{\partial^2}{\partial x^2} + \frac{\partial^2}{\partial y^2} + \frac{\partial^2}{\partial z^2} - \frac{1}{c^2}\frac{\partial^2}{\partial t^2}\right)\psi(x, y, z, t) = 0 \quad (3.344a)$$

or

$$\Box \psi(x, y, z, t) = 0 \quad (3.344b)$$

where $\Box \equiv \nabla^2 - \frac{1}{c^2}\frac{\partial^2}{\partial t^2}$ is known as a *D'Alembert* operator.

Now let us write Eq. 3.344 in cylindrical coordinate form with components of ρ, φ, and z; thus, we seek a solution in this coordinate and for the sake of simplicity, we confine the solution to an axially symmetrical one as $\psi(\rho, z, t)$, so we do not have any dependency on angle φ. Therefore, we have this mathematical relation:

$$\left(\frac{\partial^2}{\partial \rho^2} + \frac{1}{\rho}\frac{\partial}{\partial \rho} + \frac{\partial^2}{\partial z^2} - \frac{1}{c^2}\frac{\partial^2}{\partial t^2}\right)\psi(\rho, z, t) = 0 \tag{3.345}$$

In free space solution $\psi(\rho, z, t)$ can be written in terms of a Bessel-Fourier transform with reference to variable ρ, and two Fourier transforms with reference to variables z and t, as follows:

$$\psi(\rho, z, t) = \int_0^{\infty} \int_{-\infty}^{+\infty} \int_{-\infty}^{+\infty} k_\rho J_0(k_\rho \rho) e^{ik_z z} e^{-i\omega t} \overline{\psi}(k_\rho, k_z, \omega) dk_\rho dk_z d\omega \tag{3.346}$$

where $J_0(k_\rho, \rho)$ is an ordinary zero-order Bessel function and $\overline{\psi}(k_\rho, k_z, \omega)$ is the transform of $\psi(\rho, z, t)$. Substituting Eq. 3.346 into Eq. 3.345, one obtains the relation among ω, k_ρ, and k_z, given by the following form:

$$\frac{\omega^2}{c^2} = k_\rho^2 + k_z^2 \tag{3.347}$$

This equation must be satisfied. In this way, by using the condition given by Eq. 3.347 in Eq. 3.346, any solution to wave Eq. 3.345 can be written as:

$$\psi(\rho, z, t) = \int_0^{\omega/c} \int_{-\infty}^{+\infty} k_\rho J_0(k_\rho \rho) e^{i\sqrt{\omega^2/c^2 - k_\rho^2} z} e^{-i\omega t} S(k_\rho, \omega) dk_\rho d\omega \tag{3.348}$$

where $S(k_\rho, \omega)$ is the chosen spectral function.

The general integral of Eq. 3.348 yields, for instance, the non-localized Gaussian beams and pulses, to which we will refer to for illustrating the differences of the LW with respect to them [86]. A very common non-localized *beam* is the Gaussian; it corresponds to the spectrum as [87]:

$$S(k_\rho, \omega) = 2a^2 e^{-a^2 k_\rho^2} \delta(\omega - \omega_0) \tag{3.349}$$

In Eq. 3.349, a is a positive constant, which can be shown to depend on the transverse aperture of the initial pulse [86].

Figure 3.92 illustrates the interpretation of the integral solution of Eq. 3.347, with the spectral function given in Eq. 3.349, as a superposition of plane waves. Namely, from Fig. 3.92 one can easily realize that this case corresponds to plane waves propagating in all directions always with $\vec{k}_z \geq 0$; the most intense ones being those directed along positive z. Notice that in the plane-wave case $\vec{k}_z$ is the longitudinal component of the wave vector, $\vec{k} = \vec{k}_\rho + \vec{k}_y$, where $\vec{k}_\rho = \vec{k}_x + \vec{k}_y$.

3.14 Generalized Solution to Wave Equations

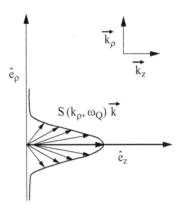

Fig. 3.92 Visual interpretation of the integral solution of Eq. 3.348 [86]

By substituting Eq. 3.349 into Eq. 3.348, and adopting the paraxial approximation, one meets the Gaussian beam wave equation as follows:

$$\psi_{\text{gauss}}(\rho,z,t) = \frac{2a^2 \exp\left(\frac{-\rho^2}{4(a^2+iz/2k_0)}\right)}{2(a^2+iz/2k_0)} e^{ik_0(z-ct)} \quad (3.350)$$

where $k_0 = \omega_0/c$. We can verify that such a beam, which suffers transverse diffraction, doubles its initial width, $\Delta\rho_0 = 2a$, after having traveled distance $z_{\text{dif}} = \sqrt{3}k_0 \Delta\rho_0^2/2$—called diffraction length. The more concentrated a Gaussian beam happens to be, the more rapidly it gets damaged [86].

It is noticeable, as well, that the most common non-localized *pulse* is the *Gaussian pulse*, which does not deviate from Eq. 3.348 by using the spectrum defined by Zamboni-Rached et al. [88] as:

$$S(k_\rho,\omega) = \frac{2ba^2}{\sqrt{\pi}} e^{-a^2 k_\rho^2} e^{-b^2(\omega-\omega_0)^2} \quad (3.351)$$

Now, by substituting Eq. 3.351 into Eq. 3.348, and once more using the paraxial approximation, one gets the Gaussian pulse [86]:

$$\psi(\rho,z,t) = \frac{a^2 \exp\left(\frac{-\rho^2}{4(a^2+iz/2k_0)}\right) \exp\left(\frac{-(z-ct)^2}{4c^2 b^2}\right)}{a^2+iz/2k_0} \quad (3.352)$$

This is endowed with speed c and temporal width $\Delta t = 2b$, and suffers a progressive enlargement of its transverse width so that its initial value already gets doubled at position $z_{\text{dif}} = \sqrt{3}k_0 \Delta\rho_0^2/2$, with $\Delta\rho_0 = 2a$.

It is noteworthy to say that the axially symmetrical Bessel beam is created by the superposition of plane waves with vectors that lay on the surface of a cone having the propagation axis as its symmetry axis and an angle equal to θ that is the *axicon angle*, as illustrated in Fig. 3.93 [86].

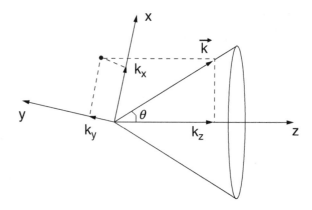

Fig. 3.93 The axially symmetrical Bessel beam

Considering that Eq. 3.343 indicates that the Bessel beam keeps its transverse shape, which is therefore invariant, while propagation with a central "spot" is given as:

$$\Delta \rho = \frac{2.405c}{(\omega \sin \theta)} \quad (3.353)$$

Now if we turn our attention to a scalar function, such as $\Phi(\vec{r}, t)$, for a discussion in quest of a generalized solution to wave equations, then we consider that an N-dimensional isotopic/homogeneous wave equation is given by [91–92]:

$$\left[\sum_{j=1}^{N} \frac{\partial^2}{\partial x_j^2} - \frac{1}{c^2} \frac{\partial^2}{\partial t^2} \right] \Phi(\vec{r}, t) = 0 \quad (3.354)$$

where $x_j (j = 1, 2, \cdots, N)$ represents rectangular coordinates in an N-dimensional space, $N \geq 1$ is an integer, $\Phi(\vec{r}, t)$ is a scalar function (e.g., sound pressure, velocity potential, or Hertz potential in EMs) of spatial variables, $\vec{r} = (x_1, x_2, \cdots, x_N)$, and time t. c is the speed of sound in a medium or the speed of light in a vacuum [89, 90].

In three-diminsional (i.e., 3D) space, we can use the *D'Alembert* operator, $\Box \equiv \nabla^2 - \frac{1}{c^2} \frac{\partial^2}{\partial t^2}$, and apply it to Eq. 3.354 to write the following forms of this equation:

$$\left(\nabla^2 - \frac{1}{c^2} \frac{\partial^2}{\partial t^2} \right) \Phi(\vec{r}, t) = 0 \quad (3.355a)$$

and

$$\Box \Phi(\vec{r}, t) \quad (3.355b)$$

3.14 Generalized Solution to Wave Equations

Fig. 3.94 Schematic of the cylindrical coordinate in 3D

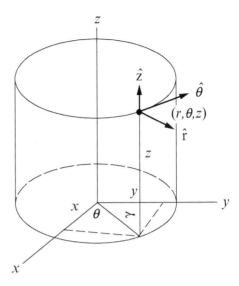

In Eq. 3.355a the symbol ∇^2 is a *Laplace* operator, while in Eq. 3.355b the symbol $\square$, as we said, is the *D' Alembert* operator. In a cylindrical coordinate, the wave equation is given by:

$$\left[\frac{1}{r}\frac{\partial}{\partial r}\left(r\frac{\partial}{\partial r}\right) + \frac{1}{r^2}\frac{\partial^2}{\partial \varphi^2} + \frac{\partial^2}{\partial z^2} - \frac{1}{c^2}\frac{\partial^2}{\partial t^2}\right]\Phi(\vec{r},t) = 0 \quad (3.356)$$

where $r = \sqrt{x^2 + y^2}$, using Fig. 3.94, is the radial distance, $\varphi = \tan^{-1}(y/x)$ is the polar angle, and z is the axial axis.

One generalized solution to the N-dimensional wave equation [66, 90, 91–92] in Eq. 3.355 is given by the following equation [91–92]:

$$\Phi(x_1, x_2, \cdots, x_N, t) = f(s) \quad (3.357)$$

where

$$s = \sum_{j=1}^{N=1} D_j x_j + D_N(x_N \pm c_1 t) \quad N \geq 1 \quad (3.358)$$

In this equation the parameter D_j is a complex coefficient and in Eq. 3.357, $f(s)$ is any well-behaved complex function of s and c_1 given by [91–92]:

$$c_1 = c\sqrt{1 + \sum_{j=1}^{N-1} D_j^2/D_N^2} \quad (3.359)$$

If c_1 is real, $f(s)$ and its linear superposition represent limited-diffraction solutions to the N-dimensional wave equation (see Eq. 3.354).

For example, if $N = 3$, $x_1 = x$, $x_2 = y$, $x_3 = z$, $D_1 = \alpha_0(k,\zeta)\cos\theta$, $D_2 = \alpha_0(k,\zeta)\sin\theta$, and $D_3 = b(k,\zeta)$, with cylindrical coordinates one obtains families of solutions to Eq. 3.356 [66, 90, 91–92]:

$$\Phi_\zeta(s) = \int_0^\infty T(k) \left[\frac{1}{2\pi} \int_{-\pi}^{+\pi} A(\theta) f(s) d\theta \right] dk \tag{3.360}$$

and

$$\Phi_K(s) = \int_{-\pi}^{+\pi} D(\zeta) \left[\frac{1}{2\pi} \int_{-\pi}^{+\pi} A(\theta) f(s) d\theta \right] d\zeta \tag{3.361}$$

where

$$s = \alpha_0(k,\zeta) r \cos(\phi - \theta) + b(k,\zeta)[z \pm c_1(k,\zeta) t] \tag{3.362}$$

and where

$$c_1(k,\zeta) = c\sqrt{1 + [\alpha_0(k,\zeta)/b(k,\zeta)]^2} \tag{3.363}$$

and $\alpha_0(k,\zeta)$, $b(k,\zeta)$, $A(\theta)$, $T(k)$, and $D(\zeta)$ are well-behaved functions, and θ, k, and ζ are free parameters. If $c_1(k,\zeta)$ is real and is not a function of k and ζ, respectively, $\Phi_\zeta(s)$ and $\Phi_K(s)$ are families of limited-diffraction solutions to the wave equation [91–92].

The following function is also a family of limited-diffraction solutions to the wave equation [66, 90, 91–92], which represents waves that can propagate to an infinite distance without changing their wave shape at the speed of c:

$$\Phi_L(r, \varphi, z - ct) = \Phi_1(r, \rho) \Phi_2(z - ct) \tag{3.364}$$

where $\Phi_2(z - ct)$ is any well-behaved function of $z - ct$ and $\Phi_1(r, \rho)$ is a solution to the transverse Laplace equation:

$$\left[\frac{1}{r} \frac{\partial}{\partial r} \left(r \frac{\partial}{\partial r} \right) + \frac{1}{r^2} \frac{\partial^2}{\partial \varphi^2} \right] \Phi_1(r, \varphi) = 0 \tag{3.365}$$

Furthermore, Lu [91–92] finds sets of solutions for Bessel beams and X-waves as follows. For Bessel beams:

$$\Phi_{B_n}(\vec{r}, t) = \Phi_{B_n}(r, \varphi, z - c_1 t) = e^{in\varphi} J_n(\alpha r) e^{i(\beta z - \omega t)} \quad \text{for} \quad n = 0, 1, 2, \cdots \tag{3.366}$$

where:

B_n = the nth-order Bessel beam
α = the scaling parameter
$J_n(\ldots)$ = the nth-order Bessel function of first kind
$c_1 = \omega/\beta$ = the phase velocity wave

For X-waves:

$$\Phi_{X_n}(\vec{r}, t) = \Phi_{X_n}(r, \varphi, z - c_1 t)$$
$$= e^{in\varphi} \int_0^\infty B(k) J_n(kr \sin \zeta) e^{-k[a_0 - i \cos \zeta (z - c_1 t)]} dk \quad \text{for} \quad n = 0, 1, 2, \cdots$$
(3.367)

where:

X_n = the subscript means nth-order X wave
$c_1 = c/\cos \geq c$ = both phase and group velocity of wave
$|\zeta| < \pi/2$ = the Axicon angle of X-wave
a_0 = a positive free parameter that determines the decaying speed of the high-frequency of a device (acoustic transducer or EM antenna) that produces the wave

Compare Eq. 3.367 with Eq. 3.366. It is easy to see the similarity, and it is easy to see the similarity and difference between a Bessel beam and an X-wave. X-waves are multiple-frequency waves while Bessel beams have a single frequency. Nevertheless, both waves have the same limited-diffraction property—that is, they are propagation invariant. Because of multiple frequencies, X-waves can be localized in both transverse space and time to form a tight wave packet. They can propagate in free space or isotropic/homogeneous media without spreading or dispersion [91–92].

References

1. https://en.wikipedia.org/wiki/Standing_wave
2. https://en.wikipedia.org/wiki/Seiche
3. G.H. Darwin, *The Tides and Kindred Phenomena in the Solar System* (John Murray, London, 1898), pp. 21–31
4. Tsunamis are normally associated with earthquakes, but landslides, volcanic eruptions, and meteorite impacts all have the potential to generate a tsunami
5. Proudman, J. (1953). Dynamical Oceanography. London: Methuen. §117 (p. 225). OCLC 223124129
6. J.R. Merian, Ueber die Bewegung tropfbarer Flüssigkeiten in Gefässen [on the motion of drippable liquids in containers] (thesis) (in German). Basel: Schweighauser. OCLC **46229431** (1828)
7. Pierce, T. (2006). Marine and coastal services abbreviations and definitions (PDF). *National Weather Service*, Office of Climate, Water, and Weather Services. Archived from *the original* (pdf) on May 17, 2008. *Retrieved April 19, 2017*

8. This behaves in a fashion similar to a tidal bore where incoming tides are funneled into a shallow, narrowing river via a broad bay. The funnel-like shape increases the height of the tide above normal, and the flood appears as a relatively rapid increase in the water level.
9. https://en.wikipedia.org/wiki/MOSE_Project
10. J.R. Reitz, F.J. Milford, R.W. Christy, *Foundations of Electromagnetic Theory*, 3rd edn. (Addison – Wesley Publishing Company, Reading, 1979)
11. J.D. Jackson, *Classical Electrodynamics*, 3rd edn. (John Wiley Publisher, New York, 1990), pp. 27–29
12. O.L. Brill, B. Goodman, Am. J. Phys. **35**, 832 (1967). for a detailed discussion of casualty in the Coulomb gauge.
13. Konstantin, Meyl, Scalar Waves. http://www.k-meyl.de/xt_shop/index.php?cat=c3_Books-in-English.html, INDEL GmbH, Verlagsabteilung ISBN 3-9802542-4-0
14. http://www.cheniere.org/books/part1/teslaweapons.htm
15. http://www.cheniere.org/images/people/moray%20pics.htm
16. http://www.prahlad.org/pub/bearden/Columbia_attack.htm
17. http://www.cheniere.org/books/excalibur/moray.htm
18. D. Griffiths, *Introduction to Electrodynamics*, 3rd edn. (Prentice Hall, Upper Saddle River, 1999)
19. H.E. Ensle, *The Electromagnetic Universe*, 2nd edn. (2013)
20. B. Zohuri, *Plasma Physics and Controlled Thermonuclear Reactions Drive Fusion Energy, 1st edn* (Springer Publisher, 2016)
21. B. Zohuri, *Dimensional Analysis and Self-Similarity Methods for Engineers and Scientists*, 1st edn. (Springer Publisher, 2015)
22. J.D. Jackson, *Classical Electrodynamics*, 2rd edn. (John Wiley Publisher, 1975), pp. 681–790
23. Y. Aharonov, D. Bohm, Significance of electromagnetic potentials in the quantum theory. Phys. Rev. **115**, 485–491 (1959)
24. M.A. Steane, *Relativity Made Relativity Easy* (Oxford Publishing Company, 2012)
25. N. Tesla, My Inventions, V. The magnifying transmitter. *Electrical Experimenter*, June 1919, p. 112. Other articles can be found in Reference 26
26. T.E. Bearden, *Solutions to Tesla's Secrets and the Soviet Tesla Weapons* (Tesla Book Company, Millbrae, 1981)
27. C.N. Yang, R.L. Mills, Conservation of isotopic spin and isotopic gauge invariance. Phys. Rev. **96**, 191–195 (1954)
28. http://www.cheniere.org/books/excalibur/moray.htm
29. J. Y. Dea, Instantaneous Interactions. (Proc. 1986 ITS)
30. S.J. Gates Jr., *Mathematica Analysis of Physical System* (Van Nostrand Reinhold, Edited by Ronald Nickkens, 1985)
31. L.D. Landau, E.N. Lifshitz, *Quantum Mechanics Non-Relativistic Theory* (Addison-Wesley, 1958)
32. D. Forster, *Hydrodynamic Fluctuations, Broken Symmetry, and Correlation Functions* (W. A. Benjamin, Inc., Reading, 1975)
33. Panofsky and Philips, *Classical Electricity and Magnetism, Second Edition (2005-01-26)* (Dover Publications, 1656)
34. R.C. Gelinas, *Cruel-Free Vector Potential Effects in Simply Connected Space* (Casner and Gelinas Co., Inc. !985, International Tesla Society, Inc), pp. 4–43
35. J.Y. Dea, Scalar fields: Their prediction from classical electromagnetism and interpretation from quantum mechanics, in *The Proceeding of the Tesla Centennial Symposium, an IEEE Centennial Activity*, (Colorado College, Colorado Springs, 1984), pp. 94–98
36. E.T. Whittaker, One the partial differential equations of mathematical physics. Math. Ann. **57**, 333–355 (1903)
37. Whittaker, E. T., 1903 and 1904, ibid. For the original basis of scalar interferometry and its creation of my kind of electromagnetic field, potential, and wave in the interference zone.

38. M. , W. Evans, P.K. Anastasovski, T.E. Bearden, et al., On Whittaker's representation of electromagnetic entity in vacuo, part V: The production of transverse field and energy by scalar interferometry. J. New Energy **4**(3). Winter 1999, p. 76 for mathematical proof of scalar interferometry. There are several other AIAS papers in the same issue of JNE, dealing with interferometry and with the implications of the Whittaker papers
39. This is a claim by T. E. Bearden and his book of Fer De Lance, *A Briefing on Soviet Scalar Electromagnetic Weapons* and not this author, so it is up to reader how they are taking this claim. "As an example, the *U.S.S. Skylark* surface companion of the *Thresher*, was beset by mysterious "jamming" of multiple types of EM systems in multiple bands. So bad was it, that it required more than an hour to get an emergency message to Naval Headquarters that the *Thresher* was in serious difficulty. Later, as the scalar interferometry field went away, all the EM systems aboard the *Skylark* resumed normal functioning. The *Skylark* was in fact in a "splatter zone" of the underwater interference zone around the *Thresher*.
40. R.G. Chambers, Physics. Rev. Lett. **5**, 3 (1960)
41. W. Tai Tsun, C.N. Yang, Concept of nonintegrable phase factors and global formulation of gauge fields. Phys. Rev. D **12**(15), 3845–3857 (1975)
42. R.P. Feynman, R.B. Leighton, M. Sands, The Feynman lectures on physics. **II**(Section 15-5), 15-8–15-14
43. J.D. Jackson, Classical electrodynamics, in *John Wiley Publisher*, 2rd edn., (1975), p. 223
44. P.A. Kossey et al., Artificial Ionospheric mirrors (AIM), in *Ionospheric Modification and its Potential to Enhance or Degrade the Performance of Military Systems*, (AGARD Conference Proceedings 485, October 1990), pp. 17A–171A
45. A. Robinson and J. Holland, Quantum Wave Theory—A Model of Unity, *Nature*, 2011
46. C. Vuille, Schrodinger's equation and general relativity. J. Math. Phys. **41**(8), 5256–5261 (August 2000)
47. E.T. Whittaker, An expression of the electromagnetic field due to electrons by means of two scalar potential functions. Proc. Lond. Math. Soc. **1**, 367–372 (1904)
48. H.A. Lorentz, A. Einstein, H. Minkowski, H. Weyl, *The Principle of Relativity: A Collection of Original Memoirs on the Special and General Theory of Relativity* (Dover, New York, 1952)
49. L. Brillouin, A. Sommerfeld, *Wave Propagation and Group Velocity* (Academic Press, New York, 1960)
50. M. Born, E. Wolf, *Principles of Optics: Electromagnetic Theory of Propagation, Interference, and Diffraction of Light*, 7th edn. (Cambridge University Press, Cambridge, 1999)
51. L.D. Landau, E.M. Lifschitz, *Electrodynamics of Continuous Media*, 2nd edn. (Pergamon Press, Oxford, 1984)
52. L.J. Wang, A. Kuzmich, A. Dogariu, Gain-assisted superluminal light propagation. Nature **406**, 277–279 (2000)
53. W.A. Rodrigues, J.Y. Lu, On the existence of undistorted progressive waves (UPWs) of arbitrary speeds in nature. Found. Phys. **27**, 435–508 (1997)
54. J.E. Maiorino, W.A. Rodrigues Jr., What is superluminal wave motion? Sci. & Tech. Mag **2** (1999). http://www.cptec.br/stm
55. http://scienceworld.wolfram.com/physics/SmarandacheHypothesis.html
56. R. Courant, D. Hilbert, Methods of Mathematical Physics, vol 2 (Wiley, New York, 1966), p. 760
57. H. Bateman, *Electrical and Optical Wave Motion* (Cambridge University Press, Cambridge, 1915)
58. A. O. Barut, G. D. Maccarrone, and E. Recami, On the shape of tachyons, *Nuovo Cimento A* **71**, 509–533 (1982), and refs. therein; see also E. Recami and G. D. Maccarrone, *Lett. Nuovo Cimento* **28**, 151–157 (1980) and **37**, 345–352 (1983); P. Caldirola, G. D. Maccarrone, and E. Recami, *Lett. Nuovo Cimento* **29**, 241–250 (1980); G. D. Maccarrone, M. Pavsic, and E. Recami, Nuovo Cimento B **73**, 91–111 (1983)

59. E. Recami, M. Zamboni-Rached, C.A. Dartora, The X-shaped, localized field generated by a superluminal electric charge, LANL archives e-print physics/0210047. Phys. Rev. E **69**, 027602 (2004)
60. A.O. Barut, A. Grant, Quantum particle-like configurations of the electromagnetic field. Found. Phys. Lett. **3**, 303–310 (1990)
61. A.O. Barut, A.J. Bracken, Particle-like configurations of the electromagnetic field: An extension of de Broglie's ideas. Found. Phys. **22**, 1267–1285 (1992)
62. J.N. Brittingham, Focus wave modes in homogeneous Maxwell's equations: Transverse electric mode. J. Appl. Phys. **54**, 1179–1189 (1983)
63. A. Sezginer, A general formulation of focus wave modes. J. Appl. Phys. **57**, 678–683 (1985)
64. H. E. Hernández-Figueroa, M. Zamboni-Rached, and E. Recami (eds.),Localized Waves, John Wiley: New York, 2008.), 386 pages
65. E. Recami, Classical tachyons and possible applications. Rivista Nuovo Cimento **9**(6), 1–178 (1986). issue 6, and refs. therein
66. J.-Y. Lu, J.F. Greenleaf, Nondiffracting X-waves: Exact solutions to free-space scalar wave equation and their finite aperture realizations. IEEE Trans. Ultrason. Ferroelectr. Freq. Control **39**, 19–31 (1992). and refs. therein
67. J.-Y. Lu, J.F. Greenleaf, Experimental verification of nondiffracting X-waves. IEEE Trans. Ultrason. Ferroelectr. Freq. Control **39**, 441–446 (1992)
68. J.A. Stratton, *"Electromagnetic Theory", Page 356, McGraw-Hill* (New York, 1941)
69. R.W. Ziolkowski, I.M. Besieris, A.M. Shaarawi, J. Phys. A Math. Gen. **33**, 7227–7254 (2000)
70. G. Shan, Quantum superluminal communication does not result in casual loop. CERN Preprint. (1999a)
71. G. Shan, Quantum superluminal communication must exist. CERN preprint. (1999b)
72. J.-Y. Lu, J. F. Greenleaf, and E. Recami, Limited diffraction solutions to Maxwell (and Schrödinger) equations, LANL archives e-print physics/9610012, Report INFN/FM–96/01, Instituto Nazionale de Fisica Nucleare, Frascati, Italy Oct. 1996; E. Recami: On localized X-shaped superluminal solutions to Maxwell equations, Physica A 252, 586–610 (1998), and refs. therein; see also R. W. Ziolkowski, I. M. Besieris, and A. M. Shaarawi, J. Opt. Soc. Am. A 10, 75 (1993); J. Phys. A Math. Gen. 33, 7227–7254 (2000)
73. R.W. Ziolkowski, I.M. Besieris, A.M. Shaarawi, Aperture realizations of exact solutions to homogeneous wave equations. J. Opt. Soc. Am. A **10**, 75–87 (1993)
74. A.M. Shaarawi, I.M. Besieris, On the superluminal propagation of X-shaped localized waves. J. Phys. A **33**, 7227–7254 (2000)
75. A.M. Shaarawi, I.M. Besieris, Relativistic causality and superluminal signaling using X-shaped localized waves. J. Phys. A **33**, 7255–7263 (2000). and refs. therein.
76. M. Zamboni-Rached, E. Recami, H.E. Hernández-Figueroa, New localized superluminal solutions to the wave equations with finite total energies and arbitrary frequencies. Eur. Phy. J. D **21**, 217–228 (2002)
77. A.P.L. Barbero, H. E. Hernández Figuerao, and E. Recami, On the propagation speed of evanescent modes , Phys. Rev. E 62, 8628 (2000), and refs. therein; see also A. M. Shaarawi and I. M. Besieris, Phys. Rev. E 62, 7415 (2000)
78. H.M. Brodowsky, W. Heitmann, G. Nimtz, Phys. Lett. A **222**, 125 (1996)
79. W.A. Rodrigues, J.Y. Lu, On the existence of undistorted progressive waves (UPWs) of arbitrary speeds in nature. Found. Phys. **27**, 435–508 (1997)
80. P. Saari, K. Reivelt, Evidence of X-shaped propagation-invariant localized light waves. Phys. Rev. Lett. **79**, 4135–4138 (1997)
81. P. Bowlan, H. Valtna-Lukner, et al., Measurement of the spatiotemporal electric field of ultrashort superluminal Bessel-X pulses. Optics and Photonics News. **20**(12), 42
82. A. Ciattoni and C. Conti, Quantum electromagnetic X-waves *arxiv.org* 0704.0442 v1
83. S., C. Frank, *Waves (Berkeley Physics Course Volume 3)*, 1st edn. (McGraw-Hill, 1968)
84. Recami, E. and Zamboni-Rached, M. (2011) Non-diffracting waves, and Frozen Waves: An introduction, 121 pages online in Geophysical Imaging with Localized Waves, Sanya, China

2011 [UCSC, S. Cruz, Cal.], and refs, therein; available at http://es.ucsc.edu/~acti/sanya/SanyaRecamiTalk.pdf. Accessed 27 Apr 2013
85. Recami, E., Zamboni-Rached, M., Hernandez-Figueroa H. E., et al. (2011) Method and Apparatus for Producing Stationary (Intense) Wavefields of arbitrary shape. Patent, Application No. US-2011/0100880
86. E. Recami, M. Zamboni-Rached and H. E. Hernández-Figueroa, Localized waves: A historical and scientific introduction., arXiv:0708.1655v2 [physics.gen-ph] 16 Aug 2007
87. A.C. Newell, J.V. Molone, *Nonlinear Optics* (Addison & Wesley, Redwood City, 1992)
88. M. Zamboni-Rached, H.E. Hernández-Figueroa, E. Recami, Chirped optical X-shaped pulses in material media. J. Opt. Soc. Am. **A21**, 2455–2463 (2004)
89. J.-Y. Lu, J. Cheng, J. Wang, High frame rate imaging system for limited diffraction array beam imaging with square-wave aperture weightings. IEEE Trans. Ultrason. Ferroelectr. Freq. Control **53**(10), 1796–1812 (2006)
90. J.-Y. Lu, J.F. Greenleaf, Diffraction-limited beams and their applications for ultrasonic imaging and tissue characterization. Proc. of SPIE **1733**, 92–119 (1992)
91. J.Y. Lu, H. Zou, J.F. Greenleaf, Biomedical ultrasound beam forming. Ultrasound Med. Biol. **20**(5), 403–428, 43, 44 (1994)
92. J.Y. Lu, Limited-diffraction beams for high-frame-rate imaging, chapter 5, in *Non-Diffracting Waves*, ed. by H. E. Hernández-Figueroa, E. Recami, M. Zamboni-Rached (Eds), (John Wiley), p. 2008

Chapter 4
The Fundamental of Electrodynamics

Electromagnetism is one of the four fundamental interactions of nature, along with strong interactions, weak interactions, and gravitation. It is the force that causes electrically charged particles to interact; the areas in which this happens are called *electromagnetic fields*. Electromagnetism is the force responsible for practically all the phenomena encountered in daily life (with the exception of gravity). Ordinary matter takes its form as a result of intermolecular forces in matter.

Electromagnetism is also the force that holds electrons and protons together inside atoms, which are the building blocks of molecules. This governs the processes involved in chemistry, which arise from interactions between the electrons orbiting atoms. Electromagnetic field theory has been and will continue to be one of the most important fundamental courses of the electrical engineering curriculum. It is one of the best-established general theories that provides explanations and solutions to intricate electrical engineering problems when other theories are no longer applicable.

4.1 Introduction

From our knowledge of college electromagnetics, we have learned that the integral form of Ampère's Law are the magnetic fields because of a current distribution that satisfies the following relationship (Fig. 4.1):

$$\oint_C \vec{H} \cdot d\vec{l} = \int_S \vec{J} \cdot \hat{n} \, da \qquad (4.1)$$

The second kind of field that enters into the study of electricity and magnetism are magnetic fields or, more appropriately, the effects of such a field. It has been known to humans since ancient times, when the effects on the naturally occurring, permanent magnetic (Fe_3O_4) were first observed. Magnetic fields even were used for

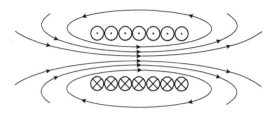

Fig. 4.1 Electricity-magnetism

medical treatment purposes in the early history of Egyptian civilization. The discovery of the orientation of North and South while seeking properties of this material had a tremendous influence on early navigation and exploration of the world via both sea and land.

Early studies done by Oersted led to his discovery of the phenomenon that an electric current does produce a magnetic field; this allows us to have a better understanding of applications of magnetism. That work, together with later efforts and research by Gauss, Henry, Faraday, and others, brought the magnetic field into better focus and a partnership with the electric field. The subject of this book is based on such a foundation.

4.2 Maxwell's Equations and the Electric Field of the Electromagnetic Wave

Although Maxwell's equations were formulated over 100 years ago, the subject of electromagnetism has not remained static. Much of modern physics (and engineering) involves time-dependent *electromagnetic* (EM) fields in which Maxwell's displacement current plays a crucial role. Maxwell's equations contain all the information necessary to characterize EM fields at any point in a medium.

To understand the behavior of materials' response to a high-power laser or radar beam one needs to consider the coupling of the laser energy or, in our case, the EM wave with materials. Therefore, we first need to know the optical reflectivity, R, and the transmissivity, T, for light incident on a surface that divides two semi-infinite media. To understand reflectivity, we must use some general results from the theory of EM waves.

For EM fields to exist they must satisfy the four Maxwell equations at the source where they are generated, at any point in a medium through which they propagate, and at the load where they are received or absorbed. Because the fields must satisfy the four coupled Maxwell equations involving four unknown variables, we first obtain an equation in terms of one unknown variable. Similar equations can then be obtained for other variables. We refer to these equations as general *wave equations*. It can be shown that the fields generated by time-varying sources propagate as *spherical waves*. In a small region far from the radiating source, however, the spherical wave may be approximated as a *plane wave*—that is, one in which all the field quantities are in a plane normal to the direction of its propagation (i.e., the

4.3 Wave Equations for Electric and Magnetic Fields

transverse plane). Consequently, a plane wave does not have any field component in its direction of propagation (i.e., the longitudinal direction).

First, we seek the solution of a plane wave in an unbounded dielectric medium and show that the wave travels with the speed of light in free space. We then consider the general case of a finitely conducting medium. We show that the wave attenuates because of a loss in energy as it travels in the conducting medium. Finally, we introduce the concept of reflection and transmission of a plane wave when it leaves one medium and enters another.

4.3 Wave Equations for Electric and Magnetic Fields

In the regions of space where no charge or current exists, Maxwell's equations read:

$$\begin{cases} \vec{\nabla} \cdot \vec{E} = 0 & \text{(i)} \\ \vec{\nabla} \cdot \vec{B} = 0 & \text{(ii)} \\ \vec{\nabla} \times \vec{E} = \dfrac{\partial \vec{B}}{\partial t} & \text{(iii)} \\ \vec{\nabla} \times \vec{B} = \mu_0 \varepsilon_0 \left(\dfrac{\partial \vec{E}}{\partial t} \right) & \text{(iv)} \end{cases} \quad (4.2)$$

where

$\vec{E}$ = the electric field
$\vec{B}$ = the magnetic field
μ_0 = the constant Biot-Savart's Law, known as permittivity of free space: $4\pi \times 10^7$ N/A^2
ε_0 = the constant Coulomb's Law, known as permittivity of free space: 8.85×10^{-12} C^2/(N · m^2)

The preceding equations constitute a set of coupled, first-order, partial differential equations for $\vec{E}$ and $\vec{B}$. They can be decoupled by applying *curl* to (iii) and (iv):

$$\vec{\nabla} \times (\vec{\nabla} \times \vec{E}) = \vec{\nabla}(\vec{\nabla} \cdot \vec{E}) - \vec{\nabla} \times \left(\frac{\partial \vec{B}}{\partial t} \right)$$
$$= -\frac{\partial}{\partial t}(\vec{\nabla} \times \vec{B}) = -\mu_0 \varepsilon_0 \frac{\partial^2 \vec{E}}{\partial t^2} \quad (4.3)$$

$$\vec{\nabla} \times (\vec{\nabla} \times \vec{B}) = \vec{\nabla}(\vec{\nabla} \cdot \vec{B}) - \nabla \times \left(\mu_0 \varepsilon_0 \frac{\partial \vec{E}}{\partial t} \right)$$
$$= \mu_0 \varepsilon_0 \frac{\partial}{\partial t}(\vec{\nabla} \times \vec{E}) = -\mu_0 \varepsilon_0 \left(\frac{\partial^2 \vec{B}}{\partial t^2} \right) \quad (4.4)$$

Or, because $\vec{\nabla} \cdot \vec{E} = 0$ and $\vec{\nabla} \cdot \vec{B} = 0$,

$$\nabla^2 \vec{E} = \mu_0 \varepsilon_0 \frac{\partial^2 \vec{E}}{\partial t^2} \quad \nabla^2 \vec{B} = \mu_0 \varepsilon_0 \frac{\partial^2 \vec{B}}{\partial t^2} \quad (4.5)$$

Equation 4.5 is a demonstration of separations between $\vec{E}$ and $\vec{B}$, but they are second order [1]. In a vacuum, then, each Cartesian component of $\vec{E}$ and $\vec{B}$ satisfies the *three-dimensional wave*, as follows:

$$\nabla^2 f = \frac{1}{v^2} \frac{\partial^2 f}{\partial t^2} \quad (4.6)$$

This supports standard wave equations in general within Cartesian form, which is like a classical wave equation of small disturbance on the string, where v represents the speed of propagation and is given by [1]:

$$v = \sqrt{\frac{T}{\mu}} \quad (4.7)$$

where μ is the mass per unit length. This equation permits as solutions all functions of the form in Eq. 4.8, for example, in z-direction of propagation [1]:

$$f(z,t) = g(z - vt) \quad (4.8)$$

This mathematical derivation can be somewhat depicted as in Fig. 4.2 at two different two times, once at $t = 0$ and later at time t. Each point on the wave form simply shifts to the right by an amount vt, where v is the velocity. Then we can see that Maxwell's equations imply that empty space supports the propagation of EM waves traveling at a speed

$$v = \frac{1}{\sqrt{\mu_0 \varepsilon_0}} = 3.00 \times 10^8 \quad m/s = 3.00 \times 10^8 \, m/s = c \quad (4.9)$$

which is precisely the velocity of light c [1]. Maxwell himself was astonished by this, and he noted that a "wave can scarcely avoid the interference that light consists in the transverse undulations of the same medium which is the cause of electric and magnetic phenomena."

Fig. 4.2 Wave propagation in z-direction [1]

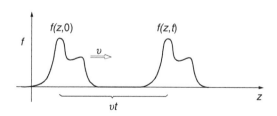

4.4 Sinusoidal Waves

Of all possible wave forms, the sinusoidal one as defined in Eq. 4.10 is, for good reason, the most familiar. Figure 4.3 shows this function at time $t = 0$ with A as the *amplitude* of the wave.

$$f(z, t) = A \cos\left[k(z - vt) + \delta\right] \tag{4.10}$$

(This is positive and represents the maximum displacement from equilibrium.) The argument of the cosine is called the *phase*, and δ is the *phase constant*; one can add any integer multiple 2π to δ without changing $f(z, t)$. Ordinarily, we use a value in the range $0 \leq \delta < 2\pi$. A point known as the *central maximum* on the curve takes place at $z = vt - \delta/k$ when the phase is zero. If $\delta = 0$, the central maximum passes the origin at time $t = 0$; more generally, δ/k is the distance by which the central maximum (and therefore the entire wave) is "delayed." Finally, k is the *wave number*. It is related to the *wavelength* λ by Eq. 4.11, for when z advances by $2\pi/k$ and the cosine executes one complete cycle:

$$\lambda = \frac{2\pi}{k} \tag{4.11}$$

As time passes, the entire wave train proceeds to the right at speed v. At any fixed point z, the string vibrates up and down, undergoing one full cycle in a *period*, that is,

$$T = \frac{2\pi}{kv} \tag{4.12}$$

If we now introduce frequency ν as the number of oscillations per unit time and show it in the form of Eq. 4.13, then we have:

$$\nu = \frac{1}{T} = \frac{kv}{2\pi} = \frac{v}{\lambda} \tag{4.13}$$

For our purposes it is better to write Eq. 4.10 in a refined form and present that in terms of *angular frequency*, ω, given in the analogous case of uniform circular motion that represents the number of radians swept out per unit time:

$$\omega = 2\pi\nu = kv \tag{4.14}$$

Fig. 4.3 A sinusoidal wave propagation [1]

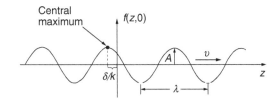

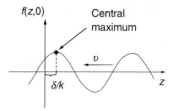

Fig. 4.4 Sinusoidal wave traveling to the left [1]

Then the new form of the sinusoidal wave in terms of ω, rather than v, is given as:

$$f(z,t) = A\cos(kz - \omega t + \delta) \tag{4.15}$$

If both k and ω travel to the left, then Eq. 4.15 can be written in this form [1]:

$$f(z,t) = A\cos(kz + \omega t - \delta) \tag{4.16}$$

This is in correspondence to and consistent with the previous convention that δ/k will represent the distance by which a wave is "delayed" because the wave is now moving to the *left*, and delay means a shift to the *right*. At $t = 0$, the wave looks like that in Fig. 4.4, and because the cosine is an *even function*, we can just as well write Eq. 4.16 in the form of Eq. 4.17:

$$f(z,t) = A\cos(-kz - \omega t + \delta) \tag{4.17}$$

Comparing Eqs. 4.16 and 4.17, we conclude that we could *simply switch the sign of k* to produce a wave with the same amplitude, phase constant, frequency, and wavelength traveling in the opposite direction.

From what we have learned in our complex variable course in college, about complex in view of Euler's formula, we have:

$$e^{i\theta} = \cos\theta + i\sin\theta \tag{4.18}$$

We can now write Eq. 4.17 as follows:

$$f(z,t)\text{Re}\left[Ae^{i(kz-\omega t+\delta)}\right] \tag{4.19}$$

where $\text{Re}[\xi]$ denotes the real part of any complex number, such as ξ. This allows for the introduction of the *complex wave function* in Eq. 4.20:

$$\tilde{f}(z,t) = \tilde{A}e^{i(kz-\omega t)} \tag{4.20}$$

With insertion of the complex amplitude, $\tilde{A} = Ae^{i\delta}$ absorbs the phase constant. The actual wave function is the real part of $\tilde{f}$:

$$f(z,t) = \text{Re}\left[\tilde{f}(z,t)\right] \tag{4.21}$$

Knowing $\tilde{f}$, it is a simple matter to find f.

Example 4.1 Combine two sinusoidal waves of f_1 and f_2.

Solution Let us write the following function f_3 as

$$f_3 = f_1 + f_2 = \text{Re}(\tilde{f}_1) + \text{Re}(\tilde{f}_2) = \text{Re}(\tilde{f}_1 + \tilde{f}_2) = \text{Re}(\tilde{f}_3).$$

with $\tilde{f}_3 = \tilde{f}_1 + \tilde{f}_2$. You may simply add the corresponding complex wave functions, and then take the real part. In particular, if they have the same frequency and wave number,

$$\tilde{f}_3 = \tilde{A}_1 e^{i(kz-\omega t)} + \tilde{A}_2 e^{i(kz-\omega t)} = \tilde{A}_3 e^{i(kz-\omega t)}$$

where

$$\tilde{A}_3 = \tilde{A}_1 + \tilde{A}_2 \quad \text{or} \quad \tilde{A}_3 e^{i\delta_3} = \tilde{A}_1 e^{i\delta_1} + \tilde{A}_2 e^{i\delta_2}$$

Now we are going to figure out what A_3 and δ_3 are as follows:

$$(A_3)^2 = (A_3 e^{i\delta_3})(A_3 e^{-i\delta_3}) = (A_1 e^{i\delta_1} + A_2 e^{i\delta_2})(A_1 e^{i\delta_1} + A_2 e^{i\delta_2})$$
$$= (A_1)^2 + (A_2)^2 + A_1 A_2 (e^{i\delta_1} e^{-i\delta_2} + e^{-i\delta_1} e^{i\delta_2})$$
$$= (A_1)^2 + (A_2)^2 + A_1 A_2 \cos(\delta_1 - \delta_2)$$

$$A_3 = \sqrt{(A_1)^2 + (A_2)^2 + A_1 A_2 \cos(\delta_1 - \delta_2)}$$

$$A_3 e^{i\delta_3} = A_2 e^{i\delta_2} + A_1 e^{i\delta_1}$$

$$A_3(\cos\delta_3 + i\sin\delta_3) = A_2(\cos\delta_2 + i\sin\delta_2) + A_1(\cos\delta_1 + i\sin\delta_1)$$
$$= (A_1\cos\delta_1 + A_2\cos\delta_2) + i(A_1\sin\delta_1 + A_2\sin\delta_2)$$

$$\tan(\delta_3) = \frac{A_3\sin(\delta_3)}{A_3\cos(\delta_3)} = \frac{A_1\sin(\delta_1) + A_2\sin(\delta_2)}{A_1\cos(\delta_1) + A_2\cos(\delta_2)}$$

$$\delta_3 = \tan^{-1}\left(\frac{A_1\sin\delta_1 + A_2\sin\delta_2}{A_1\cos\delta_1 + A_2\cos\delta_2}\right)$$

As we can see the combined wave still has the same frequency and wavelength is given by

$$f_3(z,t) = A_3 \cos(kz - \omega t + \delta_3)$$

4.5 Polarization of the Wave

Polarization, also called *wave polarization*, is an expression of the orientation of the lines of electric flux in an EM field. Polarization can be constant—that is, existing in a particular orientation at all times, or it can rotate with each wave cycle.

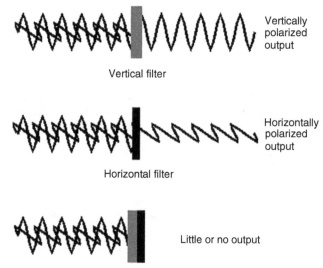

Fig. 4.5 Polarization wave

Polarization affects the propagation of EM fields at infrared (IR), visible, ultraviolet (UV), and even X-ray wavelengths. In ordinary visible light there are numerous wave components at random polarization angles. When such light is passed through a special filter, the filter blocks all light except those having a certain polarization. When two polarizing filters are placed so a ray of light passes through them both, the amount of light transmitted depends on the angle of the polarizing filters with respect to each other. The most light is transmitted when the two filters are positioned so that they polarize light in the same direction. The least light is transmitted when the filters are oriented at right angles to each other (Fig. 4.5) and in vector presentation, as in Fig. 4.6.

Electromagnetic waves are considered transverse waves in which the vibrations of electric and magnetic fields are perpendicular to each other and to the direction of propagation (Fig. 4.7). These two fields change with time and space in a sinusoid fashion. Generally, only the electric field is represented, related to the propagation direction, because it is with the electric field in which detectors interact—for example, eyes, photographic film, *charged coupled device* (CCD) .

Visible light makes up just a small part of the full EM spectrum. Electromagnetic waves with shorter wavelengths and the higher energies include ultraviolet light, X-rays, and gamma rays. Electromagnetic waves with longer wavelengths and lower energies include infrared light, microwaves, and radio and television waves. Table 4.1 is a presentation of types of radiation along with their wavelength range.

The polarization of an EM wave refers to the orientation of its electric field, $\vec{E}$. When the direction of $\vec{E}$ is randomly varying with time on a very fast scale, related to

4.5 Polarization of the Wave

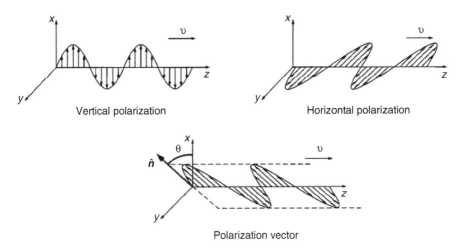

Fig. 4.6 Vector form of polarization wave [1]

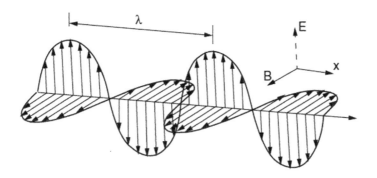

Fig. 4.7 Electromagnetic wave propagation

Table 4.1 Type of radiation and wavelength range

Type of radiation	Wavelength range
Gamma-rays	<1 pm
X-rays	1 nm-1 pm
Ultraviolet	400 nm–1 nm
Visible	750 nm–400 nm
Infrared	2.5 mm–750 nm
Microwaves	1 mm–25 mm
Radio waves	>1 mm

the direction of propagation, the wave is considered non-polarized. In case of a linearly polarized wave, the electric vector has a fixed orientation related to the propagation direction as shown in Fig. 4.8. The polarization of EM waves can be produced by absorption, scattering, reflection, and birefringence.

Fig. 4.8 Electric field orientation for polarized and non-polarized EM waves

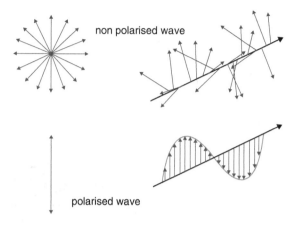

A *linear polarizer* is a device that only allows electric field components parallel to a certain direction (called the *polarization axis*) to pass through. Any EM wave that comes through such a polarizer is polarized in the direction of the polarization axis. After leaving the polarizer, the wave's polarization (i.e., $\vec{E}$-field direction) does not change, and the wave is thought to be linearly polarized. If the linearly polarized light passes through a second polarizer, the transmitted intensity, $I(\theta)$, of the waves as it leaves the second polarizer is given by the Malus Law.

The specific manner in which a beam of an EM wave at a specific polarization is reflected (and refracted), at an interface between two different media, can be used to determine the refractive index of the solid. Specifically, for a particular interface there is a specific angle of incidence (relative to the normal vector of the surface), called the *Brewster angle*, which is related to the refractive index of a material. At this angle the reflection coefficient of light polarized parallel to the plane of incidence is zero. Thus, if the incident light is non-polarized and impinges on the material at the Brewster angle, the light reflected from the solid will be polarized in the plane perpendicular to the plane of incidence. If the incident light is polarized parallel to the plane of incidence, the intensity of the reflected light will be theoretically zero at the Brewster angle.

The proposed experiment used the polarization by reflection because of its simplicity, but other polarization methods can be used too. When a light *fascicle* falls on the M mirror at a Brewster angle, the reflected fascicle is linearly polarized. Using a second rotating mirror, M', the Malus Law can be checked. If mirror M' is rotated around the PP' axe, the reflected P'S' fascicle has a variable intensity with two minimum and two maximum values. When the second fascicle falls as the first fascicle at the Brewster angle, the S'P' fascicle has a minimum value (Fig. 4.9).

The experiments with visible light and mirrors were made more than a century ago, starting with Brewster and repeated in a lot of laboratories. To date some experiments have been done in order to check the phenomena for IR, and the effect can be accepted as valid for UV and gamma rays. Related to this kind of experiment,

4.5 Polarization of the Wave

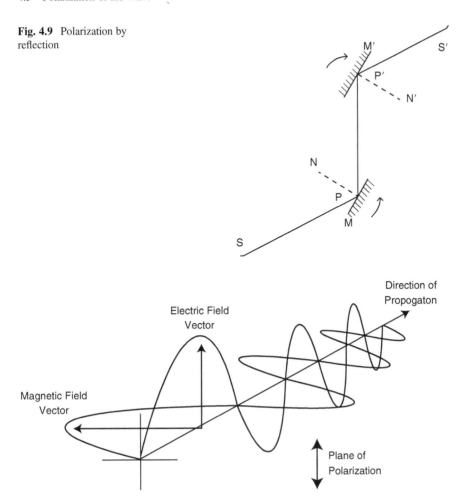

Fig. 4.9 Polarization by reflection

Fig. 4.10 Definition of a polarization vector

it is worth mentioning a paper by Lambert et al. from 1969: "Rotatable polarizers for the near infra-red" [2].

Polarization is an important optical property inherent in all laser beams. Brewster windows, reflective phase retarders, and absorbing thin-film reflectors use the advantage of polarization. On the other hand, it can cause troublesome and sometimes unpredictable results when ignored. Because virtually all laser sources exhibit some degree of polarization, understanding this effect is necessary to specify components properly. The following text gives a basic polarization definition and presents the polarization types most commonly encountered.

Light is a transverse EM wave; this means that the electric and magnetic field vectors point perpendicular to the direction of wave's travel (Fig. 4.10). When all the electric field vectors for a given wave-train lie in a plane, the wave is said to be plane or linearly polarized. The orientation of this plane is the direction of polarization.

Fig. 4.11 Unpolarized light

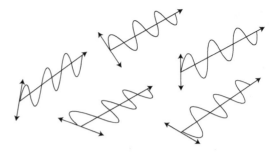

Fig. 4.12 Plane-polarized wave that points at 45° to the axes

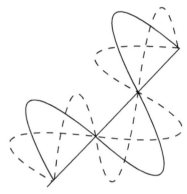

Unpolarized light refers to a wave collection that has an equal distribution of electric field orientations for all directions (Fig. 4.11). Although each individual wave-train may be linearly polarized, there is no preferred direction of polarization when all the waves are averaged together.

Randomly polarized light is exactly what it says—the light is plane-polarized, but the direction is unknown and may vary with time. Random polarization causes problems in optical systems because some components are polarization sensitive. If the polarization state changes with time, then the components' transmission, reflection, and/or absorption characteristics will also vary with time.

Polarization is a vector that has both direction and amplitude. Like any vector, it is defined in an arbitrary Coordinate system as the sum of orthogonal components. Figure 4.12 shows a plane-polarized wave that points at 45° toward the axes of our coordinate system. Thus, when described in this coordinate system, it has equal x and y components. If we then introduce a phase difference of 90° (or one-quarter wavelength) between these components, the result is a wave in which the electric field vector has a fixed amplitude but with a direction that varies as we move down the wave-train (Fig. 4.13). Such a wave is said to be circularly polarized because the tip of the polarization vector traces out a circle as it passes a fixed point.

If we have two wave-trains with unequal amplitude and with a one-quarter wave phase difference, then the result is elliptical polarization. The tip of the polarization vector will trace out an ellipse as the wave passes a fixed point. The ratio of the major to the minor axis is called the *ellipticity ratio* of the polarization. Always state the

4.5 Polarization of the Wave

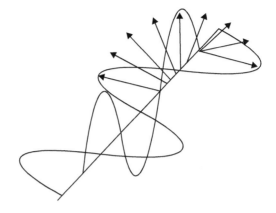

Fig. 4.13 Circularly polarized wave

polarization orientation when ordering optical coatings for use at non-normal incidences. If you are unsure about how to determine the polarization state of your source, contact applications engineers for assistance.

When light strikes an optical surface, such as a beam splitter, at a non-perpendicular angle, the reflection and transmission characteristics depend on polarization. In this case, the coordinate system we use is defined by the plane containing the input and reflected beams. Light with a polarization vector lying in this plane is called *p-polarized*, and light that is polarized perpendicular to this plane is called *s-polarized*. Any arbitrary state of input polarization can be expressed as a vector sum of these *s* and *p* components. To understand the significance of *s* and *p* polarizations, examine the graph that shows the single surface reflectance as a function of angle of incidence for these components of light at a wavelength of 10.6 μm striking a ZnSe surface (Fig. 4.14). Note that although the reflectance of the *s* component steadily increases with angle, the *p* component at first decreases to zero at 67° and then increases after that. The angle at which the *p* reflectance drops to zero is called the Brewster's angle. This effect is exploited in several ways to produce polarizing components or uncoated windows that have no transmission loss such as the Brewster windows.

The angle at which *p* reflectance drops to zero (i.e., Brewster's angle) can be calculated from $\theta_B = \tan^{-1}(n)$, where θ_B is the Brewster's angle and *n* is the material's index of refraction. Polarization state is particularly important in laser-cutting application as well as a consideration for purposes of this book and laser interaction with metal. Figure 4.15 shows a summary of polarization.

In addition, by now we know that if the waves travel down the path of say the *x*-direction, it is called *transverse* because the displacement (in the case of rope or string) is perpendicular to the direction of propagation. it is known as *longitudinal* if displacement from equilibrium is along the direction of propagation. Sound waves, which are nothing but compressed waves in air, are longitudinal although EM waves, as we will see, are transverse (see Sect. 6.2 of this book).

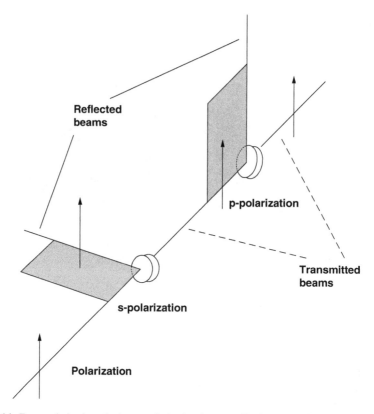

Fig. 4.14 For *s*-polarization, the input polarization is perpendicular to the plane (*shown in color*) containing the input and output beams. For *p*-polarization, the input polarization is parallel to the plane (*shown in color*) containing the input and output beams

Polarization Summary

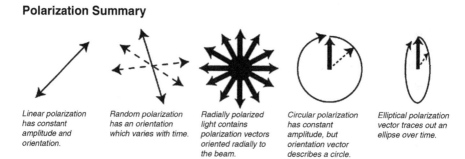

Linear polarization has constant amplitude and orientation.

Random polarization has an orientation which varies with time.

Radially polarized light contains polarization vectors oriented radially to the beam.

Circular polarization has constant amplitude, but orientation vector describes a circle.

Elliptical polarization vector traces out an ellipse over time.

Fig. 4.15 Polarization summary

4.6 Monochromatic Plane Waves

As we discussed in Sect. 3.2 of this book, we now pay attention to sinusoidal waves of frequency ω, where in visible range, each unlike frequency corresponds to different *colors*. Such waves are called *monochromatic waves* (see Table 4.2). Suppose, furthermore that the waves are traveling in the z-direction and have no x or y dependencies; these are called *plane waves* [3] because the fields are uniform over every plane perpendicular to the direction of propagation (Fig. 4.16).

Table 4.2 The electromagnetic spectrum

Frequency (Hz)	Type	Wavelength (m)
10^{22}		10^{-13}
10^{21}	Gamma rays	10^{-12}
10^{20}		10^{-11}
10^{19}		10^{-10}
10^{18}	X-rays	10^{-9}
10^{17}		10^{-8}
10^{16}	Ultraviolet	10^{-7}
10^{15}	Visible	10^{-6}
10^{14}	Infrared	10^{-5}
10^{13}		10^{-4}
10^{12}		10^{-3}
10^{11}		10^{-2}
10^{10}	Microwave	10^{-1}
10^{9}		1
10^{8}	TV, FM	10
10^{7}		10^{2}
10^{6}	AM	10^{3}
10^{5}		10^{4}
10^{4}	RF	10^{5}
10^{3}		10^{6}

The visible range

Frequency (Hz)	Color	Wavelength (m)
1.0×10^{15}	Near ultraviolet	3.0×10^{-7}
7.5×10^{14}	Shortest visible blue	4.0×10^{-7}
6.5×10^{14}	Blue	4.6×10^{-7}
5.6×10^{14}	Green	5.4×10^{-7}
5.1×10^{14}	Yellow	5.9×10^{-7}
4.9×10^{14}	Orange	6.1×10^{-7}
3.9×10^{14}	Longest visible red	7.6×10^{-7}
3.0×10^{14}	Near infrared	1.0×10^{-6}

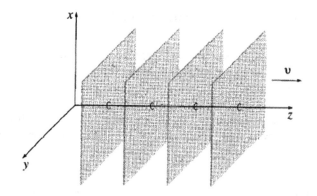

Fig. 4.16 Plane wave propagation [1]

We are interested then in complex form of both magnetic and electric fields as follows:

$$\tilde{\vec{E}}(z,t) = \tilde{\vec{E}}_0 e^{i(kz-\omega t)} \quad \text{and} \quad \tilde{\vec{B}}(z,t) = \tilde{\vec{B}}_0 e^{i(kz-\omega t)} \tag{4.22}$$

where $\tilde{\vec{E}}_0$ and $\tilde{\vec{B}}_0$ are the complex amplitudes [1]. The physical fields, of course, are the real parts of $\tilde{\vec{E}}$ and $\tilde{\vec{B}}$. Now, the wave equations for $\tilde{\vec{E}}$ and $\tilde{\vec{B}}$ (see Eq. 4.2) were derived from Maxwell's equations. Still, whereas every solution to Maxwell's equations (in empty space) must obey the wave equation, the converse is *not* true; Maxwell's equations impose extra constraints on $\tilde{\vec{E}}_0$ and $\tilde{\vec{B}}_0$. In particular, because $\vec{\nabla} \cdot \vec{E} = 0$ and $\vec{\nabla} \cdot \vec{B} = 0$, it follows that:

$$\left(\tilde{E}_0\right)_z = \left(\tilde{B}_0\right) = 0 \tag{4.23}$$

This is because the real part of $\tilde{\vec{E}}$ differs from the imaginary part only in the replacement of *sine* by *cosine*; if the former obeys Maxwell's equations, so does the latter, and thus $\tilde{\vec{E}}$ as well [1]. Equation 4.23 also indicates that EM waves *are transverse*—that is, the electric and magnetic fields are perpendicular to the direction of propagation. Moreover, Faraday's Law, $\vec{\nabla} \times \vec{E} = -\partial \vec{B}/\partial t$, implies a relationship between the electric and magnetic amplitudes in the following equation:

$$-k\left(\tilde{E}_0\right)_y = \omega\left(\tilde{B}_0\right)_x \quad \text{and} \quad -k\left(\tilde{E}_0\right)_x = \omega\left(\tilde{B}_0\right)_y \tag{4.24}$$

or in compact form complex variable presentation we have:

$$\tilde{\vec{B}}_0 = \frac{k}{\omega}\left(\hat{z} \times \tilde{\vec{E}}_0\right) \tag{4.25}$$

From this equation it is evident that $\tilde{\vec{E}}$ and $\tilde{\vec{B}}$ are *in phase and mutually perpendicular*; their (real) amplitudes are related by

4.6 Monochromatic Plane Waves

Fig. 4.17 Scalar product of $\vec{k} \cdot \vec{r}$

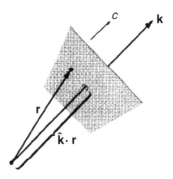

$$B_0 = \frac{k}{\omega} E_0 = \frac{1}{c} E_0 \tag{4.26}$$

The fourth of Maxwell's equations, $\vec{\nabla} \times \vec{B} = \mu_0 \varepsilon_0 (\partial \vec{E}/\partial t)$, does not yield an independent condition; it simply reduces to Eq. 4.24.

We can easily generalize to monochromatic plane waves traveling in an arbitrary direction. The notation is facilitated by the introduction of *propagation* (or *wave*) *vector* $\vec{k}$ pointing in the direction of propagation, with a magnitude that is the wave number k. The scalar product, $\vec{k} \cdot \vec{r}$, is the appropriate generalization of kz in Fig. 4.17, so we have:

$$\begin{cases} \tilde{\vec{E}}(\vec{r},t) = \tilde{E}_0 e^{i(\vec{k}\cdot\vec{r}-\omega t)} \hat{n} \\ \tilde{\vec{B}}(\vec{r},t) = \frac{1}{c}\tilde{E}_0 e^{i(\vec{k}\cdot\vec{r}-\omega t)} (\hat{k} \times \hat{n}) = \frac{1}{c}(\hat{k} \times \tilde{\vec{E}}) \end{cases} \tag{4.27}$$

where $\hat{n}$ is the polarization vector, and because $\vec{E}$ is transverse,

$$\hat{n} \cdot \hat{k} = 0 \tag{4.28}$$

The transversality of $\vec{B}$ follows automatically from Eq. 4.27. The actual (real) electric and magnetic fields in a monochromatic plane wave with propagation vector $\vec{k}$ and polarization $\hat{n}$ are [1]:

$$\vec{E}(\vec{r},t) = E_0 \cos(\vec{k}\cdot\vec{r} - \omega t + \delta) \hat{n} \tag{4.29}$$

$$\vec{B}(\vec{r},t) = \frac{1}{c} E_0 \cos(\vec{k}\cdot\vec{r} - \omega t + \delta)(\hat{k} \times \hat{n}) \tag{4.30}$$

Example 4.2 If $\vec{E}$ points in the x-direction, then $\vec{B}$ points in the y-direction Eq. 4.25. See Fig. 4.18.

Fig. 4.18 Depiction of Example 4.2 solution

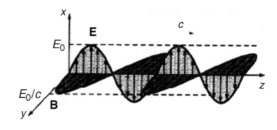

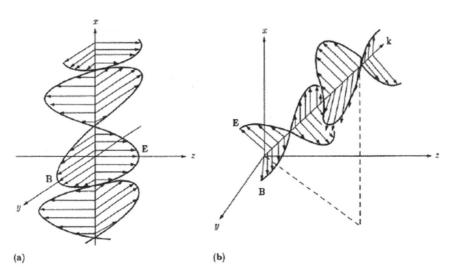

(a) (b)

Fig. 4.19 Sketch of the Example 4.3 solution

Solution Writing the following

$$\tilde{\vec{E}}(z,t) = \tilde{E}_0 e^{i(\vec{k}\cdot\vec{r}-\omega t)} \hat{x} \quad \text{and} \quad \tilde{\vec{B}}(z,t) = \frac{1}{c}\tilde{E}_0 e^{i(\vec{k}\cdot\vec{r}-\omega t)} \hat{y}$$

or taking the real part, we have

$$\begin{cases} \vec{E}(z,t) = E_0 \cos(kz - \omega t + \delta)\hat{x} \\ \vec{B}(z,t) = \frac{1}{c} E_0 \cos(kz - \omega t + \delta)\hat{y} \end{cases} \quad (4.31)$$

Example 4.3 Write down the (real) electric and magnetic fields for a monochromatic plane wave of amplitude E_0, frequency ω, and phase angle zero that is (1) traveling in the negative x-direction and polarized in the z-direction; (2) traveling in the direction from the origin to the point (1, 1, 1), with polarization parallel to the xz-plane. In each case, sketch the wave and give the explicit Cartesian components of $\vec{k}$ and $\hat{n}$. See Fig. 4.19.

4.7 Boundary Conditions: The Reflection and Refraction (Transmission)...

Solution We can start with the following for part (a):

(a) $\vec{k} = -\frac{\omega}{c}\hat{x}; \quad \hat{n} = \hat{z} \quad \vec{k} \cdot \vec{r} = \left(-\frac{\omega}{c}\hat{x}\right) \cdot (x\hat{x} + y\hat{y} + z\hat{z}) = -\frac{\omega}{c}x;$

$\vec{k} \times \hat{n} = -\hat{x} \times \hat{z} = \hat{y}$

$\vec{E}(x,t) = E_0 \cos\left(\frac{\omega}{c}x + \omega t\right)\hat{z} \quad \vec{B}(x,t) = \frac{E_0}{c}\cos\left(\frac{\omega}{c}x + \omega t\right)\hat{y}$

and we have the following for part (b):

(b) $\vec{k} = \frac{\omega}{c}\left(\frac{\hat{x} + \hat{y} + \hat{z}}{\sqrt{3}}\right) \quad \hat{n} = \frac{\hat{x} - \hat{z}}{\sqrt{2}}$ (Because $\hat{n}$ is parallel to the xz-plane, it must have the form $\alpha\hat{x} + \beta\hat{z}$; because $\hat{n} \cdot \hat{k} = 0$, $\beta = -\alpha$; and because it is a unit vector, $\alpha = 1/\sqrt{2}$):

$\vec{k} \cdot \vec{r} = \frac{\omega}{\sqrt{3}c}(\hat{x} + \hat{y} + \hat{z}) \cdot (x\hat{x} + y\hat{y} + z\hat{z}) = \frac{\omega}{\sqrt{3}c}(x + y + z);$

$\hat{k} \times \hat{n} = \frac{1}{\sqrt{6}}\begin{vmatrix} \hat{x} & \hat{y} & \hat{z} \\ 1 & 1 & 1 \\ 1 & 0 & -1 \end{vmatrix} = \frac{1}{\sqrt{6}}(-\hat{x} + 2\hat{y} - \hat{z})$

Therefore, we have the following as the final solution for part (b):

$$\begin{cases} \vec{E}(x,y,z,t) = E_0 \cos\left[\frac{\omega}{\sqrt{3}c}(x+y+z) - \omega t\right]\left(\frac{\hat{x} - \hat{z}}{\sqrt{2}}\right) \\ \vec{B}(x,y,z,t) = \frac{E_0}{c}\cos\left[\frac{\omega}{\sqrt{3}c}(x+y+z) - \omega t\right]\left(\frac{-\hat{x} + 2\hat{y} - \hat{z}}{\sqrt{6}}\right) \end{cases}$$

4.7 Boundary Conditions: The Reflection and Refraction (Transmission) Dielectric Interface

The important problem of the reflection and refraction (transmission) of a wave at the interface of two media of different dielectric constants is a boundary value one in which the physics is identical in principle to that involved in time-harmonic problems involving conductors. If, for example, one medium is vacuum, and a plane wave is incident on a second (dielectric) medium, the incident harmonic of the plane wave generates oscillating time-harmonic dipoles (or dipolar currents) that produce a field of their own [4].

The strength of the currents is not known in advance, however, and therein lays the essence of the problem. Although this is a boundary value problem, it is especially simple because of the great symmetry and simple geometry. It turns out that it can be solved by adding to the incident plane waves only two other plane waves—a *reflected* and a *transmitted* (or refracted) one. The geometry is shown Fig. 4.20 with the plane $z = 0$ taken as the interface between the two media, labeled 1 and 2.

In terms of the *vector* propagation constant, $\vec{k}_i$, the incident fields are

$$\vec{E}_i = \vec{A}e^{i\vec{k}_i \cdot \vec{r}} \quad \text{and} \quad \vec{B}_i = (\vec{k}_i/k) \times \vec{E}_i \quad \text{Incident} \quad (4.32)$$

Here $\vec{B}_i$ is derived from the assumed $\vec{E}_i$ by $\nabla \times \vec{E}_i = ik\vec{B}_i$. The normal $\hat{z}$ to the plane $z = 0$ and the vector $\vec{k}_i$ defines a *plane of incidence* that, without loss of generality, can be taken to be the x, z-plane as was shown in Fig. 4.18. We now postulate the existence of two other plane waves and will show that these suffice to solve the boundary value problem. These *reflected* and *transmitted* waves, amplitudes $\vec{E}_r$ and $\vec{E}_t$, respectively, are [4]:

$$\vec{E}_r = \vec{R}e^{i\vec{k}_r \cdot \vec{r}} \quad \text{and} \quad \vec{B} = (\vec{k}_r/k) \times \vec{E}_r \quad \text{Reflected} \quad (4.33)$$

$$\vec{E}_t = \vec{T}e^{i\vec{k}_t \cdot \vec{r}} \quad \text{and} \quad \vec{B}_t = (\vec{k}_t/k) \times \vec{E}_t \quad \text{Transmitted} \quad (4.34)$$

The vectors $\vec{k}_r$ and $\vec{k}_t$, for the moment, must be considered arbitrary in direction, for although $\vec{k}_i$ in the x, z-plane, we cannot assume the same a priori for $\vec{k}_r$ and $\vec{k}_t$. The magnitudes, $k_i \equiv |\vec{k}_i|$ and so on, of the wave vectors are, with n_1 and n_2 the indices of refraction of the two media and $k = \omega/c$,

$$k_i = k_r = n_1 k$$
$$k_t = n_2 k \quad (4.35)$$

Considering the boundary condition that states the tangential components of $\vec{E}$ and $\vec{B}$ are continuous, we have to satisfy the following two steps:

First: If the tangential components of the three fields in Eqs. 4.33, 4.34, and 4.35 are to be matched at $z = 0$, it also should be clear that the spatial dependence given by the exponents must be identical. However, this is a necessary but *not* a sufficient condition.

Second: Vector coefficients $\vec{A}$, $\vec{R}$, and $\vec{T}$ must be determined.

The first condition—that is, the spatial variation of the three fields must be identical at $z = 0$—leads to the following equation:

$$(\vec{k}_i \cdot \vec{r})_{z=0} = (\vec{k}_r \cdot \vec{r})_{z=0} = (\vec{k}_t \cdot \vec{r})_{z=0} \quad (4.36)$$

The first equality in Eq. 4.35 yields $k_{ix}x = k_{rx}x + k_{ry}y$. For this condition to hold for all x and y, we must have $k_{ry} = 0$, showing that $\vec{k}_r$ lies in the plane of incidence

4.7 Boundary Conditions: The Reflection and Refraction (Transmission)... 371

and, also that $k_{ix}x = k_{rx}x$. Similarly, from the second equality in Eq. 4.36, $\vec{k}_t$ must lie in this plane so that $\vec{k}_i$, $\vec{k}_r$, and $\vec{k}_t$ are all *co-planar*. Moreover, from the geometry of Fig. 4.19, we have from $k_{ix} = k_{rx}$ that $k_i \sin \theta_i = k_r \sin \theta_r$. Or, because of $k_i = k_r$, we can write the following:

$$\sin \theta_i = \sin \theta_r \tag{4.37}$$

The *angle of incidence equals the angle of reflection*. Similarly, the equality of k_{ix} and k_{tx} yields:

$$k_i \sin \theta_i = k_t \sin \theta_t \tag{4.38}$$

Or, using Eq. 4.35, we have:

$$\frac{\sin \theta_i}{\sin \theta_t} = \frac{n_2}{n_1} \tag{4.39}$$

Equation 4.38 is known as *Snell's Law* of refraction. The condition of Eqs. 4.37 and 4.38 are quite general ones that are independent of the detailed vectorial nature of the wave field. They hold for reflection and refraction of scalar waves (SWs). These conditions by themselves do not guarantee the continuity of tangential $\vec{E}$ and $\vec{B}$ across the boundary. To satisfy these conditions more specifications of the polarization of the fields should be analyzed [4]. For convenience we consider the general case of arbitrary incident polarization as a linear combination of a wave with polarization perpendicular to the plane of incidence and one with polarization parallel to this plane. The reflected and transmitted waves will then be similarly polarized. These two cases are sketched as parts (a) and (b) in Fig. 4.20.

Consider first the case of $\vec{E}$ perpendicular to the plane of incidence (i.e., in the y-direction). Vector coefficients $\vec{A}$, $\vec{R}$, and $\vec{T}$ become scalar ones, with a subscript $\perp$ to denote this case:

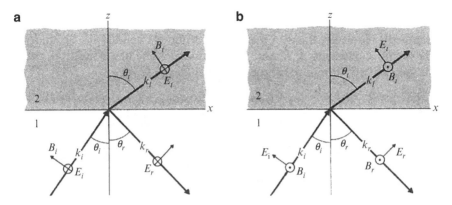

Fig. 4.20 Reflection and refraction at a dielectric interface: (**a**) Electric vector perpendicular to plane of incidence. (**b**) Electric vector parallel to plane of incidence

$$E_{i\perp} = A_\perp e^{i\vec{k}_i \cdot \vec{r}}$$
$$E_{r\perp} = A_\perp e^{i\vec{k}_r \cdot \vec{r}} \qquad (4.40)$$
$$E_{t\perp} = A_\perp e^{i\vec{k}_t \cdot \vec{r}}$$

Then from the continuity of tangential $\vec{E}$ at the boundary, we have

$$A_\perp + R_\perp = T_\perp \qquad (4.41)$$

The condition on tangential $\vec{B}$ becomes, with $B_{ix} = -A_\perp \cos\theta_i$, $B_{rx} = -R_\perp \cos\theta_r$, and $B_{tx} = -T_\perp \cos\theta_t$—also considering that $\cos\theta_i = -\cos\theta_r$—we then have:

$$n_1(A_\perp - R_\perp)\cos\theta_i = n_2 T_\perp \cos\theta_t \qquad (4.42)$$

Solving Eqs. 4.41 and 4.42 for the ratios $R_\perp/A_\perp$ and $T_\perp/A_\perp$, we find the following using Snell's Law:

$$\frac{R_\perp}{A_\perp} = \frac{1 - \dfrac{\tan\theta_i}{\tan\theta_r}}{1 + \dfrac{\tan\theta_i}{\tan\theta_r}} = -\frac{\sin(\theta_i - \theta_t)}{\sin(\theta_i + \theta_t)}$$

$$\frac{T_\perp}{A_\perp} = \frac{2}{1 + \dfrac{\tan\theta_i}{\tan\theta_r}} = \frac{2\cos\theta_i \sin\theta_t}{\sin(\theta_i + \theta_t)} \qquad (4.43)$$

For the second case, an incident wave polarized parallel to the plane of incidence, we use $A_\parallel = |\vec{A}|$ in Eq. 4.33 and, similarly, $R_\parallel = |\vec{R}|$ and $T_\parallel = |\vec{T}|$ in Eqs. 4.34 and 4.35. The boundary condition yields:

$$\cos\theta_i(A_\parallel - R_\parallel) = \cos\theta_t T_\parallel$$
$$n_1(A_\parallel - R_\parallel) = n_2 T_\parallel \qquad (4.44)$$

These equations lead to the results:

$$\frac{R_\parallel}{A_\parallel} = \frac{\tan(\theta_i - \theta_t)}{\tan(\theta_i + \theta_t)}$$

$$\frac{T_\parallel}{A_\parallel} = \frac{2\cos\theta_i \sin\theta_t}{\sin(\theta_i + \theta_t)\cos(\theta_i - \theta_t)} \qquad (4.45)$$

There are two phenomena worthy of note in connection with the preceding discussion. Consider the case of polarization in the plane of incidence. We can see from Eq. 4.43 that $R_\parallel/A_\parallel$ will be zero for $\theta_i + \theta_t = \pi/2$. Putting this condition into Snell's Law, using $\sin\theta_t = \sin(\pi/2 - \theta_i) = \cos\theta_i$, we can see that the angle of incidence, θ_B (i.e., the *Brewster angle*) for which this happens, is defined by

4.7 Boundary Conditions: The Reflection and Refraction (Transmission)...

$$\tan \theta_B = \frac{n_2}{n_1} \tag{4.46}$$

If a wave with arbitrary polarization is incident on a dielectric interface, it can be considered to be a linear combination of a wave polarized parallel to, and a wave polarized perpendicular to, the plane of incident. At the Brewster's angle, the parallel component will not be reflected, so the reflected wave will be plane-polarized in a plane perpendicular to the plane of incidence. This effect can then be made the basis of a device for polarizing an unpolarized beam of radiation.

The second phenomenon is that of *total internal reflection*. In either (a) or (b), shown earlier in Fig. 4.4, suppose that the index n_1 is greater than n_2. Then, from Snell's Law, $\sin\theta_t = (n_1/n_2) \sin \theta_i$ is always greater than θ_i. There then will be some value of θ_i, call it θ_{int}, for which $\theta_t = \pi/2$; this angle is defined by

$$\sin \theta_{\text{int}} = \frac{n_2}{n_1} \tag{4.47}$$

Because, in general,

$$\vec{E}_t = \vec{T} e^{ikn_2(x \sin \theta_t + z \cos \theta_t)} \tag{4.48}$$

for $\theta_t = \pi/2$ there is no wave in the second medium; the z-dependence vanishes. Now if $\theta_i > \theta_{\text{int}}$, then $\sin\theta_t$ is larger than unity from Snell's Law and, as a consequence, $\cos\theta_t$ is imaginary, as follows:

$$\cos \theta_t = \sqrt{1 - \left(\frac{n_1}{n_2}\right)^2 \sin^2\theta_i} = i\sqrt{\left(\frac{n_1}{n_2}\right)^2 \sin^2\theta_i - 1} \tag{4.49}$$

Equation 4.36 then becomes

$$\vec{E}_t = \vec{T} e^{-kn_2 z \sqrt{(n_1/n_2)\sin^2\theta_i - 1}} e^{ikn_2 x \sin \theta_t} \tag{4.50}$$

This corresponds to a wave that is exponentially attenuated as a function of z, and that propagates as a function of x with a propagation constant of $kn_2 \sin \theta_t$. Such a wave is the prototype of a surface wave; readers can see further discussion and more details in Eyges (1972) [4] in his Appendix 2.

The interesting question is this: What happens when a wave passes from one transparent medium into another—air to water, say, or glass to plastic? As in the case of waves on a string, we expect to get a reflected wave and a transmitted wave. The details depend on the exact nature of the electrodynamics boundary conditions [1].

$$\begin{cases} \varepsilon_1 E_1^\perp = \varepsilon_2 E_2^\perp & \text{(i)} \quad E_1^\| - E_2^\| \quad \text{(iii)} \\ B_1^\perp = B_2^\perp & \text{(ii)} \quad \frac{1}{\mu_1} B_1^\| = \frac{1}{\mu_2} B_2^\| \quad \text{(iv)} \end{cases} \tag{4.51}$$

These equations relate to the electric and magnetic fields just to the left and just to the right of the interface between two linear media. In the following sections we use them to deduce the laws governing reflection and refraction of EM waves.

4.8 Electromagnetic Waves in Matter

Some solutions to Maxwell's equations already have been discussed in previous sections. This section extends the treatment of EM waves. Because most regions of interest are free of charge, it will be assumed that charge density $\rho = 0$. Moreover, linear isotropic (invariant with respect to direction) materials will be assumed, with the following relationships:

$$\begin{cases} \vec{D} = \varepsilon \vec{E} \\ \vec{B} = \mu \vec{H} \\ \vec{J} = \sigma \vec{E} \end{cases} \quad (4.52)$$

where

$\vec{D}$ = flux density (C/m^2)
$\vec{E}$ = electric field (N/C)
ε = permittivity of the medium (C^2/N · m^2) or equivalent (F/m)
$\vec{B}$ = magnetic field (T)
$\vec{H}$ = magnetic field strength (A/m)
μ = mobility within materials (m^2/V · s)
$\vec{J}$ = current density (A/m^2)
σ = conductivity of materials (S/m)
ρ = charge density (C/m^3)

where

A = Ampère
C = Coulomb
N = Newton
F = Faraday
S = Siemens
T = Tesla

4.8.1 Propagation in Linear Media

Inside matter, but in regions where there is no *free* charge or *free current*, Maxwell's equations become

4.8 Electromagnetic Waves in Matter

$$
\begin{aligned}
&\text{(i)} \quad \nabla \cdot \vec{D} = 0 \quad \text{(iii)} \quad \nabla \times \vec{E} = -\frac{\partial \vec{B}}{\partial t} \\
&\text{(ii)} \quad \nabla \cdot \vec{B} = 0 \quad \text{(iv)} \quad \nabla \times \vec{H} = \frac{\partial \vec{D}}{\partial t}
\end{aligned}
\qquad (4.53)
$$

If the medium is *linear*, then

$$
\vec{D} = \varepsilon \vec{E} \quad \text{and} \quad \vec{H} = \frac{1}{\mu}\vec{B} \qquad (4.54)
$$

and homogeneous (so ε and μ do not vary from point to point), Maxwell's equations reduce to:

$$
\begin{aligned}
&\text{(i)} \quad \nabla \cdot \vec{E} = 0 \quad \text{(iii)} \quad \nabla \times \vec{E} = -\frac{\partial \vec{B}}{\partial t} \\
&\text{(ii)} \quad \nabla \cdot \vec{B} = 0 \quad \text{(iv)} \quad \nabla \times \vec{B} = \mu\varepsilon\frac{\partial \vec{E}}{\partial t}
\end{aligned}
\qquad (4.55)
$$

which remarkably differ from the vacuum analogs in Eq. 4.1 only in the replacement of $\mu_0\varepsilon_0$ by $\mu\varepsilon$. This is obvious mathematically, yet the physical implications are astonishing [5]. As the wave passes through the fields, they busily polarize and magnetize all the molecules, and the resulting (oscillating) dipoles create their own electric and magnetic fields. These combine with the original fields in such a way as to create a *single* wave with the same frequency but a different speed. This extraordinary conspiracy is responsible for the phenomenon of *transparency*. It is a distinctly *non-trivial* consequence of the *linearity* of the medium [5]. Looking at the set of Eq. 4.54, it is evident that EM waves propagate through a linear homogeneous medium at a speed:

$$
v = \frac{1}{\sqrt{\varepsilon\mu}} = \frac{c}{n} \qquad (4.56)
$$

where

$$
n \equiv \sqrt{\frac{\varepsilon\mu}{\varepsilon_0\mu_0}} \qquad (4.57)
$$

is the index of refraction of the material. For most materials μ is very close to μ_0, so

$$
n \cong \sqrt{\varepsilon_r} \qquad (4.58)
$$

where ε_r is the *dielectric constant*, also known as *relative permittivity* [1], and is equal to $\varepsilon_r = \varepsilon/\varepsilon_0$. Because ε_r almost always is greater than 1, light travels *more slowly* through matter—a fact that is well known from optics [1].

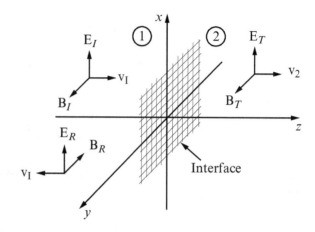

Fig. 4.21 Normal wave incidence [1]

4.8.2 Reflection and Transmission at Normal Incidence

Suppose the *xy*-plane forms the boundary between two linear media. A plane wave of frequency ω, traveling in the *z*-direction and polarized in the *x*-direction, approaches the interface from the left (see Fig. 4.21):

$$\begin{cases} \tilde{\vec{E}}_I(z,t) = \tilde{E}_{0_I} e^{i(k_1 z - \omega t)} \hat{x} \\ \tilde{\vec{B}}_I(z,t) = \dfrac{1}{v_1} \tilde{E}_{0_I} e^{i(k_1 z - \omega t)} \hat{y} \end{cases} \quad (4.59)$$

This gives rise to a reflected wave

$$\begin{cases} \tilde{\vec{E}}_R(z,t) = \tilde{E}_{0_R} e^{i(k_1 z - \omega t)} \hat{x} \\ \tilde{\vec{B}}_R(z,t) = \dfrac{1}{v_1} \tilde{E}_{0_R} e^{i(k_1 z - \omega t)} \hat{y} \end{cases} \quad (4.60)$$

which travels back to the left in medium (1), and a transmitted wave

$$\begin{cases} \tilde{\vec{E}}_T(z,t) = \tilde{E}_{0_T} e^{i(k_1 z - \omega t)} \hat{x} \\ \tilde{\vec{B}}_T(z,t) = \dfrac{1}{v_1} \tilde{E}_{0_T} e^{i(k_1 z - \omega t)} \hat{y} \end{cases} \quad (4.61)$$

which continues to the right in medium (2). Note the minus sign in $\vec{B}_R$, as required by Eq. 4.27, or, if you prefer, by the fact that Poynting's vector aims in the direction of propagation.

At $z = 0$ the combined fields on the left, $\tilde{\vec{E}}_I + \tilde{\vec{E}}_R$ and $\tilde{\vec{B}}_I + \tilde{\vec{B}}_R$, must join the fields on the right, $\tilde{\vec{E}}_T$ and $\tilde{\vec{B}}_T$, in accordance with the boundary conditions in

4.8 Electromagnetic Waves in Matter

Eq. 4.50. In this case there are no components perpendicular to the surface, so (i) and (ii) are trivial. However, (iii) requires:

$$\tilde{E}_{0_I} + \tilde{E}_{0_R} = \tilde{E}_{0_T} \tag{4.62}$$

while (iv) says

$$\frac{1}{\mu_1}\left(\frac{1}{v_1}\tilde{E}_{0_I} - \frac{1}{v_1}\tilde{E}_{0_R}\right) = \frac{1}{\mu_2}\left(\frac{1}{v_2}\tilde{E}_{0_T}\right) \tag{4.63}$$

or

$$\tilde{E}_{0_I} - \tilde{E}_{0_R} = \beta\tilde{E}_{0_T} \tag{4.64}$$

where

$$\beta \equiv \frac{\mu_1 v_1}{\mu_2 v_2} = \frac{\mu_1 n_2}{\mu_2 n_1} \tag{4.65}$$

Equations 4.61 and 4.63 are solved easily for the outgoing amplitudes in terms of the incident amplitude:

$$\tilde{E}_{0_R} = \left(\frac{1-\beta}{1+\beta}\right)\tilde{E}_{0_I} \quad \text{and} \quad \tilde{E}_{0_T} = \left(\frac{2}{1+\beta}\right)\tilde{E}_{0_I} \tag{4.66}$$

These results are strikingly similar to the ones for waves on a string. Indeed, if the permittivities μ are close to their values in vacuum (as, remember, they are for most media), then $\beta = \frac{v_1}{v_2}$, and we have [1]:

$$\tilde{E}_{0_R} = \left(\frac{v_2 - v_1}{v_2 + v_1}\right)\tilde{E}_{0_I} \quad \text{and} \quad \tilde{E}_{0_T} = \left(\frac{2v_2}{v_2 + v_1}\right)\tilde{E}_{0_I} \tag{4.67}$$

In that case, as before, the reflected wave is *in phase* (right side up) if $v_2 > v_1$ and *out of phase* (upside down) if $v_2 < v_1$; the real amplitudes are related by

$$\vec{E}_{0_R} = \left(\frac{1-\beta}{1+\beta}\right) \quad \text{and} \quad E_{0_T} = \left(\frac{2v_2}{v_2 + v_1}\right)E_{0_I} \tag{4.68}$$

or, in terms of the indices of refraction [1],

$$E_{0_R} = \left(\frac{n_1 - n_2}{n_1 + n_2}\right)E_{0_I} \quad \text{and} \quad E_{0_T} = \left(\frac{2n_2}{n_1 + n_2}\right)E_{0_I} \tag{4.69}$$

According to the definition of *intensity* (average power per unit area) that is given by Griffiths [1], as follows, we should ask these two questions: Which fraction of the incident energy is reflected? Which fraction is transmitted?

$$I = \frac{1}{2}\epsilon v E_0^2 \tag{4.70}$$

If again $\mu_1 = \mu_2 = \mu_0$, then the ratio of the reflected intensity to the incident intensity is as follows:

$$R \equiv \frac{I_R}{I_I} = \left(\frac{E_{0_R}}{E_{0_I}}\right)^2 = \left(\frac{n_1 - n_2}{n_1 + n_2}\right)^2 \quad (4.71)$$

whereas the ratio of the transmitted intensity to the incident intensity is:

$$T \equiv \frac{I_T}{I_I} = \frac{\varepsilon_2 v_2}{\varepsilon_1 v_1}\left(\frac{E_{0_T}}{E_{0_I}}\right)^2 = \frac{4 n_1 n_2}{(n_1 + n_2)^2} \quad (4.72)$$

R is called the *reflection coefficient* and T is called the *transmission coefficient*; they measure the fraction of the incident energy that is reflected and transmitted, respectively. Notice that [1]:

$$R + T = 1 \quad (4.73)$$

as conservation of energy, of course, requires. For instance, when light passes from air ($n_1 = 1$) into glass ($n_2 = 1.5$), then $R = 0.04$ and $T = 0.96$. Not surprisingly, most of the light is transmitted [1].

Example 4.4 Calculate the exact reflection and transmission coefficients, without assuming $\mu_1 = \mu_2 = \mu_0$. Confirm $R + T = 1$.

Solution From Eq. 4.65 we have $R = \left(\frac{E_{0_R}}{E_{0_I}}\right)^2$; substituting from Eq. 4.65 results, where $\beta \equiv \frac{\mu_1 v_1}{\mu_2 v_2}$ and using Eq. 4.66 results in $T = \beta\left(\frac{2}{1+\beta}\right)^2$, which is Eq. 4.60. Note the following:

$$\frac{\varepsilon_2 v_2}{\varepsilon_1 v_1} = \frac{\mu_1}{\mu_2}\frac{\varepsilon_2 v_2 v_2}{\varepsilon_1 v_1 v_1} = \frac{\mu_1}{\mu_2}\left(\frac{v_1}{v_2}\right)^2 = \frac{\mu_1 v_1}{\mu_2 v_2} = \beta$$

$$R = \frac{2}{(1+\beta)^2}\left[4\beta + (1-\beta)^2\right]$$

$$= \frac{1}{(1+\beta)^2}\left(4\beta + 1 - 2\beta + \beta^2\right)$$

$$= \frac{1}{(1+\beta)^2}\left(1 + 2\beta + \beta^2\right) = 1$$

Example 4.5 In writing Eqs. 4.60 and 4.61, the assumption was that the reflected and transmitted waves have the same *polarization* as the incident waves—along the x-direction. Prove that this must be so. [*Hint:* Let the polarization vector of the transmitted and reflected waves be $\hat{n} = \cos\theta_T \hat{x} + \sin\theta_T \hat{y}$ and $\hat{n}_R = \cos\theta_R \hat{x} + \sin\theta_R \hat{y}$, and prove from the boundary condition that $\theta_T = \theta_R = 0$.]

Solution Equation 4.52 is replaced by $\tilde{E}_{0_I}\hat{x} + \tilde{E}_{0_R}\hat{n} = \tilde{E}_{0_T}\hat{n}$ and Eq. 4.58 by $\tilde{E}_{0_I}\hat{y} - \tilde{E}_{0_R}(\hat{z} \times \hat{n}_R) = \beta\tilde{E}_{0_T}(\hat{z} \times \hat{n}_T)$. The y-component of the first equation is $\tilde{E}_{0_R}\sin\theta_R = \tilde{E}_{0_T}\sin\theta_T$; the x-component of the second is $\tilde{E}_{0_R}\sin\theta_R = -\beta\tilde{E}_{0_T}\sin\theta_T$ Comparing these two, we can conclude that $\sin\theta_R = \sin\theta_T = 0$, thus, $\theta_T = \theta_R = 0$.

4.8.3 Reflection and Transmission at Oblique Incidence

In the previous section reflection and transmission were treated at *normal* incidence—that is, when the incoming wave hits the interface head-on at 90° incident angle [1]. The more general case of *oblique* incidence, in which the incoming wave meets the boundary at an arbitrary angle θ_I, also is treated by David Griffiths (see reference [1] of this chapter). We just wrote the result of his conclusion here and encourage readers to refer to his book's Sect. 9.3.3 (p. 386) for more details.

Although the angle of incidence $\theta_I = 0$ is a special case of the oblique incidence, we did treat that separately for some cases; here we are dealing with high-power laser interactions with matter. The treatment by Griffiths is based on a monochromatic plane wave with an arbitrary angle of incidence (Fig. 4.22).

Suppose that a monochromatic plane wave (incident wave) approaches from the left as follows:

$$\tilde{E}_I(\vec{r},t) = \tilde{E}_{0_I} e^{i(\vec{k}_I \cdot \vec{r} - \omega t)} \quad \text{and} \quad \tilde{B}_I(\vec{r},t) = \frac{1}{v_1}(\hat{k} \times \tilde{E}_I) \tag{4.74}$$

This gives rise to a reflected wave of the form in Eq. 4.69:

$$\tilde{E}_I(\vec{r},t) = \tilde{E}_{0_R} e^{i(\vec{k}_R \cdot \vec{r} - \omega t)} \quad \text{and} \quad \tilde{B}_R(\vec{r},t) = \frac{1}{v_1}(\hat{k}_R \times \tilde{E}_R) \tag{4.75}$$

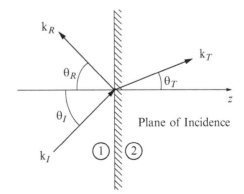

Fig. 4.22 Incidence at oblique angle [1]

and a transmitted set of wave equations of:

$$\tilde{\vec{E}}_T(\vec{r},t) = \tilde{\vec{E}}_{0_T} e^{i(\vec{k}_T \cdot \vec{r} - \omega t)} \quad \text{and} \quad \tilde{\vec{B}}_T(\vec{r},t) = \frac{1}{v_2}(\hat{\vec{k}}_T \times \tilde{\vec{E}}_T) \tag{4.76}$$

All three waves have the same frequency, ω, that is determined once and for all at the source (e.g., laser beam, flashlight, or whatever) that produces the incident beam. The three wave numbers are related per Eq. 4.11:

$$k_I v_1 = k_R v_1 = k_T v_2 = \omega \quad \text{or} \quad k_I = k_R = \frac{v_2}{v_1} k_T = \frac{n_1}{n_2} k_T \tag{4.77}$$

The combined fields in medium (1), $\tilde{\vec{E}}_I + \tilde{\vec{E}}_R$ and $\tilde{\vec{B}}_I + \tilde{\vec{B}}_R$, must now be joined to the fields $\tilde{\vec{E}}_T$ and $\tilde{\vec{B}}_T$ in medium (2), using the boundary conditions in Eq. 4.51. These all share the generic structure. Boundary conditions should hold at *all points on the plane,* and for *all times for all three plane-wave equations* when $z = 0$. Otherwise, a slight change in x would destroy the equality (see Example 4.6) [1]. The time factors in this case *already* are equal; in fact we can regard this as an independent confirmation that the transmitted and reflected frequencies must match the incidence frequency. So, for a spatial case, we have the following relation:

$$\vec{k}_I \cdot \vec{r} = \vec{k}_R \cdot \vec{r} = \vec{k}_T \cdot \vec{r} \quad \text{when} \quad z = 0 \tag{4.78}$$

or more explicitly,

$$x(k_I)_x + y(k_I)_y = x(k_R)_x + y(k_R)_y = x(k_T)_x + y(k_T)_y \tag{4.79}$$

for all x and y. Equation 4.73 can *only* hold if the components are separately equal, for if $x = 0$, we get:

$$(k_I)_y = (k_R)_y = (k_T)_y \tag{4.80}$$

While $y = 0$ gives

$$(k_I)_x = (k_R)_x = (k_T)_x \tag{4.81}$$

We may as well orient the axes so that $\vec{k}$ lies in the xy-plane (i.e., $(k_I)_y = 0$); according to Eq. 4.74, so too will $\vec{k}_R$ and $\vec{k}_T$.

Conclusion
First Law: The incident, reflected, and transmitted wave vectors form a plane (called the *plane of incidence*), which also includes the normal to the surface (here, the z-axis). Meanwhile, Eq. 4.81 implies the following:

$$k_I \sin \theta_1 = k_R \sin \theta_R = k_T \sin \theta_T \tag{4.82}$$

where θ_I is the *angle of incidence*; θ_R is the *angle of reflection*; and θ_T is the *angle of transmission*, more commonly known as the *angle of refraction*—all of them are

4.8 Electromagnetic Waves in Matter

measured with respect to normal (see Fig. 4.21). In view of Eq. 4.81, we get the second law.

Second Law: The angle of incidence is equal to the angle of reflection,

$$\theta_I = \theta_T \tag{4.83}$$

This is the law of reflection.

Third Law: The transmitted angle follows:

$$\frac{\sin \theta_T}{\sin \theta_I} = \frac{n_1}{n_2} \tag{4.84}$$

This is the law of refraction, or Snell's Law.

These are the three fundamental laws of geometrical optics. It is remarkable how little actual electrodynamics went into them; we have yet to invoke any specific boundary conditions. Now that we have taken care of the exponential factors they cancel, given Eq. 4.77 the boundary conditions in Eq. 4.51 become [1]:

$$\left.\begin{array}{ll}
\text{(i)} & \varepsilon_1 \left(\tilde{\vec{E}}_{0_I} + \tilde{\vec{E}}_{0_R} \right)_z = \varepsilon_2 \left(\tilde{\vec{E}}_{0_T} \right)_z \\
\text{(ii)} & \left(\tilde{\vec{B}}_{0_I} + \tilde{\vec{B}}_{0_R} \right)_z = \left(\tilde{\vec{B}}_{0_T} \right)_z \\
\text{(iii)} & \left(\tilde{\vec{E}}_{0_I} + \tilde{\vec{E}}_{0_R} \right)_{x,y} = \left(\tilde{\vec{E}}_{0_T} \right)_{x,y} \\
\text{(iv)} & \frac{1}{\mu_1} \left(\tilde{\vec{B}}_{0_I} + \tilde{\vec{B}}_{0_R} \right)_{x,y} = \frac{1}{\mu_2} \left(\tilde{\vec{B}}_{0_T} \right)_{x,y}
\end{array}\right\} \tag{4.85}$$

where $\tilde{\vec{B}}_0 = (1/v) \hat{\tilde{k}} \times \tilde{\vec{E}}_0$ in each case. The last two, (iii) and (iv), represent pairs of equations—one for the *x*-direction and one for the *y*-direction.

Suppose that the polarization of the incident wave is *parallel* to the plane of incidence (the *xz*-plane in Fig. 4.23); it follows that the reflected and transmitted

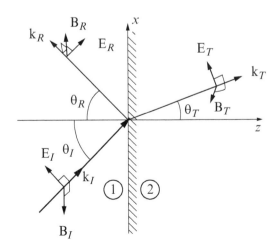

Fig. 4.23 Incident oblique wave

waves also are polarized in this plane (see Example 4.5 in this chapter). We present this in an sample to analyze the case of polarization perpendicular to the plane of incidence (see Example 4.7). Then (i) reads:

$$\varepsilon_1\left(-\tilde{\vec{E}}_{0_I} \sin\theta_I + \tilde{\vec{E}}_{0_R} \sin\theta_R\right) = \varepsilon_2\left(-\tilde{\vec{E}}_{0_T} \sin\theta_T\right) \quad (4.86)$$

Note that (ii) adds nothing (0 = 0) because the magnetic fields have no z-components; (iii) becomes [1]:

$$\tilde{\vec{E}}_{0_I} \cos\theta_I + \tilde{\vec{E}}_{0_R} \cos\theta_R = \tilde{\vec{E}}_{0_T} \cos\theta_T \quad (4.87)$$

and (iv) says,

$$\frac{1}{\mu_1 v_1}\left(\tilde{\vec{E}}_{0_I} - \tilde{\vec{E}}_{0_R}\right) = \frac{1}{\mu_2 v_2}\left(\tilde{\vec{E}}_{0_T}\right) \quad (4.88)$$

Given the laws of reflection and refraction, Eqs. 4.84 and 4.88 both reduce to the following:

$$\tilde{\vec{E}}_{0_I} - \tilde{\vec{E}}_{0_R} = \beta \tilde{\vec{E}}_{0_T} \quad (4.89)$$

where again

$$\beta \equiv \frac{\mu_1 v_1}{\mu_2 v_2} = \frac{\mu_1 n_2}{\mu_2 n_1} \quad (4.90)$$

Equation 4.87 says:

$$\tilde{\vec{E}}_{0_I} + \tilde{\vec{E}}_{0_R} = \alpha \tilde{\vec{E}}_{0_T} \quad (4.91)$$

where again

$$\alpha = \frac{\cos\theta_T}{\cos\theta_I} \quad (4.92)$$

Solving Eqs. 4.89 and 4.91 for the reflected and transmitted amplitudes, we obtain [1]:

$$\begin{cases} \tilde{\vec{E}}_{0_R} = \left(\dfrac{\alpha - \beta}{\alpha + \beta}\right) \tilde{\vec{E}}_{0_I} \\ \tilde{\vec{E}}_{0_T} = \left(\dfrac{2}{\alpha + \beta}\right) \tilde{\vec{E}}_{0_I} \end{cases} \quad (4.93)$$

These are known as Fresnel's equations, for the case of polarization in the plane of incidence. (There are two other Fresnel equations that give the reflected and transmitted amplitudes when the polarization is perpendicular to the plane of incidence— see Prob. 9.16 in Griffiths [1].) Notice that the transmitted wave is always in phase

4.8 Electromagnetic Waves in Matter

with the incident one; the reflected wave is either in phase (right side up) if $\alpha > \beta$, or 180° out of phase (upside down) if $\alpha < \beta$.

Be aware that there is an unavoidable ambiguity in the phase of the reflected wave. The convention that Griffiths [1] used in Chap. 9 of his book, which is reflected here and was adopted for Fig. 4.23, has $\tilde{\vec{E}}_R$ as positive "upward"; this is consistent with some, but not all, of the standard optics' texts. Changing the sign of the polarization vector is equivalent to a 180° phase shift.

The amplitudes of the transmitted and reflected waves depend on the angle of incidence, α, because α is a function of θ_I:

$$\alpha = \sqrt{\frac{1 - \sin^2 \theta_T}{\cos \theta_I}} = \sqrt{\frac{1 - [(n_1/n_2) \sin \theta_I]^2}{\cos \theta_I}} \tag{4.94}$$

In the case of normal incidence ($\theta_I = 0$), $\alpha = 1$ so we retrieve Eq. 4.60. At grazing incidence ($\theta_I = 90$), α diverges, and the wave is totally reflected (a fact that is painfully familiar to anyone who has driven at night on a wet road). Interestingly, there is an intermediate angle, θ_B (i.e., Brewster's angle), at which the reflected wave is completely extinguished [1].

Because waves polarized perpendicular to the plane of incidence exhibit no corresponding quenching of the reflected component, an arbitrary beam incident at Brewster's angle yields a reflected beam that is totally polarized parallel to the interface. That is why Polaroid glasses, with the transmission axis vertical, help to reduce glare off a horizontal surface [1].

According to Eq. 4.94, this occurs when $\alpha = \beta$, or

$$\sin^2 \theta_B = \frac{1 - \beta^2}{(n_1/n_2)^2 - \beta^2} \tag{4.95}$$

For the typical case $\mu_1 \cong \mu_2$, so $\beta \cong n_2/n_1$, $\sin^2 \theta_B \cong \beta^2/(1 + \beta^2)$; thus,

$$\tan \theta_B \cong \frac{n_2}{n_1} \tag{4.96}$$

Figure 4.24 shows a plot of the transmitted and reflected amplitudes as functions of θ_I, for light incident on glass, ($n_2 = 1.5$), from air, ($n_1 = 1.0$). (On the graph, a negative number indicates that the wave is 180° out of phase with the incident beam—the amplitude itself is the absolute value).

The power per unit area striking the interface is $\vec{S} \cdot \hat{z}$. Thus, the incident intensity is:

$$I_I = \frac{1}{2} \epsilon_1 v_1 E_{0_I}^2 \cos \theta_I \tag{4.97}$$

while the reflected and transmitted intensities are

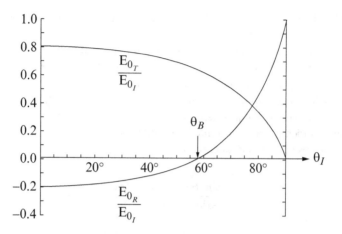

Fig. 4.24 Plot of the transmitted and reflected amplitude [1]

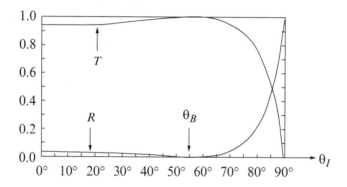

Fig. 4.25 Plot of reflection and transmission coefficient for polarized waves

$$I_I = \frac{1}{2}\varepsilon_1 v_1 E_{0_I}^2 \cos\theta_I \quad \text{and} \quad I_T = \frac{1}{2}\varepsilon_2 v_2 E_{0_I}^2 \cos\theta_T \quad (4.98)$$

(The cosines are there because we are talking about the average power per unit area of interface, and the *interface* is at an angle to the wave front [1].)

The reflection and transmission coefficients for waves polarized parallel to the plane of incidence are

$$R \equiv \frac{I_R}{I_I} = \left(\frac{E_{0_R}}{E_{0_I}}\right)^2 = \left(\frac{\alpha - \beta}{\alpha + \beta}\right)^2 \quad (4.99)$$

$$T \equiv \frac{I_T}{I_I} = \frac{\varepsilon_2 v_2}{\varepsilon_1 v_1}\left(\frac{E_{0_T}}{E_{0_I}}\right)^2 \frac{\cos\theta_T}{\cos\theta_I} = \alpha\beta\left(\frac{2}{\alpha + \beta}\right)^2 \quad (4.100)$$

They are plotted as functions of the angle of incidence in Fig. 4.25 (for the air–glass interface). R is the fraction of the incident energy that is reflected—naturally, it goes

4.8 Electromagnetic Waves in Matter

to zero at Brewster's angle; and T is the fraction transmitted—it goes to 1 at θ_B. Note that $R + T = 1$ is required by conservation of energy; the energy per unit time *reaching* a certain patch of area on the surface is equal to the energy per unit time *leaving* the patch [1].

Example 4.6 Suppose $Ae^{iax} + Be^{ibx} = Ce^{icx}$, for some non-zero constant A, B, C, a, b, c, and for all x. Prove that $a = b = c$ and $A + B = C$.

Solution $Ae^{iax} + Be^{ibx} = Ce^{icx}$, for all x, so (using $x = 0$), $A + B = C$. Differentiate: $iaAe^{iax} + ibBe^{ibx} = icCe^{icx}$, so (using $x = 0$), $aA + bB = cC$. Differentiating again: $-a^2 Ae^{iax} - b^2 Be^{ibx} = -c^2 Ce^{icx}$, so (using $x = 0$), $a^2 A + b^2 B = c^2 C$. $a^2 A + b^2 B = c(cC) = c(aA + bB)$, or $(A + B)(a^2 A + b^2 B) = (A + B)c(cC) = c(aA + bB) = cC(aA + bB)$, or $a^2 A^2 + b^2 AB + a^2 AB + b^2 B^2 = (aA + bB)^2 = a^2 A^2 + 2abAB + b^2 B^2$, or $(a^2 + b^2 - 2ab)AB = 0$, or $(a - b)^2 AB = 0$.

But A and B are non-zero, so $a = b$. Therefore, $(A + B)e^{iax} = Ce^{icx} \cdot a(A + B) = cC$, or $aC = cC$, so (because $C \neq 0$), then $a = c$. *Conclusion*: $a = b = c$ is driven.

Example 4.7 Analyze the case of polarization *perpendicular* to the plane of incidence (i.e., electric fields in the y-direction (see Fig. 4.22). Impose the boundary condition in Eq. 4.85, and obtain the Fresnel equations for $\tilde{E}_{0_R}$ and $\tilde{E}_{0_T}$. Sketch $(\tilde{E}_{0_R}/\tilde{E}_{0_I})$ and $(\tilde{E}_{0_T}/\tilde{E}_{0_I})$ as functions of $\beta = n_2/n_1 = 1.5$. (Note that for β the reflected wave is always 180° out of phase.) Show that there is no Brewster's angle for *any* n_1 and n_2: $\tilde{E}_{0_R}$ is *never* zero (unless, of course, $n_1 = n_2$ and $\mu_1 = \mu_2$, in which case the two media are optically indistinguishable). Confirm that your Fresnel equations reduce to the proper forms at normal incidence. Compute the reflection and transmission coefficients, and check that they add up to 1.

Solution We start with the following relationships:

$$\begin{cases} \tilde{\vec{E}}_I = \tilde{E}_{0_I} e^{i(\vec{k}_I \cdot \vec{r} - \omega t)} \hat{y} \\ \tilde{\vec{B}}_I = \dfrac{1}{v_1} \tilde{E}_{0_I} e^{i(\vec{k}_I \cdot \vec{r} - \omega t)} \left(-\cos\theta_1 \hat{x} + \sin\theta_1 \hat{z} \right) \end{cases}$$

$$\begin{cases} \tilde{\vec{E}}_R = \tilde{E}_{0_R} e^{i(\vec{k}_R \cdot \vec{r} - \omega t)} \hat{y} \\ \tilde{\vec{B}}_I = \dfrac{1}{v_1} \tilde{E}_{0_R} e^{i(\vec{k}_R \cdot \vec{r} - \omega t)} \left(-\cos\theta_1 \hat{x} + \sin\theta_1 \hat{z} \right) \end{cases}$$

$$\begin{cases} \tilde{\vec{E}}_T = \tilde{E}_{0_T} e^{i(\vec{k}_T \cdot \vec{r} - \omega t)} \hat{y} \\ \tilde{\vec{B}}_T = \dfrac{1}{v_2} \tilde{E}_{0_R} e^{i(\vec{k}_T \cdot \vec{r} - \omega t)} \left(-\cos\theta_2 \hat{x} + \sin\theta_2 \hat{z} \right) \end{cases}$$

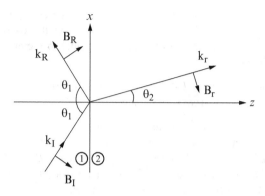

Boundary conditions:
$$\begin{cases} \text{(i)} \ \varepsilon_1 E_1^\perp = \varepsilon_2 E_2^\perp & \text{(iii)} \ E_1^\parallel = E_2^\parallel \\ \text{(ii)} \ B_1^\perp = B_2^\perp & \text{(iv)} \ \dfrac{1}{\mu_1} B_1^\parallel = \dfrac{1}{\mu_1} B_2^\parallel \end{cases}$$

Law of refractions: $\dfrac{\sin\theta_2}{\sin\theta_1} = \dfrac{v_2}{v_1}$

(Note: $\vec{k}_I \cdot \vec{r} - \omega t = \vec{k}_R \cdot \vec{r} - \omega t = \vec{k}_T \cdot \vec{r} - \omega t$ at $z = 0$, so we can drop all exponential factors in applying the boundary conditions.)

Boundary condition (i): $0 = 0$ (trivial)
Boundary condition (iii): $\tilde{E}_{0_I} + \tilde{E}_{0_R} = \tilde{E}_{0_T}$
Boundary condition (ii): $\dfrac{1}{v_1}\tilde{E}_{0_I}\sin\theta_1 + \dfrac{1}{v_1}\tilde{E}_{0_R}\sin\theta_1 = \dfrac{1}{v_2}\tilde{E}_{0_T}\sin\theta_2 \Rightarrow \tilde{E}_{0_I} + \tilde{E}_{0_R} = \left(\dfrac{v_1 \sin\theta_2}{v_2 \sin\theta_1}\right)\tilde{E}_{0_T}$

But the term in parentheses is 1, by the law of refraction, so this is the same as (ii).

Boundary condition (iv): $\dfrac{1}{\mu_1}\left[\dfrac{1}{v_1}\tilde{E}_{0_I}(-\cos\theta_1) + \dfrac{1}{v_1}\tilde{E}_{0_R}(-\cos\theta_1)\right] = \dfrac{1}{\mu_2 v_2}\tilde{E}_{0_T}(-\cos\theta_2)$

or we can write $\tilde{E}_{0_I} - \tilde{E}_{0_R} = \left(\dfrac{\mu_1 v_1 \sin\theta_2}{\mu_2 v_2 \sin\theta_1}\right)\tilde{E}_{0_T}$. Let $\alpha \equiv \dfrac{\cos\theta_2}{\cos\theta_1}$ and $\beta \equiv \dfrac{\mu_1 v_1}{\mu_2 v_2}$. Then we have $\tilde{E}_{0_I} - \tilde{E}_{0_R} = \alpha\beta\tilde{E}_{0_T}$. Solving for $\tilde{E}_{0_R}$ and $\tilde{E}_{0_T}$: $2\tilde{E}_{0_I} = (1+\alpha\beta)\tilde{E}_{0_T}$, or $\tilde{E}_{0_T} = \left(\dfrac{2}{1+\alpha\beta}\right)\tilde{E}_{0_I}$ and $\tilde{E}_{0_R} = \tilde{E}_{0_T} - \tilde{E}_{0_I} = \left(\dfrac{2}{1+\alpha\beta} - \dfrac{1+\alpha\beta}{1+\alpha\beta}\right)\tilde{E}_{0_I}$; then we have $\tilde{E}_{0_R} = \left(\dfrac{1-\alpha\beta}{1+\alpha\beta}\right)\tilde{E}_{0_I}$.

Because α and β are positive, it follows that $2/(1+\alpha\beta)$ is positive; thus, the *transmitted* wave is *in phase* with the incident wave, and the (real) amplitudes are related by $E_{0_T} = \left(\dfrac{2}{1+\alpha\beta}\right)E_{0_I}$. The *reflected* wave is in phase if $\alpha\beta < 1$ and 180° out of phase if $\alpha\beta < 1$; the (real) amplitudes are related by $E_{0_R} = \left(\dfrac{1-\alpha\beta}{1+\alpha\beta}\right)E_{0_I}$. These are the *Fresnel equations* for polarization perpendicular to the plane of incidence.

4.8 Electromagnetic Waves in Matter

To construct the graphs, note that $\alpha\beta = \beta\frac{\sqrt{2.25-\sin^2\theta}}{\cos\theta}$, where θ is the angle of incidence, so for $\beta = 1.5$ and $\alpha\beta = \beta\frac{\sqrt{1-\sin^2\theta}}{\cos\theta} = \frac{\sqrt{\beta^2-\sin^2\theta}}{\cos\theta}$.

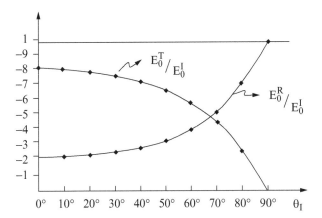

Is there a Brewster's angle? Well, $E_{0_R} = 0$ would mean that $\alpha\beta = 1$, and thus that $\alpha = \frac{\sqrt{1-(v_2/v_1)\sin^2\theta}}{\cos\theta} = \frac{1}{\beta} = \frac{\mu_2 v_2}{\mu_1 v_1}$, or $1 - \left(\frac{v_2}{v_1}\right)^2 \sin^2\theta = \left(\frac{\mu_2 v_2}{\mu_1 v_1}\right)^2 \cos^2\theta$, so $1 = \left(\frac{v_2}{v_1}\right)^2 [\sin^2\theta + (\mu_2/\mu_1)\cos^2\theta]$.

Because $\mu_1 \approx \mu_2$, this means that $1 \approx (v_2/v_1)^2$, which is only true for optically indistinguishable media in which case there is, of course, no reflection—but that would be true at any angle, not just at a special "Brewster's angle." If μ_2 were substantially different from μ_1, and the relative velocities were just right, it would be possible to get a Brewster's angle for this case, at

$$\left(\frac{v_2}{v_1}\right)^2 = 1 - \cos^2\theta + \left(\frac{\mu_2}{\mu_1}\right)^2 \cos^2\theta$$

$$\Rightarrow \cos^2\theta = \frac{(v_1/v_2)^2 - 1}{(\mu_1/\mu_2)^2 - 1} = \frac{(\mu_2\varepsilon_2/\mu_1\varepsilon_1)^2 - 1}{(\mu_2/\mu_1)^2 - 1}$$

$$= \frac{(\varepsilon_2/\varepsilon_1) - (\mu_2/\mu_1)}{(\mu_2/\mu_1) - (\mu_1/\mu_2)}$$

But the media would be very peculiar. By the same token, δ_R is either always 0, or always π, for a given interface—it does not switch over as you change θ the way it does for polarization in the plane of incidence. In particular, if $\beta = 3/2$, then $\alpha\beta > 1$, for $\alpha\beta = \left(\frac{\sqrt{1-\sin^2\theta}}{\cos\theta}\right) > 1$ if $2.25 - \sin^2\theta > \cos^2\theta$, or $2.25 > \sin^2\theta + \cos^2\theta = 1.0$. In general, for $\beta > 1$, $\alpha\beta > 1$, and thus $\delta_R = \pi$ for $\beta < 1$, $\alpha\beta < 1$, and $\delta_R = 0$. At *normal incidence*, $\alpha = 1$, so Fresnel's equations reduce to $E_{0_T} = \left(\frac{2}{1+\beta}\right)E_{0_I}$ and $E_{0_R} = \left(\frac{1-\beta}{1+\beta}\right)E_{0_I}$ consistent with Eq. 4.68.

Reflection and transmission coefficients: $R = \left(\frac{E_{0_R}}{E_{0_I}}\right)^2 = \left(\frac{1-\alpha\beta}{1+\alpha\beta}\right)^2$.

Referring to Eq. 4.100, $T = \frac{\varepsilon_2 v_2}{\varepsilon_1 v_1}\alpha\left(\frac{E_{0_T}}{E_{0_I}}\right)^2 = \alpha\beta\left(\frac{2}{\alpha+\beta}\right)^2$, therefore $R + T = \frac{(1-\alpha\beta)^2+4\alpha\beta}{(1+\alpha\beta)^2} = \frac{1-2\alpha\beta+\alpha^2\beta^2+4\alpha\beta}{(1+\alpha\beta)^2} = \frac{(1+\alpha\beta)^2}{(1+\alpha\beta)^2} = 1$.

Example 4.8 The index of refraction of a diamond is 2.42. Construct a graph analogous to Fig. 4.23 for the air–diamond interface (assume $\mu_1 = \mu_2 = \mu_0$). In particular, calculate: (1) the amplitudes at normal incidence, (2) Brewster's angle, and (3) the "crossover" angle at which the reflected and transmitted amplitudes are equal.

Solution In Eq. 4.84 we see that $\beta = 2.42$ and in Eq. 4.88 $\alpha = \frac{\sqrt{1-(\sin\theta/2.42)^2}}{\cos\theta}$
$\theta = 0 \Rightarrow \alpha = 1$. See Eq. 4.97 $\Rightarrow \left(\frac{E_{0_R}}{E_{0_I}}\right) = \frac{\alpha-\beta}{\alpha+\beta} = \frac{1-2.42}{1+2.42} = \frac{1.42}{3.42} = -0.415$,
$\left(\frac{E_{0_T}}{E_{0_I}}\right) = \frac{2}{\alpha+\beta} = \frac{2}{3.42} = 0.585$. For Eq. 4.90 $\Rightarrow \theta_B \tan^{-1}(2.42) = 67.5°$
$E_{0_R} = E_{0_T} \Rightarrow \alpha - \beta = 2; \alpha = \beta + 2 = 4.42$ and $(4.42)^2\cos^2\theta = 1 - \sin^2\theta/(2.42)^2$; $(4.42)^2(1-\sin^2\theta) = (4.42)^2 - (4.42)^2\sin^2\theta = 1 - 0.171\sin^2\theta; 19.5 - 1 = (19.5 - 017)\sin^2\theta$; $18.5 = 19.3\sin^2\theta$; $\sin^2\theta = 18.5/19.3 = 0.959 \Rightarrow \sin\theta = 0.979; \theta = 78.3°$

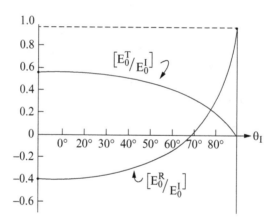

4.9 Absorption and Dispersion

Expressions for the stored energy, energy flow, and power dissipated are derived for EM waves in terms of complex permittivity $\tilde{\varepsilon}$ and permeability $\tilde{\mu}$ for a frequency-dispersive absorbing medium. This is shown to be possible when $\tilde{\varepsilon}$ and $\tilde{\mu}$ are known functions not only of frequency but also of all the loss factors (e.g., collision frequencies, etc.) The derivation is not restricted to media with small losses.

4.9 Absorption and Dispersion

In physics and electrical engineering, dispersion most often refers to frequency-dependent effects in wave propagation. Note, however, that there are several other uses of the word "dispersion" in the physical sciences. In the presence of dispersion, wave velocity is no longer uniquely defined, giving rise to the distinction of phase velocity and group velocity. A well-known effect of phase velocity dispersion is the color dependence of light refraction that can be observed in prisms and rainbows.

Dispersion relations describe the interrelations of wave properties (e.g., wavelength, frequency, velocities, refraction index, and attenuation coefficient). Besides geometry- and material-dependent dispersion relations, there are the overarching Kramers-Kronig relations that connect the frequency dependences of propagation and attenuation. Dispersion may be caused either by geometric boundary conditions (e.g., waveguides, shallow water) or by interaction of the waves with the transmitting medium.

4.9.1 Electromagnetic Waves in Conductors

Previous sections stipulated that free charge density ρ_f and free current density $\vec{J}_f$ are zero, and everything that followed was predicated on that assumption. Such a restriction is perfectly reasonable when you are talking about wave propagation through a vacuum or through insulating materials such as glass or (pure) water. But in the case of conductors we do not independently control the flow of charge and, in general, $\vec{J}_f$ is certainly not zero. In fact, according to Ohm's Law, the (free) current density in a conductor is proportional to the electric field:

$$\vec{J}_f = \sigma \vec{E} \qquad (4.101)$$

With this, Maxwell's equations for linear media assume the form

$$\begin{aligned}
&\text{(i)} \ \vec{\nabla} \cdot \vec{E} = \frac{1}{\varepsilon}\rho_f \quad \text{(iii)} \ \vec{\nabla} \times \vec{E} = -\frac{\partial \vec{B}}{\partial t} \\
&\text{(ii)} \ \vec{\nabla} \cdot \vec{B} = 0 \quad \text{(iv)} \ \vec{\nabla} \times \vec{B} = \mu\sigma\vec{E} + \mu\varepsilon\frac{\partial \vec{E}}{\partial t}
\end{aligned} \qquad (4.102)$$

Now the continuity equation for free charge becomes,

$$\vec{\nabla} \cdot \vec{J}_f = -\frac{\partial \rho_f}{\partial t} \qquad (4.103)$$

Together with Ohm's Law and Gauss's Law (i), this gives

$$\frac{\partial \rho_f}{\partial t} = -\sigma(\vec{\nabla} \cdot \vec{E}) = -\frac{\sigma}{\varepsilon}\rho_f \qquad (4.104)$$

for a homogeneous linear medium, from which it follows that

$$\rho_f(t) = e^{-(\sigma/\varepsilon)t}\rho_f(0) \tag{4.105}$$

Thus, any initial free charge density $\rho_f(0)$ dissipates in a characteristic time $\tau \equiv \varepsilon/\sigma$.

This reflects the familiar fact that if you put some free charge on a conductor, it will flow out to the edges. Time constant τ affords a measure of how "good" a conductor is: For a "perfect" conductor $\sigma = \infty$ and $\tau = 0$; for a "good" conductor, τ is much less than the other relevant times in the problem. In oscillatory systems this that means $\tau \ll 1/\omega$; for a "poor" conductor, τ is *greater* than the characteristic times in the problem ($\tau \gg 1/\omega$) [6]. Ashby points out that for good conductors, it is absurdly short (10^{-19} s for copper, whereas the time between collisions is $\tau_c = 10^{-14}$ s) [6]. The problem is that Ohm's Law itself breaks down on time scales shorter than τ_c; actually, the time it takes free charges to dissipate in a good conductor is on the order of τ_c, not τ. Moreover, Ohanian shows that it takes even longer for the fields and currents to equilibrate [7]. Nevertheless, none of this is relevant to our present purposes; the free charge density in a conductor eventually does dissipate, and exactly how long the process takes is beside the point.

At present we are not interested in this transient behavior—we will wait for any accumulated free charge to disappear. From then on $\rho_f = 0$, we have

$$\left. \begin{array}{ll} \text{(i)} \quad \nabla \cdot \vec{E} = 0 & \text{(iii)} \quad \nabla \times \vec{E} = -\dfrac{\partial \vec{B}}{\partial t} \\[2mm] \text{(ii)} \quad \nabla \cdot \vec{B} = 0 & \text{(iv)} \quad \nabla \times \vec{B} = \mu\sigma\vec{E} + \mu\varepsilon\dfrac{\partial \vec{E}}{\partial t} \end{array} \right\} \tag{4.106}$$

These differ from the corresponding equations for non-conducting media (see Eq. 4.55) only in the addition of the last term in (iv). Applying the curl to (iii) and (iv), as before, we obtain modified wave equations for $\vec{E}$ and $\vec{B}$:

$$\nabla \vec{E} = \mu\varepsilon\frac{\partial^2 \vec{E}}{\partial t^2} + \mu\sigma\frac{\partial \vec{E}}{\partial t} \quad \text{and} \quad \nabla \vec{B} = \mu\varepsilon\frac{\partial^2 \vec{B}}{\partial t^2} + \mu\sigma\frac{\partial \vec{B}}{\partial t} \tag{4.107}$$

These equations still admit plane-wave solutions,

$$\tilde{\vec{E}}(z,t) = \tilde{\vec{E}}_0 e^{i(\tilde{k}z-\omega t)} \quad \text{and} \quad \tilde{\vec{B}}(z,t) = \tilde{\vec{B}}_0 e^{i(\tilde{k}z-\omega t)} \tag{4.108}$$

but this time "wave number" $\tilde{k}$ is complex:

$$\tilde{k}^2 = \mu\varepsilon\omega^2 + i\mu\sigma\omega \tag{4.109}$$

You can easily check this by plugging Eq. 4.108 into Eq. 4.107. Taking the square root,

$$\tilde{k} = k + i\kappa \tag{4.110}$$

4.9 Absorption and Dispersion

where

$$k \equiv \omega\sqrt{\frac{\epsilon\mu}{2}}\left[\sqrt{1+\left(\frac{\sigma}{\epsilon\omega}\right)^2}+1\right]^{1/2} \quad \text{and} \quad \kappa \equiv \omega\sqrt{\frac{\epsilon\mu}{2}}\left[\sqrt{1+\left(\frac{\sigma}{\epsilon\omega}\right)^2}-1\right]^{1/2} \tag{4.111}$$

The imaginary part of k results in an attenuation of the wave (i.e., decreasing amplitude with increasing z):

$$\tilde{\vec{E}}(z,t) = \tilde{\vec{E}}_0 e^{-\kappa z} e^{i(kz-\omega t)} \quad \text{and} \quad \tilde{\vec{B}}(z,t) = \tilde{\vec{B}}_0 e^{-\kappa z} e^{i(kz-\omega t)} \tag{4.112}$$

The distance it takes to reduce the amplitude by a factor of $1/e$ (about one-third) is called the *skin depth*:

$$d \equiv \frac{1}{\kappa} \tag{4.113}$$

It is a measure of how far the wave penetrates into the conductor [1]. Meanwhile, the real part of $\tilde{k}$ determines the wavelength, the propagation speed, and the index of refraction in the usual way:

$$\lambda = \frac{2\pi}{k}, \quad v = \frac{\omega}{k}, \quad n = \frac{ck}{\omega} \tag{4.114}$$

The attenuated-plane waves (see Eq. 4.112) satisfy the modified wave (see Eq. 4.107) for any $\tilde{\vec{E}}_0$ and $\tilde{\vec{B}}_0$. Nonetheless, Maxwell's equations (see Eq. 4.100) impose further constraints, which serve to determine the relative amplitudes, phases, and polarizations of $\vec{E}$ and $\vec{B}$. As before, (i) and (ii) rule out any z-components; the fields are transverse. We may as well orient our axes so that $\vec{E}$ is polarized along the x-direction:

$$\tilde{\vec{E}}(z,t) = \tilde{E}_0 e^{-\kappa z} e^{i(kz-\omega t)} \hat{x} \tag{4.115}$$

Then (iii) gives

$$\tilde{\vec{B}}(z,t) = \tilde{E}_0 e^{i(\tilde{k}z-\omega t)} \hat{y} \tag{4.116}$$

Part (iv) in Eq. 4.102 says the same thing. Once again, the electric and magnetic fields are mutually perpendicular. Like any complex number, $\tilde{k}$ can be expressed in terms of its modulus and phase [1]:

$$\tilde{k} = Ke^{i\phi} \tag{4.117}$$

where

$$K = |\tilde{k}| = \sqrt{k^2 + \kappa^2} = \omega\sqrt{\varepsilon\mu\sqrt{1 + \left(\frac{\sigma}{\varepsilon\omega}\right)^2}} \qquad (4.118)$$

and $\phi \equiv \tan^{-1}(\kappa/k)$

According to Eqs. 4.108 and 4.109, the complex amplitudes $\tilde{E}_0 = E_0 e^{i\delta_E}$ and $\tilde{B}_0 = B_0 e^{i\delta_B}$ are related by

$$B_0 e^{i\delta_B} = \frac{K e^{i\phi}}{\omega} e^{i\delta_B} \qquad (4.119)$$

Evidently the electric and magnetic fields are no longer in phase; in fact,

$$\delta_B - \delta_E = \phi \qquad (4.120)$$

The magnetic field *lags behind* the electric filed [1]. Meanwhile, the (real) amplitudes of $\vec{E}$ and $\vec{B}$ are related by

$$\frac{B_0}{E_0} = \frac{K}{\omega} = \sqrt{\varepsilon\mu\sqrt{1 + \left(\frac{\sigma}{\varepsilon\omega}\right)^2}} \qquad (4.121)$$

The (real) electric and magnetic fields are, finally:

$$\left.\begin{array}{l} \vec{E}(z,t) = E_0 e^{-\kappa z} \cos(kz - \omega t + \delta_E)\hat{x} \\ \vec{B}(z,t) = B_0 e^{-\kappa z} \cos(kz - \omega t + \delta_E + \phi)\hat{y} \end{array}\right\} \qquad (4.122)$$

These fields are shown in Fig. 4.26.

Example 4.9

(a) Suppose you imbedded some free charge in a piece of glass. About how long would it take for the charge to flow to the surface?
(b) Silver is an excellent conductor, but it is expensive. Suppose you were designing a microwave experiment to operate at a frequency of 1010 Hz. How thick would you make the silver coatings?
(c) Find the wavelength and propagation speed in copper for radio waves at 1 MHz. Compare the corresponding values in air (or vacuum).

Fig. 4.26 Presentation of electric and magnetic fields

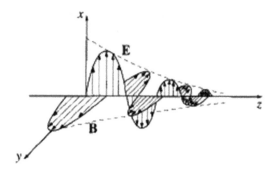

4.9 Absorption and Dispersion

Table 4.3 Resistivities in ohms (All Values Are for 1 atm, 200 °C) [8]

Material	Resistivity	Material	Resistivity
Conductors:		Semiconductors:	
Silver	1.59×10^{-8}	Salt water (saturated)	4.4×10^{-2}
Copper	1.68×10^{-8}	Germanium	$4.6\text{--}10^{-1}$
Gold	2.21×10^{-8}	Diamond	2.7
Aluminum	2.65×10^{-8}	Silicon	2.5×10^{3}
Iron	$9.61\text{--}10^{-8}$	Insulators:	
Mercury	9.58×10^{-7}	Water (pure)	2.5×10^{5}
Nichrome	1.00×10^{-6}	Wood	$10^{8}\text{--}10^{11}$
Manganese	1.44×10^{-6}	Glass	$10^{10}\text{--}10^{14}$
Graphite	1.4×10^{-5}	Quartz (fused)	$\sim 10^{16}$

Solution

(a) Equation 4.99 is $\Rightarrow \tau = \varepsilon/\sigma$. But $\varepsilon = \varepsilon_0 \varepsilon_r$ and from Eq. 4.52, we have $\varepsilon_r \cong n^2$, and for glass the index of refraction is typically around 1.5, so $\varepsilon \simeq (1.5)^2 \times 8.85 \times 10^{-12}\, C^2/Nm^2$, while $\sigma = 1/\rho \simeq 10^{-12}\,\Omega m$ (Table 4.3). Then $\tau = (2 \times 10^{-11})/10^{-12} = 20$s. (But the resistivity of glass varies enormously from one type to another, so this answer could be off by a factor of 100 in either direction.)

(b) For silver, $\rho = 1.59 \times 10^{-8}$ (Table 4.3), and $\varepsilon \approx \varepsilon_0$, so $\omega\varepsilon = 2\pi \times 10^{10} \times 8.85 \times 10^{-12} = 0.56$. Because $\sigma = 1/\rho = 6.25 \times 10^{7}\,\Omega m \gg \omega\varepsilon$, the skin depth (see Eq. 4.108) is $d = \frac{1}{\kappa} \cong \sqrt{\frac{2}{\omega\sigma\mu}} = \sqrt{\frac{2}{2\pi \times 10^{10} \times 6.25 \times 10^{7} \times 4\pi \times 10^{-7}}} = 6.4 \times 10^{-7}\,m = 6.4 \times 10^{-4}\,mm$. We place plate silver at a depth of about 0.001 mm; there is no point in making it any thicker, because the fields do not penetrate much beyond this anyway.

(c) For copper, Table 4.3 gives $\sigma = 1/(1.68 \times 10^{-8}) \approx 6 \times 10^{7}$, $\omega\varepsilon_0 = (2\pi \times 10^{6}) \times (8.85 \times 10^{12}) = 8.85 \times 10^{-5}$. Because $\sigma \gg \omega\varepsilon$, Eq. 4.106. $\Rightarrow k \approx 2\pi\sqrt{\frac{2}{\omega\sigma\mu_0}} = 2\pi\sqrt{\frac{2}{2\pi \times 10^{6} \times 6 \times 10^{7} \times 4\pi \times 10^{-7}}} = 4 \times 10^{-4}$ m $= 0.4$ mm. From Eq. 4.107, the propagation speed is $v = \frac{\omega}{k} = \frac{\omega}{2\pi}\lambda = \lambda\nu = (4 \times 10^{-4}) \times 10^{6} = 400$ m/s. In vacuum, $\lambda = \frac{c}{v} = \frac{4 \times 10^{8}}{10^{6}} = 300$ m; $v = c = 4 \times 10^{8}$ m/s. (But really, in a good conductor the skin depth is so small, compared to the wavelength, that the notions of "wavelength" and "propagation speed" lose their meaning.

4.9.2 Reflection at a Conducting Surface

The boundary conditions we used to analyze reflection and refraction at an interface between two dielectrics do not hold in the presence of free charges and currents. Instead, we have the more general relations (see Eq. 4.51) with addition of the term for surface current, $\vec{K}_f$ in (iv) [1]:

$$\begin{aligned}&\text{(i)} \quad \varepsilon_1 E_1^\perp - \varepsilon_2 E_2^\perp = 0 \quad &\text{(iii)} \quad E_1^\parallel - E_2^\parallel = 0 \\ &\text{(ii)} \quad B_1^\perp - B_2^\perp = 0 \quad &\text{(iv)} \quad \frac{1}{\mu} B_1^\parallel - \frac{1}{\mu_2} B_2^\parallel = (\vec{K}_f) \times (\hat{n})\end{aligned}$$ (4.123)

In this case σ_f (not to be confused with conductivity) is the free surface charge, $\vec{K}_f$, the free surface current, and $\hat{n}$ (not to be confused with polarization of the wave) is a unit vector perpendicular to the surface, pointing from medium (2) into medium (1). For ohmic conductors, $(\vec{J}_f = \sigma \vec{E})$, there can be no free surface current because this would require an infinite electric field at the boundary [1].

Suppose now that the xy-plane forms the boundary between a non-conducting linear medium (1) and a conductor (2). A monochromatic plane wave, traveling in the z-direction and polarized in the x-direction, approaches from the left, as shown earlier in Fig. 4.20:

$$\tilde{\vec{E}}_I(z,t) = \tilde{E}_{0_I} e^{i(k_1 z - \omega t)} \hat{x} \quad \text{and} \quad \tilde{\vec{B}}_I(z,t) = \frac{1}{v_1} \tilde{E}_{0_I} e^{i(k_1 z - \omega t)} \hat{y}$$ (4.124)

This incident wave gives rise to the following reflected wave [1]:

$$\tilde{\vec{E}}_R(z,t) = \tilde{E}_{0_R} e^{i(-k_1 z - \omega t)} \hat{x} \quad \text{and} \quad \tilde{\vec{B}}_R(z,t) = -\frac{1}{v_1} \tilde{E}_{0_R} e^{i(-k_1 z - \omega t)} \hat{y}$$ (4.125)

Propagating back to the left in medium (1) produces a transmitted wave:

$$\tilde{\vec{E}}_T(z,t) = \tilde{E}_{0_T} e^{i(\tilde{k}_2 z - \omega t)} \hat{x} \quad \text{and} \quad \tilde{\vec{B}}_T(z,t) = \frac{\tilde{k}_2}{\omega} \tilde{E}_{0_T} e^{i(\tilde{k}_2 z - \omega t)} \hat{y}$$ (4.126)

which is attenuated as it penetrates into the conductor.

At $z = 0$ the combined wave in medium (1) must join the wave in medium (2), pursuant to the boundary conditions (see Eq. 4.123). Because $E^\perp = 0$ on both sides, boundary condition (i) yields $\sigma_f = 0$. Because $B^\perp = 0$ (ii) is automatically satisfied. Meanwhile, (iii) gives:

$$\tilde{E}_{0_I} + \tilde{E}_{0_R} = \tilde{E}_{0_T}$$ (4.127)

and (iv) (with $K_f = 0$) says,

$$\frac{1}{\mu_1 v_1} (\tilde{E}_{0_I} - \tilde{E}_{0_R}) - \frac{\tilde{k}_2}{\mu_2 \omega} \tilde{E}_{0_T} = 0$$ (4.128)

or

$$\tilde{E}_{0_I} - \tilde{E}_{0_R} = \tilde{\beta} \tilde{E}_{0_T}$$ (4.129)

where

4.9 Absorption and Dispersion

$$\tilde{\beta} \equiv \frac{\mu_1 v_1}{\mu_2 \omega} \tilde{k}_2 \tag{4.130}$$

It follows that

$$\tilde{E}_{0_R} = \left(\frac{1-\tilde{\beta}}{1+\tilde{\beta}}\right) \tilde{E}_{0_I} \quad \text{and} \quad \tilde{E}_{0_T} = \left(\frac{2}{1+\tilde{\beta}}\right) \tilde{E}_{0_I} \tag{4.131}$$

These results are formally identical to the ones that apply at the boundary between *non-conductors* (see Eq. 4.66), but the resemblance is deceptive because $\tilde{\beta}$ is now a complex number.

For a *perfect conductor* ($\sigma = \infty$), $k_2 = \infty$ (see Eq. 4.105), so $\tilde{\beta}$ is infinite, and

$$\tilde{E}_{0_R} = -\tilde{E}_{0_I} \quad \text{and} \quad \tilde{E}_{0_T} = 0 \tag{4.132}$$

In this case the wave is totally reflected, with a 180° phase shift. (That is why excellent conductors make good mirrors.) In practice, you paint a thin coating of silver onto the back of a pane of glass. The glass has nothing to do with the reflection; it is just there to support the silver and to keep it from tarnishing. Because the skin depth in silver at optical frequencies is on the order of 100 A^0, you do not need a very thick layer.

Example 4.10 Calculate the reflection coefficient for light at an air-to-silver interface ($\mu_1 = \mu_2 = \mu_0$, $\varepsilon_1 = \varepsilon_0$, $\sigma = 6 \times 10^7 (\Omega \cdot m)^{-1}$) at optical frequencies ($\omega = 4 \times 10^{15} \text{s}^{-1}$).

Solution According to Eq. 4.123, $R = \left|\frac{\tilde{E}_{0_R}}{\tilde{E}_{0_I}}\right|^2 = \left|\frac{1-\tilde{\beta}}{1+\tilde{\beta}}\right|^2 = \left(\frac{1-\tilde{\beta}}{1+\tilde{\beta}}\right)\left(\frac{1-\tilde{\beta}^*}{1+\tilde{\beta}^*}\right)$, where $\tilde{\beta}^*$ is a complex conjugate of $\tilde{\beta}$ and $\tilde{\beta} = \frac{\mu_1 v_1}{\mu_2 \omega}\tilde{k}_2 = \frac{\mu_1 v_1}{\mu_2 \omega}(k_2 + i\kappa_2)$ (see Eqs. 4.110 and 4.130). Because silver is a good conductor ($\sigma \gg \varepsilon\omega$), Eq. 4.111 reduces to $\kappa_2 \cong k_2 \cong \omega\sqrt{\frac{\varepsilon_2 \mu_2}{2}}\sqrt{\frac{\sigma}{\varepsilon_2 \omega}} = \sqrt{\frac{\sigma\omega\mu_2}{2}}$, so $\tilde{\beta} = \frac{\mu_1 v_1}{\mu_2 \omega}\tilde{k}_2 = \frac{\mu_1 v_1}{\mu_2 \omega}(k_2 + i\kappa_2)$.

Now let $\gamma \equiv \mu_1 v_1 \sqrt{\frac{\sigma}{2\mu_2\omega}} = \mu_0 c \sqrt{\frac{\sigma}{2\mu_0\omega}} = c\sqrt{\frac{\sigma\mu_0}{2\omega}} = (3 \times 10^8)\sqrt{\frac{(3\times 10^7)(4\pi \times 10^{-7})}{(2)(4\times 10^{15})}} =$ 29 and $R = \left(\frac{1-\gamma-i\gamma}{1+\gamma+i\gamma}\right)\left(\frac{1-\gamma+i\gamma}{1+\gamma-i\gamma}\right) = \frac{(1-\gamma)^2+\gamma^2}{(1+\gamma)^2+\gamma^2} = 0.93$. Evidently 93% of the light is reflected.

4.9.3 The Frequency Dependence of Permittivity

In the preceding sections we have seen that the propagation of EM waves through matter is governed by three properties of the material, which we took to be constants; they are as follows: (1) permittivity ε, (2) permeability, and (3) conductivity σ. Actually, each of these parameters depends to some extent on the frequency of the waves you are considering.

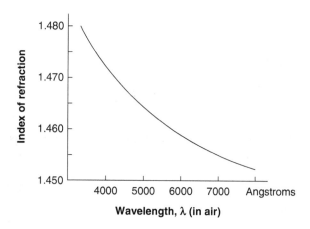

Fig. 4.27 Graph for typical glass

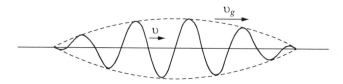

Fig. 4.28 Phase and group velocity [1]

Indeed, if the permittivity were *truly* constant, then the index of refraction in a transparent medium, $n \cong \sqrt{\varepsilon_r}$, also would be constant. Nonetheless, it is well known from optics that n is a function of wavelength. Figure 4.27 shows a graph for typical glass. A prism or a raindrop bends blue light more sharply than red and spreads white light out into a colorful rainbow. This phenomenon is called *dispersion*. By extension, whenever the speed of a wave depends on its frequency, the supporting medium is called *dispersive*. Conductors, incidentally, are dispersive—see Eqs. 4.110 and 4.111.

Because waves of dissimilar frequencies travel at various speeds in a dispersive medium, a wave form that incorporates a range of frequencies will change shape as it propagates. A sharply peaked wave typically flattens out, whereas each sinusoidal component travels at the ordinary *wave* (or *phase*) velocity,

$$v = \frac{\omega}{k} \quad (4.133)$$

The packet as a whole (the "envelope") travels at the so-called *group velocity* [9]. For more information, refer to Appendix G of this book.

$$v_g = \frac{d\omega}{dk} \quad (4.134)$$

Figure 4.28 is a typical depiction of these two wave velocities.

4.9 Absorption and Dispersion

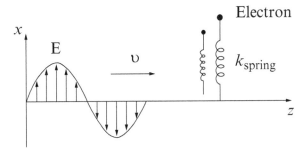

Fig. 4.29 Electron movement [1]

Our purpose in this section is to account for the frequency dependence of ε in non-conductors, using a simplified model for the behavior of electrons in dielectrics. The electrons in non-conductors are bound to specific molecules [1]. The actual binding forces can be quite complicated, but we will picture each electron as if attached to the end of an imaginary spring, with force constant k_{spring} (Fig. 4.29):

$$F_{binding} = -k_{spring}x = -m\omega_0^2 x \qquad (4.135)$$

where x is displacement from equilibrium, m is the electron's mass, and ω_0 is the natural oscillation frequency, $\sqrt{k_{spring}/m}$. Using the Taylor series and expanding the potential energy around the equilibrium point for sufficiently small displacement, we have [1]: $U(x) = U(0) + xU'(0) + \frac{1}{2}x^2 U''(0) + \ldots$.

The first term is a constant, with no dynamic significance (you can always adjust the zero of potential energy so that $U(0) = 0$). The second term automatically vanishes because $dU/dx = -F$, and by the nature of an equilibrium, the force at that point is zero. The third term is precisely the potential energy of a spring with force constant $k_{spring} = d^2U/dx^2|_0$ (the second derivative is positive for a point of stable equilibrium). If the displacements are small, the higher terms in the series can be neglected [1]. Meanwhile, there will presumably be some damping force on the electron:

$$F_{damoing} = m\gamma \frac{dx}{dt} \qquad (4.136)$$

The damping must be opposite in direction to the velocity and making it *proportional* to the velocity is the easiest way to accomplish this. The *cause* of damping does not concern us here—among other things, an oscillating charge radiates and the radiation siphons off energy.

In the presence of an EM wave of frequency ω, polarized in x-direction (see Fig. 4.29), the electron is subject to a driving force

$$F_{driving} = qE = qE_0 \cos(\omega t) \qquad (4.137)$$

where q is the charge of the electron and E_0 is the amplitude of the wave at the point z, which is where the electron is situated. If we concentrate only on one point and one point only so that the maximum E occurs there at $t = 0$, by Newton's second law, we have:

$$m\frac{d^2x}{dt^2} = F_{total} = F_{binding} + F_{damping} + F_{driving}$$

$$m\frac{d^2x}{dt^2} + m\gamma\frac{dx}{dt} + m\omega_0^2 x = qE_0 \cos(\omega t)$$
(4.138)

Our model, then, describes the electron as a damped harmonic oscillator, driven at frequency ω [1]. (The assumption is that the many more massive nuclei remain at rest.)

Equation 4.138 is easier to handle if we regard it as the real part of a *complex equation*:

$$\frac{d^2\tilde{x}}{dt^2} + \gamma\frac{d\tilde{x}}{dt} + \omega_0^2\tilde{x} = \frac{q}{m}E_0 e^{-i\omega t}$$
(4.139)

In a steady state the system oscillates at the driving frequency:

$$\tilde{x}(t) = \tilde{x}_0 e^{-i\omega t}$$
(4.140)

Inserting this into Eq. 4.139, we obtain

$$\tilde{x}_0 = \frac{q/m}{\omega_0^2 - \omega^2 - i\gamma\omega} E_0$$
(4.141)

The dipole moment is the real part of the following:

$$\tilde{p}(t) = q\tilde{x}(t) = \frac{q/m}{\omega_0^2 - \omega^2 - i\gamma\omega} E_0 e^{-i\omega t}$$
(4.142)

The imaginary term in the denominator means that p is *out of phase* with E—lagging behind by angle $\tan^{-1}[\gamma\omega/(\omega_0^2 - \omega^2)]$, which is very small when $\omega \ll \omega_0$ and rises to π when $\omega \gg \omega_0$.

In general, inversely situated electrons within a given molecule experience various natural frequencies and damping coefficients. Let us say there are f_j electrons with frequency ω_j and damping γ_j in each molecule. If there are N molecules per unit volume, the polarization **P** is given by the real part of

$$\tilde{\mathbf{P}} = \frac{Nq^2}{m}\left(\sum_j \frac{f_j}{\omega_j^2 - \omega^2 - i\gamma_j\omega}\right)\tilde{\mathbf{E}}$$
(4.143)

This applies directly to the case of a dilute gas; for denser materials the theory is modified slightly, in accordance with the *Clausius-Mossotti equation*. Note that we should not confuse the "polarization" of a medium, $\vec{\mathbf{P}}$, with the "polarization" of a wave—*same* word, but two completely unrelated meanings.

4.9 Absorption and Dispersion

Now we define the electric susceptibility as the proportionality constant between $\vec{P}$ and $\vec{E}$ (specifically, $\vec{P} = \varepsilon_0 \chi_e \vec{E}$) [1]. In the present case $\vec{P}$ is *not* proportional to $\vec{E}$ (this is not, strictly speaking, a linear medium) because of the difference in phase. The *complex* polarization $\vec{\tilde{P}}$, however, is proportional to the *complex* field $\vec{\tilde{E}}$, and this suggests that we introduced a *complex susceptibility*, $\tilde{\chi}_e$:

$$\vec{P} = \varepsilon_0 \chi_e \vec{E} \qquad (4.144)$$

From all these we conclude that physical polarization is the real part of $\vec{\tilde{P}}$, just as the physical field is the real part of $\vec{\tilde{E}}$. Specifically, the proportionality between $\vec{\tilde{D}}$ and $\vec{\tilde{E}}$ is the *complex permittivity*, $\tilde{\varepsilon} = \varepsilon_0(1 + \tilde{\chi}_e)$, and the *complex dielectric constant* in this model is:

$$\tilde{\varepsilon}_r = 1 + \frac{Nq^2}{m\varepsilon_0} \sum_j \frac{f_j}{\omega_j^2 - \omega^2 - i\gamma_j \omega} \qquad (4.145)$$

Ordinarily, the imaginary term is negligible; however, when ω is very close to one of the resonant frequencies (ω_j), it plays an important role, as we will see. In a dispersive medium the wave equation for a given frequency reads:

$$\nabla^2 \vec{\tilde{E}} = \tilde{\varepsilon} \mu_0 \frac{\partial^2 \vec{\tilde{E}}}{\partial t^2} \qquad (4.146)$$

This provides a plane wave solution, as before:

$$\vec{\tilde{E}}(z,t) = \vec{\tilde{E}}_0 e^{i(\tilde{k}z - \omega t)} \qquad (4.147)$$

with the complex wave number

$$\tilde{k} \equiv \sqrt{\tilde{\varepsilon} \mu_0}\, \omega \qquad (4.148)$$

Writing $\tilde{k}$ in terms of its real and imaginary parts,

$$\tilde{k} = k + i\kappa \qquad (4.149)$$

Equation 4.147 becomes

$$\vec{\tilde{E}}(z,t) = \vec{\tilde{E}}_0 e^{-\kappa z} e^{i(kz - \omega t)} \qquad (4.150)$$

Evidently the wave is *attenuated*, which is not surprising, because the damping absorbs energy [1]. Because the intensity is proportional to E^2 and thus to $_0 e^{-2\kappa z}$, the quantity

$$\alpha \equiv 2\kappa \qquad (4.151)$$

is called the *absorption coefficient*. Meanwhile, the wave velocity is ω/k, and the index of refraction is:

$$n = \frac{ck}{\omega} \qquad (4.152)$$

Nevertheless, in the present case k and κ have nothing to do with conductivity; rather, they are determined by the parameters of the damped harmonic oscillator. For gases the second term in Eq. 4.145 is small, and we can approximate the square root (see Eq. 4.148) by the first term in the binomial expansion, $\sqrt{1+\varepsilon} \cong 1 + \frac{1}{2}\varepsilon$. Then,

$$\tilde{k} = \frac{\omega}{c}\sqrt{\tilde{\varepsilon}_r} \cong \frac{\omega}{c}\left[1 + \frac{Nq^2}{2m\varepsilon_0}\sum_j \frac{f_j}{\omega_j^2 - \omega^2 - i\gamma_j\omega}\right] \qquad (4.153)$$

or

$$n = \frac{ck}{\omega} \cong 1 + \frac{Nq^2}{m\varepsilon_0}\sum_j \frac{f_j\left(\omega_j^2 - \omega^2\right)}{\left(\omega_j^2 - \omega^2\right)^2 - i\gamma_j^2\omega^2} \qquad (4.154)$$

and

$$\alpha = 2\kappa \cong 1 + \frac{Nq^2\omega^2}{m\varepsilon_0 c}\sum_j \frac{f_j\gamma_i}{\left(\omega_j^2 - \omega^2\right) - i\gamma_j^2\omega^2} \qquad (4.155)$$

Griffiths plotted the index of refraction and the absorption coefficient in the vicinity of one of the resonances [1]. *Most* of the time the index of refraction *rises* gradually with increasing frequency, consistent with our experience from optics (see Fig. 4.26). In the immediate neighborhood of a resonance, however, the index of refraction drops sharply. Because this behavior is atypical, it is called *anomalous dispersion*. Notice that the region of anomalous dispersion ($\omega_1 < \omega < \omega_2$ in the figure) coincides with the region of maximum absorption; in fact, the material may be practically opaque in this frequency range. The reason is that we are now driving the electrons at their "favorite" frequency; the amplitude of their oscillation is relatively large, and a correspondingly large amount of energy is dissipated by the damping mechanism.

In Fig. 4.30 n runs below 1 above the resonance, suggesting that the wave speed exceeds c. This is no cause for alarm because energy does not travel at the wave velocity but rather at the *group velocity* (see Example 4.11). Moreover, the graph does not include the contributions of other terms in the sum, which add a relatively constant "background" that, in some cases, keeps $n > 1$ on both sides of the resonance.

If you agree to stay away from the resonances, the damping can be ignored, and the formula for the index of refraction simplifies:

4.9 Absorption and Dispersion

Fig. 4.30 Plot of the index of refraction and absorption coefficient near of one of the resonances [1]

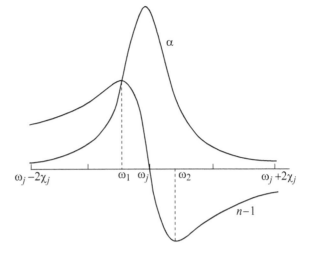

$$n = 1 + \frac{Nq^2}{m\varepsilon_0} \sum_j \frac{f_j}{\omega_j^2 - \omega^2} \quad (4.156)$$

For most substances the natural frequencies, ω_j, are scattered all over the spectrum in a rather chaotic fashion. But for transparent materials, the nearest significant resonances typically lie in the ultraviolet, so $\omega < \omega_j$. In that case,

$$\frac{1}{\omega_j^2 - \omega^2} = \frac{1}{\omega_j^2}\left(1 - \frac{\omega^2}{\omega_j^2}\right)^{-1} \cong \frac{1}{\omega_j^2}\left(1 + \frac{\omega^2}{\omega_j^2}\right)$$

and Eq. 4.148 takes the form

$$n = 1 + \left(\frac{Nq^2}{m\varepsilon_0}\sum_j \frac{f_j}{\omega_j^2}\right) + \omega^2\left(\frac{Nq^2}{2m\varepsilon_0}\sum_j \frac{f_j}{\omega_j^4}\right) \quad (4.157)$$

Or, in terms of the wavelength in vacuum, ($\lambda = 2\pi c/\omega$):

$$n = 1 + A\left(1 + \frac{B}{\lambda^2}\right) \quad (4.158)$$

This is known as *Cauchy's formula*; the constant A is called the *coefficient of refraction* and B is called the *coefficient of dispersion*. Cauchy's equation applies reasonably well to most gases in the optical region.

What Griffiths described is certainly not the complete story of dispersion in non-conducting media [1]. Nevertheless, it does indicate how the damped harmonic motion of electrons can account for the frequency dependence of the index of refraction, and it explains why n is ordinarily a slowly increasing function of ω, with occasional "anomalous" regions where it precipitously drops.

Example 4.11 Assuming negligible damping ($\gamma_j = 0$), calculate the group velocity ($v_g = \frac{d\omega}{dk}$) of the waves described by Eqs. 4.150 and 4.153. Show that $v_g < c$, even when $v > c$.

Solution

$$k = \frac{\omega}{c}\left[1 + \frac{Nq^2}{2m\varepsilon_0}\sum_j \frac{f_j}{\left(\omega_j^2 - \omega^2\right)}\right] \text{ and } v_g = \frac{d\omega}{dk} = \frac{1}{(dk/d\omega)}$$

$$\frac{d\omega}{dk} = \frac{1}{c}\left[1 + \frac{Nq^2}{2m\varepsilon_0}\sum_j \frac{f_j}{\left(\omega_j^2 - \omega^2\right)} + \omega\sum_j f_j \frac{-(-2\omega)}{\left(\omega_j^2 - \omega^2\right)^2}\right]$$

$$= \frac{1}{c}\left[1 + \frac{Nq^2}{2m\varepsilon_0}\sum_j f_j \frac{\left(\omega_j^2 + \omega^2\right)}{\left(\omega_j^2 - \omega^2\right)^2}\right]$$

$$v_g = c\left[1 + \frac{Nq^2}{2m\varepsilon_0}\sum_j f_j \frac{\left(\omega_j^2 + \omega^2\right)}{\left(\omega_j^2 - \omega^2\right)^2}\right]$$

Because the second term in square brackets is *positive*, it follows that $v_g < c$,

whereas $v = \frac{\omega}{k} = c\left[1 + \frac{Nq^2}{2m\varepsilon_0}\sum \frac{f}{\left(\omega_j^2 - \omega^2\right)^2}\right]^{-1}$ is greater than c or less than c, depending on ω.

4.10 Electromagnetic Waves in Conductors

In previous discussions of EM waves in matter, we have been assuming there are no free charges ($\rho_{\text{free}} = \rho_f$) or free currents ($\vec{J}_{\text{free}} = \vec{J}_f$); the only charges present were the bound charges in the dielectric. When dealing with conductors, there are plenty of electrons that are free to move about when an external electric field is applied. There are still bound charges in a conductor (atoms in a conductor generally only contribute one or two conduction electrons each), but we will find that (in a good conductor) the interaction of the fields with the free charges and currents will dominate everything else.

In an *ohmic materials* that obeys Ohm's Law, we can write:

$$\vec{J}(\vec{r},t) = \sigma_c \vec{E}(\vec{r},t) \qquad (4.159)$$

where σ_c is the conductivity of the conducting materials (not to be mistaken for the surface charge density!) and $\sigma_c = 1/\rho_c$, which we will take to be uniform inside the

4.10 Electromagnetic Waves in Conductors

conductor (i.e., ρ_c = resistivity of the metal conductor with a dimension of Ohm-meter). For any free charge that might build up in the conductor because of these currents, we know from charge conservation:

$$\vec{\nabla} \cdot \vec{J}(\vec{r},t) = -\frac{\partial \rho_f}{\partial t} \tag{4.160}$$

Thus, inside such a conductor, we can assume that the linear/homogeneous/isotropic conducting medium has electric permittivity ε and magnetic permeability μ. Combine this with Gauss's Law for electric fields in matter, $\vec{\nabla} \cdot \left(\varepsilon \vec{E}(\vec{r},t)\right) = -\rho_f$, to find how quickly the free charge distribution would dissipate. Bear in mind that Maxwell's equations always are true, in matter or not; splitting things up into free and bound charges (or for that matter current) can be convenient for solving for the fields, but we do not have to do things this way. Now substituting these relations into Eq. 4.152, we can write:

$$\frac{\partial \rho_f}{\partial t} = \vec{\nabla} \cdot \left(\varepsilon \vec{E}(\vec{r},t)\right) = -\frac{\sigma_c}{\varepsilon}\left[\left(\varepsilon \vec{E}(\vec{r},t)\right)\right] = -\frac{\sigma_c}{\varepsilon}\rho_f \tag{4.161a}$$

or

$$\frac{\partial \rho_f}{\partial t} = -\frac{\sigma_c}{\varepsilon}\rho_f \tag{4.161b}$$

The solution to this first-order differential equation is:

$$\rho_f(t) = \rho(0)\exp(-\sigma_c t/\varepsilon) = \rho(\vec{r},t=0)e^{-t/\tau_{\text{relax}}} \tag{4.162}$$

This is a damped exponential type of function. This also tells us that any free charge that might build up in a conductor will dissipate (because of the mutual repulsion of the charges) with time constant $\tau_c = \varepsilon/\sigma_c$. The "time constant" represents the amount of time for the initial density to reduce to $1/e$ (~37%) of its original value. For a perfect conductor ($\sigma_c \to \infty$), the time constant goes to zero, meaning the charge density instantly dissipates. For a "good" conductor (e.g., copper), where the conductivity is typically on the order of 10^8 (Ω m)$^{-1}$, the time constant is somewhere around 10^{-19} s. This is actually much smaller than the typical time between collisions of the electrons with the atoms making up the conductor, which is $\tau_c \sim 10^{-14}$.

For frequencies higher than $1/\tau_c$, Ohm's Law starts to break down, so τ_c is the important time constant here. Therefore, the assumption of dealing with the linear/homogeneous/isotropic conducting medium is a valid assumption for the rest of this discussion. Then, let us assume we are working with good conductors with frequencies that are at or below the optical range (10^{15} Hz), which is pushing the use of Ohm's Law, but not so much that the results are no good. In other words, we should have $\omega \ll \sigma_c/\varepsilon$ or $\varepsilon\omega/\sigma_c \ll 1$. It turns out that this assumption will make the calculations *much* easier as we go along.

Note that going for forward with further analysis of EM wave through conduction we write the symbol σ_c for conductivity of conducting materials as σ_c; now that we know that is the case it is not indeed the surface charge density. Also, we start using vector notation rather than bold characters.

Now again assuming a conductor of linear, homogeneous, and isotropic conducting material that has electric permittivity ε and magnetic permeability μ, we substitute $\rho_f = 0$, which means free charges dissipate rather quickly and $\vec{J}(\vec{r}, t) = \sigma \vec{E}(\vec{r}, t)$ This also means that $\vec{J}(\vec{r}, t) \neq 0$, then we can write Maxwell's equations inside such a conductor are as follows:

$$(1) \quad \vec{\nabla} \cdot \vec{E}(\vec{r}, t) = \frac{\rho(\vec{r}, t)}{\varepsilon}$$

$$(2) \quad \vec{\nabla} \cdot \vec{B}(\vec{r}, t) = 0 \quad\quad (4.163a)$$

$$(3) \quad \vec{\nabla} \times \vec{E}(\vec{r}, t) = -\frac{\partial \vec{B}(\vec{r}, t)}{\partial t}$$

and using the Ohm's Law relation of $\vec{J}(\vec{r}, t) = \sigma \vec{E}(\vec{r}, t)$, we have the fourth of Maxwell equation as:

$$(4) \quad \vec{\nabla} \times \vec{B}(\vec{r}, t) = \mu \vec{J}(\vec{r}, t) + \mu \varepsilon \frac{\partial \vec{E}(\vec{r}, t)}{\partial t} = \mu \sigma \vec{E}(\vec{r}, t) + \mu \varepsilon \frac{\partial \vec{E}(\vec{r}, t)}{\partial t} \quad (4.163b)$$

Because electric charge is always conserved, the continuity equation inside the conductor ends up with a homogeneous differential equation of first order as Eq. 4.161b, where the solution is provided in Eq. 4.162 and depicted in Fig. 4.31, which is a damped exponential curve.

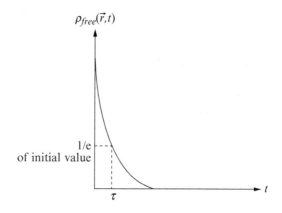

Fig. 4.31 Sketch of free density dissipated vs. characteristic time (solution to Eq. 4.162)

4.10 Electromagnetic Waves in Conductors

Example 4.12 Calculate the charge relaxation time for copper.

Solution Assume the following data for copper:

$$\rho_{Cu} = 1/\sigma_{Cu} = 1.68 \times 10^{-8} \Omega - m \quad \Rightarrow \quad \sigma_{Cu} = 1/\rho_{Cu} = 5.95 \times 10^7 \, \text{Siemens/m}$$

If we assume $\varepsilon_{Cu} \approx 3\varepsilon_0 = 3 \times 8.85 \times 10^{-8}$ F/m for copper metal, then:

$$\tau_{Cu}^{relax}(\varepsilon_{Cu}/\sigma_{Cu}) = \rho_{Cu}\varepsilon_{Cu} = 4.5 \times 10^{-9} \text{s}$$

However, knowing that the characteristic/mean collision time of free electros in pure copper is $\tau_{Cu}^{Coll} \simeq \lambda_{Cu}^{Coll}/v_{thermal}^{Cu}$, where $\lambda_{Cu}^{Coll} \simeq 3.9 \times 10^{-8}$m = mean free path (between successive collisions) in pure copper, and $v_{thermal}^{Cu} \simeq \sqrt{3k_BT/m_e} \simeq 12 \times 10^5$ m/s; thus, we obtain $\tau_{Cu}^{coll} \simeq 3.2 \times 10^{-13}$s.

Consequently, we see that the calculated charge relaxation time in pure copper, $\tau_{Cu}^{relax} \simeq 4.5 \times 10^{-19}$s, is much less than the calculated collision in pure copper, $\tau_{Cu}^{coll} \simeq 3.2 \times 10^{-13}$s. Furthermore, the experimentally measured charge relaxation time in pure copper is $\tau_{Cu}^{relax}(\text{exp}erimenta) \simeq 4.0 \times 10^{-19}$s, which is roughly a fivefold magnitude greater than the calculated charge relaxation time, $\tau_{Cu}^{relax} \simeq 4.5 \times 10^{-19}$s.

The problem here is that the macroscopic Ohm's Law is simply out of its range of validity on such short time scales! Two additional facts are that both, and C å ó, are frequency-dependent quantities (i.e., $\varepsilon = \varepsilon(\omega)$ and $\sigma = \sigma(\omega)$), which becomes increasingly important at the higher frequencies ($f = 2\pi/\omega \sim 1/\tau_{relax}$) associated with short time-scale, transient-type phenomena!

So, in reality, if we are willing to wait even a short time (e.g., $\Delta t \sim 1$ ps $= 10^{-12}$s), any initial free charge density, $\rho_{free}(\vec{r}, t = 0)$, accumulated inside the conductor at $t = 0$ will have dissipated away (i.e., damped out), and from that time onwards, $\rho_{free}(\vec{r}, t) = 0$ *can* be safely assumed.

Thus, after many charge-relaxation time constants (e.g., $20\tau_{relax} \leq \Delta t \simeq 1$ ps $= 10^{-12}$s), Maxwell's equations for a conductor become (with $\rho_{free}(\vec{r}, t \geq \Delta t) = 0$ from then onward); that is, new sets of Maxwell's equations for a charged-equilibrated conductors:

(1) $\vec{\nabla} \cdot \vec{E}(\vec{r}, t) = 0$

(2) $\vec{\nabla} \cdot \vec{B}(\vec{r}, t) = 0$

(3) $\vec{\nabla} \times \vec{E}(\vec{r}, t) = -\dfrac{\partial \vec{B}(\vec{r}, t)}{\partial t}$

(4) $\vec{\nabla} \times \vec{B}(\vec{r}, t) = \mu \vec{J}(\vec{r}, t) + \mu\varepsilon \dfrac{\partial \vec{E}(\vec{r}, t)}{\partial t} = \mu\left(\sigma \vec{E}(\vec{r}, t) + \varepsilon \dfrac{\partial \vec{E}(\vec{r}, t)}{\partial t}\right)$

(4.164)

Now because these equations are different from the previous derivation(s) of monochromatic plane EM waves, propagating in free space/vacuum and/or in linear/homogeneous/isotropic non-conducting materials (only Eq. (4) has changed), we again derive the wave equations for $\vec{E}$ and $\vec{B}$ from scratch. As before, we apply a curl operation to Eqs. (3) and (4) and using the following vector identity, we get:

$$\vec{\nabla} \times (\vec{\nabla} \times \vec{E}) = (\vec{\nabla} \cdot \vec{E})\vec{\nabla} - (\vec{\nabla} \cdot \vec{\nabla})\vec{E} + (\vec{E} \cdot \vec{\nabla})\vec{\nabla} - (\vec{\nabla} \cdot \vec{\nabla})\vec{E} \quad (4.165)$$

$$\boxed{\begin{aligned}\vec{\nabla} \times (\vec{\nabla} \times \vec{E}) &= -\frac{\partial}{\partial t}(\vec{\nabla} \times \vec{B}) \\ = \vec{\nabla}(\cancel{\vec{\nabla} \cdot \vec{E}})^{0} - \nabla^{2}\vec{E} &= -\frac{\partial}{\partial t}\left(\mu\sigma\vec{E} + \mu\varepsilon\frac{\partial \vec{E}}{\partial t}\right) \\ = \nabla^{2}\vec{E} &= \mu\varepsilon\frac{\partial^{2}\vec{E}}{\partial t^{2}} + \mu\sigma\frac{\partial \vec{E}}{\partial t}\end{aligned}}$$

(4.166)

$$\boxed{\begin{aligned}\vec{\nabla} \times (\vec{\nabla} \times \vec{B}) &= \mu\left[\sigma(\vec{\nabla} \times \vec{E})\right] + \varepsilon\frac{\partial}{\partial t}(\vec{\nabla} \times \vec{E}) \\ = \vec{\nabla}(\vec{\nabla} \cdot \vec{B}) - \nabla^{2}\vec{B} &= \mu\sigma\frac{\partial \vec{B}}{\partial t} - \mu\varepsilon\frac{\partial^{2}\vec{B}}{\partial t^{2}} \\ = \nabla^{2}\vec{B} &= \mu\varepsilon\frac{\partial^{2}\vec{B}}{\partial t^{2}} + \mu\sigma\frac{\partial \vec{B}}{\partial t}\end{aligned}}$$

Again, we can write:

$$\nabla^{2}\vec{E}(\vec{r},t) = \mu\varepsilon\frac{\partial^{2}\vec{E}(\vec{r},t)}{\partial t^{2}} + \mu\sigma\frac{\partial \vec{E}(\vec{r},t)}{\partial t} \quad \text{and}$$

$$\nabla^{2}\vec{B}(\vec{r},t) = \mu\varepsilon\frac{\partial^{2}\vec{B}(\vec{r},t)}{\partial t^{2}} + \mu\sigma\frac{\partial \vec{B}(\vec{r},t)}{\partial t}$$

(4.167)

Note that these four-dimensional wave equations for $\vec{E}$ and $\vec{B}$ in a conductor have an additional term that has a single time derivative, which is analogous to a velocity-dependent *damping term* (e.g., for a mechanical harmonic oscillator).

The general solution(s) to the preceding wave equations are usually in the form of an oscillatory function, *, a damping term (i.e., a decaying exponential) in the direction of the propagation of the EM wave; for example, complex plane-wave type solutions for $\vec{E}$ and $\vec{B}$ associated with the preceding wave equation(s) are of the general form:

$$\tilde{\vec{E}}(z,t) = \tilde{\vec{E}}_{0}e^{i(\tilde{k}x - \omega t)} \quad (4.168a)$$

and

4.10 Electromagnetic Waves in Conductors

$$\tilde{\vec{B}}(z,t) = \tilde{\vec{B}}_0 e^{i(\tilde{k}z - \omega t)} = \left(\frac{\tilde{k}}{\omega}\right)\hat{k} \times \tilde{\vec{E}}(z,t) = \frac{1}{\omega}\hat{k} \times \tilde{\vec{E}}(z,t) \quad (4.168b)$$

With a frequency-dependent complex wave number, $\tilde{k}(\omega) = k(\omega) + i\kappa(\omega)$, where $k(\omega) = \Re(\tilde{k}(\omega))$ and $\kappa(\omega) = \Im(\tilde{k}(\omega))$; thus, the corresponding complex vector in the positive $+\hat{z}$-direction is $\vec{\tilde{k}}(\omega) = \tilde{k}(\omega)\hat{k} = \tilde{k}(\omega)\hat{z}$ —that is, $\vec{\tilde{k}}(\omega) = [k(\omega) + i\kappa(\omega)]\hat{z}$.

We plugged $\tilde{\vec{E}}(z,t) = \tilde{\vec{E}}_0 e^{i(\tilde{k}x - \omega t)}$ and $\tilde{\vec{B}}(z,t) = \tilde{\vec{B}}_0 e^{i(\tilde{k}z - \omega t)}$ into their respective wave equations before, and obtained from each the same/identical *characteristic equation* (i.e., dispersion relation) between complex $\tilde{k}(\omega)$ and ω; thus, we get the following relation:

$$\tilde{k}^2(\omega) = \mu\varepsilon\omega^2 + i\mu\sigma\omega \quad (4.169)$$

Thus, because $\tilde{k}(\omega) = k(\omega) + i\kappa(\omega)$, then:

$$\boxed{\tilde{k}^2(\omega) = [k(\omega) + i\kappa(\omega)]^2 = k^2(\omega) - \kappa^2(\omega) + 2ik(\omega)\kappa(\omega) = \mu\varepsilon\omega^2 + i\mu\sigma\omega}$$
$$(4.170)$$

If we temporarily suppress the ω-dependence of complex $\tilde{k}(\omega)$, this relation becomes as:

$$\tilde{k}^2(\omega) = (k + i\kappa)^2 = k^2 - \kappa^2 + 2ik\kappa = \mu\varepsilon\omega^2 + i\mu\sigma\omega \quad (4.171)$$

We can solve this relation to determine $(\omega) = \Re(\tilde{k}(\omega))$ and $\kappa(\omega) = \Im(\tilde{k}(\omega))$ as follows.

First, separate this relation into two relations by separating its real and imaginary parts:

$$k^2 - \kappa^2 = \mu\varepsilon\omega^2 \quad \text{and} \quad 2ik\kappa = i\mu\sigma\omega \quad 2k\kappa = \mu\sigma\omega \quad (4.172)$$

Now we have two separate independent equations, $k^2 - \kappa^2 = \mu\varepsilon\omega^2$ and $2k\kappa = \mu\sigma\omega$, and we have two unknowns (i.e., k and κ). Therefore, solving these two equations simultaneously, we find the following results:

$$\kappa = \frac{1}{2}\mu\sigma\omega/k$$
$$(4.173)$$
$$k^2 - \kappa^2 = k^2 - \left(\frac{1}{2}\mu\sigma\omega/k\right)^2 = k^2 - \frac{1}{k^2}\left(\frac{1}{2}\mu\sigma\omega\right)^2 = \mu\varepsilon\omega^2$$

Then multiply by k^2 and rearrange the terms to obtain the following relation:

$$k^4 - (\mu\varepsilon\omega^2)k^2 - \left(\frac{1}{2}\mu\sigma\omega\right)^2 = 0 \quad (4.174)$$

To solve this equation, let $x \equiv k^2$, $a \equiv 1$, $b \equiv -(\mu\varepsilon\omega^2)$, and $c \equiv -(\frac{1}{2}\mu\sigma\omega)^2$, then Eq. 4.170 reduces to the form of a quadratic equation of $ax^2 + bx + c = 0$ with solution roots of

$$x = \frac{-b \pm \sqrt{b^2 - 4ac}}{2a} \quad \text{or} \quad k^2 = \frac{1}{2}\left[+(\mu\varepsilon\omega^2) \mp \sqrt{(\mu\varepsilon\omega^2)^2 + 4\left(\frac{1}{2}\mu\sigma\omega\right)^2}\right]$$

$$k^2 = \frac{1}{2}(\mu\sigma\omega^2)\left[1 \mp \sqrt{1 + 4\frac{(\mu^2\sigma^2\omega^2)}{4(\mu^2\varepsilon^2\omega^4)}}\right] = \frac{1}{2}(\mu\sigma\omega^2)\left[1 \mp \sqrt{1 + \frac{(\sigma^2)}{(\varepsilon^2\omega^2)}}\right]$$

$$= \frac{1}{2}(\mu\sigma\omega^2)\left[1 \mp \sqrt{1 + \left(\frac{\sigma}{\varepsilon\omega}\right)^2}\right]$$

Now we can see that on physical grounds ($k^2 > 0$), we *must* select the + sign, thus:

$$k^2 = \frac{1}{2}(\mu\sigma\omega^2)\left[1 + \sqrt{1 + \left(\frac{\sigma}{\varepsilon\omega}\right)^2}\right]$$

and thus:

$$k = \sqrt{k^2} = \omega\sqrt{\frac{\varepsilon\mu}{2}}\left[1 + \sqrt{1 + \left(\frac{\sigma}{\varepsilon\omega}\right)^2}\right]^{1/2} = \omega\sqrt{\frac{\varepsilon\mu}{2}}\left[1 + \sqrt{1 + \left(\frac{\sigma}{\varepsilon\omega}\right)^2}\right]^{1/2} \quad (4.175)$$

Having solved for k (or equivalently k^2), we can use either of the original two relations to solve for κ (e. g. $k^2 - \kappa^2 = \mu\varepsilon\omega^2$), then:

$$\kappa^2 = k^2 - \mu\varepsilon\omega^2 = \frac{1}{2}(\mu\varepsilon\omega^2)\left[\sqrt{1 + \left(\frac{\sigma}{\varepsilon\omega}\right)^2}\right] - \mu\varepsilon\omega^2\left[\sqrt{1 + \left(\frac{\sigma}{\varepsilon\omega}\right)^2} - 1\right] \quad (4.176)$$

Thus, we obtain:

$$\begin{cases} k(\omega) = \Re(\tilde{k}(\omega)) = \omega\sqrt{\frac{\varepsilon\mu}{2}}\left[\sqrt{1 + \left(\frac{\sigma}{\varepsilon\omega}\right)^2} + 1\right]^{1/2} \\ \kappa(\omega) = \Im(\tilde{k}(\omega)) = \omega\sqrt{\frac{\varepsilon\mu}{2}}\left[\sqrt{1 + \left(\frac{\sigma}{\varepsilon\omega}\right)^2} - 1\right]^{1/2} \end{cases} \quad (4.177)$$

Note that the imaginary part of $\tilde{k}$, $\kappa(\omega) = \Im(\tilde{k}(\omega))$, results in an *exponential* attenuation and damping of the monochromatic plane EM wave, increasing z as:

$$\begin{cases} \tilde{E}(z,t) = \tilde{E}_0 e^{-\kappa z} e^{i(kz-\omega t)} \text{ for Electric Field} \\ \tilde{B}(z,t) = \tilde{B}_0 e^{-\kappa z} e^{i(kz-\omega t)} = \frac{1}{\omega}\tilde{k} \times \tilde{E}_0 e^{-\kappa z} e^{i(kz-\omega t)} \text{ for Magnetic Filed} \end{cases} \quad (4.178)$$

4.10 Electromagnetic Waves in Conductors

These solutions (Eq. 4.178) satisfy the preceding wave equations for *any* choice of $\tilde{\vec{E}}_0$.

The characteristic distance over which $\vec{E}$ and $\vec{B}$ are attenuated, and reduced to $1/e = e^{-1} = 0.3679$ of their initial value at $z = 0$, is known as the *skin depth* and is shown as $\delta_{\text{skin depth}} = \delta_{\text{sc}} \equiv 1/\kappa(\omega)$ and the SI unit has a dimension of the meter:

$$\delta_{\text{sc}}(\omega) = \frac{1}{\kappa(\omega)} = \frac{1}{\omega\sqrt{\frac{\varepsilon\mu}{2}\left[\sqrt{1+\left(\frac{\sigma}{\varepsilon\omega}\right)^2}-1\right]^{1/2}}} \Rightarrow \boxed{\begin{array}{l}\tilde{\vec{E}}(z=\delta_{\text{sc}},t) = \tilde{\vec{E}}_0 e^{-1} e^{i(kz-\omega t)} \\ \tilde{\vec{B}}(z=\delta_{\text{sc}},t) = \tilde{\vec{B}}_0 e^{-1} e^{i(kz-\omega t)}\end{array}}$$

(4.179)

The real part of $\tilde{k}$ (i.e. $k(\omega) = \Re(\tilde{k}(\omega))$) determines spatial wavelength $\lambda(\omega)$, propagation speed $v(\omega)$ of the monochromatic EM plane wave (see Fig. 4.25) in the conductor, and also the index of refraction:

$$\lambda(\omega) = \frac{2\pi}{k(\omega)} = \frac{2\pi}{\Re(\tilde{k}(\omega))}$$

$$v(\omega) = \frac{\omega}{k(\omega)} = \frac{\omega}{\Re(\tilde{k}(\omega))} \quad (4.180)$$

$$n(\omega) = \frac{\omega}{v(\omega)} = \frac{ck(\omega)}{\omega} = \frac{c\Re(\tilde{k}(\omega))}{\omega}$$

The preceding plane-wave solutions satisfy the wave equations for any choice of $\tilde{\vec{E}}_0$. As we also have seen, it can similarly be shown here that Maxwell's equations (1) and (2), $\{\vec{\nabla}\cdot\vec{E} = 0$ and $\vec{\nabla}\cdot\vec{B} = 0\}$, rule out the presence of any (longitudinal) z-components for $\vec{E}$ and $\vec{B}$ for EM waves propagating in the $+\hat{z}$-direction: $\Rightarrow \vec{E}$ and $\vec{B}$ are purely transverse waves.

If we consider, for example, linearly polarized monochromatic plane EM waves propagating in the $+\hat{z}$-direction in a conducting medium (e.g., $\tilde{\vec{E}}(z,t) = \tilde{\vec{E}}_0 e^{-\kappa z} e^{i(kz-\omega t)}\hat{x}$), then:

$$\tilde{\vec{B}}(z,t) = \frac{1}{\omega}\tilde{\vec{k}} \times \tilde{\vec{E}}(z,t) = \left(\frac{\tilde{k}}{\omega}\right)\tilde{\vec{E}}_0 e^{-\kappa z} e^{i(kz-\omega t)}\hat{y} = \left(\frac{k+i\kappa}{\omega}\right)\tilde{\vec{E}}_0 e^{\kappa z} e^{i(kz-\omega t)}\hat{y}$$

$$\Rightarrow \tilde{\vec{E}}(z,t) \perp \tilde{\vec{B}}(z,t) \perp \hat{z} \quad (+\hat{z} = \text{propagation direction})$$

(4.181)

The complex wave number is $\tilde{k} = k + i\kappa = Ke^{i\phi_k}$, where $K \equiv |\tilde{k}| = \sqrt{k^2 + \kappa^2}$ and $\phi_k \equiv \tan^{-1}(\kappa/k)$ (Fig. 4.32). Then, we see that: $\tilde{\vec{E}}(z,t) = \tilde{\vec{E}}_0 e^{-\kappa z} e^{i(kz-\omega t)}\hat{x}$ has $\tilde{\vec{E}}_0 = E_0 e^{i\delta_E}$ and that:

Fig. 4.32 In the complex $\tilde{k}$-plane

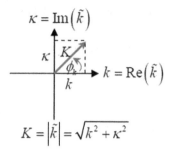

$$\vec{\tilde{B}}(z,t) = \tilde{B}_0 e^{-\kappa z} e^{i(kz-\omega t)} \hat{y} = \frac{\tilde{k}}{\omega} \tilde{E}_0 e^{-\kappa z} e^{i(kz-\omega t)} \hat{y}$$

has $\tilde{B}_0 = B_0 e^{-i\delta_B} = \dfrac{\tilde{k}}{\omega} \tilde{E}_0 = \dfrac{K e^{i\phi_k}}{\omega} E_0 e^{i\delta_E}$. Thus, we see that:

$$B_0 e^{-i\delta_B} = \frac{K e^{i\phi_k}}{\omega} E_0 e^{i\delta_E} = \frac{K}{\omega} E_0 e^{i(\delta_E + \phi_k)} = \frac{\sqrt{k^2 + \kappa^2}}{\omega} E_0 e^{i(\delta_E + \phi_k)}$$

That is, inside a conductor $\vec{E}$ and $\vec{B}$ are no longer in phase with each other. Phases of $\vec{E}$ and $\vec{B}$ are $\boxed{\delta_B = \delta_E + \phi_k}$. We also see that $\delta\varphi_{B-E} \equiv \delta_B - \delta_E = \phi_k$ magnetic field *lags* behind the electric field. We also see that:

$$\boxed{\frac{B_0}{E_0} = \frac{K}{\omega} = \left[\varepsilon\mu\sqrt{1 + \left(\frac{\sigma}{\varepsilon\omega}\right)^2}\right]^{1/2} \neq \frac{1}{c}}$$

The real physical $\vec{E}$ and $\vec{B}$ fields associated with linearly polarized monochromatic plane EM waves propagating in a conducting medium are *exponentially* damped:

$$\vec{E}(z,t) = \mathfrak{Re}\left(\vec{\tilde{E}}(z,t)\right) = E_0 e^{-\kappa z} \cos(kz - \omega t + \delta_E)\hat{x} \qquad (4.182a)$$

$$\vec{B}(z,t) = \mathfrak{Re}\left(\vec{\tilde{B}}(z,t)\right) = B_0 e^{-\kappa z} \cos(kz - \omega t + \delta_B)\hat{y}$$
$$= B_0 \cos(kz - \omega t + \{\delta_E + \phi_k\})\hat{y} \qquad (4.182b)$$

$$\frac{B_0}{E_0} = \frac{K(\omega)}{\omega}\left[\varepsilon\mu\sqrt{1 + \left(\frac{\sigma}{\varepsilon\omega}\right)^2}\right]^{1/2} \qquad (4.183a)$$

where,

$$K(\omega) \equiv |\tilde{k}(\omega)| = \sqrt{k^2(\omega) + \kappa^2(\omega)} = \omega\left[\varepsilon\mu\sqrt{1 + \left(\frac{\sigma}{\varepsilon\omega}\right)^2}\right]^{1/2} \qquad (4.183b)$$

4.10 Electromagnetic Waves in Conductors

$$\delta_B = \delta_E + \phi_k \quad \tilde{\vec{k}}(\omega) = [k(\omega) + i\kappa(\omega)]\hat{z} \quad (4.184c)$$

$$\phi_k(\omega) \equiv \tan^{-1}\left(\frac{\kappa(\omega)}{k(\omega)}\right) \quad \tilde{k}(\omega) = \left|\tilde{\vec{k}}(\omega)\right| = k(\omega) + i\kappa(\omega) \quad (4.184d)$$

The definition of the skin depth in a conductor is:

$$\boxed{\delta(\omega)_{\text{skin depth}} \equiv \frac{1}{\kappa(\omega)} = \frac{1}{\omega\sqrt{\frac{\varepsilon\mu}{2}}\left[\sqrt{1 + \left(\frac{\sigma}{\varepsilon\omega}\right)^2} - 1\right]^{1/2}} = \begin{array}{l}\text{Distance over which,}\\ \text{the } \vec{E} \text{ and } \vec{B} \text{ fields fall to}\\ \text{their initial values}\end{array}}$$

$$(4.185)$$

Example 4.13 What is the skin depth of fine silver at a microwave frequency of 10^{10} Hz, which is the common microwave region; assume the silver has a conductivity of $g = 10^{10}$ Hz.

Solution The skin depth is presented by the following equation:

$$\delta = \sqrt{\frac{2}{\mu_0 \omega g}} = \sqrt{\frac{2}{(2\pi \times 10^{10})(4\pi \times 10^{-7})(3 \times 10^3)}} = 9.2 \times 10^{-5}\text{cm}$$

Thus, at microwave frequencies the skin depth in silver is very small, and, consequently, the difference in performance between a pure silver component and a silver-plated brass component would be expected to be negligible.

Example 4.14 For a seawater case we calculate the frequency at which the skin depth is 1 m. For seawater use $\mu = \mu_0$ and $g \approx 4.3$ S/m.

Solution The expression for the frequency corresponding to given skin depth δ is:

$$\omega = \frac{2}{g\mu_0\delta^2} = \frac{2}{4.3 \times 4\pi \times 10^{-7}\delta^2} = \frac{3.70 \times 10^5}{\delta^2}(\text{s})^{-1}$$

which yields $f = 58 \times 10^3$ HZ, or frequency of 60 kHz for a skin depth of 1 m. If a submarine is equipped with a very sensitive receiver, and if a very powerful transmitter is used, it is possible to communicate with a submerged submarine. However, a low radio frequency must be used, and even then, an extremely severe attenuation of the signal occurs. At five skin depths (5 m in the case calculated here), only 1% of the initial electric field remains and only 0.01% of the incident power.

References

1. D. Griffiths, *Introduction to Electrodynamics*, 3rd edn. (Prentice Hall, New York, 1999)
2. R.M. Lambert et al., J. Phys. E: Sci. Instrum., 2799–2801. https://doi.org/10.1088/0022-3735/2/9/311
3. J.R. Reitz, F.J. Milford, R.W. Christy, *Foundations of Electromagnetic Theory*, 3rd edn. (Addison Wesley, Reading, 1979). Section 17-5
4. L. Eyges, *The Classical Electromagnetic Field* (Dover Publication, New York, 1972)
5. M.B. James, D.J. Griffiths, Am. J. Physics **60**, 309 (1992)
6. N. Ashby, Am. J. Phys. **43**, 553 (1975)
7. H.C. Ohanian, Am. J. Phys. **51**, 1020 (1983)
8. W. M. Hayes (ed.), *Handbook of Chemistry and Physics*, 78th edn. (CRC Press, Inc., Boca Raton, 1997)
9. F.A. Jenkins, *Fundamentals of Optics*, 4th edn. (McGraw-Hill Science/Engineering/Math, New York, 2001)

Chapter 5
Deriving the Lagrangian Density of an Electromagnetic Field

As part of understanding relativistic particles and electromagnetic fields, we need to have some awareness of the kinematic special theory of relativity, and then turn our attention to the dynamic aspect of charged particle motion in external electromagnetic fields. Thus, in this chapter we take a Lagrangian approach to the equations of motion and deal with the total energy involved with the motion of a particle. We also introduce the Hamiltonian in relation to the total energy of particle motion; analogous to classical mechanics, it relates to the corresponding *kinetic* and *potential* energy of the particle or system of concern. The Lagrangian approach is based in electrodynamics; thus, equations of motion are presented mainly as an avenue to introduce the concept of a Lorentz invariant action to a Hamiltonian, with the definition of the canonical momentum discussed in this chapter.

5.1 Introduction

Lagrangian field theory is a formalism in classical field theory. It is the *field-theoretic* analogue of Lagrangian mechanics, which is used for discrete particles each with a finite number of degrees of freedom. Lagrangian field theory applies to continua and fields that have an infinite number of degrees of freedom.

The relativistic aspects of Hamiltonian and Lagrangian theories, which are rather abstract constructions in *classical mechanics*, have a very simple interpretation in *relativistic quantum mechanics*. They both are proportional to the number of phase changes per unit of time. The Hamiltonian runs over the time axis, whereas the Lagrangian runs over the trajectory of the moving particle, the t'-axis, as illustrated in Fig. 5.1.

The figure shows the relativistic *de Broglie wave* in a *Minkowski diagram*. The triangle represents the relation between the Lagrangian ($\mathcal{L}$) and the Hamiltonian ($\mathcal{H}$), which holds in both relativistic and non-relativistic physics; it can be written with classical mechanics notation:

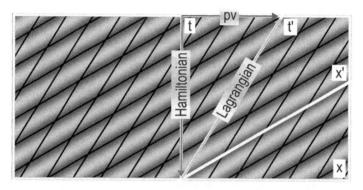

Fig. 5.1 The Hamiltonian and Lagrangian depictions scenario

$$\mathcal{L} = pv - \mathcal{H} \tag{5.1}$$

The Hamiltonian counts the phase changes per unit of time on the vertical axis, whereas the term pv counts the phase changes per unit on the horizontal axis; v is the distance traveled per unit of time, whereas p is proportional with the phase changes per unit of distance—therefore the term pv. We now can understand the classical relation with $\dot{q} = \dot{x} = v$ and write the following relation:

$$\frac{\partial \mathcal{L}}{\partial \dot{q}} = p \tag{5.2}$$

Thus, for the free classical relativistic particles we have the Hamiltonian energy and the pv as follows:

$$\mathcal{H} = \frac{mc^2}{\sqrt{1-\frac{v^2}{c^2}}} \quad \text{and} \quad pv = \frac{mv^2}{\sqrt{1-\frac{v^2}{c^2}}} \tag{5.3}$$

Calculating the Lagrangian, we can see that the Hamiltonian is proportional to γ, whereas the Lagrangian is proportional to $1/\gamma$,

$$\mathcal{L} = -(\mathcal{H} - pv) = -\left[\frac{c^2 - v^2}{\sqrt{1-\frac{v^2}{c^2}}}\right] m = -\left(\sqrt{1-\frac{v^2}{c^2}}\right) mc^2 \tag{5.4}$$

This is what we expect from time dilation. The moving particle has fewer clock-ticks by a factor γ because of time dilation; we now check that:

$$\frac{\partial \mathcal{L}}{\partial \dot{q}} = \frac{\partial}{\partial v}\left\{-\left(\sqrt{1-\frac{v^2}{c^2}}\right) mc^2\right\} = \frac{mv}{\sqrt{1-\frac{v^2}{c^2}}} = p \tag{5.5}$$

5.2 How the Fields Transform

We have not yet discussed potential energy, which we need to expand on. To obtain the equation for motion of a relativistic particle in a potential field, we have to add the *potential energy* term $V(q)$. In the non-relativistic case we have

$$\mathcal{H} = T(\dot{q}) + V(q)$$
$$\mathcal{L} = T(\dot{q}) - V(q) \tag{5.6}$$

where $T(\dot{q}) = \frac{1}{2}mv^2$ is kinetic energy.

The relativistic Hamiltonian and Lagrangian we have discussed, however, also include rest mass energy. The *rest mass energy* can be considered part of potential energy. Kinetic part T in the relativistic case can be obtained as follows:

$$\mathcal{H} + \mathcal{L} = 2T = pv \Rightarrow T = \frac{1}{2}pv = \frac{1}{2}mv^2 \left\{ \sqrt{1 - \frac{v^2}{c^2}} \right\}^{-1} \approx \frac{1}{2}mv^2 \tag{5.7}$$

The approximation in this equation is valid when $v \ll c$. Using the term $\mathcal{L} = \frac{1}{2}mv^2$ in Eq. 5.5 provides the result for $p = mv$ in the non-relativistic momentum, the same as we got here.

5.2 How the Fields Transform

As part of the derivation of the Lagrangian density of electromagnetic (EM) fields, we need to have some fundamental understanding of what field transformations and field tensors are all about; these two subjects are presented in this and the following sections. What we need to get our mind around is the *general* transformation rules for EM fields so that we can answer, for example, this question:

1. Given the fields in *system S*, what are the magnetic fields in *system S'*?

To answer this question, our first guess might be that electric field $\vec{E}$ is the spatial part of one 4-vector, and the magnetic field $\vec{B}$ is the spatial part of another. If this is true, then our intuition of one person's electric field as another's magnetic field is an incorrect assumption. Thus, let us begin by making an explicit assumption that we very implicitly can express in relativistic electrodynamics—that is, charge itself is an invariant case.

Covariance	The property of a function to retain its form when the variables are linearly transformed ....under a covariance condition.
Invariance	In mathematics an invariance is a property held by a class of mathematical objects that remains unchanged when transformations of a certain type are applied to the objects.
Invariant	A function, quantity, or property that remains unchanged when a specified transformation is applied.

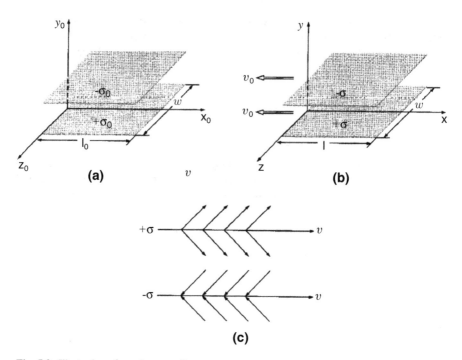

Fig. 5.2 Illustration of moving capacitor

This includes like mass, but unlike energy; the charge of the particle is a fixed number, independent of how fast the particle happens to travel from one point to another. Thus, we assume that the transformation rules are the same no matter how the fields are produced—that is, electric fields get generated by moving currents and transform the same way as those set up by stationary charges. If this is not the case, then we would need to abandon the field formulation altogether because what it is essential for a field theory is that the fields at a given point provide *all the information* one needs to know. Yet, in case of EM fields, one does not have to have extra information, electromagnetically, regarding their source [1].

Bearing the preceding argument in mind, we can consider one of the *simplest possible* electric fields, which is uniform in the region between the *two-large parallel-capacitor plates*, as shown in Fig. 5.2a, by assuming that the capacitor is at rest in S_0 and carries surface charges $\pm\sigma_0$. Then, we can write the electric field $\vec{E}_0$ in terms of *surface charge density* σ_0 in free space and $\varepsilon_0 = 8.85 \times 10^{-12}$ $C^2/N \times m^2$, this being the *permittivity of free space*, as well as a unit vector $\hat{y}$ in the y-direction, as follows:

$$\vec{E}_0 = \frac{\sigma_0}{\varepsilon_0}\hat{j} \qquad (5.8)$$

5.2 How the Fields Transform

If we consider the same capacitor from system S moving to the right at speed v_0, as shown in Fig. 5.2b, however, we can see that these system plates are moving to the left, but the field takes the form of the following relationship [2]:

$$\vec{E} = \frac{\sigma}{\varepsilon_0}\hat{y} \tag{5.9}$$

The only difference between Eqs. 5.8 and 5.9 is the value of the surface charge σ, but we need to see whether that is the only difference. Equation 5.9 is written for a parallel plate capacitor, which came from *Gauss's Law*, whereas Gauss's Law is perfectly valid for moving charges, where this application totally relies on the symmetry issue of the problem. If the plates tilt, however, as we can see in Fig. 5.2c, is it possible to say that electric filed $\vec{E}$ still is perpendicular to the moving direction of the plate? The answer is that it does not matter, and the field between the plates is the creation of the superposition of the surface charge of $+\sigma$. The one from the $-\sigma$ nevertheless would run perpendicular to the plates that are in motion, even in a tilted condition. For the surface charge, $-\sigma$, field would orient as indicated in Fig. 5.2c; changing the sign of the charges reverses the direction of the field, and the vector sum kills off the parallel components [2].

Now we are at the point where we can consider the total charge on each plate and claim that as *invariant* and the *width* (w) is unchanged; however, the *length* (l) is *Lorentz contracted* by a factor of:

$$\frac{1}{\gamma_0} = \sqrt{1 - v_0^2/c^2} \tag{5.10}$$

So that the charge per unit area is *increased* by a factor of γ_0 as indicated here:

$$\sigma = \gamma_0 \sigma_0 \tag{5.11}$$

Thus, accordingly, we can say that:

$$\vec{E}^{\perp} = \gamma_0 \vec{E}_0^{\perp} \tag{5.12}$$

The symbol $\perp$ is clearly an indication of a *perpendicular* component of electric field $\vec{E}$, while symbol $\|$ is a presentation of a *parallel* component of same field in the direction of the motion of system S, considering that the capacitor lined up with the yz-plane, as shown in Fig. 5.3. At this time, it is plate separation d that is Lorentz connected, whereas l and w, consequently, σ, are the same in both frames. Because the field does not depend on d, it follows that:

$$\vec{E}_{\|} = \gamma_0 \vec{E}_{\|}^{0} \tag{5.13}$$

What we have learned so far from the previous discussion we can put into perspective with the following examples given by Griffiths [2]. Here, we are going to show a simple problem of an electric field with a point charge in uniform motion.

Fig. 5.3 Two parallel charged plates

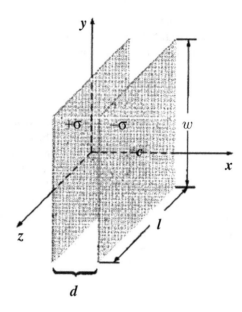

Example 5.1 A point charge q is at the origin in system S_0.

Question What is the electric field of this same charge in system S, which moves to the right at speed v_0 relative to system S_0?

Solution In system S_0, the field is:

$$\vec{E}_0 = \frac{1}{4\pi\varepsilon_0} \frac{q}{r_0^2} \hat{r}_0$$

$$\begin{cases} E_{x0} = \dfrac{1}{4\pi\varepsilon_0} \dfrac{qz_0}{\left(x_0^2 + y_0^2 + z_0^2\right)^{3/2}} \\ E_{y0} = \dfrac{1}{4\pi\varepsilon_0} \dfrac{qx_0}{\left(x_0^2 + y_0^2 + z_0^2\right)^{3/2}} \\ E_{z0} = \dfrac{1}{4\pi\varepsilon_0} \dfrac{qy_0}{\left(x_0^2 + y_0^2 + z_0^2\right)^{3/2}} \end{cases}$$

From the transformation rules of Eqs. 5.12 and 5.13, we can state that:

$$\begin{cases} E_x = E_{x0} = \dfrac{1}{4\pi\varepsilon_0} \dfrac{qx_0}{\left(x_0^2 + y_0^2 + z_0^2\right)^{3/2}} \\ E_y = \gamma_0 E_{y0} = \dfrac{1}{4\pi\varepsilon_0} \dfrac{\gamma_0 qy_0}{\left(x_0^2 + y_0^2 + z_0^2\right)^{3/2}} \\ E_z = \gamma_0 E_{z0} = \dfrac{1}{4\pi\varepsilon_0} \dfrac{\gamma_0 qz_0}{\left(x_0^2 + y_0^2 + z_0^2\right)^{3/2}} \end{cases}$$

5.2 How the Fields Transform

Fig. 5.4 Illustration of point charge in system S_0

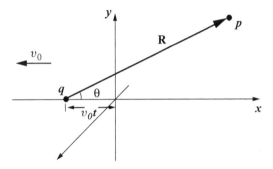

The expression is still in terms of the system S_0 coordinates (x_0, y_0, z_0) of the field point (P), and preferably we need to write them in terms of the system S coordinate of point (P). From the Lorentz transformations, or actually, the inverse transformations point of view, we can write:

$$\begin{cases} x_0 = \gamma_0(x + v_0 t) = \gamma_0 R_x \\ y_0 = y = R_y \\ z_0 = z = R_z \end{cases}$$

where $\vec{R}$ is the vector from charge q to point P, as depicted in Fig. 5.4. Thus, we can write:

$$\vec{E} = \frac{1}{4\pi\varepsilon_0} \frac{\gamma_0 q \vec{R}}{\left[\gamma_0^2 R^2 \cos^2(\theta) + R^2 \sin^2(\theta)\right]^{3/2}}$$

$$= \frac{1}{4\pi\varepsilon_0} \frac{q\left(1 - v_0^2/c^2\right)}{\left[1 - \left(v_0^2/c^2\right)\sin^2(\theta)\right]^{3/2}} \frac{\widehat{R}}{R^2} \quad (5.14)$$

This equation then is identified as the field charge in uniform motion; thus, we get the same result if we use a retarded potential equation, known as *Lienard-Wiechert potential* for point charges if they move on a specific trajectory [2].

This is a significant and efficient derivation, and shows that the field points away from the instantaneous as opposed to the retarded position of the charge; E_x gets a factor of γ_0 from the *Lorentz transformation* of the *coordinates*, and E_y as well as E_z pick up their factors from the transformation of the *field*. It is the balancing of these two γ_0s that leaves $\vec{E}$ parallel to $\vec{R}$.

Nonetheless, Eqs. 5.12 and 5.13 are not the most general transformation laws that we can express because we start with system S_0 in which the charges were at rest and where, consequently, there was no magnetic field. For us to derive the *general* rule, what we need to start with is out in a system with both electric and magnetic fields simultaneously. For this purpose, S itself will serve very nicely.

In addition to the *electric* filed,

$$E_y = \frac{\sigma}{\varepsilon_0} \quad (5.15)$$

there exists a *magnetic* field because of the surface currents, as shown earlier in Fig. 5.2b:

$$\vec{K}_\mp = \mp \sigma v_0 \hat{x} \quad (5.16)$$

Using the right-hand side (RHS) rule, as it was mentioned in Chap. 1 of this book, this field points in the negative z-direction and its magnitude is given by Ampère's Law as:

$$B_z = -\mu_0 \sigma v_0 \quad (5.17)$$

Using a figure, such as depicted Fig. 5.5, for a *third* system $\bar{S}$, traveling to the right with speed v relative to S, the field would be:

$$\bar{E}_y = \frac{\bar{\sigma}}{\varepsilon_0} \bar{B}_z = -\mu_0 \bar{\sigma} \bar{v} \quad (5.18)$$

where $\bar{v}$ is the velocity of system $\bar{S}$ relative to system S_0, and it can be written in the following form:

$$\bar{v} = \frac{v + v_0}{1 + vv_0/c^2}$$
$$\bar{\gamma} = \frac{1}{\sqrt{1 - \bar{v}^2/c^2}} \quad (5.19)$$

and

$$\bar{\sigma} = \bar{\gamma} \sigma_0 \quad (5.20)$$

Fig. 5.5 Illustration of systems S_0, S_0, S_0 and their motions

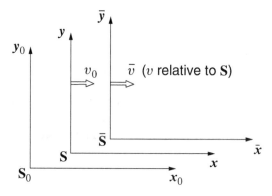

5.2 How the Fields Transform

The only thing that remains is to express electric filed $\vec{\bar{E}}$ and magnetic field $\vec{\bar{B}}$, as shown in Eq. 5.18, in terms of $\vec{E}$ and $\vec{B}$ (see Eqs. 5.15 and 5.17). In view of Eqs. 5.11 and 5.20, we have the following:

$$\bar{E}_y = \left(\frac{\bar{\gamma}}{\gamma_0}\right)\frac{\sigma}{\varepsilon_0}$$

$$\bar{B}_z = -\left(\frac{\bar{\gamma}}{\gamma_0}\right)\mu_0\sigma\bar{v} \tag{5.21}$$

With a little manipulation of algebra, we find the following result:

$$\frac{\bar{\gamma}}{\gamma_0} = \frac{\sqrt{1-v_0^2/c^2}}{\sqrt{1-\bar{v}^2/c^2}} = \frac{1+vv_0/c^2}{\sqrt{1-v^2/c^2}} = \gamma\left(1+\frac{vv_0}{c^2}\right) \tag{5.22}$$

where

$$\gamma = \frac{1}{\sqrt{1-v^2/c^2}} \tag{5.23}$$

Thus, we can write:

$$\bar{E}_y = \gamma\left(1+\frac{vv_0}{c^2}\right)\frac{\sigma}{\varepsilon_0} = \gamma\left(E_y - \frac{v}{c^2\varepsilon_0\mu_0}B_z\right) \tag{5.24}$$

whereas

$$\bar{B}_2 = -\gamma\left(1+\frac{vv_0}{c^2}\right)\mu_0\sigma\left(\frac{v+v_0}{1+vv_0/c^2}\right)$$
$$= \gamma(B_z - \mu_0\varepsilon_0 vE_y) \tag{5.25}$$

Or, because $\mu_0\varepsilon_0 = 1/c^2$, then Eqs. 5.24 and 5.25 reduce to the following form:

$$\begin{cases} \bar{E}_y = \gamma(E_y - vB_z) \\ \bar{B}_z = \gamma\left(B_z - \frac{v}{c^2}E_y\right) \end{cases} \tag{5.26}$$

This equation indicates how E_y and B_z transform—for E_z and B_y simply align the same capacitor parallel to the xy-plane instead of xz-plane, as illustrated in Fig. 5.6. Thus, the fields in system S are written:

$$E_z = \frac{\sigma}{\varepsilon_0}$$
$$B_y = \mu_0\sigma v_0 \tag{5.27}$$

Fig. 5.6 Two capacitors parallel to the xy-plane

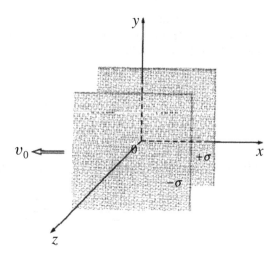

Using the RHS rule of Chap. 1 again, we get B_y. The rest of the argument is identical—everywhere we had E_y before reads E_z, and everywhere that we had B_z reads $-B_y$; thus, we can write:

$$\begin{cases} \bar{E}_z = \gamma(E_z + vB_y) \\ \bar{B}_y = \gamma\left(B_y + \dfrac{v}{c^2}E_z\right) \end{cases} \quad (5.28)$$

As for the x-components, we already have seen, by orienting the capacitor parallel to the yz-plane, that:

$$\bar{E}_x = E_x \quad (5.29)$$

Because in this case there is no accompanying magnetic field, we cannot conclude and deduce the transformation rule for B_x. Nevertheless, another configuration, such as the one in Fig. 5.7, which is an illustration of a long *solenoid* aligned parallel to the x-axis and at rest in system S. The magnetic field within the coils is then [2]:

$$B_x = \mu_0 n I \quad (5.30)$$

where n is the number of turns per unit length and I is the current in the coil. In system $\bar{S}$ the length contracts, so n *increases*:

$$\bar{n} = \gamma n \quad (5.31)$$

On the other hand, time *dilates*. Thus, the system S clock that rides along with the solenoid runs slow, so the current (i.e., change *per unit time*) in $\bar{S}$ is given by:

5.2 How the Fields Transform

Fig. 5.7 Depiction of long solenoid aligned parallel to the x-axis

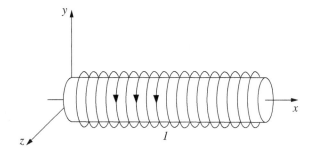

$$\bar{I} = \frac{1}{\gamma} I \quad (5.32)$$

The two factors of γ exactly cancel, and we conclude that:

$$\bar{B}_x = B_x \quad (5.33)$$

Like $\vec{E}$, component $\vec{B}$, which is *parallel* to the motion, is unchanged [1].

Let us now collect all the components of transformation in order to have a complete set of transformation rules:

$$\boxed{\begin{array}{lll} \bar{E}_x = E_x & \bar{E}_y = \gamma(E_y - vB_z) & \bar{E}_z = \gamma(E_z + vB_y) \\ \bar{B}_x = B_x & \bar{B}_y = \gamma\left(B_y + \dfrac{v}{c^2}E_z\right) & \bar{B}_z = \gamma\left(B_z - \dfrac{v}{c^2}E_y\right) \end{array}} \quad (5.34)$$

The following are two special cases that we need to pay close attention to:

1. If $\vec{B} = 0$ in system S, then we can write:

$$\vec{\bar{B}} = \gamma \frac{v}{c^2}\left(E_z\hat{y} - E_y\hat{z}\right) = \frac{v}{c^2}\left(\bar{E}_z\hat{y} - \bar{E}_y\hat{z}\right) \quad (5.35)$$

or, since $\vec{v} = v\hat{x}$, then

$$\vec{\bar{B}} = -\frac{1}{c^2}\left(\vec{v} \times \vec{\bar{E}}\right) \quad (5.36)$$

2. If $\vec{E} = 0$ in system S, then

$$\vec{\bar{E}} = -\gamma v\left(B_z\hat{y} - B_y\hat{z}\right) = -v\left(\bar{B}_z\hat{y} - \bar{B}_y\hat{z}\right) \quad (5.37)$$

or

$$\vec{\bar{E}} = \vec{v} \times \vec{\bar{B}} \quad (5.38)$$

5.3 The Field Tensor

As part of electrodynamics and relativity, we can expand our knowledge into the *field tensor*, where the classical electrodynamics *already* is consistent with *special relativity* that can be extended and be applied to *Maxwell's equations* and the *Lorentz force law* It is possible, legitimately, in any inertial system, even in an invariant *Lorentz transformation*, to show how the field transforms. See Appendix B for more details [2].

To understand what we discussed as part of developing Eq. 5.6 and the scalar $\left(\|\vec{E}\|^2 - c^2\|\vec{B}\|^2\right)$, one of the two *Lorentz invariants*, we consider the following tensor field analyses; refer to Appendix B for more essential information and derivations. Still, bear in mind that a 4-vector transforms by the rule shown in the following equation:

$$a^{\mu'} = \Lambda^\mu_\nu a^\nu \tag{5.39}$$

Note that the summation over ν implies that a Λ is the *Lorentz transformation* matrix. If S' is moving in the x-direction at speed $\vec{v}$, then Λ has the form,

$$\Lambda = \begin{pmatrix} \gamma & -\gamma\beta & 0 & 0 \\ -\gamma\beta & \gamma & 0 & 0 \\ 0 & 0 & 0 & 0 \\ 0 & 0 & 0 & 0 \end{pmatrix} \tag{5.40}$$

Λ^μ_ν is the entry in row μ and column ν. A second-rank tensor is an object with *two* indices that transforms with *two* factors of Λ, which is one for each index as demonstrated in the following equation, where we have:

$$t^{\mu\nu} = \Lambda^\mu_\lambda \Lambda^\nu_\sigma t^{\lambda\sigma} \tag{5.41}$$

A tensor in four dimensions has $4 \times 4 = 16$ components, which we can display in a 4×4 matrix array:

$$t^{\mu\nu} = \begin{bmatrix} t^{00} & t^{01} & t^{02} & t^{03} \\ t^{10} & t^{11} & t^{12} & t^{13} \\ t^{20} & t^{21} & t^{22} & t^{23} \\ t^{30} & t^{31} & t^{32} & t^{33} \end{bmatrix} \tag{5.42}$$

But, the 16 elements need not all to be different. For instance, a *symmetric* tensor has the property as follows:

$$t^{\mu\nu} = t^{\nu\mu} \quad \text{(Symmetric Tensor)} \tag{5.43}$$

5.3 The Field Tensor

In this case, there are 10 distinct components and 6 out of 16 represent the repeats ($t^{01} = t^{10}$, $t^{02} = t^{20}$, $t^{03} = t^{30}$, $t^{12} = t^{21}$, $t^{13} = t^{31}$, and $t^{23} = t^{32}$). Similarly, and *antisymmetric* tensor obeys the following relation as well:

$$t^{\mu\nu} = -t^{\nu\mu} \quad \text{(Antisymmetric Tensor)} \tag{5.44}$$

Such an object has just 6 distinct elements of the original 16 components, where 6 are repeats of the same ones as before, only this time with a minus sign in front of each element, and 4 of them are zeros; they are (t^{00}, t^{11}, t^{22}, and t^{33}). Thus, the general form of an antisymmetric tensor has the following format:

$$t^{\mu\nu} = \begin{bmatrix} 0 & t^{01} & t^{02} & t^{03} \\ -t^{10} & 0 & t^{12} & t^{13} \\ -t^{20} & -t^{21} & 0 & t^{23} \\ -t^{30} & -t^{31} & -t^{32} & 0 \end{bmatrix} \tag{5.45}$$

Now let us see how the transformation of the rule defined by Eq. 5.41 works for the 6 distinct components of an anti-symmetric tensor. Thus, starting with component $-t^{10}$, we have the following relation:

$$-t^{01} = \Lambda^0_\lambda \Lambda^1_\sigma t^{\lambda\sigma} \tag{5.46}$$

According to Eq. 5.40, $\Lambda^0_\lambda = 0$; however, unless $\lambda = 0$ or 1, and $\Lambda^1_\sigma = 0$ unless $\sigma = 0$ or 1. Thus, there are four terms in the sum:

$$-t^{01} = \Lambda^0_0 \Lambda^1_0 t^{00} + \Lambda^0_0 \Lambda^1_1 t^{01} + \Lambda^0_1 \Lambda^1_0 t^{10} + \Lambda^0_1 \Lambda^1_1 t^{11} \tag{5.47}$$

On the other hand, $t^{00} = t^{11} = 0$, while $t^{01} = -t^{10}$, then we can write the following relations:

$$-t^{01} = \left(\Lambda^0_0 \Lambda^1_1 - \Lambda^0_1 \Lambda^1_0\right) t^{01} = \left[\gamma^2 - (\gamma\beta)^2\right] t^{01} = t^{01} \tag{5.48}$$

Working out the other components, similarly, we obtain the following results:

$$\begin{cases} -t^{01} = t^{01} & -t^{02} = \gamma(t^{02} - \beta t^{12}) & -t^{03} = \gamma(t^{03} - \beta t^{31}) \\ -t^{23} = t^{23} & -t^{31} = \gamma(t^{31} - \beta t^{03}) & -t^{12} = \gamma(t^{12} - \beta t^{02}) \end{cases} \tag{5.49}$$

These are precisely the rules we can derive on physical grounds for the EM field in Sect. 5.2 [1].

In fact, we can construct *field tensor* $F^{\mu\nu}$ by direct comparison as:

$$\begin{array}{ccc} F^{01} = \dfrac{E_x}{c}, & F^{02} = \dfrac{E_y}{c}, & F^{03} = \dfrac{E_z}{c} \\ F^{12} = B_z, & F^{31} = B_y, & F^{23} = B_x \end{array} \tag{5.50}$$

Written in Eq. 5.50 as a matrix array, we can produce the following:

$$F^{\mu\nu} = \begin{bmatrix} 0 & E_x/c & E_y/c & E_z/c \\ -E_x/c & 0 & B_z & -B_y \\ -E_y/c & -B_z & 0 & B_x \\ -E_x/c & B_y & -B_x & 0 \end{bmatrix} \quad (5.51)$$

With the results obtained in Eq. 5.51, relativity completes and makes a perfect job by Oersted, combining the electric and magnetic fields into a single entity as $F^{\mu\nu}$. Note that some authors prefer the other notation convention—$F^{01} = E_x$, $F^{12} = cB_z$, and so on—and some use the opposite signs. Accordingly, most of the equations from here on will have a little different appearance, depending on the text you are reading.

If we follow the argument that yields Eq. 5.51 results, with some cautious care, we can notice that there is a *different* way of relating an anti-symmetric tensor with electric field $\vec{E}$ and magnetic field $\vec{B}$. Instead of comparing the first line of Eq. 5.51 with the first line of Eq. 5.49, and the second with the second, we can relate the first line of Eq. 5.51 to the *second* line of Eq. 5.49 and vice versa. This leads to and results in the *dual tensor*, $G^{\mu\nu}$, as follows:

$$G^{\mu\nu} = \begin{bmatrix} 0 & B_x & B_y & B_z \\ -B_x & 0 & -E_z/c & -E_y/c \\ -B_y & E_z/c & 0 & -E_x/c \\ -B_z & -E_y/c & E_x/c & 0 \end{bmatrix} \quad (5.52)$$

$G^{\mu\nu}$ can be obtained directly from $F^{\mu\nu}$ by the substitution of $\vec{E}/c \to \vec{B}$ and $\vec{B} \to \vec{E}/c$. Notice that this operation leaves Eq. 5.51 unchanged; this is why both tensors generate the correct transformation rules for $\vec{E}$ and $\vec{B}$.

5.4 The Electromagnetic Field Tensor

The transformation of electric and magnetic fields under a Lorentz boost was established even before Einstein developed the theory of relativity. We know the $\vec{E}$-fields can transform into $\vec{B}$-fields and vice versa. For example, a point charge at rest gives an *electric* field. If we boost to a frame in which the charge is moving, there is an *electric* and a *magnetic* field. This means that the $\vec{E}$-field cannot be a *Lorentz vector*. We need to put the electric and magnetic fields into one *tensor object* to properly handle a *Lorentz transformation* and to write the equations in a covariant way.

5.4 The Electromagnetic Field Tensor

The simplest way, and the correct way, to do this is to make the electric and magnetic fields components of a rank 2 anti-symmetric tensor as follows:

$$F_{\mu\nu} = \begin{pmatrix} 0 & B_z & -B_y & -iE_x \\ -B_z & 0 & B_x & -iE_y \\ B_y & -B_x & 0 & -iE_z \\ iE_x & iE_y & iE_z & 0 \end{pmatrix} \quad (5.53)$$

The fields can be written simply in terms of the vector potential, which is Lorentz vector $A_\mu = (\vec{A}, i\phi)$, thus Eq. 5.53 can be expressed in the following form:

$$F_{\mu\nu} = \frac{\partial A_\nu}{\partial x_\mu} - \frac{\partial A_\mu}{\partial x_\nu} \quad (5.54)$$

Note that this is automatically anti-symmetric under the interchange of the indices. As before, the first two source-less *Maxwell equations* are automatically satisfied for fields derived from potential. We may write the other two Maxwell equations in terms of 4-vector $j_\mu = (\vec{j}, ic\rho)$ in the following form:

$$\frac{\partial F_{\mu\nu}}{\partial x_\nu} = \frac{j_\mu}{c} \quad (5.55)$$

Comparing Eqs. 5.54 and 5.55, we can write the following:

$$\frac{\partial}{\partial x_\nu}\left(\frac{\partial A_\nu}{\partial x_\mu} - \frac{\partial A_\mu}{\partial x_\nu}\right) = \frac{j_\mu}{c} \quad (5.56)$$

Of course, we have not yet quantized the theory in Eq. 5.56.

For some peace of mind, let us verify a few terms in the equations. Clearly, all the diagonal terms in the field tensor are zero by anti-symmetry. Taking some examples of off-diagonal terms in the field tensor, check the old definition of the fields in terms of potential:

$$\vec{B} = \vec{\nabla} \times \vec{A}$$

$$\vec{E} = -\vec{\nabla}\phi - \frac{1}{c}\frac{\partial \vec{A}}{\partial t}$$

$$F_{12} = \frac{\partial A_2}{\partial x_1} - \frac{\partial A_1}{\partial x_2} = (\vec{\nabla} \times \vec{A})_z = B_z$$

$$F_{13} = \frac{\partial A_3}{\partial x_1} - \frac{\partial A_1}{\partial x_3} = (\vec{\nabla} \times \vec{A})_y = -B_y$$

$$F_{4i} = \frac{\partial A_i}{\partial x_4} - \frac{\partial A_4}{\partial x_i} = \frac{1}{ic}\frac{\partial A_i}{\partial t} - \frac{\partial(i\phi)}{\partial x_i} = -i\left(\frac{1}{c}\frac{\partial A_i}{\partial t} + \frac{\partial \phi}{\partial x_i}\right) = -i\left(\frac{\partial \phi}{\partial x_i} + \frac{1}{c}\frac{\partial A_i}{\partial t}\right)$$

$$(5.57)$$

Let us also check what the Maxwell equation says for the last row in the tensor, as follows:

$$\frac{\partial F_{4v}}{\partial x_v} = \frac{j_4}{c}$$

$$\frac{\partial F_{4i}}{\partial x_i} = \frac{ic\rho}{c}$$

$$\frac{\partial (iE_i)}{\partial x_i} = i\rho \qquad (5.58)$$

$$\frac{\partial E_i}{\partial x_i} = \rho$$

$$\vec{\nabla} \cdot \vec{E} = \rho$$

We will not bother to check the Lorentz transformation of the field here.

5.5 The Lagrangian and Hamiltonian for Electromagnetic Fields

There are not many ways to make a *scalar Lagrangian* from the field tensor. We already, from Eq. 5.55, know that

$$\frac{\partial F_{\mu v}}{\partial x_v} = \frac{j_\mu}{c} \qquad (5.59)$$

In addition, we need to make our Lagrangian out of the field, not just the current again; x_μ cannot appear explicitly because it violates symmetries. Also, we want a linear equation, so higher powers of the field should not occur. Term $mA_\mu A^\mu$ is a mass term and would cause fields to fall off faster than $1/r$ (i.e., we later show how the term $1/r$ comes into play). Thus, the only reasonable choice is:

$$F_{\mu v}F^{\mu v} = 2(B^2 - E^2) \qquad (5.60)$$

One might consider the following relationship:

$$e_{\mu v}\lambda_\sigma F_{\mu v}F_{\lambda\sigma} = \vec{B} \cdot \vec{E} \qquad (5.61)$$

But that is a pseudo-scalar, not a scalar completely—that is, it changes its sign under a parity transformation. The EM interaction is known to conserve parity, so this is not a real option. As with the scalar field, we need to add an interaction with a source term. Of course, we know electromagnetism well, so finding the right Lagrangian is not really guess work. The source of the field is the vector $\vec{j}_\mu$; thus, the simple scalar that we can write is $j_\mu A^\mu$.

5.5 The Lagrangian and Hamiltonian for Electromagnetic Fields

The Lagrangian for *classical electricity and magnetism* we will try is:

$$\mathcal{L}_{em} = -\frac{1}{4} F_{\mu\nu} F^{\mu\nu} + \frac{1}{c} \vec{j}_\mu A^\mu \tag{5.62}$$

In working with this Lagrangian, we will treat each component of A as an independent field.

The next step is to check what the *Euler-Lagrange equation* provides:

$$\frac{\partial}{\partial x_\nu} \left(\frac{\partial \mathcal{L}}{\partial (\partial A^\mu / \partial x_\nu)} \right) - \frac{\partial \mathcal{L}}{\partial A^\mu} = 0 \tag{5.63}$$

where

$$\mathcal{L} = -\frac{1}{4} F_{\mu\nu} F^{\mu\nu} + \frac{1}{c} j_\mu A^\mu = -\frac{1}{4} \left(\frac{\partial A^\nu}{\partial x_\mu} - \frac{\partial A^\mu}{\partial x_\nu} \right) \left(\frac{\partial A^\nu}{\partial x_\mu} - \frac{\partial A^\mu}{\partial x_\nu} \right) \tag{5.64}$$

$$\frac{\partial \mathcal{L}}{\partial (\partial A^\mu / \partial x_\nu)} = -\frac{1}{4} \frac{\partial}{\partial (\partial A^\mu / \partial x_\nu)} \left(\frac{\partial A^\sigma}{\partial x_\lambda} - \frac{\partial A^\lambda}{\partial x_\sigma} \right) \left(\frac{\partial A^\sigma}{\partial x_\lambda} - \frac{\partial A^\lambda}{\partial x_\sigma} \right)$$

$$= -\frac{1}{4} \frac{\partial}{\partial (\partial A^\mu / \partial x_\nu)} \left(2 \frac{\partial A^\sigma}{\partial x_\lambda} \frac{\partial A^\sigma}{\partial x_\lambda} - 2 \frac{\partial A^\sigma}{\partial x_\lambda} \frac{\partial A^\lambda}{\partial x_\sigma} \right) \tag{5.65}$$

$$= -\frac{1}{4} 4 \left(\frac{\partial A^\lambda}{\partial x_\nu} - \frac{\partial A^\nu}{\partial x_\mu} \right)$$

$$= -F_{\nu\mu} = F_{\mu\nu}$$

$$\begin{cases} \dfrac{\partial}{\partial x_\nu} F^{\mu\nu} - \dfrac{\partial \mathcal{L}}{\partial A_\mu} = 0 \\[4pt] \dfrac{\partial}{\partial x_\nu} F^{\mu\nu} - \dfrac{j_\mu}{c} = 0 \\[4pt] \dfrac{\partial}{\partial x_\nu} F^{\mu\nu} = \dfrac{j_\mu}{c} \end{cases} \tag{5.66}$$

Now that we have four independent components of A^μ as autonomous fields, we have four equations or one 4-vector equation. The Euler-Lagrange equation gets us back to Maxwell's equations with this choice of the Lagrangian. Clearly, this justifies the choice of $\mathcal{L}$.

It is important to emphasize that we have a Lagrangian-based, formal classical field theory for electricity and magnetism, which has the four components of the 4-vector potential as the independent fields. We could not treat each component of $F_{\mu\nu}$ as independent because they obviously are correlated. We could have tried using the six independent components of the anti-symmetric tensor, but it would not have given the correct answer. Using the 4-vector potentials as the fields does give the correct answer. Electricity and magnetism is a theory of 4-vector field A^μ.

5 Deriving the Lagrangian Density of an Electromagnetic Field

We also can calculate the free field Hamiltonian density—that is, the *Hamiltonian density* in regions with no source term. We use the standard definition of the Hamiltonian in terms of the Lagrangian:

$$\mathcal{H} = \left(\frac{\partial \mathcal{L}}{\partial(\partial A^\mu/\partial t)}\right)\frac{\partial A^\mu}{\partial t} - \mathcal{L} = \left(\frac{\partial \mathcal{L}}{\partial(\partial A^\mu/\partial x_4)}\right)\frac{\partial A^\mu}{\partial x_4} - \mathcal{L} \qquad (5.67)$$

We calculated before that:

$$\frac{\partial \mathcal{L}}{\partial(\partial A^\mu/\partial x_\nu)} = F_{\mu\nu} \qquad (5.68)$$

which we can use to get:

$$\frac{\partial \mathcal{L}}{\partial(\partial A^\mu/\partial x_4)} = F_{\mu 4} \qquad (5.69)$$

Thus, the free field Hamiltonian density becomes:

$$\begin{aligned}\mathcal{H} &= (F_{\mu 4})\frac{\partial A^\mu}{\partial x_4} - \mathcal{L} \\ &= F_{\mu 4}\frac{\partial A^\mu}{\partial x_4} + \frac{1}{4}F_{\mu\nu}F^{\mu\nu}\end{aligned} \qquad (5.70)$$

Therefore,

$$\boxed{\mathcal{H} = F_{\mu 4}\frac{\partial A^\mu}{\partial x_4} + \frac{1}{4}F_{\mu\nu}F^{\mu\nu}} \qquad (5.71)$$

We will use this once we have written the radiation field in a convenient form. In the meantime, we can check what this gives us in general in a region with no sources:

$$\begin{aligned}\mathcal{H} &= F_{\mu 4}\left(F_{4\varpi} + \frac{\partial A^4}{\partial x_\mu}\right) + \frac{1}{4}F_{\mu\nu}F^{\mu\nu} \\ &= -F_{\mu 4}\left(F_{4\mu} + \frac{\partial A^4}{\partial x_\mu}\right) + \frac{1}{4}F_{\mu\nu}F^{\mu\nu} \\ &= -F_{\mu 4}F_{4\mu} - F_{\mu 4}+\frac{\partial A^4}{\partial x_\mu} + \frac{1}{4}F_{\mu\nu}F^{\mu\nu} \\ &= E^2 - F_{4i}\frac{\partial A^4}{\partial x_i} + \frac{1}{2}(B^2 - E^2) \\ &= \frac{1}{2}(E^2 + B^2) - iE_i\frac{\partial(i\phi)}{\partial x_i} \\ &= \frac{1}{2}(E^2 + B^2) + E_i\frac{\partial \phi}{\partial x_i}\end{aligned} \qquad (5.72)$$

If we integrate the last term by parts (and the fields fall to zero at infinity), then that term contains a $\vec{\nabla} \cdot \vec{E}$, which is zero with no sources in the region. We therefore can drop it and are left with:

$$\mathcal{H} = \frac{1}{2}\left(E^2 + B^2\right) \qquad (5.73)$$

This is the result we expected: the energy density and an EM field. (Remember the fields have been decreased by a factor of $\sqrt{4\pi}$ compared to centimeter-gram-second, CGS, units.)

We will study the interaction between electrons and the EM field with the *Dirac equation* later. Until then, the Hamiltonian used for non-relativistic quantum mechanics will be sufficient. We have derived the Lorentz force law from the Hamiltonian.

5.6 Introduction to Lagrangian Density

In considering the *special relativistic EM field*, we understand that assuming a *Lagrangian density* of the form results in:

$$\mathcal{L} = -\frac{1}{4} F_{\mu\nu} F^{\mu\nu} + \frac{1}{c} \vec{j}_\mu A^\mu \qquad (5.74)$$

and following the *Euler-Lagrange equation* recovers Maxwell's equation. In this equation $F^{\mu\nu}$ is the Maxwell field tensor and c is the speed of light, $c^2 = 1/\varepsilon\mu$ (not necessarily in a vacuum). The four-current is $\vec{j}_\mu = (\rho c, \vec{j})$; the four-potential is $\vec{A}_\mu = (\phi/c, \vec{A})$; the Compton wave number for a photon with mass (m) takes the form $k = 2\pi mc/h$; and h is *Plank's constant*.

In the following we will prove that a compatible *Lagrangian density* for the EM field in empty space is,

$$\mathcal{L}_{em} = \varepsilon_0 \left\{ \frac{\|\vec{E}\|^2 - c^2 \|\vec{B}\|^2}{2} \right\} - \rho\phi + \vec{j} \cdot \vec{A} \qquad (5.75)$$

which will be proof of Eq. 5.74. In following section, we take a very general approach to the more *classical electrodynamics* of Lagrangian density for the EM field.

Equation 5.75 is the *Euler-Lagrange equation* produced from this Lagrangian and are Maxwell's equations for the EM field. Lagrangian density is derived by a trial-and-error procedure, not by guessing, and one also can find the Lagrangian in a more difficult and complicated case. (See the proof provided in Law [3], where the interaction between a moving mirror and radiation pressure is a Hamiltonian formulation.)

5 Deriving the Lagrangian Density of an Electromagnetic Field

Maxwell's differential equations of the EM field in empty space are:

$$\vec{\nabla} \times \vec{E} = -\frac{\partial \vec{B}}{\partial t} \tag{5.76a}$$

$$\vec{\nabla} \times \vec{B} = \mu_0 \vec{j} + \frac{1}{c^2}\frac{\partial \vec{E}}{\partial t} \tag{5.76b}$$

$$\vec{\nabla} \cdot \vec{E} = \frac{\rho}{\varepsilon_0} \tag{5.76c}$$

$$\vec{\nabla} \cdot \vec{B} = 0 \tag{5.76d}$$

where $\vec{E}$ is the electric field intensity vector, $\vec{B}$ is the magnetic-flux density vector, ρ is the electric charge density, and $\vec{j}$ is the electric current-density vector. All quantities are functions of the three space coordinates $(x_1, x_2, x_3) \equiv (x, y, z)$ and time $t = x_4$ for time coordinates of the fourth dimension.

From Eq. 5.76d, magnetic-flux vector $\vec{B}$ may be expressed as the *curl* of vector potential $\vec{A}$ and is written as follows:

$$\vec{B} = \vec{\nabla} \times \vec{A} \tag{5.77}$$

In addition, from Eqs. 5.77 and 5.76a, we can conclude that:

$$\vec{\nabla} \times \left(\vec{E} + \frac{\partial \vec{A}}{\partial t} \right) = 0 \tag{5.78}$$

So the parentheses term may be expressed as the gradient of a scalar function:

$$\vec{E} + \frac{\partial \vec{A}}{\partial t} = -\nabla \phi \tag{5.79a}$$

Equation 5.78 can be rearranged as follows as well:

$$\vec{E} = -\nabla \phi - \frac{\partial \vec{A}}{\partial t} \tag{5.79b}$$

Therefore, the six scalar variables, the components of vectors $\vec{E}$ and $\vec{B}$, can be expressed as functions of four *scalar variables*, the scalar potential ϕ, and three components of vector potential $\vec{A}$. Inserting the expressions of $\vec{E}$ and $\vec{B}$, Eqs. 5.77, 5.79a, and 5.79b, respectively, into Eqs. 5.76b and 5.76c, we obtain the following result:

$$\vec{\nabla} \times (\vec{\nabla} \times \vec{A}) = \mu_0 \vec{j} + \frac{1}{c^2}\frac{\partial}{\partial t}\left(-\nabla \phi - \frac{\partial \vec{A}}{\partial t} \right) \tag{5.80}$$

and

$$-\nabla^2 \phi - \frac{\partial}{\partial t}(\vec{\nabla} \cdot \vec{A}) = \frac{\rho}{\varepsilon_0} \quad (5.81)$$

Given that the vector identity is of this form:

$$\vec{\nabla} \times (\vec{\nabla} \times \vec{A}) = \vec{\nabla}(\vec{\nabla} \cdot \vec{A}) - \nabla^2 \vec{A} \quad (5.82)$$

Per Eqs. 5.82, and 5.80, we find the following result:

$$\frac{1}{c^2}\frac{\partial^2 \vec{A}}{\partial t^2} - \nabla^2 \vec{A} + \nabla\left(\nabla \cdot \vec{A} + \frac{1}{c^2}\frac{\partial \phi}{\partial t}\right) = \mu_0 \vec{j} \quad (5.83)$$

5.7 Euler-Lagrange Equation of the Electromagnetic Field

Now the main goal and task is to find a Lagrangian density $\mathcal{L}$ function of the four "field coordinates" and their first-order derivatives as follows:

$$\mathcal{L} = \mathcal{L}(\eta_j, \dot{\eta}_j, \vec{\nabla}\eta_j) \quad j = 1, 2, 3 \quad (5.84)$$

such that the four scalar EM fields, Eqs. 5.81 and 5.83, are derived from the Lagrange equations:

$$\frac{\partial}{\partial t}\left[\frac{\partial \mathcal{L}}{\partial \left(\frac{\partial \eta_j}{\partial t}\right)}\right] + \sum_{k=1}^{k=3} \frac{\partial}{\partial x_k}\left[\frac{\partial \mathcal{L}}{\partial \left(\frac{\partial \eta_j}{\partial x_k}\right)}\right] - \frac{\partial \mathcal{L}}{\partial \eta_j} = 0 \quad j = 1, 2, 3 \quad (5.85)$$

This can be simplified in notation to:

$$\frac{\partial}{\partial t}\left(\frac{\partial \mathcal{L}}{\partial \dot{\eta}_j}\right) + \vec{\nabla} \cdot \left[\frac{\partial \mathcal{L}}{\partial (\vec{\nabla}\eta_j)}\right] - \frac{\partial \mathcal{L}}{\partial \eta_j} = 0 \quad j = 1, 2, 3 \quad (5.86)$$

Here the *Lagrangian density* $\mathcal{L}$ is a function of the following:

1. The four "field coordinates"

$$\eta_1 = A_1(x_1, x_2, x_3, t) \quad (5.87a)$$

5 Deriving the Lagrangian Density of an Electromagnetic Field

$$\eta_2 = A_2(x_1, x_2, x_3, t) \tag{5.87b}$$

$$\eta_3 = A_3(x_1, x_2, x_3, t) \tag{5.87c}$$

$$\eta_4 = A_4(x_1, x_2, x_3, t) \tag{5.87d}$$

2. Their time derivatives

$$\dot{\eta}_1 \equiv \frac{\partial \eta_1}{\partial t} = \frac{\partial A_1}{\partial t} \equiv \dot{A}_1 \tag{5.88a}$$

$$\dot{\eta}_2 \equiv \frac{\partial \eta_2}{\partial t} = \frac{\partial A_2}{\partial t} \equiv \dot{A}_2 \tag{5.88b}$$

$$\dot{\eta}_3 \equiv \frac{\partial \eta_3}{\partial t} = \frac{\partial A_3}{\partial t} \equiv \dot{A}_3 \tag{5.88c}$$

$$\dot{\eta}_4 \equiv \frac{\partial \eta_4}{\partial t} = \frac{\partial \phi}{\partial t} \equiv \dot{\phi} \tag{5.88d}$$

3. Their gradients

$$\nabla \eta_1 = \nabla A_1, \quad \nabla \eta_2 = \nabla A_2, \quad \nabla \eta_3 = \nabla A_3, \quad \nabla \eta_4 = \nabla \phi \tag{5.89}$$

We express Eqs. 5.81 and 5.83 in forms that are similar to Lagrange Eq. 5.86:

$$\boxed{\frac{\partial}{\partial t}\left(\vec{\nabla} \cdot \vec{A}\right) + \vec{\nabla} \cdot (\nabla \phi) - \left(-\frac{\rho}{\varepsilon_0}\right) = 0} \tag{5.90}$$

Finally, we can write:

$$\boxed{\frac{\partial}{\partial t}\left(\frac{\partial A_k}{\partial t} + \frac{\partial \phi}{\partial x_k}\right) + \vec{\nabla} \cdot \left[c^2\left(\frac{\partial \vec{A}}{\partial x_k} - \nabla A_k\right)\right] - \frac{j_k}{\varepsilon_0} = 0} \tag{5.91}$$

The Lagrange Eq. 5.86 for $j = 4$—that is, for $\eta_4 = \phi$ is given as:

$$\frac{\partial}{\partial t}\left(\frac{\partial \mathcal{L}}{\partial \dot{\phi}}\right) + \vec{\nabla} \cdot \left[\frac{\partial \mathcal{L}}{\partial (\nabla \phi)}\right] - \frac{\partial \mathcal{L}}{\partial \phi} = 0 \tag{5.92}$$

Comparing Eqs. 5.90 and 5.2, we note that the first could be derived from the second if,

$$\frac{\partial \mathcal{L}}{\partial \dot{\phi}} = \vec{\nabla} \cdot \vec{A}, \quad \frac{\partial \mathcal{L}}{\partial (\nabla \phi)} = \nabla \phi, \quad \frac{\partial \mathcal{L}}{\partial \phi} = -\frac{\rho}{\varepsilon_0} \tag{5.93}$$

so that Lagrangian density $\mathcal{L}$ must contain, respectively, the terms:

5.7 Euler-Lagrange Equation of the Electromagnetic Field

$$\mathcal{L}_{\alpha_1} \equiv (\nabla \cdot \vec{A})\dot{\phi}, \quad \mathcal{L}_{\alpha_2} \equiv \frac{1}{2}\|\nabla\phi\|^2, \quad \mathcal{L}_{\alpha_3} \equiv -\frac{\rho\phi}{\varepsilon_0} \tag{5.94}$$

Consequently, their sum becomes:

$$\mathcal{L}_\alpha = \mathcal{L}_{\alpha_1} + \mathcal{L}_{\alpha_2} + \mathcal{L}_{\alpha_3} = (\vec{\nabla} \cdot \vec{A})\dot{\phi} + \frac{1}{2}\|\nabla\phi\|^2 - \frac{\rho\phi}{\varepsilon_0} \tag{5.95}$$

We suppose that an appropriate Lagrangian density $\mathcal{L}$ would be of the form:

$$\mathcal{L} = \mathcal{L}_\alpha + \mathcal{L}_\beta \tag{5.96}$$

and because $\mathcal{L}_\alpha$ produces Eq. 5.90, we expect that $\mathcal{L}_\beta$—to be determined—will produce Eq. 5.91. This expectation would be correct if Eqs. 5.90 and 5.91 were decoupled; for example, if the first contains only ϕ-terms and the second contains only $\vec{A}$-terms. But, here this is not the case: $\mathcal{L}_\alpha$ containing $\vec{A}$-terms would participate to the production of Eq. 5.91; moreover, $\mathcal{L}_\beta$ would participate in the production of Eq. 5.90. It is possible that this could mutually destroy the production of the equations as we had expected. Nonetheless, here we follow a trial-and-error procedure, which will direct us to the correct answer, as we can see in the following.

Now Lagrangian Eq. 5.86 for $j = k = 1, 2, 3$—that is, for $\eta_k = A_k$—is:

$$\frac{\partial}{\partial t}\left(\frac{\partial \mathcal{L}}{\partial \dot{A}_k}\right) + \nabla \cdot \left[\frac{\partial \mathcal{L}}{\partial (\nabla A_k)}\right] - \frac{\partial \mathcal{L}}{\partial A_k} = 0 \tag{5.97}$$

Comparing Eqs. 5.91 and 5.97, we note that the first could be derived from the second if,

$$\frac{\partial \mathcal{L}}{\partial \dot{A}_k} = \dot{A}_k + \frac{\partial \phi}{\partial x_k}, \quad \frac{\partial \mathcal{L}}{\partial (\nabla A_k)} = c^2\left(\frac{\partial \vec{A}}{\partial x_k} - \nabla A_k\right), \quad \frac{\partial \mathcal{L}}{\partial A_k} = \frac{j_k}{\varepsilon_0} \tag{5.98}$$

From the first part of Eq. 5.98 the $\mathcal{L}_\beta$ of Lagrange density $\mathcal{L}$ must contain the following terms:

$$\frac{1}{2}\|\dot{A}_k\|^2 + \frac{\partial \phi}{\partial x_k}\dot{A}_k, \quad k = 1, 2, 3 \tag{5.99}$$

So their sum with respect to k is:

$$\mathcal{L}_{\beta_1} \equiv \frac{1}{2}\|\dot{\vec{A}}\|^2 + \nabla\phi \cdot \dot{\vec{A}} \tag{5.100}$$

From the second part in Eq. 5.98, the $\mathcal{L}_\beta$ of Lagrange density $\mathcal{L}$ must contain the terms as follows:

$$\frac{1}{2}c^2\left[\frac{\partial \vec{A}}{\partial x_k} \cdot \nabla A_k - \|\nabla A_k\|^2\right] \quad k = 1, 2, 3 \tag{5.101}$$

So their sum with respect to k turns into this format:

$$\mathcal{L}_{\beta_2} \equiv \frac{1}{2}c^2 \sum_{k=1}^{k=3} \left[\frac{\partial \vec{A}}{\partial x_k} \cdot \nabla A_k - \|\nabla A_k\|^2 \right] \quad (5.102)$$

From the third part in Eq. 5.98, the $\mathcal{L}_\beta$ of Lagrange density $\mathcal{L}$ must contain the terms:

$$\frac{j_k A_k}{\varepsilon_0} \quad k = 1, 2, 3 \quad (5.103)$$

Therefore, their sum, with respect to k, becomes:

$$\mathcal{L}_{\beta_3} \equiv \frac{\vec{j} \cdot \vec{A}}{\varepsilon_0} \quad (5.104)$$

From Eqs. 5.101, 5.103, and 5.104 the $\mathcal{L}_\beta$ part of Lagrange density $\mathcal{L}$ is:

$$\begin{aligned}\mathcal{L}_\beta &= \mathcal{L}_{\beta_1} + \mathcal{L}_{\beta_2} + \mathcal{L}_{\beta_3} \\ &= \frac{1}{2}\|\dot{\vec{A}}\|^2 + \nabla \phi \cdot \dot{\vec{A}} + \frac{1}{2}c^2 \sum_{k=1}^{k=3}\left[\frac{\partial \vec{A}}{\partial x_k} \cdot \nabla A_k - \|\nabla A_k\|^2\right] + \frac{\vec{j} \cdot \vec{A}}{\varepsilon_0}\end{aligned} \quad (5.105)$$

Finally, from Eqs. 5.96 and 5.104 for densities $\mathcal{L}_\alpha$ and $\mathcal{L}_\beta$, Lagrange density $\mathcal{L} = \mathcal{L}_\alpha + \mathcal{L}_\beta$ is given as:

$$\begin{aligned}\mathcal{L} &= \mathcal{L}_\alpha + \mathcal{L}_\beta \\ &= (\nabla \cdot \vec{A})\dot{\phi} + \frac{1}{2}\|\nabla \phi\|^2 - \frac{\rho\phi}{\varepsilon_0} \\ &+ \frac{1}{2}\|\dot{\vec{A}}\|^2 + \nabla \phi \cdot \dot{\vec{A}} + \frac{1}{2}c^2 \sum_{k=1}^{k=3}\left[\frac{\partial \vec{A}}{\partial x_k} \cdot \nabla A_k - \|\nabla A_k\|^2\right] + \frac{\vec{j} \cdot \vec{A}}{\varepsilon_0}\end{aligned} \quad (5.106)$$

Yet, this is a "wrong" Lagrange density; thus, we need to perform the following error-trial-final success procedure (see next section).

5.7.1 Error-Trial-Final Success

Insertion of this Lagrange density expression into the Lagrange equation with respect to ϕ (i.e., Eq. 5.92) does not yield Eq. 5.81 but gets:

$$-\nabla^2 \phi - \frac{\partial}{\partial t}(2\nabla \cdot \vec{A}) = \frac{\rho}{\varepsilon_0} \quad \text{(wrong)} \quad (5.107)$$

5.7 Euler-Lagrange Equation of the Electromagnetic Field

The appearance of an extra $(\nabla \cdot \vec{A})$ is because of the term $(\nabla \phi \cdot \dot{\vec{A}})$ of $\mathcal{L}_\beta$ and that is why the Lagrange density given by Eq. 5.106 is not an appropriate one. To resolve this problem, we must look at Eq. 5.90 (i.e., Eq. 5.81) from a different point of view as follows:

$$\nabla \cdot \left(\nabla \phi + \dot{\vec{A}} \right) - \left(-\frac{\rho}{\varepsilon_0} \right) = 0 \tag{5.108}$$

Comparing Eqs. 5.108 and 5.93, we note that the first could be derived from the second if in place of Eq. 5.93 we have the following result:

$$\frac{\partial \mathcal{L}}{\partial \dot{\phi}} = 0, \quad \frac{\partial \mathcal{L}}{\partial (\nabla \phi)} = \nabla \phi + \dot{\vec{A}}, \quad \frac{\partial \mathcal{L}}{\partial \phi} = -\frac{\rho}{\varepsilon_0} \tag{5.109}$$

So in place of Eqs. 5.94 and Eq. 5.95, respectively, the new form of the equations is as follows:

$$\mathcal{L}'_{\alpha_1} \equiv 0, \quad \mathcal{L}'_{\alpha_2} \equiv \frac{1}{2}\|\nabla \phi\|^2 + \nabla \phi \cdot \dot{\vec{A}}, \quad \mathcal{L}'_{\alpha_3} \equiv \mathcal{L}_{\alpha_3} \equiv -\frac{\rho \phi}{\varepsilon_0} \tag{5.110}$$

$$\mathcal{L}'_\alpha = \mathcal{L}'_{\alpha_1} + \mathcal{L}'_{\alpha_2} + \mathcal{L}'_{\alpha_3} = \frac{1}{2}\|\nabla \phi\|^2 + \left(\nabla \phi \cdot \dot{\vec{A}} \right) - \frac{\rho \phi}{\varepsilon_0} \tag{5.111}$$

Now it is necessary to omit from $\mathcal{L}_{\beta_1}$, Eq. 5.100, the second term $\left(\nabla \phi \cdot \dot{\vec{A}} \right)$ because it appears in $\mathcal{L}'_{\alpha_2}$ (see the second term of Eq. 5.110). Thus, we have in place of Eq. 5.100, the following form:

$$\mathcal{L}'_{\beta_1} \equiv \frac{1}{2}\|\dot{\vec{A}}\|^2 \tag{5.112}$$

Although $\mathcal{L}_{\beta_2}$ and $\mathcal{L}_{\beta_3}$ remain unchanged, as in Eqs. 5.102 and 5.104, we have:

$$\mathcal{L}'_{\beta_2} = \mathcal{L}_{\beta_2} \equiv \frac{1}{2}c^2 \sum_{k=1}^{k=3} \left[\frac{\partial \vec{A}}{\partial x_k} \cdot \nabla A_k - \|\nabla A_k\|^2 \right] \tag{5.113}$$

$$\mathcal{L}'_{\beta_3} = \mathcal{L}_{\beta_3} \equiv \frac{\vec{j} \cdot \vec{A}}{\varepsilon_0} \tag{5.114}$$

In place of Eq. 5.105 we now can write:

$$\mathcal{L}'_\beta = \mathcal{L}'_{\beta_1} + \mathcal{L}'_{\beta_2} + \mathcal{L}'_{\beta_3}$$

$$= +\frac{1}{2}\|\dot{\vec{A}}\|^2 + \frac{1}{2}c^2 \sum_{k=1}^{k=3} \left[\frac{\partial \vec{A}}{\partial x_k} \cdot \nabla A_k - \|\nabla A_k\|^2 \right] + \frac{\vec{j} \cdot \vec{A}}{\varepsilon_0} \tag{5.115}$$

Finally, for the new Lagrangian density we have the following in place of Eq. 5.106:

$$\mathcal{L}' = \mathcal{L}'_\alpha + \mathcal{L}'_\beta$$
$$= \frac{1}{2}\|\nabla\phi\|^2 + \nabla\phi \cdot \dot{\vec{A}} - \frac{\rho\phi}{\varepsilon_0} \qquad (5.116)$$
$$\frac{1}{2}\|\dot{\vec{A}}\|^2 + \frac{1}{2}c^2 \sum_{k=1}^{k=3}\left[\frac{\partial \vec{A}}{\partial x_k} \cdot \nabla A_k - \|\nabla A_k\|^2\right] + \frac{\vec{j}\cdot\vec{A}}{\varepsilon_0}$$

Density $\mathcal{L}'$ of Eq. 5.116 is obtained from density $\mathcal{L}$ of Eq. 5.106 if we omit the term $(\nabla \cdot \vec{A})\dot{\phi}$. Thus, $\mathcal{L}'$ is *independent* of $\dot{\phi}$.

In the following equations the brace over the left three terms groups that part of density $\mathcal{L}'$, which essentially participates in the production of electromagnetic Eq. 5.81 from the Lagrange equation with respect to ϕ, Eq. 5.92. Although the brace under the right four terms groups that part of density $\mathcal{L}'$, which essentially participates in the production of electromagnetic Eq. 5.83 from the Lagrange equations with respect to A_1, A_2, A_3, Eq. 5.87. Thus, we have:

$$\overbrace{\mathcal{L}' = \frac{1}{2}\|\nabla\phi\|^2 - \frac{\rho\phi}{\varepsilon_0} + \nabla\phi\cdot\dot{\vec{A}}}^{\text{with respect to } \phi} + \frac{1}{2}\|\dot{\vec{A}}\|^2 + \frac{1}{2}c^2\sum_{k=1}^{k=2}\left[\frac{\partial\vec{A}}{\partial x_k}\cdot\nabla A_k - \|\nabla A_k\|^2\right] + \frac{\vec{j}\cdot\vec{A}}{\varepsilon_0}$$

$$\mathcal{L}' = \frac{1}{2}\|\nabla\phi\|^2 - \frac{\rho\phi}{\varepsilon_0} + \underbrace{\nabla\phi\cdot\dot{\vec{A}} + \frac{1}{2}\|\dot{\vec{A}}\|^2 + \frac{1}{2}c^2\sum_{k=1}^{k=2}\left[\frac{\partial\vec{A}}{\partial x_k}\cdot\nabla A_k - \|\nabla A_k\|^2\right] + \frac{\vec{j}\cdot\vec{A}}{\varepsilon_0}}_{\text{with respect to } \vec{A}}$$

$$(5.117)$$

Note the common term $(\nabla\phi\cdot\dot{\vec{A}})$.

Recording the terms in Eq. 5.116 of density $\mathcal{L}'$, we have:

$$\mathcal{L}' = \underbrace{\frac{1}{2}\|\dot{\vec{A}}\|^2 + \frac{1}{2}\|\nabla\phi\|^2 + \nabla\phi\cdot\dot{\vec{A}}}_{\frac{1}{2}\|-\nabla\phi-\frac{\partial\vec{A}}{\partial t}\|^2} - \underbrace{\frac{1}{2}c^2\sum_{k=1}^{k=3}\left[\|\nabla A_k\|^2 - \frac{\partial\vec{A}}{\partial x_k}\cdot\nabla A_k\right]}_{\|\nabla\times\vec{A}\|^2} + \frac{1}{\varepsilon_0}(-\rho\phi + \vec{j}\cdot\vec{A})$$

$$(5.118)$$

that is,

$$\mathcal{L}' = \frac{1}{2}\left\|-\nabla\phi - \frac{\partial\vec{A}}{\partial t}\right\|^2 - \frac{1}{2}c^2\|\nabla\times A\|^2 + \frac{1}{\varepsilon_0}(-\rho\phi + \vec{j}\cdot\vec{A}) \qquad (5.119)$$

or

$$\mathcal{L}' = \frac{\|\vec{E}\|^2 - c^2\|\vec{B}\|^2}{2} + \frac{1}{\varepsilon_0}\left(-\rho\phi + \vec{j}\cdot\vec{A}\right) \qquad (5.120)$$

Now if density $\mathcal{L}'$ must have dimensions of energy per unit volume, we define $\mathcal{L}_{em} = \varepsilon_0 \mathcal{L}'$; therefore, we can write the following result:

$$\boxed{\mathcal{L}_{em} = \varepsilon_0 \left[\frac{\|\vec{E}\|^2 - c^2\|\vec{B}\|^2}{2}\right] - \rho\phi + \vec{j}\cdot\vec{A}} \qquad (5.121)$$

keeping in mind that

$$\|\vec{E}\|^2 = \left\|-\nabla\phi - \frac{\partial \vec{A}}{\partial t}\right\|^2 = \|\dot{\vec{A}}\|^2 + \|\nabla\phi\|^2 + 2(\nabla\phi \cdot \dot{\vec{A}}) \qquad (5.122a)$$

$$\|\vec{B}\|^2 = \|\vec{\nabla}\times\vec{A}\|^2 = \sum_{k=1}^{k=3}\left[\|\nabla A_k\|^2 - \frac{\partial \vec{A}}{\partial x_k}\cdot \nabla A_k\right] \qquad (5.122b)$$

The scalar $\left(\|\vec{E}\|^2 - c^2\|\vec{B}\|^2\right)$ is one of the two *Lorentz invariants* (e.g., see subsection in the preceding description of the field); the other one is $\vec{E}\cdot\vec{B}$. The scaler essentially is equal to constant time $\varepsilon_{\mu\nu}e^{\mu\nu}$, where $e^{\mu\nu}$ is the anti-symmetric field tensor (see next subsection as well). (More details can be found in Appendix A under the topic Relativity and Electromagnetism.)

On the other hand, the scalar $\left(-\rho\phi + \vec{j}\cdot\vec{A}\right)$ is essentially the inner product $J_\mu A^\mu$ of two 4-vectors in *Minkowski space*. The four-current density $J^\mu = (c\rho, \vec{J})$ and the four-potential $A^\mu = (\phi/c, \vec{A})$ is a Lorentz invariant scalar too. Therefore, Lagrangian density $\mathcal{L}_{em}$ in Eq. 5.6 is a Lorentz invariant.

5.8 Formal Structure of Maxwell's Theory

Magnetic theory and its set of equations, in essence, now stands before us. To make the connection with its usual three-dimensional formulation in terms of *electric field* $\vec{E}$ and *magnetic field* $\vec{B}$, we define $\vec{E}$ and $\vec{B}$ in each inertial frame by the first of the following equations:

$$\mathcal{E}_{\mu\nu} = \begin{bmatrix} 0 & E_1 & E_2 & E_3 \\ -E_1 & 0 & -cB_3 & cB_2 \\ -E_2 & cB_3 & 0 & -cB_1 \\ -E_3 & -cB_2 & cB_1 & 0 \end{bmatrix}$$

so (5.123)

$$\mathcal{E}^{\mu\nu} = \begin{bmatrix} 0 & -E_1 & -E_2 & -E_3 \\ E_1 & 0 & -cB_3 & cB_2 \\ E_2 & cB_3 & 0 & -cB_1 \\ E_3 & -cB_2 & cB_1 & 0 \end{bmatrix}$$

which, by making the duality replacement $\vec{E} \to c\vec{B}$ and $c\vec{B} \to \vec{E}$, yields in the following tensor set:

$$\mathcal{B}_{\mu\nu} = \begin{bmatrix} 0 & -cB_1 & -cB_2 & -cB_3 \\ cB_1 & 0 & -E_3 & E_2 \\ cB_2 & cE_3 & 0 & -E_1 \\ cB_3 & -E_2 & E_1 & 0 \end{bmatrix}$$

so (5.124)

$$\mathcal{B}^{\mu\nu} = \begin{bmatrix} 0 & cB_1 & cB_2 & -cB_3 \\ E_1 & 0 & -cB_3 & cE_2 \\ E_2 & cB_3 & 0 & -E_1 \\ E_3 & -cB_2 & cB_1 & 0 \end{bmatrix}$$

The two *invariants* of $\mathcal{E}^{\mu\nu}$—immediately recognizable as such from their mode of formation—can be expressed as follows:

$$\begin{cases} X = \dfrac{1}{2}\mathcal{E}_{\mu\nu}\mathcal{E}^{\mu\nu} = -\dfrac{1}{2}\mathcal{B}_{\mu\nu}\mathcal{B}^{\mu\nu} = c^2\|\vec{B}\|^2 - \|\vec{E}\|^2 \\ Y = \dfrac{1}{2}\mathcal{B}_{\mu\nu}\mathcal{E}^{\mu\nu} = c\vec{B}\cdot\vec{E} \end{cases} \quad (5.125)$$

But if we expand the first equation into the form of *Stueckelberg-Lagrangian density* in the following equation form, we have [1]:

$$\mathcal{L} = -\frac{\varepsilon c^2}{4}F_{\mu\nu}F^{\mu\nu} + J_\mu A^\mu - \frac{\gamma\varepsilon c^2}{2}\left(\partial_\mu A^\mu\right)^2 - \frac{\varepsilon c^2 k^2}{2}\left(A_\mu A^\mu\right) \quad (5.126)$$

Note that this equation is known as the Lagrangian density equation, or what we call the *More Complete Electromagnetic* (MCE) theory in order to generate a *scalar longitudinal wave* (SLW); more information about such matters is addressed in Chap. 6 of this book.

Equation 5.126 will be written in a general form based on the following argument. That is, the *Helmholtz theorem* (e.g., Griffiths, 2007 [2]) states that any three-

5.8 Formal Structure of Maxwell's Theory

dimensional vector (electrical current density $\vec{J}$, in this case) can be uniquely decomposed into circulating- and gradient-driven components, thus $\vec{J} = \nabla \kappa + \vec{\nabla} \times \vec{G}$. Here κ and $\vec{G}$ are scalar and vector functions of spacetime, respectively.

Using the result of Woodside [4] provides a unique decomposition for a smooth *Minkowski field*, and using only *4-vetor geometry* will produce a new form of Maxwell's equations as follows:

$$\vec{E} = -\nabla \phi - \frac{\partial \vec{A}}{\partial t} \tag{5.127}$$

$$\vec{B} = \vec{\nabla} \times \vec{A} \tag{5.128}$$

$$C \equiv \nabla \cdot \vec{A} + \frac{1}{c^2} \frac{\partial \phi}{\partial t} \tag{5.129}$$

$$\nabla \times \vec{B} - \frac{1}{c^2} \frac{\partial \vec{E}}{\partial t} \underbrace{-\vec{\nabla} \cdot \vec{A} + \frac{1}{c^2} \frac{\partial \phi}{\partial t}}_{C} = \mu \vec{J} \tag{5.130}$$

$$\vec{\nabla} \cdot \vec{E} + \frac{\partial}{\partial t} \underbrace{\left\{ \vec{\nabla} \cdot \vec{A} + \frac{1}{c^2} \frac{\partial \phi}{\partial t} \right\}}_{C} = \frac{\rho}{\varepsilon} \tag{5.131}$$

$\vec{A}$ and ϕ are the classical vector and scalar potentials, respectively. $\vec{B}$ and $\vec{E}$ are the magnetic and electric fields, respectively, as well. In the preceding set of Maxwell's equations, constant C is a non-zero dynamic field with the *same dimensions* as the magnetic field.

Bear in mind that in classical electrodynamics, $C = 0$ is known as the *Lorentz gauge*. Also, in the previous set of equations ρ and $\vec{J}$ are electric charge density and current charge density, respectively, while ε and μ represent the permittivity and permeability of media, not necessarily of the vacuum, respectively.

The homogeneous equations $\nabla \cdot \vec{B} = 0$ and $\vec{\nabla} \times \vec{E} + (\partial \vec{B}/\partial t) = 0$ are identical to the classical model, yielding Eqs. 5.129 and 5.130. Using Eqs. 5.127 through 5.131, the most general Lagrangian density for a mass-less, 4-vector field (A^μ), which is no more than quadratic in its variables and derivatives, is given by the following equation based on Eq. 2 in Stueckelberg's 1938 article [1]—that is, Stueckelberg-Lagrangian density:

$$\mathcal{L} = \frac{\varepsilon c^2}{2} \left[\frac{1}{c^2} \left(\nabla \phi + \frac{\partial \vec{A}}{\partial t} \right)^2 - (\nabla \times \vec{A})^2 \right] - \rho \phi + \vec{J} \cdot \vec{A} - \frac{\gamma \varepsilon c^2}{2} \left(\vec{\nabla} \cdot \vec{A} + \frac{1}{c^2} \frac{\partial \phi}{\partial t} \right)^2 \tag{5.132}$$

The result we obtained in Eq. 5.126 will be applied to generate a SLW, where we use it as a Lagrangian density equation for what we call the MCE theory.

References

1. F.C.G. Stueckelberg, Interaction forces in electrodynamics and in field theory of nuclear forces. Helv. Phys. Acta. **11**, 299–328 (1938)
2. D. Griffiths, *Introduction to electrodynamics*, 3rd edn. (Prentice Hall, Upper Saddle River, 1999)
3. C.K. Law, Interaction between a moving mirror and radiation pressure: A Hamiltonian formulation. Phys. Rev. A **51**, 2537 (1995)
4. D.A. Woodside, Three-vector and scalar field identities and uniqueness theorems in Euclidean and Minkowski spaces. Am. J. Phys. **77**, 438–446 (2009)

Chapter 6
Scalar Waves

There is wide confusion about what "scalar waves" are both in serious and less serious literature on electrical engineering. This chapter explains that this type of wave is a longitudinal wave of potentials. It has been shown that a longitudinal wave is a combination of a vector potential with a scalar potential. There is a full analogue to acoustic waves. Transmitters and receivers for longitudinal electromagnetic waves are discussed. Scalar waves were found and used at first by Nikola Tesla in his wireless energy transmission experiments. The SW is the extension of the Maxwell's equation part that we call the More Complete Electromagnetic equation, as described herein.

6.1 Introduction

It is the purpose of this chapter to discuss a new unified field theory based on the work of Tesla. This *unified field* and particle theory describes quantum and classical physics, mass, gravitation, the constant speed of light, neutrinos, waves, and particles—all of which can be explained by vortices [1]. In addition, we discuss unique and various recent inventions and their possible modes of operation in order to convince those studying this of their value for hopefully directing a future program geared toward the rigorous clarification and certification of the specific role the electroscalar domain, and how it might play a role in shaping a future consistent with classical electrodynamics. Also, by extension, to perhaps shed light on inconsistencies that do exist within current conceptual and mathematical theories in the present interpretation of relativistic quantum mechanics. In this regard, we anticipate that by incorporating this more expansive electrodynamic model, the source of the extant problems with gauge invariance in quantum electrodynamics and the subsequent unavoidable divergences in energy/charge might be identified and ameliorated.

Not only does the electroscalar domain have the potential to address such lofty theoretical questions surrounding fundamental physics, but another aim in this chapter is to show that the protocol necessary for generating these field effects may not be present only in exotic conditions involving large field strengths and specific frequencies involving expensive infrastructures such as the *large hadron collider* (LHC). As recent discoveries suggest, however, SWs may be present in the physical manipulation of everyday objects. We also will explain that nature has been and may be engaged in the process of using *scalar longitudinal waves* (SLWs) in many ways yet unsuspected and undetected by humanity. Some of the modalities of SW generation we will investigate include chemical bond-breaking, particularly as a precursor to seismic events (i.e., illuminating the study and development of earthquake early warning systems); solar events (i.e., related to eclipses); and sunspot activity and how it affects the Earth's magnetosphere. Moreover, this overview of the unique aspects of the electroscalar domain suggests that many of the currently unexplained anomalies—for example, overunity power observed in various energy devices and exotic energy effects associated with *low-energy nuclear reactions* (LENRs)—may find some basis in fact.

In regard to the latter, cold fusion or LENR fusion-type scenarios, the *electroscalar wave* might be the actual agent needed to reduce the nuclear Coulomb barrier, thereby providing the long sought for viable theoretical explanation of this phenomenon [2]. Longitudinal electrodynamic forces (e.g., in exploding wires) actually may be because of the operation of electroscalar waves at subatomic levels of nature. For instance, the extraordinary energies produced by Ken Shoulder's charge clusters (i.e. Particles of like charge repel each other - that is one of the laws describing the interaction between single sub-atomic particles) perhaps may be because of electroscalar mechanisms.

Moreover, these observations, spanning as they do many cross-disciplines of science, beg the question as to the possible universality of the SLW—that is, the concept of the longitudinal electroscalar wave, not present in current electrodynamics, may represent a general key, overarching principle, leading to new paradigms in other sciences besides physics. This idea also will be explored in the chapter, showing the possible connection of scalar–longitudinal (i.e., electroscalar) wave dynamics to biophysical systems. Admittedly, we are proposing quite an ambitious agenda in reaching for these goals, but we think you will see that recent innovations may have proved equal to the task of supporting this quest.

6.2 Descriptions of Transverse and Longitudinal Waves

As you know from a classical physics point of view, typically the following are three kinds of waves—the *soliton wave* is an exceptional case and should be addressed separately—and wave equations that we can talk about:

1. Mechanical waves (i.e., waves on string)
2. Electromagnetic (EM) waves (i.e., $\vec{E}$ and $\vec{B}$ fields from Maxwell's equations to deduce the wave equations, where these waves carry energy from one place to another)
3. Quantum mechanical waves (i.e., using Schrödinger equations to study particles' movements)

6.2 Descriptions of Transverse and Longitudinal Waves

The second one is our subject of interest in terms of te two types of waves involved in EM waves: (1) transverse waves and (2) *longitudinal pressure waves* (lpws), also known as *scalar longitudinal waves* (slws).

From the preceding two waves, the SLW is of interest in directed energy weapons (DEWs) [3] and here is why. First, we briefly describe SLWs and their advantages for DEW purposes as well as communication within non-homogeneous media such as seawater with different *electrical primitivity* ε and *magnetic permeability* μ at various ocean depths (see Chap. 4 of this book).

A wave is defined as a disturbance that travels through a certain medium. The medium is material through which a wave moves from one to another location. If we take as an example a slinky coil that can be stretched from one end to the other, a static condition then develops . This static condition is called the wave's *neutral condition* or equilibrium state.

In the slinky coil the particles are moved up and down then come into their equilibrium state. This generates disturbance in the coil that is moved from end one to the other. This is the movement of a *slinky pulse*, which is a *single disturbance in medium* from one to another location. If it is done *continuously* and in a *periodical manner*, then it is called a *wave,* also known as an *energy transport medium*. They are found in diverse shapes, show a variety of behaviors, and have characteristic properties. On this basis, they are classified mainly as longitudinal, transverse, and surface waves. Here we discuss the properties of LWs and provide examples. The movement of waves is parallel to the medium of the particles in them.

1. **Transverse Waves**

For TWs the medium is displaced perpendicular to the wave's direction of propagation. A ripple on a pond and a wave on a string are visualized easily as TWs (Fig. 6.1).

TWs cannot propagate in a gas or a liquid because there is no mechanism for driving motion perpendicular to the propagation of the wave. In summary, a transverse wave (TW) is a wave in which the oscillation is perpendicular to the direction of wave propagation. Electromagnetic waves (and secondary waves, S-waves or shear waves, sometimes called elastic S-waves) in general are TWs.

2. **Longitudinal Waves**

In a LW the displacement of the medium is parallel to the propagation of the wave. A wave in a slinky is a good visualization. Sound waves in air are LWs (Fig. 6.2).

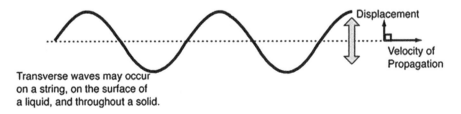

Fig. 6.1 Depiction of a transverse wave

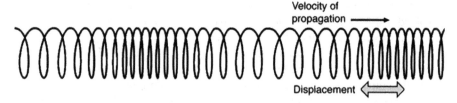

Fig. 6.2 Depiction of a longitudinal wave

In summary, an LW is a wave in which the oscillation is in an opposite direction to the direction of wave propagation. Sound waves—that is, primary waves, or P-waves, in general—are LWs. On the other hand, a wave with a motion that occurs through the particles of the medium oscillating about their mean positions in the direction of propagation is called an LW.

We use our knowledge to expand the subject of the *longitudinal wave* before we go deeper into the subject of the *scalar longitudinal wave*; for an LW the particles of the medium vibrate in the direction of wave propagation. An LW proceeds in the form of compression and rarefaction, which is stretch and compression in the same direction as the wave moves. For an LW at places of compression the pressure and density tend to be maximal, whereas at places where rarefaction takes place, the pressure and density are minimal. In gases only an LW can propagate; LWs are known as *compression waves*.

An LW travels through a medium in the form of compressions or condensations, C, and rarefaction, R. A compression is a region of the medium in which particles are compressed (i.e., particles come closer); in other words, the distance between the particles becomes less than the normal distance between them. Thus, volume temporarily decreases and, therefore the density of the medium increases in the region of compression. A *rarefaction* is a region of the medium in which particles are rarefied (i.e., particles get farther apart than what they normally are). Thus, volume temporary increases and, consequently, the density of the medium decreases in the region of rarefaction.

The distance between the centers of two consecutive rarefactions and two consecutive compressions is called *wavelength*. Examples of LWs are sound waves, tsunami waves, earthquake P-waves, ultra-sounds, vibrations in gas, and oscillations in springs internal water waves, waves in slink, and so on.

(a) *Longitudinal waves*

Examples of the various types of waves are:

1. Sound wave
2. Earthquake *P*-wave
3. Tsunami wave
4. Waves in a slinky
5. Glass vibrations

6.2 Descriptions of Transverse and Longitudinal Waves

6. Internal water waves
7. Ultra-sound
8. Spring oscillations

(b) *Sound waves*

Now the question is: *Are sound waves longitudinal?* The answer is *Yes*. A sound wave travels as an LW in nature. It behaves as a TW in solids. Through gases, plasma, and liquids the sound travels as an LW. Through solids the wave can be transmitted as a TW or an LW.

A material medium is mandatory for the propagation of the sound waves. They mostly are longitudinal in common nature. The speed of sound in air is 332 m/s at normal temperature and pressure. Vibrations of an air column above the surface of water in the tube of a resonance apparatus are longitudinal. Vibrations of an air column in organ pipes are longitudinal. Sound is audible only between 20 Hz and 20 KHz. Sound waves cannot be polarized.

(i) *Propagation of sound waves in air*: Sound waves are classified as LWs. Let us now see how sound waves propagate. Take a tuning fork, vibrate it, and concentrate on the motion of one of its prongs, say prong A. The normal position of the tuning fork and the initial condition of air particles is shown in Fig. 6.3a. As prong A moves toward the right, it compresses air particles near it, forming a compression as shown in Fig. 6.3b. Because of vibrating air layers, this compression moves forward as a disturbance.

As prong A moves back to its original position, the pressure on its right decreases, thereby forming a rarefaction. This rarefaction moves forward like compression as a disturbance. As the tuning fork goes on vibrating, waves consisting of alternate compressions and rarefactions spread in the air as shown in Fig. 6.3c,d. The direction of motion of the sound waves is the same as that of air particles, thus they are classified as LWs. The LWs travel in the form of compressions and rarefactions.

The main parts of the sound wave follow, along with descriptions:

1. *Amplitude*: The maximum displacement of a vibrating particle of the medium from the mean position. A shows amplitude in $y = A \sin(\omega t)$. The maximum height of the wave is called its amplitude. If the sound is more, then the amplitude is more.
2. *Frequency*: The number of vibrations made per second by the particles and is denoted by f, which is given as $f = 1/T$ and its unit is Hz. We also can get the expression for angular frequency.
3. *Pitch*: It is characteristic of sound with the help of which we can distinguish between a *shrill* note and a note that is *grave*. When a sound is shriller, it is said to be of higher pitch and is found to be of greater frequency, as $\omega = 2\pi f$. On the other hand, a grave sound is said to be of low pitch and is of low frequency. Therefore the pitch of a sound depends on its frequency. It should be made clear that pitch is not the frequency but changes with frequency.

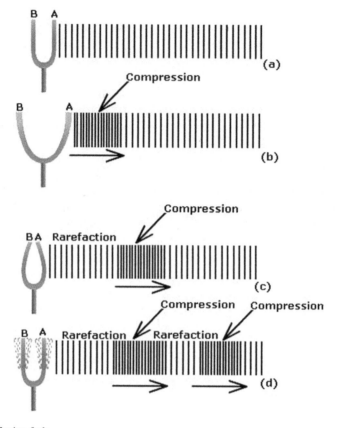

Fig. 6.3 Tuning fork

4. *Wavelength*: The distance between two consecutive particles in the same phase or the distance traveled by the wave in one periodic time and denoted by lambda, Λ.
5. *Sound wave*: This is a LW with regions of compression and rarefaction. The increase of pressure above its normal value may be written as:

$$\sum p = \sum p_0 \sin \omega \left(t - \frac{c}{v}\right) \tag{6.1}$$

where

$\sum p$ = increase in pressure at x position at time t
$\sum p_0$ = maximum increase in pressure
$\omega = 2\pi f$ where f is frequency

If $\sum p$ and $\sum p_0$ are replaced by P and P_0, then Eq. 6.1 has the following form:

$$P = P_0 \sin \omega \left(t - \frac{c}{v}\right) \tag{6.2}$$

6.2 Descriptions of Transverse and Longitudinal Waves

(ii) *Sound intensity*: Loudness of sound is related to the intensity of it. The sound's intensity at any point may be defined as the amount of sound energy passing per unit time per unit area around that point in a perpendicular direction. It is a physical quantity and is measured in W/m² in S.I. units.

The sound wave falling on the ear drum of the observer produces the sensation of hearing. The sound's sensation, which enables us to differentiate between a loud and a faint sound, is called *loudness*, and we can designate it by symbol L. It depends on the intensity of the sound I and the sensitivity of the ear of the observer at that place. The lowest intensity of sound that can be perceived by the human ear is called the *threshold of hearing*, denoted by I_0. The mathematical relation between intensity and loudness is:

$$L = \log \frac{I}{I_0} \qquad (6.3)$$

The intensity of sound depends on:

Amplitude of vibrations of the source
Surface area of the vibrating source
Distance of the source from the observer
Density of the medium in which sound travels from the source
Presence of other surrounding bodies
Motion of the medium

(iii) *Sound reflection*: When a sound wave gets reflected from a rigid boundary, the particles at the boundary are unable to vibrate. Thus, the generation of a reflected wave takes place, which interferes with the oncoming wave to produce zero displacement at the rigid boundary. At the points where there is zero displacement, the variation in pressure is at a maximum. This shows that the phase of the wave has been reversed, but the nature of the sound wave does not change (i.e., on reflection the compression is reflected as compression and rarefaction as rarefaction. Let the incident wave be represented by the given equation:

$$Y = a \sin(\omega t - kx) \qquad (6.4)$$

Then the Eq. 6.4 of reflected wave takes the form

$$Y = a \sin(\omega t + kx + \pi) = -a \sin(\omega t + kx) \qquad (6.5)$$

Here in both Eqs. 6.4 and 6.5 the symbol of a is basically the designation of the amplitude of the reflected wave.

A sound wave also is reflected if it encounters a rarer medium or free boundary or low-pressure region. A common example is the traveling of a sound wave in a narrow open tube. On reaching an open end, the wave gets reflected. So the force exerted on the particles there because of the outside air is quite small and, therefore, the particles vibrate with increasing amplitude. Because of this the pressure there tends to remain at the average value. This means that there is no alteration in the phase of the wave, but the ultimate nature of the wave has been altered (i.e., on the reflection of the wave the compression is reflected as rarefaction and vice versa).

The amplitude of the reflected wave would be a' this time and Eq. 6.4 becomes:

$$y = a' \sin(\omega t + kx) \tag{6.6}$$

(c) *Wave interface*: When listening to a single sine wave, amplitude is directly related to loudness and frequency and directly related to pitch. When there are two or more simultaneously sounding sine waves, wave interference takes place. There are basically two types of wave interference: (1) constructive and (2) destructive.

(d) *Decibel*: A smaller and practical unit of loudness is a decibel (dB) and is defined as follows:

$$1 \text{ Decibel} = \frac{1}{10} \text{ bel} \tag{6.7}$$

In dB the loudness of a sound of intensity I is given by

$$L = 10 \log \left(\frac{I}{I_0}\right) \tag{6.8}$$

(e) *Timber*: Timber can be called the property that distinguishes two sounds and makes them different from each other, even when they have the same frequency. For example, when we play violin and guitar on the same note and same loudness, the sound is still different. It also is denoted as *tone color*.

(f) *S-waves*: An *S*-wave is a wave in an elastic medium in which the restoring force is provided by shear. *S*-waves are divergence-less,

$$\nabla \cdot \vec{u} = 0 \tag{6.9}$$

where $\vec{u}$ is the displacement of the wave and comes in two polarizations: (1) SV (vertical) and (2) SH (horizontal).

The speed of an *S*-wave is given by:

$$v_s = \sqrt{\frac{\mu}{\rho}} \tag{6.10}$$

where μ is the shear modulus and ρ is the density.

(g) *P-waves:* Primary waves also are called P-waves. These are compressional waves and are longitudinal in nature. They are a type of pressure wave. The speed of P-waves is greater than other waves. These are called the primary waves

6.2 Descriptions of Transverse and Longitudinal Waves

because they are the first to arrive during an earthquake. This is because of great their velocity. The propagation of these waves knows no bounds and thus can travel through any type of material, including fluids.

P-waves, also called pressure waves, are longitudinal waves (i.e. the oscillation occurs in the same direction, and the opposite direction of wave propagation). The restoring force for P-waves is provided by the medium's bulk modulus. In an elastic medium with rigidity or shear modules being zero ($\mu = 0$), a harmonic plane wave has the form:

$$S(z,t) = S_0 \cos(kz - \omega t + \phi) \quad (6.11)$$

where S_0 is the amplitude of displacement, k is the wave number, z is the distance along the axis of propagation, ω is the angular frequency, t is the time, and ϕ is a phase offset. From the definition of bulk modulus (K), we can write:

$$K = -V \frac{dP}{dV} \quad (6.12)$$

where V is the volume and dP/dV is the derivative of pressure with respect to volume. The bulk modulus gives the change in volume of a solid substance as the pressure on it is changed, then we can write:

$$\begin{aligned} K &\equiv -V\left(\frac{dP}{dV}\right) \\ &\equiv \rho\left(\frac{\partial P}{\partial \rho}\right) \end{aligned} \quad (6.13)$$

Consider a wavefront with surface area A, then the change in pressure of the wave is given by the following relationship:

$$\begin{aligned} dP &= -K\frac{dV}{V} = -K\frac{A[S(z+\Delta z) - S(z)]}{A\Delta z} \\ &= -K\frac{S(z+\Delta z) - S(z)}{\Delta z} = -K\frac{\partial S}{\partial z} \end{aligned} \quad (6.14)$$

where ρ is the density. The bulk modulus has units of pressure.

6.2.1 Pressure Waves and More Details

As we mentioned in the preceding, the pressure waves present the behavior and concept of LWs, thus many of the important concepts and techniques used to analyze TWs on a string, as part of mechanical waves components, also can be applied to longitudinal pressure waves.

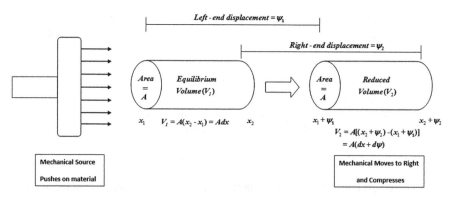

Fig. 6.4 Displacement and compression of a segment of materials

You can see an illustration of how a pressure wave works in Fig. 6.4. As the mechanical wave source moves through the medium, it pushes on a nearby segment of the material, and that segment moves away from the source and is compressed (i.e., the same amount of mass is squeezed into a smaller volume, so the density of the segment increases). That segment of increased density exerts pressure on adjacent segments, and in this way a pulse wave (if the source gives a single push) or a harmonic wave (if the source oscillates back and forth) is generated by the source and propagates through the material.

The "disturbance" of such waves involves three things: (1) the longitudinal displacement of material, (2) changes in the density of the material, and (3) variation of the pressure within the material. So pressure waves also could be called "density waves" or even "longitudinal displacement waves," and when you see graphs of wave disturbance in physics and engineering textbooks, you should make sure you understand which of these quantities is being plotted as the displacement of the wave.

As you can see in Fig. 6.4, we are still considering one-dimensional wave motion (i.e., the wave propagates only along the χ-axis). But pressure waves exist in a three-dimensional medium, so instead of considering linear mass density μ (as we did for the string in the previous section), in this case it is the volumetric mass density ρ that will provide the inertial characteristic of the medium. Nonetheless, just as we restricted the string motion to small angles and considered only the transverse component of the displacement, in this case we will assume that the pressure and density variations are small relative to the equilibrium values and consider only longitudinal displacement (i.e., the material is compressed or rarefied only by changes in the segment length in the χ-direction).

The most straightforward route to finding the wave equation for this type of wave is very similar to the approach used for TWs on a string, which means you can use Newton's second law to relate the acceleration of a segment of the material to the sum of the forces acting on that segment. To do that, start by defining the pressure (P) at any location in terms of the equilibrium pressure (P_0) and the incremental change in pressure produced by the wave (dP):

6.2 Descriptions of Transverse and Longitudinal Waves

$$P = P_0 + dP \tag{6.15}$$

Likewise, density (ρ) at any location can be written in terms of equilibrium density (ρ_0) and the incremental change in density produced by the wave (dP):

$$\rho = \rho_0 + d\rho \tag{6.16}$$

Before relating these quantities to the acceleration of material in the medium using Newton's second law, it is worthwhile to familiarize yourself with the terminology and equations of volume compressibility. As you might imagine, when external pressure is applied to a segment of material, how much the volume (thus the density) of that material changes depends on the nature of the material. To compress a volume of air by 1% requires a pressure increase of about 1000 pascals (Pa or N/m^2), but to compress a volume of steel by 1% requires a pressure increase of more than one billion Pa. The compressibility of a substance is the inverse of its "bulk modulus" (usually written as K or B, with units of Pa), which relates an incremental change in pressure (dP) to the fractional change in density ($d\rho/\rho_0$) of the material:

$$K \equiv \frac{dP}{d\rho/\rho_0} \tag{6.17}$$

or

$$dP = K\frac{d\rho}{\rho_0} \tag{6.18}$$

With this relationship in mind, you are ready to consider Newton's second law for the segment of material being displaced and compressed (or rarefied) by the wave. To do that consider the pressure from the surrounding material acting on the left and on the right side of the segment, as shown in Fig. 6.5. Notice that the pressure (P_1) on the left end of the segment is pushing in the positive χ-direction and the pressure on

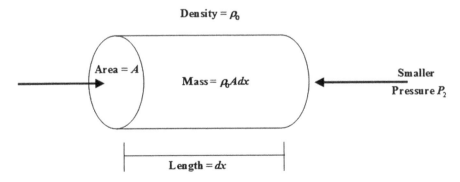

Fig. 6.5 Pressure on a segment of material

the left end of the segment is pushing in the negative χ-direction. Setting the sum of the χ-direction forces equal to the acceleration in the χ-direction gives:

$$\sum F_x = P_1 A - P_2 A = m a_x \tag{6.19}$$

where m is the mass of the segment.

If the cross-sectional area of the segment is A and the length of the segment is dx, the volume of the segment is Adx, and the mass of the segment is this volume times the equilibrium density of the material:

$$m = \rho_0 A dx \tag{6.20}$$

Notice also that the pressure on the right end of the segment is smaller than the pressure on the left end because the source is pushing on the left end, which means that the acceleration at this instant will be toward the right.

Using the symbol ψ to represent the displacement of the material because of the wave, the acceleration in the χ-direction can be written:

$$a_x = \frac{\partial^2 \psi}{\partial t^2} \tag{6.21}$$

Substituting these expressions for m and a_x into Newton's second law, Eq. 6.19 gives:

$$\sum F_x = P_1 A - P_2 A = \rho_0 A dx \frac{\partial^2 \psi}{\partial t^2} \tag{6.22}$$

Writing the pressure P_1 at the left end as $P_0 + dP_1$ and the pressure P_2 at the right end as $P_0 + dP_2$ means that

$$P_1 A - P_2 A = (P_0 + dP_1)A - (P_0 + dP_2)A$$
$$= (dP_1 - dP_2)A \tag{6.23}$$

But the change in dP (i.e., the change in the overpressure, or under-pressure, produced by the wave) over the distance dx can be written:

$$\text{Change in over-pressure} = dP_2 - dP_1 = \frac{\partial (dP)}{\partial x} dx \tag{6.24}$$

which means

$$-\frac{\partial (dP)}{\partial x} dx A = \rho_0 A dx \frac{\partial^2 \psi}{\partial t^2} \tag{6.25}$$

or

6.2 Descriptions of Transverse and Longitudinal Waves

$$\rho_0 \frac{\partial^2 \psi}{\partial t^2} = -\frac{\partial (dP)}{\partial x} \tag{6.26}$$

But $dP = d\rho K/\rho_0$, so

$$\rho_0 \frac{\partial^2 \psi}{dt^2} = -\frac{\partial \left[\left(\frac{K}{\rho_0}\right) d\rho\right]}{\partial x} \tag{6.27}$$

The next step is to relate the change in density ($d\rho$) to the displacements of the left and right ends of the segment (ψ_1 and ψ_2). To do that note that the mass of the segment is the same before and after the segment is compressed. That mass is the segment's density times its volume ($m = \rho V$), and the volume of the segment can be seen to be $V_1 = A dx$ before compression (see Fig. 6.4) and $V_2 = A(dx + d\psi)$ after compression. Thus,

$$\begin{aligned} \rho_0 V_1 &= (\rho_0 + d\rho) V_2 \\ \rho_0 (A dx) &= (\rho_0 + d\rho) A (dx + d\psi) \end{aligned} \tag{6.28}$$

The change in displacement ($d\psi$) over distance dx can be written:

$$d\psi = \frac{\partial \psi}{\partial x} dx \tag{6.29}$$

so

$$\rho_0 (A dx) = (\rho_0 + d\rho) A \left(dx + \frac{\partial \psi}{\partial x} dx \right)$$

$$\rho_0 = (\rho_0 + d\rho) \left(1 + \frac{\partial \psi}{\partial x} \right) \tag{6.30}$$

$$= \rho_0 + d\rho + \rho_0 \left(\frac{\partial \psi}{\partial x}\right) + d\rho \left(\frac{\partial \psi}{\partial x}\right)$$

Because we are restricting our consideration to the cases in which the density change ($d\rho$) produced by the wave is small relative to the equilibrium density (ρ_0), the term $d\rho(\partial \psi/\partial x)$ must be small compared with the term $\rho_0(d\psi/dx)$. Thus, to a reasonable approximation we can write:

$$d\rho = -\rho_0 \frac{\partial \psi}{\partial x} \tag{6.31}$$

which we can insert into Eq. 6.27, giving the following:

$$\rho_0 \left(\frac{\partial^2 \psi}{\partial t^2}\right) = -\frac{\partial\left\{\left(\frac{K}{\rho_0}\right)\left[-\rho_0\left(\frac{\partial \psi}{\partial x}\right)\right]\right\}}{\partial x}$$

$$= \frac{\partial\left[K\left(\frac{\partial \psi}{\partial x}\right)\right]}{\partial x} \quad (6.32)$$

Rearranging makes this into an equation with a familiar form of wave equation in one dimension:

$$\rho_0 \frac{\partial^2 \psi}{\partial t^2} = K \frac{\partial^2 \psi}{\partial x^2} \quad (6.33)$$

or

$$\frac{\partial^2 \psi}{\partial x^2} = \left(\frac{\rho_0}{K}\right) \frac{\partial^2 \psi}{\partial t^2} \quad (6.34)$$

As in the case of TWs on a string, you can determine the phase speed of a pressure wave by comparing the multiplicative term in the classical wave equation of Eq. 6.35 with that in Eq. 6.34.

$$\frac{\partial^2 \psi}{\partial x^2} = \frac{1}{v^2} \frac{\partial^2 \psi}{\partial t^2} \quad (6.35)$$

Setting these factors equal to one another gives the result:

$$\frac{1}{v^2} = \frac{\rho_0}{K} \quad (6.36)$$

or

$$v = \frac{K}{\rho_0} \quad (6.37)$$

As expected, the phase speed of the pressure wave depends both on the elastic (K) and on the inertial (ρ_0) properties of the medium. Specifically, the higher the bulk modulus of the material (i.e., the stiffer the material), the faster the components of the wave will propagate (because K is in the numerator), and the higher the density of the medium, the slower those components will move (because ρ_0 is in the denominator).

6.2.2 What Are Scalar Longitudinal Waves?

Scalar longitudinal waves (SLW) are conceived of as LWs because they are sound waves. Unlike the TWs of electromagnetism, which move up and down perpendicularly to the direction of propagation, LWs vibrate in line with the direction of propagation. Transverse waves can be observed in water ripples: the ripples move up and down as the overall waves move outward, such that there are two actions—one moving up and down and the other propagating in a specific direction outward.

Technically speaking, SWs have magnitude but no direction because they are imagined to be the result of two EM waves that are 180° out of phase with one another, which leads to both signals being canceled out. This results in a kind of "pressure wave." Mathematical physicist James Clerk Maxwell, in his original mathematical equations concerning electromagnetism, established the theoretical existence of SWs. After his death, however, later physicists assumed these equations were meaningless because SWs had not been empirically observed, and they had not been repeatedly verified among the scientific community at large.

Vibrational or subtle energetic research, however, has helped advance our understanding of SWs. One important discovery states that there are many different types of them, not just those of the EM variety. For example, there are vital SWs (corresponding with the vital or "Qi" body, described next), emotional SWs, mental SWs, causal SWs, and so forth. In essence, as far as we are aware, all "subtle" energies are made up of various types of scalar waves.

> **Qi Body**
> *Qi* can be interpreted as the "life energy" or "life force" that flows within us. Sometimes, it is known as the "vital energy" of the body. In traditional Chinese medicine (TCM) theory, *qi* is the vital substance constituting the human body. It also refers to the physiological functions of organs and meridians.

Some of the general properties of SWs (of the beneficial kind) include that they:

- Travel faster than the speed of light
- Seem to transcend space and time
- Cause the molecular structure of water to become coherently reordered
- Positively increase immune function in mammals
- Are involved in the formation process in nature

More details about SLW applications are discussed in next section.

6.2.3 Scalar Longitudinal Wave Applications

The possibility of developing a means of establishing communication through something non-homogeneous is looking very promising via use of the *More Complete Electrodynamic* (MCE) theory [4]. This theory reveals that the SLW, which is created by a *gradient-driven* current, has *no* magnetic field and, thus, is *not* constrained by the skin effect. The SLW is slightly attenuated by the non-linearities in electrical conductivity as a function of electrical field magnitude. The SLW does not interfere with classical *transverse electromagnetic* (TEM) transmission or vice versa. By contrast, TEM waves are severely attenuated in conductive media because of magnetically driven eddy currents that arise from the skin effect. Consequently, only very-low and ultra-low frequency TEM waves can be successfully used for long-distance underwater communications. The SLW also has immediate implications for the efficient redesign and optimization of existing TEM-based electronic technologies because both TEM and SLW are created simultaneously with present electronic technologies.

The goal of application of SLW-based ($\geq$150 kb/s) digital data propagation over distances of many kilometers (km) to address strategic, tactical, surveillance, and undersea warfare missions of an organization such as the Navy. With this goal in mind, the optimization of SLW underwater-antenna design will be guided by development of a first principles SLW simulator from the MCE theory because all existing simulators model only circulating current-based TEM waves.

A proof-of-principle demonstration of the prototype antenna through freshwater will be conducted in-house, followed by controlled tests at typical government underwater test range(s). These tests would include characterization of wave attenuation versus frequency, modulation bandwidth, and beam-width control. The deliverable will be an initial prototype for SLW communications over tactical distances or more, followed by a field-deployable prototype, as dictated by Navy performance needs.

The unique properties of the SLW lead to more sophisticated application areas, with implications for ocean surveillance systems, underwater imaging, energy production, power transmission, transportation, guidance, and national security. This disruptive technology has the potential to transform communications, as well as electrodynamic applications in general.

As far as *low energy nuclear reactions* are concerned, it is certain that most of us have heard of scalar electrodynamics. Nevertheless, we probably have many questions about this phenomenon. Because up to now it has been mostly shrouded in mystery, we may even wonder whether electrodynamics exists at all; and if it exists, do we need exotic conditions to produce and use it? In addition, will it require a drastic transformation in our current understanding of classical electrodynamics? How much of an impact will it have on future modes of power generation and conversion? Which applications: weaponry, medical, or a low-energy fusion-driven source of energy—a D + D reaction? Most of these were mentioned in the preceding.

There is also a possibility of applying a scalar electrodynamic wave (SEW) in applications such as developing and demonstrating of an *all-electronic* (AE) engine that replaces *electromechanical* engines for vehicle propulsion. As far as other applications of SLW are concerned, a few are discussed next.

6.2.3.1 Medical Applications for Scalar Longitudinal Waves

Not all SWs, or subtle energies, are beneficial to living systems. Electromagnetism of the 60 Hz AC variety, for example, emanates a secondary longitudinal–scalar wave that is typically detrimental to living systems. To use the SLW as an application in biofield technology effectively, however, we need to cancel the detrimental aspect of wave scale and transform it into a beneficial wave; therefore, this innovative approach qualifies as a medical application of a SLW, where we can approach biomedical people to suggest such an invention and to ask for funding as well. Currently, there seems to be an interest in a biofield approach application of SLWs.

6.2.3.2 A Genuine Application of SLW for Low-Temperature Fusion Energy

In the case low-temperature fusion interactions of D + D, by lowering the nuclear potential barrier for purposes of "cold fusion," for lack of better words, we know that in low-energy heavy-ion fusion, the term "Coulomb barrier" commonly refers to the barrier formed by the repulsive Coulomb and the attractive "nuclear" (nucleus-nucleus) interactions in a central (S-wave) collision. This barrier frequently is called a *fusion barrier* (for light and medium mass heavy-ion systems) or capture barrier (for heavy systems). In general, there is a centrifugal component to such a barrier (non-central collisions).

Experimenters may use the term *Coulomb barrier* to the define the nominal value of the "Coulomb barrier distribution" when either coupled-channel effects operate or (at least) a collision partner is deformed because the barrier features depend on orientation. To this author's knowledge, the terminology "transfer barrier" has not been used much. In my view it could be applied to the transfer of charged particles/clusters. There is a vast literature on methods for calculating Coulomb barriers. For instance, the double-folding method is broadly used in the low-energy nuclear physics community. Based on this technique, there is a potential, called the "Sao-Paulo potential," because it was developed by theorists in Sao Paulo, Brazil.

The Coulomb barrier is calculated theoretically by adding the nuclear and Coulomb contributions of the interaction potential. For fusion there are other contributions coming from the different degrees of freedom such as the angular momentum (centrifugal potential), the vibrational and rotational states in both interacting nuclei, in addition to the transfer contribution. This is an area for which we could approach the DOE or NRC with some Requests for Proposals—for example, the Idaho National Laboratory (INL).

6.2.3.3 Application of SLW for Directed-Energy Weapons

Scalar beam weapons were originally invented in 1904 by Nicola Tesla, an American immigrant from Yugoslavia (1856 or 1857–1943). Since his death in 1943, many nations have secretly developed his beam weapons, which now have been further refined to be so powerful that just by satellite one can create the following: a nuclear-like destruction, an earthquake, a hurricane, a tidal wave, as well as cause an instant freeze—killing every living thing instantly over many miles. An SLW also can cause intense heat, like a burning fireball over a wide area; induce hypnotic mind control of a whole population; or even remotely read anyone on the planet's mind.

Because of the nature of a pressure wave's behavior and its ability to carry tremendous energy, SLWs can remove something right out of its place in time and space faster than the speed of light, without any detectable warning, by crossing two or more beams with each other. Moreover, any target can be aimed at or even right through to the opposite side of the Earth. If either of the major scalar weapons' armed countries (e.g., U.S. or Russia) were to fire a nuclear missile to attack the other, it possibly may not even reach the target because it could be destroyed with scalar technology before it even left its place or origin. The knowledge via radio waves that it was about to be fired could be eavesdropped on and the target could be destroyed in the bunker if fired at from space by a satellite.

Above 60 Hz AC frequency, this wave can be very detrimental in nature. A scalar beam can be sent from a transmitter to the target, coupled with another sent from another transmitter, and as they cross an explosion can be initiated. This interference grid method could enable scalar beams to explode the missile before launch, as well as en route by knowing the right coordinates. If the target does manage to launch, what are known as Tesla globes, or Tesla hemispheric shields, can be sent to envelop a missile or aircraft. These are made of luminous plasma, which emanates physically from crossed scalar beams and can be created in any size, even more than 100 miles across.

Initially detected and tracked as it moves on the scalar interference grid, a continuous electromagnetic pulse (EMP) Tesla plasma globe could kill the electronics of the target. More intensely hot Tesla "fireball" globes could vaporize the missile. Tesla globes also could activate a missile's nuclear warhead en route by creating a violent low-order nuclear explosion. Various parts of flying debris can be subjected to smaller and more intense Tesla globes where the energy density to destroy is more powerful than the larger globe first encountered. This can be done in pulse mode with any remaining debris given maximum continuous heating to vaporize metals and materials. If anything still rains down on Russia or America, either could have already made a Tesla shield over the targeted area to block it from entering its airspace.

Other useful aspects of SLWs in military applications is for a community in the U.S. that believes the SWs are realizable in nature by a mathematical approach. In recent conferences sponsored by the *Institute of Electrical and Electronics Engineers* (IEEE), these were discussed openly and a *Proceedings* report of the conference exists. Dedicated to Nicola Tesla and his work, the conference's papers

6.2 Descriptions of Transverse and Longitudinal Waves

presented claims that some of Tesla's work used SW concepts. Thus, there is an implied "Tesla Connection" in all of this. As was stated in the preceding, these are unconventional waves that are not necessarily a contradiction to Maxwell's equations as some have suggested; they might represent an extension to Maxwell's understanding at the time. If realizable, SLWs could represent a new form of wave propagation that may well penetrate seawater (knowing the permeability, permittivity of salt water, and consequently skin depth), resulting in a new method of submarine communications and possibly a new form of technology for *anti-submarine warfare* (ASW). This technology also helps those in the Naval Special Warfare (NSW) community, such as the Navy SEALs, to be able to communicate with each other even in murky water conditions.

The following are some mathematical notations and physics involved with this aspect of SLWs:

1. The SW, as it is understood, is not an *electromagnetic (EM) wave*. An EM wave has both electric $(\vec{E})$ and magnetic $(\vec{B})$ fields and power flow in EM waves is by means of the Poynting vector, as written in Eq. 6.38:

$$\vec{S} = \vec{E} \times \vec{B} \text{ Watts/m}^2 \tag{6.38}$$

The energy per second crossing a unit area with a normal that is pointed in the direction of $\vec{S}$ is the energy in the EM wave.

A SW has no time-varying $\vec{B}$ field. In some cases, it also has no $\vec{E}$ field. Thus, it has no energy propagated in EM wave form. It must be realized, however, that any vector could be added that may be integrated into zero over a closed surface and the *Poynting theorem* still applies. Thus, there is some ambiguity in even stating the relationship that is given by Eq. 6.38—that is, the total EM energy flow.

2. The SW could be accompanied by a vector potential $\vec{A}$, $\vec{E}$, and yet $\vec{B}$ remains zero in the far field. From EM theory, we can write as follows:

$$\begin{cases} \vec{E} = -\vec{\nabla}\phi - \dfrac{1}{c}\dfrac{\partial \vec{A}}{\partial t} \\ \vec{B} = \vec{\nabla} \times \vec{A} \end{cases} \tag{6.39}$$

In this case ϕ is the scalar (electric) potential and $\vec{A}$ is the (magnetic) vector potential. Maxwell's equations then predict the following mathematical relation:

$$\nabla^2 \phi - \frac{1}{c^2}\frac{\partial^2 \phi}{\partial t^2} = 0 \quad \text{(Scalar Potential Waves)} \tag{6.40}$$

$$\nabla^2 \vec{A} - \frac{1}{c^2}\frac{\partial^2 \vec{A}}{\partial t^2} = 0 \quad \text{(Vector Potential Waves)} \tag{6.41}$$

A solution appears to exist for the special case of $\vec{E} = 0$, $\vec{B} = 0$, and $\nabla \times \vec{A} = 0$, for a new wave satisfying the following relations:

$$\begin{cases} \vec{A} = \vec{\nabla} s \\ \phi = -\frac{1}{c}\frac{\partial s}{\partial t} \end{cases} \quad (6.42)$$

s then stratifies the following relationship:

$$\nabla^2 s - \frac{1}{c^2}\frac{\partial^2 s}{\partial t^2} \quad (6.43)$$

Note that quantity c represents the speed of light. Mathematically, s is a "potential" with a wave equation, one that suggests propagation of this wave even through $\vec{E} = \vec{B} = 0$ and the Poynting theorem indicates no EM power flow.

3. From paragraph 2 in the preceding, there is the suggestion of a solution to Maxwell's equations involving a SW with potential s that can propagate without a *Poynting vector* EM power flow. The question arises, however, as to where the energy is drawn from to sustain such a flow of energy. Some suggest a vector that integrates to zero over a closed surface might be added in the theory. Another is the possibility of drawing energy from the vacuum, assuming net energy could be drawn from "free space."

Quantum electrodynamics allows random energy in free space but conventional EM theory has not allowed this to date. Random energy in free space that is built of force fields that sum to zero is a possible approach. If so, these might be a source of energy to drive the s waves drawn from "free space." A number of engineers and scientists in the community suggested, as stated earlier in a statement within this discussion, that, if realizable, the SW could represent a new form of wave propagation that could penetrate seawater or be used as a new approach for DEWs.

This author suggests considering another scenario where on3 may need to look at equations of the MCE theory and obtain new predictions for producing energy that way; thus, generate an SLW where the *Lagrangian density equation* for MCE can be defined as:

$$\mathcal{L} = -\frac{\varepsilon c^2}{4} F_{\mu\nu} F^{\mu\nu} + J_\mu A^\mu - \frac{\gamma \varepsilon c^2}{2}\left(\partial_\mu A^\mu\right) - \frac{\varepsilon c^2 k^2}{2}\left(A_\mu A^\mu\right) \quad (6.44)$$

where the Lagrangian density equation written in terms of the potentials $\vec{A}$ and ϕ as follows:

$$\mathcal{L}_{EM} = \frac{\varepsilon c^2}{2}\left[\frac{1}{c^2}\left(\vec{\nabla}\phi + \frac{\partial \vec{A}}{\partial t}\right)^2 - (\vec{\nabla}\times\vec{A})^2\right]$$

$$-\rho\phi + \vec{J}\cdot\vec{A} - \frac{\varepsilon c^2}{2}\underbrace{\left(\frac{1}{c^2}\frac{\partial \phi}{\partial t} + \vec{\nabla}\cdot\vec{A}\right)}_{C}^2 \quad (6.45)$$

6.2 Descriptions of Transverse and Longitudinal Waves

The proof of the equation was given in Chap. 5 of this book (see Eq. 5.126).

This is the area where there is a lot of speculation among scientists, around the community of EM and the use of the SW as a weapons application, and you will find a lot of good, as well as nonsense, approaches on the Internet by various authors. The present approach uses several approaches:

1. Acoustic signals that travel slowly (1500 m per second in seawater)
2. Blue-green laser light that has a typical range of 270 m and is readily scattered by seawater particulate
3. High-frequency radio waves that are limited to a range of 7–10 m in seawater at high frequencies
4. Extremely low-frequency radio signals that are long range (worldwide) but transmit only a few characters per second for one-way, bell-ring calls to individual submarines

The new feature of this proposed work is the use of a novel electrodynamic waves that have no magnetic field, and thus are not so severely constrained by the high conductivity of seawater, as regular radio waves are. We have demonstrated the low loss property of this novel (scalar–longitudinal) wave experimentally by sending a video signal through 2 mm of solid copper at 8 GHz.

If you have a background in physics or electrical engineering, you know that, unquestionably, our knowledge of the properties and dynamics of EM systems is believed to be the most solid and firmly established in all classical physics. By its extension, the application of quantum electrodynamics, describing accurately the interaction of light and matter at the subatomic realms, has resulted in the most successful theoretical scientific theory to date, agreeing with corresponding experimental findings to astounding levels of precision. Accordingly, these developments have led to the belief among physicists that the theory of classical electrodynamics is complete and that it is essentially a closed subject.

Nevertheless, at least as far back to the era of *Nikola Tesla*, there have been continual rumblings of discontent stemming from occasional physical evidence from both laboratory experimental protocols and knowledge obtained from observation of natural phenomena (e.g., the dynamics of atmospheric electricity, etc.). This may suggest that in extreme situations involving the production of high energies at specific frequencies, there might be some cracks exposed in the supposed impenetrable monolithic fortress of classical/quantum electrodynamics, implying possible key missing theoretical and physical elements. Unfortunately, some of these experimental phenomena have been difficult to replicate and produce on-demand. Moreover, some have been shown apparently to violate some of the established principles underlying classical thermodynamics.

On top of that, many of those courageous individuals promoting the study of this phenomenon have couched their understanding of the limited reliable experimental evidence available from the sources in a language unfamiliar to the legion of mainstream technical specialists in electrodynamics, preventing clear communication of these ideas. Also, the various sources that have sought to convey this information have at times delivered contradictory statements.

It is therefore no wonder that for many decades such exotic claims have been disregarded, ignored, and summarily discounted by mainstream physics. Because of important developments over the past 2 years, however, there has been a welcome resurgence of research in this area, bringing back renewed interest toward the certifying the existence of these formerly rejected anomalous energy phenomena. Consequently, this renaissance of the serious enterprise in searching for specific weaknesses, which currently plague a fuller understanding of electrodynamics, has propelled the proponents of this research to more systematically outline their ideas in a clearer fashion. The possible properties of these dynamics and how inclusion of them could change our current understanding of electricity and magnetism, as well as suggest implications for potential vast, practical ramifications, may change the disciplines of physics, engineering, and energy generation.

It is the purpose of this book, and particularly this chapter, not only to report on these unique, various recent inventions and their possible modes of operation, but also to convince readers of their value for hopefully directing a future program geared toward the rigorous clarification and certification of the specific role the electroscalar domain might play in shaping a future consistent, classical electrodynamics. Also, by extension, to perhaps shed light on thorny conceptual and mathematical inconsistencies that do exist in the present interpretation of relativistic quantum mechanics. In this regard, it is anticipated that, by incorporating this more expansive electrodynamic model, the source of the extant problems with gauge invariance in quantum electrodynamics and the subsequent unavoidable divergences in energy/charge might be identified and ameliorated. Not only does the electroscalar domain have the potential to address such lofty theoretical questions surrounding fundamental physics, but it also aims to show that the protocol necessary for generating these field effects may not be present in exotic conditions relating to large field strengths and specific frequencies involving an expensive infrastructure such as the *large hadron collider* (LHC).

Insight into the incompleteness of classical electrodynamics can begin with the Helmholtz theorem, which states that any sufficiently smooth three-dimensional vector field can be uniquely decomposed into two parts. By extension, a generalized theorem exists that was certified through the scholarly work of physicist-mathematician Dale Woodside (2009) [5] (see Eq. 6.44 in the preceding) for unique decomposition of a sufficiently smooth, Minkowski 4-vector field (three spatial dimensions, plus time) into *four-irrotational* and *four-solenoidal* parts, together with the tangential and normal components on the bounding surface.

With this background, the theoretical existence of the electroscalar wave can be attributed to failure to include certain terms in the standard, general four-dimensional EM Lagrangian density related to the four-irrotational parts of the vector field. Here ε is electrical permittivity, not necessarily of the vacuum. Specifically, the electroscalar field becomes incorporated into the structure of electrodynamics when we get Eq. 6.44 in the preceding for $\gamma = 1$ and $k = (2\pi m c/h) = 0$. As we can see in this representation, as written in the equation, it is the presence of the third term that describes these new features.

6.2 Descriptions of Transverse and Longitudinal Waves

We can see more clearly how this term arises by writing Lagrangian density in terms of the standard EM scalar (ϕ) (see Eq. 6.45) and magnetic vector potentials (A), without the electroscalar representation included. This equation has zero divergence of the potentials (formally called solenoidal) consistent with classical electromagnetics, as we can see here. The second class of 4-vector fields has zero curl of the potentials (i.e., irrotational vector field), which will emerge once we add this scalar factor. Here this is represented by the last term, which is usually zero in standard classical EMs. The expression in the parentheses, when set equal to zero, describes what is known as the Lorentz condition that makes the scalar potential and the vector potential in their usual form mathematically *dependent* on each other.

Accordingly, the usual EM theory then specifies that the potentials may be chosen arbitrarily based on the specific so-called gauge that is chosen for this purpose. The MCE theory, however, allows for a non-zero value for this scalar-valued expression, essentially making the potentials *independent* of each other, where this new scalar-valued component (*C* in Eq. 6.45 that can be called *Lagrangian density*) is a dynamic function of space and time. It is this new idea of the independence of the potentials, out of which the scalar value (*C*) is derived, and from which the unique properties and dynamics of the *scalar longitudinal electrodynamic* wave arises.

To put all these in perspective, a more complete electrodynamic model can be derived from this last equation of the Lagrangian density. The Lagrangian expression is important in physics because invariance of the Lagrangian under any transformation gives rise to a conserved quantity. Now, as is well known, conservation of charge-current is a fundamental principle of physics and nature. Conventionally, in classical electrodynamics charged matter creates an $\vec{E}$ field. Motion of charged matter creates a magnetic $\vec{B}$ field from an electrical current that in turn influences the $\vec{B}$ and $\vec{E}$ fields.

Before, we continue further, let us write the following equations:

$$\vec{E} = -\nabla\phi - \frac{\partial \vec{A}}{\partial t} \quad \text{Relativistic Covariance} \tag{6.46}$$

$$\vec{B} = \nabla \times \vec{A} \quad \begin{array}{l}\text{Classical Fields } (\vec{B} \text{ and } \vec{E})\\ \text{in terms of usual classical}\\ \text{potentials } (\vec{A} \text{ and } \vec{\phi})\end{array} \tag{6.47}$$

$$C = \frac{1}{c^2}\frac{\partial \vec{\phi}}{\partial t} + \nabla \cdot \vec{A} \quad \text{Classical wave equations for } \vec{A}, \vec{B} \tag{6.48}$$

$$\nabla \times \vec{B} - \frac{1}{c^2}\frac{\partial \vec{E}}{\partial t} - \nabla C = \mu \vec{J} \quad \vec{E} \text{ and } \vec{\phi} \text{ without the use of a gauge} \tag{6.49}$$

$$\nabla \cdot \vec{E} + \frac{\partial C}{\partial t} = \frac{\rho}{\varepsilon} \quad \begin{array}{l}\text{Condition (the MCE theory produces}\\ \text{cancellation of } \partial C/\partial t \text{ and } -\vec{\nabla}C \text{ in the}\\ \text{classical wave equation for } \vec{\phi} \text{ and } \vec{A},\\ \text{thus eliminating the need for a gauge}\\ \text{condition)}\end{array} \tag{6.50}$$

These effects can be modeled by Maxwell's equations. Now, exactly how and to what degree do these equations change when the new scalar-valued C field is incorporated? Those of you who have knowledge of Maxwellian theory will notice that the two homogeneous Maxwell's equations—representing *Faraday's Law* and $\vec{\nabla} \cdot \vec{B}$, the standard Gauss's Law equation for a divergence-less magnetic field—are both unchanged from the classical model. Notice the last three equations incorporate this new scalar component that is labeled C.

This formulation, as defined by Eq. 6.48, creates a somewhat revised version of Maxwell's equations, with one new term, $-\vec{\nabla}C$, in Gauss's Law (Eq. 6.50), where ρ is the charge density, and one new term ($\partial C/\partial t$) in Ampère's Law (Eq. 6.49), where J is the current density. We can see that these new equations lead to some important conditions. First, relativistic covariance is preserved. Second, unchanged are the classical fields $\vec{E}$ and $\vec{B}$ in terms of the usual classical potentials ($\vec{A}$ and ϕ). We have the same classical wave equations for $\vec{A}$, ϕ, $\vec{E}$, and $\vec{B}$ *without* the use of a gauge condition (and its attendant incompleteness) because the MCE theory shows cancellation of $\partial C/\partial t$ and $-\vec{\nabla}C$, the classical wave equations for ϕ and $\vec{A}$; and a SLW is revealed, composed of the scalar and longitudinal-electric fields.

A wave equation for C is revealed by use of the time derivative of Eq. 6.50, added to the divergence of Eq. 6.49. Now, as is known, matching conditions at the interface between two different media are required to solve Maxwell's equations. The divergence theorem on Eq. 6.51 will yield an interface matching in the normal component ("$\hat{n}$") of $\nabla C/\mu$, as shown in Eq. 6.51:

$$\frac{\partial^2 C}{\partial c^2 t^2} - \nabla^2 C \equiv \Box^2 C = \mu \left(\frac{\partial \rho}{\partial t} + \nabla \cdot \vec{J} \right) \tag{6.51}$$

$$\left(\frac{\nabla C}{\mu} \right)_{1n} = \left(\frac{\nabla C}{\mu} \right)_{1n} \tag{6.52}$$

$$C = C_0 \exp[j(kr - \omega t)]/r \tag{6.53}$$

Note: The preceding, Eqs. 6.51 and 6.52, present a wave equation for *scalar factor C* matching condition in the normal component of $\nabla C/\mu$, spherically *symmetric wave solution* for C, and the operator $\Box^2 = \frac{\partial^2}{\partial c^2 t^2} - \nabla^2$, called the *d'Alembert operator*.

The subscripts in Eq. 6.51 denote $\nabla C/\mu$ in medium 1 or medium 2, respectively; (μ) is magnetic permeability—again not necessarily that of the vacuum. In this regard, with the vector potential $(\vec{A})$ and scalar potential (ϕ) now stipulated as independent of each other, it is the surface charge density at the interface that produces a discontinuity in the gradient of the scalar potential, rather than the standard discontinuity in the normal component of $\vec{E}$ (see Hively [4]).

Notice, also from Eq. 6.51, the source for scalar factor C implies a violation of charge conservation on the non-zero right-hand side (RHS), a situation that we noted cannot exist in macroscopic nature. Nevertheless, this will be compatible with standard Maxwellian theory if this violation occurs at very short time scales, such

6.2 Descriptions of Transverse and Longitudinal Waves

as in subatomic interactions. Now, interestingly, with the stipulation of charge conservation on large time scales, giving zero on the RHS of Eq. 6.51., longitudinal wave-like solutions are produced with the lowest order form in a spherically symmetric geometry at a distance (r), $C = C_0 \exp[j(kr - \omega t)]/r$. Applying the boundary condition $C \to 0$ as $r \to \infty$ is thus trivially satisfied. The C wave therefore is a *pressure* wave, similar to that in acoustics and hydrodynamics.

This is unique under the new MCE model because, although classical electrodynamics forbids a spherically symmetric TW to exist, this constraint will be absent under MCE theory. Also, an unprecedented result is that these longitudinal C waves will have energy but no momentum. But then again, this is not unlike charged particle–anti-particle fluctuations that also have energy but no net momentum.

Now that we are here so far, the question of why this constraint prohibiting a spherically symmetric wave is lifted in MCE can be resolved in the following sets of Eq. 6.54 for the wave equation for the vertical magnetic field:

$$\begin{cases} \dfrac{1}{c^2}\dfrac{\partial^2 \vec{B}}{\partial t^2} - \nabla^2 \vec{B} = \mu_0(\nabla \times \vec{J}) \\ \nabla \times \vec{J} = 0 \to J = \nabla k \\ \text{Gradient-driven Current} \to \text{SLW} \end{cases} \quad (6.54)$$

The sets of this equation are established for the $\vec{B}$ wave equation, resulting in s gradient-driven current in MCE for generating the SLW.

Notice again that the source of the magnetic field on the RHS is a non-zero value of $\nabla \times \vec{J}$, which signifies solenoidal current density, as is the case in standard Maxwellian theory. When $\vec{B}$ is zero, so is $\nabla \times \vec{J}$. This is an important result. Thus, the current density is irrotational, which implies that $J = \nabla \kappa$. Here κ is a scalar function of space and time. Therefore, in contrast to closed current paths generated in ordinary Maxwell theory that result in classical waves that arise from a solenoidal current density $(\nabla \times \vec{J} \neq 0)$, J for the SLW is gradient-driven and may be uniquely detectable.

We also can see from this result that a zero value of the magnetic field is a necessary and sufficient condition for this gradient-driven current. Now, because in linearly conductive media, the current density $(\vec{J})$ is directly proportional the electric field intensity $(\vec{E})$ that produced it, where σ is the conductivity, this gradient-driven current will then produce a longitudinal $\vec{E}$-field. Based on calculations so far, we can establish a wave equation for the $\vec{E}$ solution for longitudinal $\vec{E}$ in MCE *spherically symmetric wave solutions* for $\vec{E}$ and $\vec{J}$ in *linearly conductive media*, as follows:

$$\dfrac{\partial^2 \vec{E}}{\partial c^2 t^2} - \nabla^2 \vec{E} = \left(\dfrac{\partial^2}{\partial c^2 t^2} - \nabla^2\right)\vec{E} \equiv \Box^2 \vec{E} = \mu \dfrac{\partial \vec{J}}{\partial t} - \dfrac{\nabla \rho}{\varepsilon} \quad (6.55)$$

$$E = E_r \, \hat{r} \, \exp[j(kr - \omega t)]/r \tag{6.56}$$

$$\vec{J} = \sigma \vec{E} \rightarrow \Box^2 \vec{J} = 0 \tag{6.57}$$

We also can see this from examining the standard vectorial wave equation for the electric field. The wave equation for $\vec{E}$ (Eq. 6.50) arises from the curl of Faraday's Law, use of $\nabla \cdot \vec{B}$ from Ampère's Law (Eq. 5.49), and the substitution of $\nabla \cdot \vec{E}$ from Eq. 6.50 with cancellation of the terms $\nabla(\partial C/\partial t) = (\partial/\partial t)\nabla C$. When the RHS of Eq. 13 is zero, the lowest order outgoing spherical wave is $E = E_r \, \hat{r} \, \exp[j(kr - \omega t)]/r$, where $\hat{r}$ represents the unit vector in the radial direction and r represents the radial distance. The electrical field is also longitudinal. Substitution of $\vec{J} = \sigma \vec{E}$ into $\Box^2 \vec{E} = 0$ results in $\Box^2 \vec{J} = 0$, meaning that the current density is also radial. The SLW equations for E and J are remarkable for several reasons.

First, the vector SLW equations for $\vec{E}$ and $\vec{J}$ are fully captured in one wave equation for the scalar function (κ), $\Box^2 \kappa = 0$. Second, these forms are like $\Box^2 C = 0$. Third, these equations have zero on the RHS for propagation in conductive media. This occurs because $\vec{B} = 0$ for the SLW, implying no back EM field from $\partial \vec{B}/\partial t$ in Faraday's Law, which in turn gives no circulating eddy currents. Experimentation has shown that the SLW is not subject to the skin effect in media with linear electric conductivity and travels with minimum resistance in any conductive media.

This last fact affords some insight into another related ongoing conundrum in condensed matter physics—the mystery surrounding high-temperature superconductivity (HTS). As we know, the physical problem of HTS is one of the major unsolved problems in theoretical condensed matter physics—in part, because the materials are somewhat complex, multilayered crystals. Here the MCE theory may provide an explanation on the basis of gradient-driven currents between (or among) the crystal layers. The new MCE Hamiltonian (Eq. 6.58) includes the SLW because of gradient-driven currents among the crystalline layers as an explanation for HTS. The electrodynamic Hamiltonian for MCE is written:

$$\mathcal{H}_{EM} = \left(\frac{\varepsilon E^2}{2} + \frac{B^2}{2\mu}\right) + (\rho - \varepsilon \nabla \cdot \vec{E})\phi - \vec{J} \cdot \vec{A} + \frac{C^2}{2\mu} + \frac{C\nabla \cdot \vec{A}}{\mu} \tag{6.58}$$

In conclusion we can build an antenna based on the preceding concept within a laboratory environment and use a simulation software such as Multi-Physics COMSOL© or ANSYS computer code to model such an antenna. We believe, however, that we have done adequate analysis in this chapter to show the field of electrodynamics (classical and quantum), although considered to be totally understood, with any criticisms of incompleteness on the part of dissenters essentially taken as veritable heresy; nevertheless, it needs reevaluation in terms of apparent unfortunate sins of omission in the failure to include an electroscalar component.

Anomalies previously not completely understood may get a boost of new understanding from the operation of electroscalar energy. We have seen in the three instances examined—the mechanism of generation of seismic precursor electrical

6.2 Descriptions of Transverse and Longitudinal Waves

signals because of the movement of the Earth's crust, the ordinary peeling of adhesive tape, and irradiation by the special TESLAR chip—the common feature of the breaking of chemical bonds. In fact, we ultimately may find that any phenomenon requiring the breaking of chemical bonds, in either inanimate or biological systems, actually may be mediated by SWs.

Thus, we may discover that the scientific disciplines of chemistry or biochemistry may be more closely related to physics than is currently thought. Accordingly, the experimental and theoretical reevaluation of even the simplest phenomena in this regard, such as triboelectrification processes, is the absolute essence for those researchers knowledgeable of the necessity for this reassessment of EMs. As this author said in the introduction, it may even turn out that the gradient-driven current and associated SLW could be the umbrella concept under which many of the currently unexplained electrodynamic phenomena are discussed frequently at conferences, yielding a satisfying explanation.

The new SLW patent itself, which is the centerpiece of this chapter, is a primary example of the type of invention that probably would not have seen the light of day even 10 years ago. As previously mentioned, we are seeing more of this inspired breakthrough technology based on operating principles, formerly viewed with rank skepticism bordering on haughty derision by mainstream science, now surfacing to provide an able challenge to the prevailing worldview by reproducible corroborating tests by independent sources. This revolution in the technological witnessing of the overhaul of current orthodoxy is definitely a harbinger of the rapidly approaching time when many of the encrusted and equally ill-conceived, but still accepted, paradigms of science—thought to underpin our sentient reality—will fall by the wayside.

On a grander panoramic scale, our expanding knowledge gleaned from further examining the electroscalar wave concept, as applied to areas of investigation (e.g., cold fusion research, overunity power sources, etc.), explicitly will shape the future of society as well as science, especially concerning our openness to phenomena that challenge current belief systems.

To this point the incompleteness in established understanding of the properties of electrodynamical systems can be attributed to the failure to properly incorporate what can be termed the electroscalar force into the structural edifice of electrodynamics. Unbeknownst to most specialists in the disciplines mentioned, over the last decade in technological development circles, there has quietly, but inexorably, emerged bona fide physical evidence of the demonstration of the existence of SLW dynamics in recent inventions and discoveries.

As technology leads to new understanding, we are rapidly approaching a time in which these findings no longer can be pushed aside or ignored by orthodox physics, and physics must come to terms with their potential physical and philosophical impacts on our world. By the time you read this book, this author thinks you might agree with the fact that many could be on the brink of a new era in science and technology, the likes of which the current generation has never seen before. Despite what mainstream physics may claim, the study of electrodynamics is by no means a closed book. Further details are provided in the following sections.

6.3 Description of the $\vec{B}^{(3)}$ Field

During the investigation of the theory optically induced line shifts in *nuclear magnetic resonance* (NMR), people have come across the result that the anti-symmetric part of the intensity tensor of light is directly proportional in free space to an entirely novel, phase-free, magnetic field of light, which was identified as the $\vec{B}^{(3)}$ *field*, as defined in Eq. 6.59a. The presence of $\vec{B}^{(3)}$ in free space shows that the usual, propagating TWs of EM radiation are linked geometrically to spin field $\vec{B}^{(3)}$, which indeed emerges directly from the fundamental, classic equation of motion of a single electron in a circularly polarized light beam [6]:

$$\vec{B}^{(1)} \times \vec{B}^{(2)} = iB^{(0)} \vec{B}^{(3)} \quad (6.59a)$$

$$\vec{B}^{(2)} \times \vec{B}^{(3)} = iB^{(0)} \vec{B}^{(1)*} \quad (6.59b)$$

$$\vec{B}^{(3)} \times \vec{B}^{(1)} = iB^{(0)} \vec{B}^{(2)*} \quad (6.59c)$$

Note that the symbol * means conjugate form of the field, and superscripts (1), (2), and (3) can be permuted to give the other two equations, Eqs. 6.1; thus, the fields $\vec{B}^{(1)}$, $\vec{B}^{(2)}$, and $\vec{B}^{(3)}$ are simply components of the *magnetic flux density* of *free space* electromagnetism on a circular, rather than on a Cartesian, basis. In quantum field theory, longitudinal component $\vec{B}^{(3)}$ becomes the fundamental photomagnetic of light, and the operator is defined by the following relationship [7–12]:

$$\widehat{B}^{(3)} = B^{(0)} \frac{\widehat{P}}{\hbar} \quad (6.60)$$

where $\widehat{P}$ is the angular momentum operator of one photon. The existence of the longitudinal $\widehat{B}^{(3)}$ in free space is indicated experimentally by optically induced nuclear magnetic resonance (NMR) shifts and by several well-known phenomena of magnetization by light—for example, the *inverse Faraday effects*.

The core logic of Eq. 6.59a asserts that there exists a novel cyclically symmetric field algebra in free space, implying that the usual transverse solutions of Maxwell's equations are tied to the longitudinal, non-zero, real, and physical magnetic flux density $\vec{B}^{(3)}$, which we name the *spin field*. This deduction fundamentally changes our current appreciation of electrodynamics and therefore the principles on which the old quantum theory was derived—for example, the Planck Law [13] and the light quantum hypothesis proposed in 1905 by Einstein.

The belated recognition of $\vec{B}^{(3)}$ implies that there is a magnetic field in free space that is associated with the longitudinal space axis, z, which is labeled (3) in the circular basis. Conventionally, the radiation intensity distribution is calculated using only two transverse degrees of freedom, right and left circular, corresponding to (1) and (2) on the circular basis. The $\vec{B}^{(3)}$ field of vacuum electromagnetism introduces a new paradigm of the field theory, summarized in the cyclically symmetric equations linking it to the usual transverse magnetic plane wave components, $\vec{B}^{(1)} = \vec{B}^{(2)*}$ [6, 14, 15].

In January 1992 at Cornell University the $\vec{B}^{(3)}$ field was first and obliquely inferred from a careful reexamination of known magneto-optics phenomena [16, 17] that had previously been interpreted by convention through the conjugate product $\vec{E}^{(1)} = \vec{E}^{(2)}$ of electric plane-wave components $\vec{E}^{(1)} = \vec{E}^{(2)*}$. In the intervening three-and-a-half years its understanding developed substantially into monographs and papers [6, 14, 15] covering several fundamental aspects of field theory.

The $\vec{B}^{(3)}$ field produces magnetization in an electron plasma that is proportional to the square root of the power density dependence of the circularly polarized EM radiation—conclusive evidence for the presence of the phase-free $\vec{B}^{(3)}$ in the vacuum. There are many experimental consequences of this finding, some of which are of practical utility (e.g., optical NMR). Nevertheless, the most important theoretical consequence is that there exist longitudinal components in free space of EM radiation—a conclusion that is strikingly reminiscent of that obtained from the theory of finite photon mass.

The two ideas are interwoven throughout this book. The characteristic square root light intensity dependence of $\vec{B}^{(3)}$ dominates and theoretically is observable at low cyclotron frequencies when intense, circularly polarized EM radiation interacts with a single electron, or in practical terms an electron plasma or beam. The magnetization induced in such an electron ensemble by circularly polarized radiation therefore is expected to be proportional to the square root of the power density (i.e., the intensity in W/m^2) of the radiation. This result emerges directly from the fundamental, classic equation of motion of one electron in the beam, the *relativistic Hamilton-Jacobi equation*.

To establish the physical presence of $\vec{B}^{(3)}$ in the vacuum therefore requires the observation of this magnetization as a function of the beam's power density, a critically important experiment. Other possible experiments to detect $\vec{B}^{(3)}$, such as the optical equivalent of the *Aharonov-Bohm effect*, are suggested throughout the book.

More details about this section's subject in can be found in the references listed at the end of this chapter. Further details are beyond the scope of this book; thus, we encourage readers to refer to [4, 6–12, 14, 17].

6.4 Scalar Wave Description

What is a "scalar wave" exactly? A *scalar wave* (SW) is just another name for a *longitudinal wave* (LW). The term "scalar" is sometimes used instead because the hypothetical source of these waves is thought to be a "scalar field" of some kind, similar to the Higgs field, for example. In general, the definition of the LW falls into the following description: "A wave motion, in which the particles of the medium oscillate about their mean positions in the direction of propagation of the wave, is called longitudinal wave."

For LWs the vibration of the particles of the medium are in the direction of wave propagation. An LW proceeds in the form of compression and rarefaction, which is the stretch and compression in the same direction as the wave moves. For an LW at places of compression the pressure and density tend to be maximum, while at places where rarefaction takes place, the pressure and density are minimum. In gases only, LWs can propagate. Longitudinal waves also are known as compression waves.

There is nothing particularly controversial about LWs in general. They are a ubiquitous and well-acknowledged phenomenon in nature. Sound waves traveling through the atmosphere (or underwater) are longitudinal, as are plasma waves propagating through space (also known as *Birkeland currents*). Longitudinal waves moving through the Earth's interior are known as *telluric currents*. They can all be thought of as pressure waves of sorts.

Scalar and longitudinal waves are quite different from "transverse" waves. You can observe a *transverse wave* by plucking a guitar string or watching ripples on the surface of a pond. They oscillate (i.e., vibrate, move up and down or side-to-side) perpendicular to their arrow of propagation (i.e., directional movement), as shown in Fig. 6.6.

In modern-day electrodynamics (both classical and quantum), EM waves traveling in "free space" (e.g., photons in vacuum) generally are considered to be TW. Nonetheless, this was not always the case. When the preeminent mathematician James Clerk Maxwell first modeled and formalized his unified theory of electromagnetism in the late nineteenth century, neither the EM SW or LW nor the EM TW had been experimentally proved, but he had postulated and calculated the existence of both.

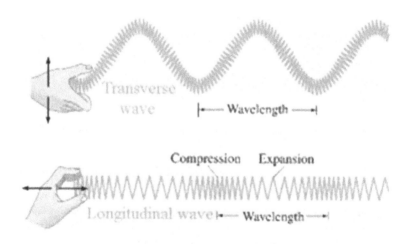

Fig. 6.6 Illustration of a transverse wave and a longitudinal wave

6.4 Scalar Wave Description

After Heinrich Hertz demonstrated experimentally the existence of transverse radio waves in 1887, theoreticians (e.g., Heaviside, Gibbs, and others) went about revising Maxwell's original equations; he was now deceased and could not object. They wrote out the SW/LW component from the original equations because they felt the mathematical framework and theory should be made to agree only by experimenting. Obviously, the simplified equations worked; they helped make the AC/DC electrical age engineerable. But at what expense?

Then in the 1889 Nikola Tesla, a prolific experimental physicist and inventor of the *alternating current* (AC) theory, threw a proverbial wrench in the works when he discovered experimental proof for the elusive electric SW. This seemed to suggest that SW/LW, opposed to TW, could propagate as pure electric waves or as pure magnetic waves. Tesla also believed these waves carried a hitherto-unknown form of excess energy he referred to as "radiant." This intriguing and unexpected result was said to have been verified by Lord Kelvin and others soon after.

Instead of merging their experimental results into a unified proof for Maxwell's original equations, however, Tesla, Hertz, and others decided to bicker and squabble over who was more correct. In actuality they all derived correct results. Nevertheless, because humans (even "rational" scientists) are fallible and prone to fits of vanity and self-aggrandizement, each side insisted dogmatically that they were correct and the other side was wrong.

The issue was allegedly settled after the dawn of the twentieth century when: (1) the concept of the mechanical (passive/viscous) ether was purportedly disproved by Michelson-Morley and replaced by Einstein's Relativistic Spacetime Manifold, and (2) detection of SW/LWs proved much more difficult than initially thought (mostly because of the waves' subtle densities, fluctuating frequencies, and orthogonal directional flow). As a result, the truncation of Maxwell's equations was upheld.

The SW and LW in free space, however, are quite real. Beside Tesla, empirical work carried out by electrical engineers (e.g., Eric Dollard, Konstantin Meyl, Thomas Imlauer, and Jean-Louis Naudin, to name only some) clearly have demonstrated their existence experimentally. These waves seem able to exceed the speed of light, pass through EM shielding (also known as Faraday cages), and produce overunity (more energy out than in) effects. They seem to propagate in a yet unacknowledged counterspatial dimension (also known as hyper-space, pre-space, false-vacuum, Aether, implicit order, etc.).

Because the concept of an all-pervasive material ether was discarded by most scientists, the thought of vortex-like electric and/or magnetic waves existing in free space, without the support of a viscous medium, was thought to be impossible. Nevertheless, later experiments carried out by Dayton Miller, Paul Sagnac, E. W. Silvertooth, and others have contradicted the findings of Michelson and Morley. More recently Italian mathematician-physicist Daniele Funaro, American physicist-systems theorist Paul LaViolette, and British physicist Harold Aspden have all conceived of (and mathematically formulated) models for a free space ether that is dynamic, fluctuating, self-organizing, and allows for the formation and propagation of SWs and LWs.

With the appearance of experiments on the non-classical effects of electrodynamics, authors often speak of EM waves not being based on oscillations of electric and magnetic fields. For example, it is claimed that there is an effect of such waves on biological systems and the human body. Even medical devices are sold that are assumed to work on the principle of transmitting any kind of information via "waves" that have a positive effect on human health. In all cases the explanation of these effects is speculative, and even the transmission mechanism remains unclear because there is no sound theory about such waves, often subsumed under the notion SWs. We have tried to give a clear definition of certain types of waves that can explain the observed effects [18].

Before analyzing the problem in more detail, we must distinguish between SWs that contain fractions of ordinary electric and magnetic fields and such waves that do not and therefore appear even more obscure. Often SWs are assumed to consist of longitudinal fields. In ordinary Maxwellian electrodynamics such fields do not exist, and EM radiation is said always to be transversal. In modern unified physics' approaches (e.g., *Einstein-Cartan-Evans theory* [19, 20], however, it has been shown that the polarization directions of EM fields do exist in all directions of four-dimensional space. So in the direction of transmission, an ordinary EM wave has a longitudinal magnetic component, the so-called $\vec{B}^{(3)}$ field of Evans [21]. (See Sect. 6.4 for more details about the $\vec{B}^{(3)}$ field.)

The $\vec{B}^{(3)}$ field is detectable by the so-called inverse Faraday effect that has been known experimentally since the 1960s [22]. Some experimental setups (e.g., the "magnifying transmitter" of Tesla [16, 17]) claim to use these longitudinal components. They can be considered to consist of an extended resonance circuit where the capacitor plates each have been displaced to the transmitter and receiver site (Fig. 6.7). In an ordinary capacitor (or cavity resonator), a very high-frequency

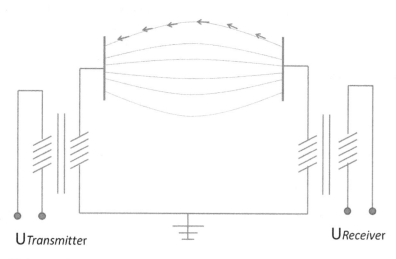

Fig. 6.7 Propagation of longitudinal electric waves according to Tesla

6.5 Longitudinal Potential Waves

wave (GHz or THz range) leads to significant runtime effects of the signal so that the quasistatic electric field can be assumed to be cut into pulses.

These represent the near-field of an EM wave and may be thought of as longitudinal. For lower frequencies the electric field between the capacitor plates remains quasistatic and therefore longitudinal too. We do not want to go deeper into this subject here. Having given hints for the possible existence of longitudinal electric and magnetic fields, we leave this area of concentration on mechanisms that allow transmission of signals, even without any detectable EM fields, to readers.

Before we move on with more details about the SW, we need to lay groundwork about the types of waves and where the SW falls under that category; thus, we need to have some idea about TWs and LTWs and what their descriptions are. This subject was discussed in a previous section of this chapter quite extensively; however, we describe the subject of longitudinal potential waves in the next section.

6.5 Longitudinal Potential Waves

In the following we develop the theory of EM waves with vanishing field vectors. Such a field state normally is referred to as a *vacuum state* and was described in full relativistic detail by Eckardt and Lindstrom [22]. Vacuum states also play a role in the microscopic interaction with matter. Here we restrict consideration to ordinary electrodynamics to give engineers a chance to fully understand the subject.

With $\vec{E}$ and $\vec{B}$ designating the classical electric and magnetic field vectors, a vacuum state is defined by:

$$\vec{E} = 0 \tag{6.61}$$

$$\vec{B} = 0 \tag{6.62}$$

The only possibility to find EM effects then is by the potentials. These are defined as vector and scalar potentials to constitute the "force" fields $\vec{E}$ and $\vec{B}$:

$$\vec{E} = -\nabla U - \dot{\vec{A}} \tag{6.63}$$

$$\vec{B} = \vec{\nabla} \times \vec{A} \tag{6.64}$$

with electric scalar potential U and magnetic vector potential $\vec{A}$. The dot above the $\vec{A}$ in Eq. 6.63 denotes the time derivative. The vacuum conditions, as stated in Eqs. 6.61 and 6.62, will lead to the following sets of equations:

$$\nabla U = -\dot{\vec{A}} \tag{6.65}$$

$$\vec{\nabla} \times \vec{A} = 0 \tag{6.66}$$

From Eq. 6.66 it follows immediately that the *vector potential* is vortex-free, representing a laminar flow. The gradient of the scalar potential is coupled to the time derivative of the vector potential, so both are not independent of one another. A general solution of these equations was derived by Eckardt and Lindstrom [22]. This is a wave solution where $\vec{A}$ is in the direction of propagation (i.e., this is a LW). Several wave forms are possible, which may even result in a propagation velocity different from the speed of light c. As a simple example, we assume a sine-like behavior of vector potential $\vec{A}$:

$$\vec{A} = \vec{A}_0 \sin\left(\vec{k} \cdot \vec{x} - \omega t\right) \tag{6.67}$$

with direction of propagation $\vec{k}$ (wave vector), space coordinate vector $\vec{x}$, and time frequency ω. Then it follows from Eq. 6.66 that

$$\nabla U = \vec{A}_0 \omega \cos\left(\vec{k} \cdot \vec{x} - \omega t\right) \tag{6.68}$$

This condition must be met for any potential U. We make the approach as follows:

$$U = U_0 \sin\left(\vec{k} \cdot \vec{x} - \omega t\right) \tag{6.69}$$

to find that:

$$\nabla U = k U_0 \cos\left(\vec{k} \cdot \vec{x} - \omega t\right) \tag{6.70}$$

which, compared to Eq. 6.68, defines the constant $\vec{A}_0$ to be:

$$\vec{A}_0 = \vec{k}\left(\frac{U_0}{\omega}\right) \tag{6.71}$$

Obviously, the waves of $\vec{A}$ and U have the same phase.

Next, we consider the energy density of such a combined wave. This is given in general by:

$$w = \frac{1}{2}\varepsilon_0 \vec{E}^2 + \frac{1}{2\mu_0}\vec{B}^2 \tag{6.72}$$

From Eqs. 6.65 and 6.66 it can be seen that the magnetic field disappears identically, but the electric field is a vanishing sum of two terms that are different from zero.

These two terms evoke an energy density of space where the wave propagates. This cannot be obtained out of the force fields (these are zero) but must be computed from the constituting potentials. As discussed in a paper by Eckardt and Lindstrom [20], we must write:

6.5 Longitudinal Potential Waves

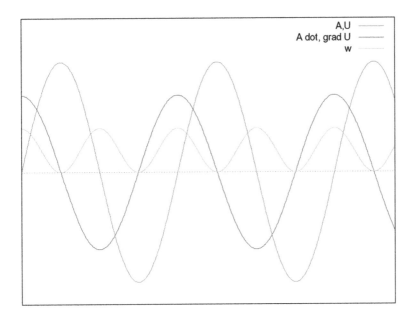

Fig. 6.8 Phases of potentials $\vec{A}$ and U and energy density w

$$w = \frac{1}{2}\varepsilon_0 \left(\dot{\vec{A}}^2 + (\nabla U)^2 \right) \tag{6.73}$$

With Eq. 6.67 and Eq. 6.69, it follows that:

$$w = \varepsilon_0 k^2 U_0^2 \cos^2\left(\vec{k} \cdot \vec{x} - \omega t \right) \tag{6.74}$$

This is an oscillating function, meaning that the energy density varies over space and time in phase with the propagation of the wave. All quantities are depicted in Fig. 6.8. Energy density is maximal where the potentials cross the zero axis. There is a phase shift of 90° between both plots that can be observed in the figure.

There is an analogy between longitudinal potential waves and acoustic waves. It is well known that acoustic waves in air or solids are mainly longitudinal too. The elongation of molecules is in the direction of wave propagation, as shown in Fig. 6.9. This is a variation in velocity. Therefore, the magnetic vector potential can be compared with a velocity field. The differences in elongation evoke a local pressure difference. Where the molecules are pressed together, the pressure is enhanced and vice versa. From conservation of momentum, the force $\vec{F}$ in a compressible fluid is given by:

$$\vec{F} = \dot{\vec{u}} + \frac{\nabla p}{\rho} \tag{6.75}$$

In this equation the term $\vec{u}$ is the velocity field, p is the pressure, and ρ is the density of the medium.

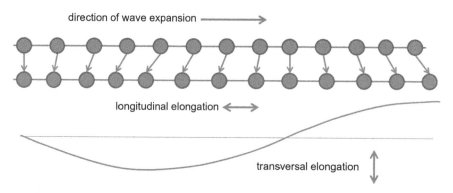

Fig. 6.9 Schematic representation of longitudinal and transversal waves

This is a full analysis of Eq. 6.63. In particular, we can see that in the EM case spacetime must be "compressible"; otherwise, there is no gradient of the scalar potential. As a consequence, space itself must be compressible, leading us to the principles of general relativity.

6.6 Transmitters and Receivers for Longitudinal Waves

A sender for longitudinal potential waves must be a device that avoids producing $\vec{E}$ and $\vec{B}$ fields but sends out oscillating potentials waves. We discuss two propositions about how this can be achieved technically. In the first case, we use two ordinary transmitter antennas (with directional characteristics) with a distance of half a wavelength (or an odd number of half waves). This means that ordinary EM waves cancel out, assuming that the near-field is not disturbed significantly. Because the radiated energy cannot disappear, it must propagate in space and is transmitted in the form of potential waves. This is depicted in Fig. 6.10.

A more common example is a bifilar flat coil (e.g., from the patent of Tesla [23])—see second drawing in Fig. 6.10. The currents in opposite directions effect an annihilation of the magnetic field component, while an electric part may remain because of the static field of the wires, as shown in the Fig. 6.11.

Construction of a receiver is not so straightforward. In principle no magnetic field can be retrieved directly from $\vec{A}$ because of Eq. 6.66. The only way is to obtain an electrical signal by separating both contributing parts in Eq. 6.63 so that the equality is outweighed and an effective electric field remains that can be detected by conventional devices [22]. A very simple method would be to place two plates of a capacitor at a distance of half a wavelength (or odd multiples of it). Then the voltage in space should influence the charge carriers in the plates, leading to the same effect as if a voltage had been applied between the plates. The real voltage in the plates or the compensating current can be measured (Fig. 6.12). The *tension of space* operates directly on the charge carriers while no electric field is induced. The $\vec{A}$ part

6.6 Transmitters and Receivers for Longitudinal Waves

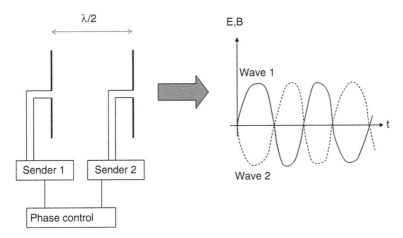

Fig. 6.10 Suggestion for a transmitter of longitudinal potential waves

Fig. 6.11 Tesla coils according to the patent [23]

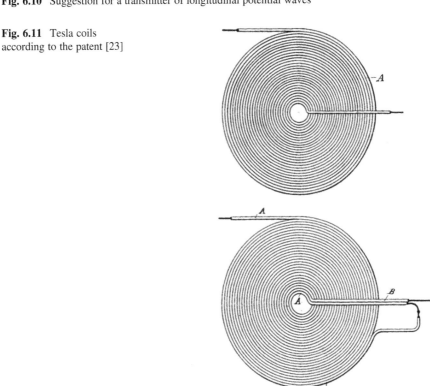

is not contributing because the direction of the plates is perpendicular to it (i.e., no significant current can be induced).

Another possibility of a receiver is to use a screened box (*Faraday cage*). If the mechanism described for the capacitor plates is valid, the electrical voltage part of

Fig. 6.12 Suggestion for a receiver of longitudinal potential waves (capacitor)

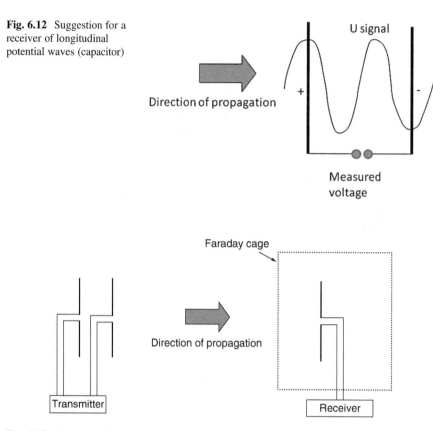

Fig. 6.13 Suggestion for a receiver of longitudinal potential waves (Faraday cage)

the wave creates charge effects that are compensated immediately because of the high conductivity of the material. As is well known, the interior of a Faraday cage is free of electric fields. The potential is constant because it is constant on the box's surface. Therefore, only the magnetic part of the wave propagates in the interior where it can be detected by a conventional receiver (Fig. 6.13).

Another method of detection is using vector potential effects in crystalline solids. As is well known from solid-state physics, the vector potential produces excitations within the quantum mechanical electronic structure, provided the frequency is near to the optical range. Crystal batteries work in this way. They can be engineered through chemical vapor deposition of carbon. In the process you get strong lightweight crystalline shapes that can handle lots of heat and stress by high currents. For detecting LWs, the excitation of the electronic system must be measured (e.g., by photoemission or other energetic processes in the crystal).

All these are suggestions for experiments with LWs. Additional experiments can be performed for testing the relationship between wave vector k and frequency ω to check whether this type of wave propagates with the ordinary velocity of light c:

6.6 Transmitters and Receivers for Longitudinal Waves

$$c = \frac{\omega}{k} \qquad (6.76)$$

where k is defined from the *wavelength* λ by the following relation:

$$k = \frac{2\pi}{\lambda} \qquad (6.77)$$

As pointed out in the paper by Eckhardt and Lindstrom [22], the speed of propagation depends on the form of the waves.

This even can be a non-linear step-function. The experimental setup shown earlier in Fig. 6.11 can be used directly for finding the $\omega(\vec{k})$ relationship because the wavelength and frequency are measured at the same time. There are rumors that Eric P. Dollard [24] found a propagation speed of LWs of $(\pi/2) \cdot c$, which is 1.5 times the speed of light; however, no reliable experiments on this have been reported in the literature.

The ideas worked out in this section may not be the only way LWs can be explained and technically handled. As mentioned in the introduction, electrodynamics derived from a unified field theory (Evens et al. [19]) predicts effects of polarization in all space and time dimensions and may lead to a discovery of even richer and more interesting effects.

6.6.1 Scalar Communication System

The basic scalar communication system indicates that the communications antenna does not make any sense according to normal EM theory. The goal of a scalar antenna is to create powerful repulsion and/or attraction between two magnetic fields to create large scalar bubbles/voids. This is done by using an antenna with two opposing EM coils that effectively cancel out as much of each other's magnetic field as possible. An ideal scalar antenna will emit no EM field (or as little as possible) because all power is being focused into the repulsion–attraction between the two opposing magnetic fields. Normal EM theory suggests that because such a device emits no measurable EM field, it is useless and will only heat up.

A scalar signal reception antenna similarly excludes normal EM waves and only measures changes in magnetic field attraction and repulsion. This typically will be a two-coil powered antenna that sets up a static opposing or attracting magnetic fields between the coils, and the coils are counterwound so that any normal radio frequency (RF) signal will be picked up by both coils and effectively canceled out.

It has been suggested that scalar fields do not follow the same rules as EM waves and can penetrate through materials that would normally slow or absorb them. If true, a simple proving method is to design a scalar signal emitter and a scalar signal receiver and encase each inside separate shielded and grounded metal boxes, known as Faraday cages. These boxes will absorb all normal EM energy and will prevent any regular non-scalar signal transmissions from passing from one box to the other.

Some people have suggested that organic life may make use of scalar energies in ways that we do not yet understand. Therefore, caution is recommended when experimenting with this fringe technology. Nevertheless, keep in mind that if scalar fields do exist, we are likely already deeply immersed in an unseen field of scalar noise all the time, generated anywhere two magnetic fields oppose or attract. Common scalar field noise sources include AC electrical cords, powerlines carrying high currents, and electric motors that operate on the principle of powerful spinning regions of repulsion and attraction.

6.7 Scalar Waves Experiments

It can be shown that SWs normally remain unnoticed and are very interesting in practical use for information and energy technology for reasons of their special attributes. The mathematical and physical derivations are supported by practical experiments. Such demonstrations show the following:

1. Wireless transmission of electrical energy
2. Reactions of the receiver to the transmitter
3. Free energy with an overunity effect of about 3
4. Transmission of SWs with 1.5 times the speed of light
5. Inefficiency of a Faraday cage to shield SWs

6.7.1 Tesla Radiation

Shown here are five extraordinary science experiments that are incompatible with textbook physics. Short courses, that were given by Meyl [25] show the transmission of longitudinal electric waves.

This is a historical experiment because 100 years ago the famous experimental physicist Nikola Tesla measured the same wave properties as this author. From him stems a patent concerning the wireless transmission of energy (1900) [26]. Because he also had to find out that very much more energy arrives at the receiver than the transmitter takes up, he spoke of a *magnifying transmitter*.

By the effect back on the transmitter Tesla sees that he has found the resonance of the Earth that lies according to his measurement at 12 Hz. Because the Schumann resonance of a wave, which goes with the speed of light lies at 7.8 Hz, however, Tesla came to the conclusion that his wave was 1.5 times the speed of light c [27]. As founder of diathermy Tesla, already had pointed to the biological effectiveness and to the possible use in medicine. The diathermy of today has nothing to do with Tesla radiation; it uses the incorrect wave and, as a consequence, hardly has any medical importance.

6.7 Scalar Waves Experiments

The discovery of the Tesla radiation has been denied and is not mentioned in textbooks anymore. For that there are two reasons:

1. No high school ever has rebuilt a "magnifying transmitter," The technology simply was too costly and too expensive. In that way the results have not been reproduced, as it is imperative to acknowledge. This author has solved this problem using modern electronics by replacing the spark a gap generator with a function generator and it operates with high-tension with 2–4 V low-tension. Meyl [25] sells the experiment as a demonstration-set so that it is reproduced as often as possible. It fits in a case and has been sold more than 100 times. Some universities already could confirm the effects. The measured degrees of effectiveness lie between 140 and 1000%.
2. The other reason why this important discovery could fall into oblivion can be seen in the missing suitable field description. Maxwell's equations in any case only describe TWs, for which the field pointers oscillate perpendicular to the direction of propagation.

The vectorial part of the wave equation derived from Maxwell's equations are presented here:

$$\begin{cases} \vec{\nabla} \times \vec{E} = -\dfrac{\partial \vec{B}}{\partial t} \\ \vec{\nabla} \times \vec{H} = \vec{J} + \dfrac{\partial \vec{D}}{\partial t} \\ \vec{B} = \mu \vec{H} \\ \vec{D} = \varepsilon \vec{E} \\ \vec{J} = 0 \end{cases} \Rightarrow \text{In Linear Media} \quad (6.78)$$

and

$$\vec{\nabla} \times (\vec{\nabla} \times \vec{E}) = -\mu \dfrac{\partial (\vec{\nabla} \times \vec{H})}{\partial t} = -\mu\varepsilon \left(\dfrac{\partial^2 \vec{E}}{\partial t^2} \right) \quad (6.79)$$

Then, from the result of Eqs. 6.78 and 6.79, we obtain the wave equation:

$$\begin{cases} \nabla^2 \vec{E} = \vec{\nabla}(\vec{\nabla} \cdot \vec{E}) - \vec{\nabla} \times (\vec{\nabla} \times \vec{E}) = \dfrac{1}{c^2} \dfrac{\partial^2 \vec{E}}{\partial t^2} \\ \mu\varepsilon = \dfrac{1}{c^2} \end{cases} \quad (6.80)$$

See Chap. 4 of this book for more details on derivation of wave equations from Maxwell's equations. Note that in all these calculations, the following symbols apply:

$\vec{E}$ = electric filed or electric force
$\vec{H}$ = auxiliary field or magnetic field
$\vec{D}$ = electric displacement ($\vec{D} = \varepsilon \vec{E}$ in linear medium)
$\vec{B}$ = magnetic intensity or magnetic induction
$\vec{J}$ = current density

Now breaking down the first equation of in the sets of Eq. 6.80 will be as follows:

$$\underbrace{\nabla^2 \vec{E}}_{\text{Laplace operator over } \vec{E}} = \underbrace{\vec{\nabla}(\vec{\nabla} \cdot \vec{E}) - \vec{\nabla} \times (\vec{\nabla} \times \vec{E})}_{\begin{array}{l} \text{if } \vec{\nabla} \cdot \vec{E} = 0 \text{ then we have Transversal Wave} \\ \text{if } \vec{\nabla} \times \vec{E} = 0 \text{ then we have Longitudinal Wave} \end{array}} = \underbrace{\frac{1}{c^2} \frac{\partial^2 \vec{E}}{\partial t^2}}_{c \text{ is speed of light}} \quad (6.81)$$

Note that in the equation that if $\vec{\nabla} \cdot \vec{E} \neq 0$, then we have a situation that provides the SW conditions, while the following relationships apply as well:

$$\vec{E} = -\vec{\nabla}\phi : \begin{cases} (1) \not{\vec{\nabla}}(\vec{\nabla} \cdot \vec{E}) = \not{\vec{\nabla}}\left[\frac{1}{C^2} \frac{\partial^2 \phi}{\partial T^2}\right] \\ (2) \quad \vec{\nabla} \cdot \vec{E} = -\vec{\nabla} \cdot \vec{\nabla}\phi \end{cases} \quad (6.82)$$

$$\vec{\nabla} \cdot \vec{D} = \rho : \left\{ (3) \quad \vec{\nabla} \cdot \vec{E} = \frac{\rho}{\varepsilon} \right.$$

From this equation we also can conclude that the *plasma wave* is:

$$\nabla^2 \phi = \frac{1}{c^2} \cdot \left(\frac{\partial^2 \phi}{\partial t^2}\right) - \frac{\rho}{\varepsilon} \quad (6.83)$$

The results found in Eqs. 6.81 and Eq. 6.82 are the scalar part of the wave equation describing longitudinal electric waves, which end up with a deviation of plasma waves, as can be seen in Eq. 6.83. In these equations symbol ϕ represents a *scalar field*, as described in Chap. 4.

If we derive the field vector from a scalar potential ϕ, then this approach immediately leads to an inhomogeneous wave equation, which is called a *plasma wave*. Solutions are known, such as the electron plasma waves, that are *longitudinal oscillations* of the electron density—Langmuir waves.

6.7.2 Vortex Model

The Tesla experiment and this author's historical rebuild, however, show more. Such LWs obviously exist even without plasma in the air and even in vacuum. The questions thus are:

6.7 Scalar Waves Experiments

I. What does divergence $\vec{E}$ describe in this case?
II. How is the impulse passed on so that a longitudinal standing wave can form?
III. How should a shock wave come about if there are no particles that can push each other?

We have solved these questions by extending Maxwell's field theory for vortices of the electric field. These so-called potential vortices are able to form structure and propagate in space for reason of their particle nature as a longitudinal shock wave. The model concept is based on the ring vortex model of Hermann von Helmholtz, which Lord Kelvin made popular. In Volume 3 of the Meyl book, *Potential Vortex* [1], the mathematical and physical derivation is described.

In spite of the field theoretical set of difficulties every physicist at first will seek a conventional explanation. We will try three approaches as follows: (1) resonant circuit interpretation, (2) ld interpretation, and (3) vortex interpretation. The details of these approaches are given in the following subsections.

6.7.2.1 Resonant Circuit Interpretation

Tesla presented his experiment to, among others, Lord Kelvin, and 100 years ago Tesla spoke about a vortex transmission. In the opinion of Kelvin, however, vortex transmission by no means concerns a wave but rather radiation. Kelvin recognized clearly that every radio-technical interpretation had to fail because alone the course of the field lines is a completely different one.

It presents itself assuming a resonant circuit, consisting of a capacitor and an inductance (Fig. 6.14). If both electrodes of the capacitor are pulled apart, then between both stretches an electric field. The field lines start at one sphere, the transmitter, and they bundle up again at the receiver. In this manner a higher degree of effectiveness and a very tight coupling can be expected. In this way, without doubt, some but not all, of the effects can be explained.

The inductance is split up in two air transformers, which are wound in a completely identical fashion. If a field in sinusoidal tension voltage is transformed up in the transmitter, then it again is transformed down at the receiver. The output voltage should be smaller or, at most, equal to the input voltage, but it is substantially higher!

An alternative wiring diagram can be drawn and calculated, but in no case does the measurable result that light-emitting diodes at the receiver glow brightly ($U > 2$ V), whereas at the same time the corresponding light-emitting diodes at the transmitter go out ($U < 2$ V)! To check this result, both coils are exchanged.

The measured degree of effectiveness lies, despite the exchange, at 1000%. If the law of conservation of energy is not to be violated, then only one interpretation is left: The open capacitor withdraws field energy from its environment. Without consideration of this circumstance, the error deviation of every conventional model calculation lies at more than 90%. In this case, one should do without the calculation.

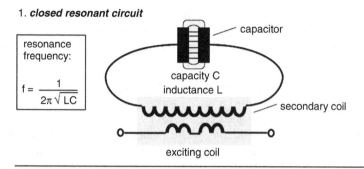

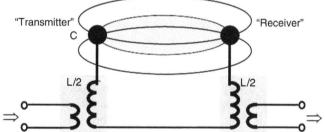

Fig. 6.14 Interpretation as an open resonant circuit

The calculation concerns oscillating fields because the spherical electrodes are changing in polarity with a frequency of approximately 7 MHz. They are operated in resonance. The condition for resonance reads as: identical frequency and opposite phase. The transmitter obviously modulates the field in its environment, while the receiver collects everything that fulfills the condition for resonance. Also, in the open question regarding the transmission velocity of the signal, the resonant circuit interpretation fails. But a HF-technician still has another explanation on the tip of his or her tongue.

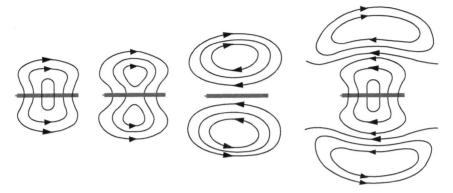

Fig. 6.15 The coming off of the electric field lines of the dipole

6.7.2.2 Near-Field Interpretation

At the antenna of a transmitter in the near-field (a fraction of the wavelength) only SWs (*potential vortex*) exist. They decompose into EM waves in the far-field and further. The near-field is not described by Maxwell's equations and the theory only is postulated. It is possible to pick up only SWs from radio transmissions. Receivers that pickup EM waves actually are converting those waves into potential vortices that are conceived as "standing waves."

This presents itself assuming a resonant circuit consisting of a capacitor and an inductance (Fig. 6.15). If both electrodes of the capacitor are pulled apart, then between both stretches an electric field. The field lines start at one sphere, the transmitter, and they bundle up again at the receiver. In this manner, a higher degree of effectiveness and a very tight coupling can be expected. In this way, without doubt, some but not all, of the effects can be explained.

In the near-field of an antenna effects are measured, on the one hand, as inexplicable because they evade the normally used field theory; on the other hand, can be shown as very close to SW effects. Everyone knows a practical application—for example, at the entrance of department stores, where the customer has to go through or between SW detectors.

In Meyl's experiment [25] the transmitter is situated in the mysterious near zone. Also, Tesla always worked in the near zone. But one who asks for the reasons will discover that the near-field effect is nothing else but the SW part of the wave equation. Meyl's explanation goes as follows: "The charge carriers which oscillate with high frequency in an antenna rod form longitudinal standing waves. As a result, also, the fields in the near zone of a *Hertzian dipole* are *longitudinal scalar wave fields*. The picture shows clearly how vortices are forming and how they come off the dipole."

Like for charge carriers in the antenna rod the phase angle between current and tension voltage amounts to 90°, and occurs in the near-field; likewise, electric and magnetic fields phase shifted for 90°. In the far-field, however, the phase angle is zero. In this author's interpretation the vortices are breaking up, they decay, and transverse radio waves are formed.

6.7.2.3 Vortex Interpretation

The vortex decay, however, depends on the velocity of propagation. Calculated at the speed of light the vortices already have decayed within half the wavelength. The faster the velocity, the more stable they get to remain stable above 1.6 times the velocity. These very fast vortices contract in the dimensions. They now can tunnel. Therefore, speed faster than light occurs at the tunnel effect. Therefore, no *Faraday cage* can shield fast vortices.

Because these field vortices with a particle nature following the high-frequency oscillation permanently change their polarity from positive to negative and back, they do not have a charge on the average over time. As a result, they almost are unhindered penetrate solids. Particles with this property are called *neutrinos* in physics. The field energy that is collected in this experiment stems from the neutrino radiation that surrounds us. Because the source of this radiation, all the same if the origin is artificial or natural, is far away from the receiver, every attempt of a near-field interpretation goes wrong. After all, does the transmitter installed in the near-field zone supply less than 10% of the received power. The 90%, however, which is of concern here, cannot stem from the near-field zone!

6.7.3 Meyl's Experiment

In Meyl's experimental set up he also takes a few other steps in order to conduct his research that is reported here [25]:

At the function generator he adjusts frequency and amplitude of the sinusoidal signal, with which the transmitter is operated. At the frequency regulator I turn so long, until the light-emitting diodes at the receiver glow brightly, whereas those at the transmitter go out. Now an energy transmission takes place.

If the amplitude is reduced so far, until it is guaranteed that no surplus energy is radiated, then in addition a gain of energy takes place by energy amplification.

If he takes down the receiver by pulling out the earthing, then the lighting up of the LED's signals, the mentioned effect back on the transmitter. The transmitter thus feels, if its signal is received.

The self-resonance of the Tesla coils, according to the frequency counter, lies at 7 MHz. Now the frequency is running down and see there, at approx. 4.7 MHz the receiver again glows, but less bright, easily shieldable and without discernible effect back on the transmitter. Now we unambiguously are dealing with the transmission of the Hertzian part and that goes with the speed of light. Because the wavelength was not changed, does the proportion of the frequencies determine the proportion of the velocities of propagation? The SW according to that goes with (7/4.7=) 1.5 times the speed of light.

If Meyl puts the transmitter into the aluminum case and closes the door, then nothing should arrive at the receiver. Expert laboratories for EM compatibility in this case indeed cannot detect anything and, although in spite of that, the receiver lamps glow! By turning the receiver coil, it can be verified that an electric and not a magnetic coupling is present although the Faraday cage should shield electric fields. The SW obviously overcomes the cage with a speed faster than light by *tunneling*. We can summarize what we have discussed so far in respect to the SW in next subsection as follows.

6.7.4 Summary

German professor Konstantin Meyl developed a new unified field and particle theory based on the work of Tesla. Meyl's theory describes quantum and classical physics, mass, gravitation, the constant speed of light, neutrinos, waves, and particles—all explained by vortices. The subatomic particle characteristics are calculated accurately by this model. Well-known equations also are derived by the unified equation. He provides tools replicating one of Tesla's experiments, which demonstrates the existence of SWs. Scalar waves are simply energy vortices in the form of particles. Here is a summary of an interview with Konstantin Meyl on his theory and technologies.

The *unified field theory* describes the electromagnetic, eddy current, potential vortex, and special distributions. This combines an extended wave equation with a Poisson equation. Maxwell's equations can be derived as a special case where Gauss's Law for magnetism is not equal to zero. This means that magnetic charges do exist in Meyl's theory [25]. That electric and magnetic fields always are generated by motion is the fundamental idea that this equation is derived from. The unipolar generator and transformer have conflicting theories under standard theories. Meyl splits them into the equations of transformation of the electric and magnetic fields separately, which describes unipolar induction and the equation of convection, relatively.

Meyl says that the field is always first, which generates particles by decay or conversion. Classical physics does not recognize energy particles (i.e., potential vortices), so they were not included in the theory. Quantum physics effectively tried to explain everything with vortices, which is why it is incomplete. The derivation of Schrodinger's equation from the extended Maxwell's equations means they are vortices. For example, photons are light as particle vortices and EM light is in wave form, which depends on the detection method that can change the form of light.

Gravitation is from the speed of light difference caused by proximity that, proportional to field strength, decreases the distance of everything for the field strength. This causes the spin of the Earth or another mass to move quicker farther away from the greatest other field's influence, thus orbit the Sun or larger mass. The closest parts of the bodies have smaller distances because of larger total fields and thus slower speeds of light. These fields are generated by closed field lines of vortices and largely are matter. Matter does not move as energy because the speed of light is zero in the field of the vortex because of infinite field strength within the closed field. The more mass in proximity to something, the greater the field strength and the shorter the distances, which causes larger groups of subatomic particles to individually have smaller sizes.

The total field energy in the Universe is exactly zero, but particle and energy forms of vortices divide the energy inside and outside the vortex boundary. When particles are destroyed, no energy is released. No energy was produced when large amounts of matter was destroyed at MIT with accelerated natrium atoms. This is what Tesla predicted but contradicts Einstein's $E = MC^2$. Einstein's equation is correct as long as the number of subatomic particles is only divided; energy comes from mass defect, not from destruction.

There are various kinds of waves. Electromagnetic waves are fields, scalar electric or eddy currents or a magnetic vortex, which Tesla started with, and magnetic scalar or the potential vortex, which Meyl focuses on and is used in nature. The EM is fixed at the speed of light at that specific closed field strength. Scalar vortices can be any speed. Neutrinos travel at 1.6c or higher and do not decay to EM. Tesla-type SWs are between c and 1.6c and decay at distances proportional to their speed (used in the traditional radio near-field). Under the speed c, the scalar vortex acts as an electron.

Black holes may produce and emit neutrinos by condensing and transforming matter into massive fast particles with apparently no mass or charge because of their very high frequency of fluctuation. Neutrinos oscillate in mass and charge. When neutrinos hit matter and have a precise charge or mass, they produce one of three effects: a gain in mass, a production of EM, or emission of slower neutrinos.

Resonance requires the same frequency, same modulation, and opposite phase angle. Once (scalar) resonance is reached, a direct connection is created from the transmitter to the receiver. Signal and power will pass through a *Faraday cage*.

References

1. http://www.k-meyl.de/xt_shop/index.php?cat=c3_Books-in-English.html
2. B. Zohuri, *Plasma Physics and Controlled Thermonuclear Reactions Driven Fusion Energy* (Springer, Cham, 2016)
3. B. Zohuri, *Directed Energy Weapons: Physics of High Energy Lasers (HEL)*, 1st edn. (Springer Publishing Company, Cham, 2016)
4. L.M. Hively, G.C. Giakos, Toward a more complete electrodynamic theory. Int. J. Signals Imaging Syst. Engr. **5**, 3–10 (2012)
5. D.A. Woodside, Three vector and scalar field identities and uniqueness theorems in Euclidian and Minkowski space. Am. J. Phys. **77** (May 2009)
6. M.W. Evans, J.P. Vigier, *The Enigmatic Photon, Volume 1: The Field $\vec{B}^{(3)}$* (Springer, Dordrecht, 1994).; Softcover Reprint of the Originated. Edition
7. M.W. Evans, *Phys. B* **182**, 237 (1992).; **183**, 103 (1993)
8. M.W. Evans, in *Waves and Particles in Light and Matter*, ed. by A. Garuccio, A. Van der Merwe (Eds), (Plenum, New York, 1994)
9. M.W. Evans, *The Photon's Magnetic Field* (World Scientific, Singapore, 1992)

10. A.A. Hasanein, M.W. Evans, *Quantum Chemistry and the Photomagnetic* (World Scientific, Singapore, 1992)
11. M.W. Evans, *Mod. Phys. Lett.* **7**, 1247 (1993).; *Found. Phys. Lett.* **7**, 67 (1994); *Found. Phys*
12. M.W. Evans, S. Kielich, in *Modern Nonlinear Optics, Vols. 85(1), 85(2), 85(3) of Advances in Chemical Physics*, ed. by I. Prigogine, S. A. Rice (Eds), (Wiley Interscience, New York, 1993/1994). Volume 85(2) contains a discussion of the cyclic algebra
13. A. Einstein, Zur Elektrodynamik bewegter Körper. *Annalen der Physik* **17**(1), 891–921 (1905). pp. 910–911
14. M.W. Evans, J.-P. Vigier, *The Enigmatic Photon Volume 2: Non-Abelian Electrodynamics* (Kluwer Academic, Dordrecht, 1995)
15. M.W. Evans, J.-P. Vigier, *The Enigmatic Photon Volume 3: $\vec{B}^{(3)}$ Theory and Practice* (Kluwer Academic, Dordrecht, 1995)
16. J.P. van der Ziel, P.S. Pershan, L.D. Malmstrom, Optically-induced magnetization resulting from the inverse faraday effect. Phys. Rev. Lett. **15**(5), 190–193 (1965)
17. W. Happer, Rev. Mod. Phys. **44**, 169 (1972). T. W. Barrett, H. Wohltjen, A. Snow, *Nature* **301**, 694 (1983)
18. H. Eckardt, What are scalar waves?, A.I.A.S. and UPITEC, www.aias.us, www.atomicprecision.com, www.upitec.org
19. M.W. Evans et al., *Generally Covariant Unified Field Theory* (Abramis, Suffolk, 2005). vol. 1–7 (see also www.aias.us, section UFT papers)
20. H. Eckardt, D.W. Lindstrom, Reduction of the ECE Theory of Electromagnetism to the Maxwell-Heaviside Theory, part I–III., www.aias.us, section publications
21. M.W. Evans, *The Enigmatic Photon*, vol 1–5 (Kluwer Academic Publishers, Dordrecht, 1994). (see also www.aias.us, section omnia opera)
22. H. Eckardt, D.W. Lindstrom, *Solution of the ECE vacuum equations, in Generally Covariant Unified Field Theory*, vol 7 (Abramis, Suffolk, 2011), pp. 207–227 (see also www.aias.us, section publications)
23. U.S. Patent 512,340, *Coil for Electro-Magnets* (Nikola Tesla, 1894)
24. http://ericpdollard.com/free-papers/
25. *Ing Konstantin Meyl: Scalar Waves, Theory and Experiments*, http://www.k-meyl.de/go/Primaerliteratur/Scalar-Waves.pdf
26. U.S. Patent 645,576, Apparatus for Transmission of Electrical Energy, (Nikola Tesla 1900); see also http://en.wikipedia.org/wiki/Wardenclyffe_Tower#Theory_of_wireless_transmission
27. Nikola Tesla: Art of transmitting electrical energy through the natural medium, (US Patent No. 787,412, N.Y. 18.4.1905)

Appendix A: Relativity and Electromagnetism

In this appendix we examine and, relativistically, modify Newtonian particle mechanics; thus, it would be natural to look with the same intention at Maxwell's electrodynamics, at first in a vacuum. That theory, however, turns out to be already "special-relativistic." Unlike Newtonian mechanics, classical electrodynamics is consistent with special relativity. Maxwell's equations and Lorentz's force law can be applied legitimately in any inertial system (e.g., system S). The only two assumptions we need to make about the electromagnetic force are that it is *pure* force—in other words, rest mass is preserving—and that it acts on particles in proportion to point charge q, which they carry. Beyond that only "simplicity" and some analogies with Newton's gravitational theory will guide us, which what this appendix is all about.

A.1 Introduction

With corrected Newtonian mechanics, we are now in a position to develop a complete and consistent formulation of relativistic electrodynamics. We will not be changing the rules of electrodynamics at all, however; rather, we will be *expressing* these rules in a notation that exposes and illustrates their relativistic character.

As we said the at the beginning of this appendix, the intention is to look at Maxwell's electrodynamics and its basic laws, as they have been summarized by the four Maxwell's equations, plus *Lorentz's force law* , are for an *invariant condition* under Lorentz transformations from one inertial frame to another. As also mentioned before, by inertial frame with only two assumptions: (1) the force induced from electromagnetics (EM) will be a pure force; and (2) the rest mass of moving particles persevere, and that it acts on particles in proportion to point charge q, which the particle carries. Indeed, it was the problem of finding a transformation that leaves

Maxwell's equations invariant, which led Lorentz to the discovery of the equations now associated with his name.

Nevertheless, even though relativity did not *modify* Maxwell's theory in a vacuum, it added immeasurably to our understanding of it and also gave us a new means for working in it. On the other hand, the theory of electromagnetism in moving media, originally because of Minkowski, is purely a relativistic development [1].

Maxwell's theory, within its old *Galilean framework*, seems quite far-fetched. Galileo was the first one to observe that a brick on a table, to which is applied a shoving force, soon comes to a stop when a certain time elapses after the initial shove because of friction between the two surfaces (i.e., brick and table). He also concluded, however, that the more you polish the surface of a brick, or a table, the farther the brick travels. Moreover, if the table *itself* is moving, the brick comes to rest with respect to the table, not the floor across which the table is moving or standing. Galileo concluded that the brick stops because of the *friction* that may exist between the brick and the table, as we said. If you can eliminate *all* existing friction by some means of polishing the surfaces to perfect smoothness, then the brick would keep moving forever. Newton, of course, incorporated this discovery into his first law of motion into the classical mechanics of motion [2].

In the real world and in practice, some friction always exists; however, this does not bring an object to an "absolute" state of rest, but merely to rest with respect to whatever it is rubbing against. Within relativity, on the other hand, it is one of the two or three simplest possible theories of a field force such as Yukawa-Klein-Gordon's scalar meson field theory and Nordström's attempt at a special-relativistic theory of gravity.

In fact, rather than *verifying* that Maxwell's equations are *Lorentz invariant*, we will use a synthetic approach that will parallel what we know about classical mechanics and highlight once more the "human-made" aspect of physical laws. We will construct a relativistic field theory that is consistent with just a very few of the basic facts of electromagnetism, and we intend to find that it is Maxwell's. Thus, the main purpose of this appendix is to provide a deeper understanding of the structure of electrodynamics laws that had seemed arbitrary and unrelated before taking on a kind of coherence and inevitability when approached from the point of view of relativity.

To begin with we will show why there needs to be such a thing as magnetism, given electrostatics and relativity, and how, in particular, we can calculate the magnetic force between a current-carrying wire and a moving charge without ever invoking the laws of magnetism. Using Fig. A.1, suppose you had a string of positive charges moving along to the right at speed v. We will assume the charges are close enough together so that we can regard them as a continuous line charge λ.

Superimposed on this positive string is a negative one, $-\lambda$, proceeding to the left at the same speed v. We then have a net current to the right of magnitude:

$$I = 2\lambda v \qquad (A.1)$$

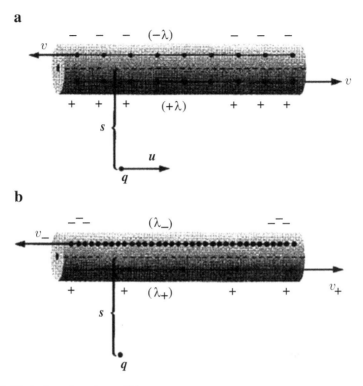

Fig. A.1 Illustration of moving particle q

Meanwhile, distance s away there is a point charge q traveling to the right at speed $u < v$, as shown in Fig. A.1a. Because, the two line charges cancel each other, there is *no electrical force* on charge q in this system S.

We now examine the same situation from another system's point of view—namely, system $\bar{S}$, which moves to the right with speed u, as illustrated in Fig. A.1b. In this frame of reference charge particle q is at rest. According to Einstein's theory and his velocity addition rule, however, the velocities of the positive and negative lines are now defined by the following equation:

$$v_{\pm} = \frac{v \mp u}{1 \mp vu/c^2} \quad (A.2)$$

Because v_- has a greater value than v_+, the *Lorentz contraction* of the space between negative charges is more severe than that between positive charges; *in this frame*, therefore, the *wire carries net negative charge*s [2]. In fact,

$$\lambda_{\pm} = \pm(\gamma_{\pm})\lambda_0 \quad (A.3)$$

where

$$\gamma_\pm = \frac{1}{\sqrt{1 - v_\pm^2/c^2}} \quad (A.4)$$

and λ_0 is the charge density of the positive line in its own rest system. Still, that is not the same as λ; indeed, in system S they already are moving at speed v, thus:

$$\lambda = \gamma \lambda_0 \quad (A.5)$$

It takes some algebra to put $\gamma_\pm$ into a simple form as:

$$\gamma_\pm = \frac{1}{\sqrt{1 - \frac{1}{c^2}(v \mp u)^2 (1 \mp vu/c^2)^{-2}}} = \frac{c^2 \mp uv}{\sqrt{(c^2 \mp uv)^2 - c^2(v \mp u)^2}}$$

$$= \frac{c^2 \mp uv}{\sqrt{(c^2 - v^2) - (c^2 - u^2)}} = \gamma \left\{ \frac{1 \mp uv/c^2}{\sqrt{1 - u^2/c^2}} \right\} \quad (A.6)$$

Naturally, then, the net line change in frame system $\bar{S}$ is going to be:

$$\lambda_{\text{total}} = \lambda_+ + \lambda_- = \lambda_0 (\gamma_+ - \gamma_-) = \frac{-2\lambda uv}{c^2 \sqrt{1 - u^2/c^2}} \quad (A.7)$$

At the conclusion, as the result of unequal Lorentz contraction of the positive and negative lines, we can see that a electrically neutral current-carrying wire in one inertial system will be charged in another.

Now a line charge λ_{total} sets up an electric field as:

$$E = \frac{\lambda_{\text{total}}}{2\pi\varepsilon_0 s} \quad (A.8)$$

But if there is a force on charge particle q in system $\bar{S}$, there must be one in system S as well; in fact, we can *calculate* it by using the transformation rules for forces. Because charge q is at rest in system $\bar{S}$ and force $\bar{F}$ is perpendicular to u, the force in system S is given by a set of equations as follows:

$$\begin{cases} \vec{\bar{F}}_\perp = \frac{1}{\gamma} \vec{F}_\perp \\ \bar{F}_\parallel = F_\parallel \end{cases} \quad (A.9)$$

This equation is valid under the condition that particle q is *instantaneously* at rest in system S, as we mentioned in the preceding as part of the two conditions and assumptions. Equation A.9 is the component of vector force $\vec{F}$ *parallel* to the motion of system $\bar{S}$ and is unchanged, whereas components *perpendicular* are divided by γ [2].

Note that perhaps it may occur to us that we could avoid the bad transformation behavior of force $\vec{F}$ by introducing a "proper" force, analogous to proper velocity,

Appendix A: Relativity and Electromagnetism

that would be the derivative of momentum with respect to *proper* time; it can be written as:

$$K^\mu \equiv \frac{dp^\mu}{d\tau} \tag{A.10}$$

Equation A.10 is a presentation of what is called *Minkowski force*, and it is plainly a *4-vector* because p^μ is a 4-vector and proper time is invariant. The spatial components of K^μ are related to the "ordinary" force by:

$$\vec{K} = \left(\frac{dt}{d\tau}\right) \frac{d\vec{p}}{\sqrt{1 - u^2/c^2}} \vec{F} \tag{A.11}$$

while the zeroth component is provided as:

$$K^0 = \frac{dp^0}{d\tau} = \frac{1}{c} \frac{dE}{d\tau} \tag{A.12}$$

Remember that γ and β pertain to the motion of system $\bar{S}$ with respect to system S, and they are constant values and that $\vec{u}$ is the velocity of the particle q with respect to S. See Chap. 4 of this book, section "Electrodynamics and Relativity," for some further details.

Additionally, in Eq. A.12, apart from factor $1/c$, shows the *proper rate* at which the *energy* of particle q increases—that is, the *proper power* delivered to the particle. Thus, considering all the preceding explanations, we can deduce that

$$F = \left(\sqrt{1 - u^2/c^2}\right) \bar{F} = -\frac{\lambda v}{\pi \varepsilon_0 c^2} \frac{qu}{s} \tag{A.13}$$

Moreover, the charge is attached to the wire by a force that is purely electrical in $\bar{S}$, where the wire is charged and particle charge q is at rest; however, it is distinctly *non-electrical* in system S, where the wire is neutral. Taking them into consideration together, then electrostatics and relativity imply the existence of another force. This so-called "*other force*" is for sure a *magnetic* force. In fact, we can cast Eq. A.13 into more familiar form by using the relation of $c^2 = (\varepsilon_0 \mu_0)^{-1}$ and expressing λv in terms of the current Eq. A.1, and achieve the following form:

$$F = -qu\left(\frac{\mu_0 I}{2\pi s}\right) \tag{A.14}$$

The term contained in parentheses in Eq. A.15 is the magnetic field of a long, straight wire, and the force is precisely what we would have obtained by using the Lorentz force law in system S.

A.2 The Formal Structure of Maxwell's Theory

As we stated in the introduction of the appendix, the only two assumptions that we specifically took into consideration were related to electromagnetic (EM) force. The two assumptions were: (1) pure force, which means that the rest mass is preserving; and (2) that it acts on particles in proportion to the charge particle q, which they carry. Beyond these two assumptions only "simplicity" and some analogies with *Newton's gravitational* theory will guide us to the formal structure of *Maxwell's theory*.

We can start our analysis of this subject by considering and rejecting certain simple possibilities from a Newtonian gravitational force point of view by taking a field of *three-force* $\vec{f} = m\vec{a}$ into consideration where, like the *Newtonian gravitational force*, it acts on a particle independently of its velocity in some frame system S. If we use the set of transformation equations for $\vec{f}$, and its designated components f_1, f_2, and f_3, by showing the new transformed components as f'_1, f'_2, and f'_3 in frame system $\bar{S}$, with a *Lorentz factor* of $\gamma = \gamma(v)$, as:

$$\begin{cases} f'_1 = \dfrac{f_1 - v(dm/dt)}{1 - u_1(v/c^2)} & f'_1 = \dfrac{f_1 - v(\vec{f} \cdot \vec{u}/c^2)}{1 - u_1(v/c^2)} & \text{if } m_0 = \text{constant} \\ f'_2 = \dfrac{f_2}{\gamma[1 - u_1(v/c^2)]} & f'_3 = \dfrac{f_3}{\gamma[1 - u_1(v/c^2)]} \end{cases} \quad (A.15)$$

The second formula in the first row of this equation applies when m_0 is constant. Bear in mind that velocity transformation from two inertial frame systems, S and $\bar{S}$, in a standard configuration of special relativity can be derived easily if we assume vector velocity $\vec{u}$ is the instantaneous velocity in frame S of a particle; or that it is simply a geometrical point charge q, as long as we do not exclude the possibility of $u \geq c$, where c is the speed of light. Then, we can deduce the vector velocity $\vec{u}'$ in frame system $\bar{S}$. In classical kinematics, however, we define the following two equations in a Cartesian coordinate and *Euclidian space* for both $\vec{u}$ and $\vec{u}'$ as:

$$\vec{u} = (u_1, u_2, u_3) = (dx/dt, dy/dt, dz/dt) \quad (A.16)$$

$$\vec{u}' = (u'_1, u'_2, u'_3) = (dx'/dt', dy'/dt', dz'/dt') \quad (A.17)$$

Yet, from properties of the *Lorentz transformation* and utilization of the *Lorentz factor* γ, the following relationship in a movement of x-direction, as illustrated in Fig. A.2, is also a valid one:

$$\Delta t' = \gamma[\Delta t - v(\Delta x/c^2)] \quad \Delta x' = \gamma(\Delta x - v\Delta t) \quad \Delta y' = \Delta y \quad \Delta z' = \Delta z \quad (A.18)$$

$$dt' = \gamma[dt - v(dx/c^2)] \quad dx' = \gamma(dx - vdt) \quad dy' = dy \quad dz' = dz \quad (A.19)$$

Now, substituting Eq. A.19 into Eq. A.17 and dividing each numerator and denominator by factor dt, and comparing the result with Eq. A.16, we immediately

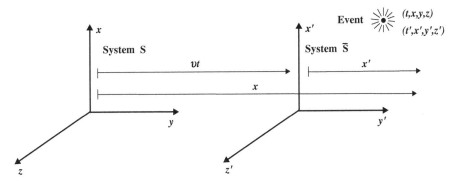

Fig. A.2 Illustration of Lorentz transformation between system S and system $\bar{S}$

can see the results of the velocity transformation components of the vector velocity $\vec{u}'$ formula [1] as:

$$\begin{cases} u'_3 = \dfrac{u_1}{\gamma[1 - u_1(v/c^2)]} \\ u'_2 = \dfrac{u_2}{\gamma[1 - u_2(v/c^2)]} \\ u'_3 = \dfrac{u_3}{\gamma[1 - u_1(v/c^2)]} \end{cases} \quad (A.20)$$

Note that in order to derive Eq. A.20, no assumption as to the uniformity of $\vec{u}$ was made, and these formulas apply equally to the instantaneous velocity in a non-uniform motion as well. We also can note how these sets of equations (i.e., Eq. A.20) will reduce to the classical formation as Eq. A.21 when formally either $v \ll c$ or $c \rightarrow \infty$ are established, knowing these classical formations are in the *standard Galilean transformation*, as discussed in the following.

To establish such a standard Galilean transformation, we first briefly recall from a classical mechanics point of view the relationship between these two sets of coordinates of system $\vec{S}$ and $\vec{S}'$, under a Newtonian condition as:

$$t' = t \quad x' = x - vt \quad y' = y \quad z' = z \quad (A.21)$$

Again using Fig. A.2 and Newton's axiom of universality, or absoluteness, of time and differentiating Eq. A.21 with respect to $t' = t$, we instantly can observe that the classical velocity transformation, which relates the velocity components of an arbitrarily moving system S with those in system $\bar{S}$ as follows:

$$u'_1 = u_1 - v \quad u'_2 = u_2 \quad u'_3 = u_3 \quad (A.22)$$

where

$$(u_1, u_2, u_3) = \left(\frac{dx}{dt}, \frac{dy}{dt}, \frac{dz}{dt}\right) \quad (u'_1, u'_2, u'_3) = \left(\frac{dx'}{dt'}, \frac{dy'}{dt'}, \frac{dz'}{dt'}\right) \quad \text{(A.23)}$$

Thus, from Eqs. A.22 and A.23 we conclude that if a particle or signal has velocity $u'_1 = c$ along the x'-axis in system $\bar{S}$, it has velocity $u_1 = c + v$ in system S. This is known as the *"Common Sense" law of velocity addition*. From a relativity point of view, however, this law conflicts with *Einstein's Law of light propagation*, according to which $u'_1 = c$ must imply that $u_1 = c$. Obviously, by differentiating sets of Eq. A.22 with respect to time, we obtain the transformation of the particle's acceleration components as well [1]:

$$a'_1 = a_1 \quad a'_2 = a_2 \quad a'_3 = a_3 \quad \text{(A.24)}$$

Now, if we get back to the discussion with respect to velocity transformation and the sets of Eq. A.20, with either $v \ll c$ or $c \to \infty$ circumstances, we immediately can obtain the reverse form of Eq. A.20 without extra effort by applying a *"v-reversal transformation"* using the following relationships; thus, we find a set of equations similar to the sets of Eq. A.20. The *inverse transformation* sets are:

$$t = \gamma\left[t' + v(x'/c^2)\right] \quad x = \gamma(x' + vt') \quad y = y' \quad z = z' \quad \text{(A.25)}$$

Consequently, we can write the following sets of equations for reversal transformation of velocity v by taking Eq. A.20 into consideration as:

$$\begin{cases} u_1 = \dfrac{u'_1}{\gamma\left[1 + u'_1(v/c^2)\right]} \\[2pt] u_2 = \dfrac{u'_2}{\gamma\left[1 + u'_1(v/c^2)\right]} \\[2pt] u_3 = \dfrac{u'_3}{\gamma\left[1 + u'_1(v/c^2)\right]} \end{cases} \quad \text{(A.26)}$$

The sets of Eq. A.26 can be considered to provide the resultant vector velocity $\vec{u}$ of the two velocities, $\vec{v} = (v, 0, 0)$ and $\vec{u}'$. Therefore, occasionally we refer to the sets of Eq. A.26 as the *relativistic velocity additional* formula. Considering Eq. A.26, we notice, in particular, that the first member gives the result of two collinear velocities, v and u'_1; thus, it is the same form as the velocity parameter in the result of two successive *Lorentz transformations*.

Gathering all the information we have learned so far with the sets of Eq. A.15, which is a transformation for three-force $\vec{f}$, we can see that, in another frame $\bar{S}$, such a force will depend on velocity $\vec{u}$ of the particle on which it acts. Thus, velocity independence is not a *Lorentz invariant* condition we can impose on a *three-force* field.

Similarly, the *four-force* $\vec{F}$ calculation can be found, for which we refer the reader to Rindler [1] and Woodside [3] references at the end of the appendices. The next

Appendix A: Relativity and Electromagnetism

one is uncomplicated and actually applies in Maxwell's theory; however, it is a force that everywhere depends *linearly* on the velocity of the particles on which it acts.

In special relativity it is natural to make this requirement on the respective 4-vector force $\vec{F}$ and 4-vector velocity $\vec{U}$ [1] as:

$$F_\mu = \frac{q}{c} E_{\mu\nu} U^\nu \qquad (A.27)$$

where coefficient $E_{\mu\nu}$ in this linear relation must be tensorial to make it through Lorentz invariant, if as is natural, we take charge particle q in question to be *scalar invariant*. The inclusion of light speed c is for convenience. In this theory and approach we assume no "moving" charge analogous to moving mass, $m = \gamma m_0$, will arise. We regard $E_{\mu\nu}$ as the field tensor, which can be determined in practice by use of a *test charge*, so it does not change in Eq. A.27. Whereas in Newtonian gravitational theory, for example, field g and force $\vec{f} = m\vec{g}$ are very closely related and in fact often are confused. If we next seek to confirm that the force given by Eq. A.27 is a *pure* force, then we need to write:

$$F_\mu U^\mu = (q/c) E_{\mu\nu} U^\mu U^\nu = 0 \qquad (A.28)$$

for all U^μ, and thus,

$$E_{\mu\nu} = -E_{\mu\nu} \qquad (A.29)$$

In that case the field tensor must be *anti-symmetric* and easily can be observed by giving U^μ only a zeroth component, then a zeroth and first, zeroth and second, and so on. Note that we are presenting conventional vector velocity $\vec{V}$ with symbol $\vec{U}$. See Chap. 2, Sect. 2.7, and Appendix C for classical and relativistic mechanics.

Further investigation of Eq. A.27 is an indication of the field effects and the charges in accordance with this equation. Nonetheless, the question is: How do the charges, reciprocally, affect the field? The answer is provided by the *field equations*. One of the characteristics of a field theory is that the action of the sources can spread throughout the field at a finite speed, although it does not need to. It is the *field* that eventually acts on the test particles rather than the sources themselves by some "action at a distance."

The field equations are therefore *differential* equations, explaining how the sources affect the field in their vicinity, and how one part of the field affects a neighboring part as well. In classical mechanics Newton's theory, too, can be regarded as a field theory; however, the effects spread instantaneously through the field with field equation div $\cdot \vec{g} = -4\pi G \rho$, where ρ presents the mass density and G is the constant of gravity [1]. Further details of the preceding analysis can be found in Rindler [1], where he shows for a four-potential $\vec{\Phi}_\mu$, along with the *Lorentz gauge condition*; the field equation he defined as decoupling, which simplifies to the following equation:

$$\Box \vec{\Phi}_\mu = k J_\mu \qquad (A.30)$$

where k is some universal constant and

$$\Box \equiv \frac{1}{c^2}\frac{\partial^2}{\partial t^2} - \frac{\partial^2}{\partial x^2} - \frac{\partial^2}{\partial y^2} - \frac{\partial^2}{\partial z^2} \qquad (A.31)$$

and J_μ is a region charge density and is in a charge-free region, where $J_\mu = 0$ (see Eq. A.30) reduces to $\Box \vec{\Phi}_\mu = 0$.

This is nothing more than a wave equation with speed of light c, showing that disturbances of the potential in a vacuum are propagated at the speed of light. The potential, however, often is regarded as an "unphysical" auxiliary. Yet, when $J_n = 0$, the field $E_{\mu\nu}$ also satisfies the wave equation as:

$$\Box E_{\mu\nu} = 0 \qquad (A.32)$$

Thus, in the present theory disturbances of the *field* propagate in a vacuum at the speed of light too.

If we define some scalar vector element, such as $\vec{\Psi}$, where $\vec{\Psi} = \vec{\Phi}_\mu - \tilde{\Phi}_\mu$ two different potentials, $\vec{\Phi}_\mu$ and $\tilde{\Phi}_\mu$, can give rise to the same field $E_{\mu\nu}$, then, by the theory of differential equations, Eq. A.31 is known to be solvable in general for $\vec{\Psi}$ relative to any inertial frame S and can be shown as:

$$\Psi(P) = \frac{1}{4\pi}\iiint \frac{[F]dV}{r} \Rightarrow \Box\Psi = F \qquad (A.33)$$

where $F = F(ct, x, y, z)$ is any integrable function and it must be sufficiently small at infinity, $[F]$ denotes the value of F, a retarded value by the high-level time to origin P of position vector $\vec{r}$ of element volume dV, where the volume integral extends over the entire three-space in system frame S. Thus, we can assume, without loss of generality, that the potential satisfies the *Lorentz gauge condition*, where $\Phi^\mu_{\mu\nu} = 0$ and the decoupled field and simple version of Eq. A.30 is established.

Maxwell's theory now will make the connection with its usual three-dimensional formulation in terms of *electric field* $\vec{E}$ and *magnetic field* $\vec{B}$, where they both can be defined in each inertial frame by the first of the following equations:

$$E_{\mu\nu} = \begin{pmatrix} 0 & E_1 & E_2 & E_3 \\ -E_1 & 0 & -cB_3 & cB_2 \\ -E_2 & cB_3 & 0 & -cB_1 \\ -E_3 & -cB_2 & cB_1 & 0 \end{pmatrix}$$

$$E^{\mu\nu} = \begin{pmatrix} 0 & -E_1 & -E_2 & -E_3 \\ E_1 & 0 & -cB_3 & cB_2 \\ E_2 & cB_3 & 0 & -cB_1 \\ E_3 & -cB_2 & cB_1 & 0 \end{pmatrix} \qquad (A.34)$$

while the second results from the first by raising the indices.

Appendix A: Relativity and Electromagnetism

At this point, the four-force law of Eq. A.27 takes the familiar form of *Lorentz's force law* as:

$$\vec{F} = q(\vec{E} + \vec{U} \times \vec{B}) \tag{A.35}$$

When we substitute into Eq. A.35 the expressions $F_\mu = \gamma([\vec{F} \cdot \vec{U}/c] - \vec{F})$, along with Eq. A.33 and $U^\mu = \gamma(c, \vec{U})$, thus by equation the spatial components; equating the temporal components as well yields a mere corollary of Eq. A.35, where $\vec{F} \cdot \vec{U} = q\vec{E} \cdot \vec{U}$. We can note parenthetically that *any* pure three-force that is velocity-independent in one frame S must be a *Lorentz-type force*. The corresponding four-force can then be expressed as in Eq. A.27 with $E_{\mu\nu}$ defined by its components as system frame S—namely, an array like Eq. A.34 with zero Bs in first set, and in other frames there will be non-zero Bs.

Now, we can determine and elaborate on *four-current density* J^μ by writing the following:

$$J^\mu = \rho_0 U^\mu = \rho_0 \gamma(c, \vec{U}) = (c\rho, \vec{J}) \tag{A.36}$$

where we have introduced the charge density ρ and the *three-current density* $\vec{J}$; thus, we can write:

$$\rho = \rho_0 \gamma \quad \vec{J} = \rho \vec{U} \tag{A.37}$$

In this equation ρ is the charge per unit volume in the frame of observation because in that frame a unit's proper volume has a measure of $1/\gamma$ by length contraction, while $\vec{J}$ is the amount of charge crossing a fixed unit area at right angles to its motion per unit. The equation of continuity takes the form [1]:

$$\frac{\partial \rho}{\partial t} + \mathrm{div}\vec{J} = 0 \tag{A.38}$$

The meaning of Eq. A.38 is very clear. Because $\mathrm{div}\vec{J}$ measures the outflux of charge from a unit volume in unit time, it indicates the precise extent that a charge leaves a small region, so the total charge inside that region must decrease. This is an expression of *charge conservation* [1].

If we now translate the electric field $E_{\mu\nu}$ and in SI units assume the universal constant k is equal to $1/c\varepsilon_0$ and assume the value of ν holding 0,1,2,3 in turn, then we find that the equivalent Maxwell's equations to the 3-vector equations are as follows:

$$\begin{cases} \mathrm{div}\vec{E} = \dfrac{1}{\varepsilon_0}\rho \\ \mathrm{curl}\vec{B} = \dfrac{1}{c^2\varepsilon_0}\vec{J} + \dfrac{1}{c^2}\dfrac{\partial \vec{E}}{\partial t} \end{cases} \tag{A.39}$$

The equation sets of A.39, as you can see, are the two familiar *Maxwell equations* [1]. Note that in Gaussian units the universal constant k is equal to $4\pi/c$, while permittivity of free space is $\varepsilon_0 = 1/4\pi$.

The other two Maxwell equations are the 3-vector equivalents of the "potential condition" of $E_{\mu\nu,\,\sigma} + E_{\nu\sigma,\,\mu} + E_{\sigma\mu,\,\nu} = 0$—for values 1,2,3; 2,3,0; 3,0,1; 0,1,2—in turn to indices μ, ν, σ in all other combinations either reproduce one of the equations so obtained, or $0 = 0$, we find:

$$\begin{cases} \text{div}\,\vec{B} = 0 \\ \text{curl}\,\vec{E} = -\dfrac{\partial \vec{B}}{\partial t} \end{cases} \tag{A.40}$$

In summary, with all the information in the preceding, we can finally establish and define a *three-scalar potential* φ and a *3-vector potential* $\vec{a}$ in each inertial frame by the following equation; see Chap. 2, Sect. 2.7, and Appendix C.

$$\Phi_\mu = (\rho, -c\vec{a}) \tag{A.41}$$

In Gaussian units we would write $\vec{a}$ for $c\vec{a}$ and then use the relationship of $E_{\mu\nu} = \Phi_{\nu\mu} - \Phi_{\mu\nu}$ between the field and the potential translates into:

$$\begin{cases} \vec{E} = -\text{grad}\,\varphi - \dfrac{\partial \vec{a}}{\partial t} \\ \vec{B} = \text{curl}\,\vec{a} \end{cases} \tag{A.42}$$

and the Lorentz gauge condition will translate into:

$$\frac{\partial \varphi}{\partial t} + c^2 \text{div}\,\vec{a} = 0 \tag{A.43}$$

Thus, field Eq. A.30 for the potential translates into:

$$\begin{cases} \Box\varphi = \dfrac{1}{\varepsilon_0}\rho \\ \Box\vec{a} = \dfrac{1}{c^2\varepsilon_0}\vec{j} \end{cases} \tag{A.44}$$

The remainder of Maxwell's theory in a vacuum "merely" consists of working out the implications of the basic equation that we have developed here. Readers need to refer to Rindler [1] for detailed information about this appendix.

Appendix B: The Schrödinger Wave Equation

At the beginning of the twentieth century, experimental evidence suggested that atomic particles also were wave-like in nature. For example, electrons were found to give diffraction patterns when passed through a double slit in a similar way to light waves. Therefore, it was reasonable to assume that a wave equation could explain the behavior of atomic particles. The Schrödinger equation plays the role of Newton's laws and conservation of energy in classical mechanics (i.e., it predicts the future behavior of a dynamic system). It is a wave equation in terms of wave function, which the Schrödinger equation predicts analytically and precisely based on the probability of events or outcomes. The detailed outcome is not strictly determined, but given a large number of events, the Schrödinger equation will predict the distribution of results.

B.1 Introduction

Schrödinger developed a differential equation for the time development of a wave function. Because the *energy* operator has a time derivative, the *kinetic energy* operator has space derivatives, and we expect the solutions to be traveling waves; it is natural to try an energy equation. The Schrödinger equation is the operator statement that kinetic energy plus potential energy is equal to total energy.

The kinetic and potential energies in classical mechanics are transformed into Hamiltonian values, which act on the wave function to generate the evolution of the wave in time and space. The *Schrödinger equation* gives the quantized energies of a system and gives the form of wave function so that other properties can be calculated.

Conservation of energy in classical mechanics, using Newton's law, can be written as a harmonic oscillator example:

$$\underbrace{\frac{1}{2}mv^2}_{\text{Kinetic Energy}} + \underbrace{\frac{1}{2}kx^2}_{\text{Potential Energy}} = \underbrace{E}_{\substack{\text{Energy for the}\\ \text{Particle}}} \qquad (B.1)$$

where

$$F = ma = -kx \qquad (B.2)$$

In both Eqs. B.1 and B.2, quantity m represents the mass of a particle in orbit, whereas x represents the position and a represents particle acceleration in Newtons; force is represented by F.

Conservation of energy from a quantum mechanics point of view will be presented in terms of the Schrödinger equation; the energy becomes the Hamiltonian operator, where we can write:

$$H\Psi = E\Psi \qquad (B.3)$$

In this equation Ψ is the wave function, E is the eigenvalue of energy for the system, and quantity H is given as follows:

$$\begin{cases} H = \dfrac{p^2}{2m} + \dfrac{1}{2}kx^2 \\ p \Rightarrow \dfrac{\hbar}{i}\dfrac{\partial}{\partial x} \end{cases} \qquad (B.4)$$

In making the transition to a wave equation, physical variables take the form of "operators" and quantity H converts to:

$$H \rightarrow \frac{-\hbar^2}{2m}\frac{\partial^2}{\partial x^2} + \frac{1}{2}kx^2 \qquad (B.5)$$

where $\hbar = h/2\pi$ and h is Planck's constant and is equal to 6.626070×10^{-34} J·s (m² kg/s); therefore, the $\hbar$ value becomes 1.054573×10^{-34} J·s (m² kg/s).

Both a classical harmonic oscillator and a quantum harmonic oscillator can be depicted in the forms of Figs. B.1 and B.2, respectively. The first illustration is a transition from classical mechanics of simple harmonic oscillation to quantum mechanics oscillation of harmonic particles.

B.2 The Schrödinger Equation

Schrödinger was the first person to write down a wave equation. Much discussion then centered on what the equation meant. The eigenvalues of the wave equation were shown to be equal to the energy levels of the quantum mechanical system, and

Appendix B: The Schrödinger Wave Equation

Fig. B.1 Classical harmonic oscillator

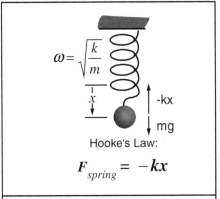

A mass on a spring has a single resonant frequency determined by its spring constant k and the mass m

$$mg - kx = ma = m\frac{d^2x}{dt^2}$$

$$x = A\sin(\omega t - \varphi) + B$$

$$\omega^2 = \frac{k}{m} \Rightarrow \omega = \sqrt{\frac{k}{m}}$$

Fig. B.2 Quantum harmonic oscillator

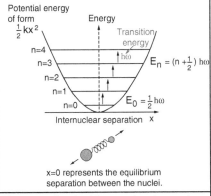

x=0 represents the equilibrium separation between the nuclei.

The energy levels of the quantum harmonic oscillator are:

$$E_n = (n+\frac{1}{2})\hbar\omega$$

$\omega = 2\pi$ (frequency)

$\hbar$ = Planck's constant / 2π

Fig. B.3 The Schrödinger equation

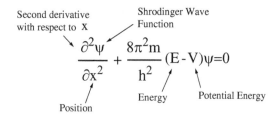

the best test of the equation was when it was used to solve for the energy levels of the hydrogen atom, and the energy levels were found to be in accordance with *Rydberg's Law*.

The Schrödinger equation is shown in Fig. B.3.

The solution to this equation is a wave that describes the quantum aspects of a system. Physically interpreting the wave, however, is one of the main philosophical problems of quantum mechanics.

Thus, the solution to Schrödinger's equation, the wave function for the system, was replaced by the wave functions of the individual series, natural harmonics of each other, an infinite series. Schrödinger discovered that the replacement waves described the individual states of the quantum system and their amplitudes gave the relative importance of that state to the whole system.

Schrödinger's equation shows all of the wave-like properties of matter and was one of greatest achievements of twentieth-century science. It is used in physics and most of chemistry to deal with problems about the atomic structure of matter. It is an extremely powerful mathematical tool and the whole basis of wave mechanics.

In summary, the *Schrodinger equation* is the name of the basic non-relativistic wave equation used in one version of quantum mechanics to describe the behavior of a particle in a field of force. There is the time-dependent equation used for describing progressive waves, applicable to the motion of free particles. The time-independent form of this equation, however, is used for describing standing waves.

Schrödinger's time-independent equation can be solved analytically for a number of simple systems. The time-dependent equation is of the first order in time but of the second order with respect to the coordinates; thus, it is not consistent with relativity. The solutions for bound systems give three quantum numbers, corresponding to three coordinates, and an approximate relativistic correction is possible by including a fourth spin quantum number.

B.3 Time-Dependent Schrödinger Equation Concept

Schrödinger was the first person to write down such a wave equation. Much discussion then centered on what the equation meant. The eigenvalues of the wave equation were shown to be equal to the energy levels of the quantum mechanical

Appendix B: The Schrödinger Wave Equation

system, and the best test of the equation was when it was used to solve for the energy levels of the hydrogen atom, and the energy levels were found to agree with *Rydberg's Law*.

It was initially much less obvious what the wave function of the equation was. After much debate, the wave function is now accepted to be a probability distribution. The Schrödinger equation is used to find the allowed energy levels of quantum mechanical systems (e.g, atoms or transistors). The associated wave function gives the probability of finding the particle at a certain position.

Imagine a particle of mass m constrained to move along the x-axis, subject to some specific force $F(x, t)$, as illustrated in Fig. B.4. The task of *classical mechanics* is to determine the position of the particle at a given time, $x(t)$. We know that we can figure out the velocity $v = dx/dt$, the momentum $p = mv$, the kinetic energy K. E. $= E_{KE} = (1/2)mv^2$, or any other dynamical variable of interest. Additionally, how do we go about determining $x(t)$?

We apply *Newton's second law*, which is $F = ma$. For conservative systems—the only kind we would consider, and fortunately, the only kind that *occur* at the microscopic level—the force can be expressed as the derivative of a potential energy function, $F = -\partial V/\partial x$, and Newton's law reads $m(d^2x/dt^2) = -\partial V/\partial x$. Note that magnetic forces are an exception, but we are not concerned about them for the time being. The preceding argument together with appropriate initial conditions, which is typically the position and velocity at $t = 0$, determines $x(t)$.

Quantum mechanics approaches this same problem quite differently. In this case what we are looking for is the *wave function* $\psi(x, t)$ in a one-dimensional case in direction x of the particle, and we get it by solving *Schrödinger equation*:

$$i\hbar \frac{\partial \Psi(x,t)}{\partial t} = -\frac{\hbar^2}{2m}\frac{\partial^2 \Psi(x,t)}{\partial x^2} + V\Psi(x,t) \tag{B.6}$$

Here quantity $i = \sqrt{-1}$ and $\hbar$ is Plank's constant—or rather, his original constant h divided by 2π:

$$\hbar = \frac{h}{2\pi} = 1.054573 \times 10^{-34} \text{ Js} \tag{B.7}$$

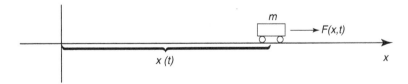

Fig. B.4 A "particle" constrained to move in one dimension under the influence of a specified force

The Schrödinger equation plays a role logically analogous to *Newton's second law*. Providing suitable initial conditions, typically $\psi(x, 0)$, the Schrödinger equation determines $\psi(x, t)$ for all future time, just as in classical mechanics, and Newton's Law determines $x(t)$ for all future times. In summary, the Schrödinger equation is:

$$\frac{\partial^2 \Psi}{\partial x^2} + \frac{8\pi^2 m}{h^2}(E - V)\Psi = 0 \tag{B.8}$$

where:

$\frac{\partial^2}{\partial x^2}$ = second derivative with respect x
Ψ = Schrödinger wave function
x = position
E = energy
V = potential energy

The solution to this equation is a wave that describes the quantum aspects of a system. Physically interpreting the wave, however, is one of the main philosophical problems of quantum mechanics.

The solution to the equation is based on the method of eigenvalues devised by Fourier. This is where any mathematical function is expressed as the sum of an infinite series of other periodic functions. The trick is to find the correct functions that have the right amplitudes so that when added together by superposition they give the desired solution.

So the solution to *Schrödinger's equation*, the wave function for the system, was replaced by the wave functions of the individual series, natural harmonics of each other, an infinite series. Schrödinger has discovered that the replacement waves described the individual states of the quantum system and their amplitudes gave the relative importance of that state to the whole system.

Schrödinger's equation shows all the wave-like properties of matter and was one of greatest achievements of twentieth-century science. It is used in physics and most of chemistry to deal with problems about the atomic structure of matter. It is an extremely powerful mathematical tool and the whole basis of wave mechanics.

The Schrödinger equation is the name of the basic non-relativistic wave equation used in one version of quantum mechanics to describe the behavior of a particle in a field of force. There is the time-dependent equation used for describing progressive waves, applicable to the motion of free particles. Moreover, the time-independent form of this equation is used for describing standing waves.

Schrodinger ½ s time-independent equation can be solved analytically for several simple systems. The time-dependent equation is of the first order in time but of the second order with respect to the coordinates; thus, it is not consistent with relativity. The solutions for bound systems give three quantum numbers, corresponding to three coordinates, and an approximate relativistic correction is possible by including a fourth spin quantum number.

B.4 Time-Independent Schrödinger Equation Concept

In previous section we talked about the wave function and how one can use it to calculate various quantities of interest. Now we need to ask the question: How do we get $\Psi(x, t)$ in the first place—do you go about solving the *Schrödinger equation*? We assume that potential V is *independent* of time t in order to continue. In that case, the Schrödinger equation can be solved by the separation of variables method for solving this partial differential equation. We look for solutions that are simple products such as:

$$\Psi(x,t) = \psi(x) T(t) \tag{B.9}$$

where $\psi(x)$ is purely a function of x, and $T(t)$ is a function of t. For separable solutions, we have:

$$\begin{cases} \dfrac{\partial \Psi(x,t)}{\partial t} = \psi(x) \dfrac{T(t)}{dt} \\ \dfrac{\partial^2 \Psi(x,t)}{\partial x^2} = T(t) \dfrac{d^2 \psi(x)}{dx^2} \end{cases} \tag{B.10}$$

Thus, the Schrödinger equation as an ordinary derivative of Eq. B.6 and results in:

$$i\hbar \psi(x) \frac{dT(t)}{dt} = -\frac{\hbar^2}{2m} \frac{d^2 \psi(x)}{dx^2} T(t) + V \psi(x) T(t) \tag{B.11}$$

Or, dividing through by $\psi(x)T(t)$, we get:

$$i\hbar \frac{1}{T(t)} \frac{dT(t)}{dt} = -\frac{\hbar^2}{2m} \frac{1}{\psi(x)} \frac{d^2 \psi(x)}{dx^2} + V \tag{B.12}$$

The left side of this equation is a function of t alone, and the right side is a function of x alone; however, note that this will not be true if potential V was a function of t as well as x. Therefore, the only possible way for Eq. B.12 to be true is that both sides are in fact *constant*; otherwise, by varying t, we could change the left side without the left side impacting the right side, and the two would no longer be equal to each other. For the time being we call the separation constant E, which explains why we have:

$$\begin{cases} i\hbar \dfrac{1}{T(t)} \dfrac{dT(t)}{dt} = E \\ \dfrac{dT(t)}{dt} = -\dfrac{iE}{\hbar} T(t) \end{cases} \tag{B.13}$$

and

$$\begin{cases} -\dfrac{\hbar^2}{2m}\dfrac{1}{\psi(x)}\dfrac{d^2\psi(x)}{dx^2} + V = E \\ -\dfrac{\hbar^2}{2m}\dfrac{d^2\psi(x)}{dx^2} + V = E\psi(x) \end{cases} \quad (\text{B.14})$$

Separation of variables has turned a *partial differential equation* into two *ordinary differential equations*—namely, the sets of Eqs. B.13 and B.14. The first of these is easy to solve by just multiplying through by dt and integrating; the general solution is $C\exp(-iEt/\hbar)$, but we might as well absorb the constant C into ψ because the quantity of interest is the product of ψT; then we can write the following mathematical notation:

$$T(t) = e^{-iEt/\hbar} \quad (\text{B.15})$$

The second part of Eq. B.14 is called the *time-independent Schrödinger equation*; we can go no further with it until potential $V(x)$ is specified.

Three solutions can be offered to Eq. B.14, as follows:

1. *They are stationary states.* Although the wave function itself is,

$$\Psi(x,t) = \psi(x)e^{-iEt/\hbar} \quad (\text{B.16})$$

obviously depends on t, the probability density is then:

$$|\Psi(x,t)|^2 = \Psi^*\Psi = \psi^* e^{+iEt/\hbar}\psi e^{-iEt/\hbar} = |\psi(x)|^2 \quad (\text{B.17})$$

The time dependence does not cancel out. The same thing happens in calculating the expression value of any dynamical variable; then we can write:

$$\langle Q(x,p) \rangle = \int \psi^* Q\left(x, \dfrac{\hbar}{i}\dfrac{d}{dx}\right)\psi\, dx \quad (\text{B.18})$$

Every expectation value is constant in time; we might as well drop factor $T(t)$ altogether, and simply use ψ in place of Ψ.

2. *They are states of definite total energy.* In classical mechanics the total energy (i.e., kinetic plus potential) is called the Hamiltonian:

$$H(x,p) = \dfrac{p^2}{2m} + V(x) \quad (\text{B.19})$$

The corresponding Hamiltonian operator, obtained by the canonical substitution $p \to (\hbar/i)(\partial/\partial x)$, is therefore

$$H = -\dfrac{\hbar^2}{2m}\dfrac{\partial^2}{\partial x^2} + V(x) \quad (\text{B.20})$$

Appendix B: The Schrödinger Wave Equation

Thus, the time-independent Schrödinger Eq. B.14 second term can be written as follows, which is the exact expression in Eq. B.3:

$$H\psi = E\psi \tag{B.21}$$

Moreover, the expectation value of the total energy is

$$\langle H \rangle = \int \psi^* H \psi \, dx = E \int |\psi|^2 dx = E \tag{B.22}$$

Note that the normalization of Ψ enables the normalization of ψ. Moreover,

$$H^2 \psi = H(H\psi) = H(E\psi) = E(H\psi) = E^2 \psi \tag{B.23}$$

thus,

$$\langle H^2 \rangle = \int \psi^* H^2 \psi \, dx = E^2 \int |\psi|^2 dx = E^2 \tag{B.24}$$

So the standard deviation in H is given by

$$\sigma_H^2 = \langle H^2 \rangle - \langle H \rangle^2 = E^2 - E^2 = 0 \tag{B.25}$$

Still, remember if $\sigma = 0$, then every member of the sample must share the same value, where the distribution has zero spread. As a conclusion, we can say that a separable solution has the property that *every measurement* of the total energy *is certain* to return the value, such as E, and that is why the letter E is chosen as the separation constant.

3. The *general solution* is a *linear combination* of separable solutions. As we determine, time-independent Schrödinger Eq. B.6 yields an infinite collection of solutions, $\{\psi_1(x), \psi_2(x), \psi_3(x), \cdots\}$, each with its associated value of the separation constant $\{E_1, E_2, E_3, \cdots\}$; thus, there is a different wave function for each *allowed energy*:

$$\begin{cases} \Psi_1(x,t) = \psi_1(x) e^{-E_1 t/\hbar} \\ \Psi_2(x,t) = \psi_2(x) e^{-E_2 t/\hbar} \\ \Psi_3(x,t) = \psi_3(x) e^{-E_3 t/\hbar} \\ \vdots \end{cases} \tag{B.26}$$

Now we easily can see that the time-dependent Schrödinger equation has the following format and has the property that any linear combination of the solution is itself a solution:

$$i\hbar \frac{\partial \Psi(x,t)}{\partial t} = -\frac{\hbar^2}{2m} \frac{\partial^2 \Psi(x,t)}{\partial x^2} + V\Psi(x,t) \tag{B.27}$$

Note: A linear combination of the function $f_1(z), f_2(z), f_3(z) \cdots$ is an expression of the form $f(z) = c_1 f_1(z) + c_2 f_2(z) + c_3 f_3(z) \cdots$, where $c_1, c_2, c_3, \cdots$ are any complex constants.

Once we have found the separable solutions, we immediately can construct a much more general solution of the form:

$$\Psi(x,t) = \sum_{n=1}^{\infty} c_n \psi_n e^{-iE_n t/\hbar} \tag{B.28}$$

It so happens that *every solution* to the time-dependent *Schrödinger equation* can be written in the form of Eq. B.28, and it is simply a matter of finding the right constants $c_1, c_2, c_3, \cdots$ to fit the initial conditions for the problem at hand.

B.5 A Free Particle Inside a Box and Density of State

We now consider a free particle of mass m inside a cube of volume L^3. Because inside the box the potential energy is zero, the time-independent Schrödinger equation will be given as:

$$\nabla^2 \Psi(x,y,z) + \frac{2m\varepsilon}{\hbar^2} \Psi(x,y,z) = 0 \left. \begin{array}{l} 0 < x < L \\ 0 < y < L \\ 0 < z < L \end{array} \right\} \tag{B.29}$$

The particle cannot go outside the box and therefore we must solve Eq. B.29 subject to the boundary condition that wave function $\psi(x,y,z)$ should vanish everywhere on the surface of the cube. We use the method of separation of variables to solve Eq. B.29 and write:

$$\Psi(x,y,z) = X(x)Y(y)Z(z) \tag{B.30}$$

Substituting Eq. B.30 into Eq. B.29 and dividing by $\psi(x,y,z)$, we obtain:

$$\frac{1}{X}\frac{d^2 X}{dx^2} + \frac{1}{Y}\frac{d^2 Y}{dy^2} + \frac{1}{Z}\frac{d^2 Z}{dz^2} + k^2 = 0 \tag{B.31}$$

where

$$k^2 = \frac{2m3}{\hbar^2} \tag{B.32}$$

The variables have indeed separated out and we can write:

$$\frac{1}{X}\frac{d^2 X}{dx^2} = -k_x^2 \tag{B.33}$$

Appendix B: The Schrödinger Wave Equation

$$\frac{1}{Y}\frac{d^2Y}{dy^2} = -k_y^2 \tag{B.34}$$

$$\frac{1}{Z}\frac{d^2Z}{dZ^2} = -k_z^2 \tag{B.35}$$

where k_x^2, k_y^2, and k_z^2 are constants subject to the condition:

$$\begin{cases} k_x^2 + k_y^2 + k_z^2 = k^2 \\ \vec{k} = \hat{i}k_x + \hat{j}k_y + \hat{w}k_z \end{cases} \tag{B.36}$$

In Eqs. B.33, B.34, and B.35 we have a set of each term equal to a negative constant; otherwise, the boundary conditions cannot be satisfied. The solution of Eq. B.33 is given as:

$$X(x) = A \sin k_x x + B \cos k_x x \tag{B.37}$$

and because Ψ has to vanish at *all* points on surfaces $x = 0$ and $x = L$ we must have $B = 0$ and

$$k_x = n_x \pi / L \quad n_x = 1, 2, 3, \cdots \tag{B.38}$$

Similarly, the allowed values of k_y and k_z are:

$$k_y = n_y \pi / L \quad n_y = 1, 2, 3, \cdots \tag{B.39}$$

$$k_z = n_z \pi / L \quad n_z = 1, 2, 3, \cdots \tag{B.40}$$

It may be noted that negative values of n_x, n_y, and n_z do not lead to an independent solution. The wave function remains the same except for a trivial change of sign. Using Eq. B.32, the permitted values of ε are given by

$$\varepsilon = \frac{\pi^2 \hbar^2}{2mL^2}\left(n_x^2 + n_y^2 + n_z^2\right) \tag{B.41}$$

Now the number of states with x component of $\vec{k}$ lies between k_x and $k_x + dk_x$ simply would be the number of integers lying between $k_x L/\pi$ and $(k_x + dk_x)L/\pi$. This number would be approximately equal to Ldk_x/π. Similarly, the number of states with y and z components of $\vec{k}$ that lie between k_y and $k_y + dk_y$ as well as k_z and $k_z + dk_z$ would be, respectively, $k_y L/\pi$ and $k_z L/\pi$. Thus, there will be

$$\left(\frac{L}{\pi}dk_x\right)\left(\frac{L}{\pi}dk_y\right)\left(\frac{L}{\pi}dk_z\right) = \frac{V}{\pi^3} dk_x dk_y dk_z \tag{B.42}$$

states in the range $dk_x dk_y dk_z$ of $\vec{k}$; here $V(=L^3)$ represents the volume of the box, but not to be mistaken for potential energy as before in terms of the Schrödinger equation case. Because $\vec{p} = \hbar\vec{k}$ is the number of states per unit volume in $\vec{p}$ space, which is considered to be the momentum space, the form would be:

$$\frac{V}{\pi^3\hbar^3} \tag{B.43}$$

If $P(p)dp$ represents the number of states with $|\vec{p}|$ that lies between p and $p + dp$, then we have

$$P(p)dp = \frac{1}{8} \times \frac{V}{\pi^3\hbar^3} 4\pi p^2 dp = \frac{V}{\hbar^3} 4\pi p^2 dp \tag{B.44}$$

where the factor $4\pi p^2 dp$ represents the volume element in $\vec{p}$ for $|\vec{p}|$ to lie between p and $p + dp$ as well as the factor 1/8 because of the fact that k_x, k_y, and k_z can take only positive values, as per Eqs. B.38, B.39, and B.40. Thus, while counting the state in the $\vec{k}$ or $\vec{p}$ space, we must consider only the positive constant. Now if $g(\varepsilon)d\varepsilon$ represents the number of states with energy $(=p^2/2m)$ lying between ε and $\varepsilon + d\varepsilon$, then

$$g(\varepsilon)d\varepsilon = P(p)dp = \frac{V}{8\pi^3\hbar^3} 4\pi p^2 dp \tag{B.45}$$

Using the relation $p^2 = 2m\varepsilon$, we get

$$g(\varepsilon)d\varepsilon = \frac{(2m)^{3/2}V}{4\pi^2\hbar^3} \varepsilon^{1/2} d\varepsilon \tag{B.46}$$

for the density of states. If there are degeneracies associated with each energy state, then the density of states will be given by:

$$g(\varepsilon)d\varepsilon = G \frac{(2m)^{3/2}V}{4\pi^2\hbar^3} \varepsilon^{1/2} d\varepsilon \tag{B.47}$$

where G is known as the *degeneracy parameter*. For example, for an electron gas each electron has two additional degrees of freedom because of its spin and therefore there will be twice as many states as given by Eq. B.46 and G will be equal to 2.

Note that it is of interest to indicate that instead of using the boundary condition that ψ vanishes at the surface of the cube, it often is more convenient to use periodic boundary conditions according to which

$$\begin{cases} \psi(x+L,y,z) = \psi(x,y,z) \\ \psi(x,y+L,z) = \psi(x,y,z) \\ \psi(x,y,z+L) = \psi(x,y,z) \end{cases} \tag{B.48}$$

Appendix B: The Schrödinger Wave Equation

Then, as an alternative to choosing the sine and cosine functions as a solution as we did in Eq. B.37, it is more appropriate to choose exponential terms as solutions, like this:

$$e^{ik_x x}, e^{ik_y y} \text{ and } e^{ik_z z} \tag{B.49}$$

Using Eq. B.48, we get

$$e^{ik_x x} = e^{ik_y y} = e^{ik_z z} = 1 \tag{B.50}$$

Thus, the allowed values of k_x, k_y, and k_z are given as:

$$\begin{cases} k_x = \dfrac{2n_x \pi}{L} & n_x = 0, \pm 1, \pm 2, \cdots \\ k_y = \dfrac{2n_y \pi}{L} & n_y = 0, \pm 1, \pm 2, \cdots \\ k_z = \dfrac{2n_x \pi}{L} & n_z = 0, \pm 1, \pm 2, \cdots \end{cases} \tag{B.51}$$

Notice that a reversal of sign k_x changes the wave function, thus $n_x = 2$ and $n_x = -2$ correspond to two different states although they both correspond to the same energy eigenvalue, which is now, given by:

$$\varepsilon = \frac{2\pi\hbar^2}{mL^2}\left(n_x^2 + n_y^2 + n_z^2\right) \tag{B.52}$$

Proceeding as before we will now have, as per Eq. B.42:

$$\left(\frac{L}{\pi}dk_x\right)\left(\frac{L}{\pi}dk_y\right)\left(\frac{L}{\pi}dk_z\right) = \frac{V}{\pi^3}dk_x dk_y dk_z \tag{B.53}$$

which states the range $dk_x dk_y dk_z$. Further, the number of states per unit volume in the $\vec{p}$ space as per Eq. B.43 would be:

$$\frac{V}{8\pi^3 \hbar^3}\left(=\frac{V}{\hbar^3}\right) \tag{B.54}$$

This equation implies that a volume h^3 in the phase space corresponds to one state. It may be noted that the number of states with $|\vec{p}|$ that lies between p and $p + dp$ would be:

$$P(p)dp = \frac{V}{8\pi^3 \hbar^3}4\pi p^2 dp \tag{B.55}$$

which is identical to Eq. B.44; thus, we would get the same expression for $g(\varepsilon)$ similar to Eq. B.46. We note that in Eq. B.55 we do not have an extra 1/8 factor because p_x, p_y, and p_z can now also take negative values. Thus, we see that the two boundary conditions lead to the same expression for the density of states.

B.6 Relativistic Spin Zero Parties: Klein-Gordon Equation

It is a non-trivial problem to generalize quantum mechanics to describe relativistic particles, and as we shall see, the solution contains some rather unusual features. First, we will study spin-less, or scalar, particles and then later particles with spin. The reason is that the properties of spin, an angular momentum, are closely subject to the requirements of special relativity and, as a consequence, the form of the wave equation for a particle depends on its spin. Common relativistic spin-less particles are, for example, π and K mesons [4].

A non-relativistic free particle has an energy–momentum relation such as:

$$E = \frac{p^2}{2m} \tag{B.56}$$

which is invariant under a *Galilean transformation* to a new coordinate system traveling with velocity $-\vec{v}$ with respect to the first. Energy E' and momentum $\vec{p}'$ of the particle in the new coordinate system are related to E and p, non-relativistically, by:

$$E' = E + \vec{p}\cdot\vec{v} + \frac{mv^2}{2} \tag{B.57}$$

and

$$\vec{p}' = \vec{p} + m\vec{v} \tag{B.58}$$

then $E' = p'/2m$. The transformation laws of Eqs. B.57 and B.58 are not restricted to a free particle. For a *Galilean transformation* on a general closed system, E is the total energy, $\vec{p}$ the total momentum, and m the total mass of the system.

Note that within an additive term independent of $\vec{v}$, the energy–momentum relation in Eq. B.56 is the only form that is invariant under a Galilean transformation. Quantum mechanically the energy and momentum of a particle are related to the change of the phase of its wave function in time and space, respectively. For a free particle to have the energy–momentum relation in Eq. B.56, its wave function is:

$$\Psi(\vec{r},t) = \exp\left\{\frac{i(\vec{p}\cdot\vec{r} - Et)}{\hbar}\right\} \tag{B.59}$$

Equation B.59 *must* obey the Schrödinger equation as:

$$i\hbar\frac{\partial \Psi(\vec{r},t)}{\partial t} = \frac{1}{2m}\left(\frac{\hbar}{i}\nabla\right)^2 \Psi(\vec{r},t) \tag{B.60}$$

Thus, the Schrödinger equation may be written from the momentum relation of Eq. B.56 by making the correspondences:

Appendix B: The Schrödinger Wave Equation

$$E \to i\hbar \frac{\partial}{\partial t}$$
$$p \to \frac{\hbar}{i} \nabla$$
(B.61)

In Eq. B.56 the result acts on Ψ, as in Eq. B.60. Furthermore, one can add in the coupling of a charged particle to an electromagnetic (EM) field described by a vector potential $\vec{A}(\vec{r},t)$ and a scalar potential $\phi(\vec{r},t)$ by letting the following occur:

$$\begin{cases} E \to E - e\phi \\ \vec{p} \to \vec{p} - \frac{e}{c}\vec{A} \end{cases}$$
(B.62)

or

$$\begin{cases} i\hbar \frac{\partial}{\partial t} \to i\hbar \frac{\partial}{\partial t} - e\phi \\ \frac{\hbar}{i}\vec{\nabla} \to \frac{\hbar}{i}\vec{\nabla} - \frac{e}{c}\vec{A} \end{cases}$$
(B.63)

into Eq. B.60, where e is charge of the particle.

This connection between the energy–momentum relation and the wave equation provides a guide for writing down a relativistic: wave equation that will give the correct relativistic energy–momentum relation:

$$E = c\sqrt{p^2 + m^2 c^2}$$
(B.64)

This simple guess is to say that the momentum space amplitude is given as:

$$\Psi_{\vec{p}}(t) = \int d^3 \vec{r}\, e^{-i\vec{p}\cdot\vec{r}/\hbar} \psi(\vec{r},t)$$
(B.65)

This equation obeys the following one that is written as:

$$i\hbar \frac{\partial}{\partial t} \Psi_{\vec{p}}(t) = c\sqrt{p^2 + m^2 c^2}\, \Psi_{\vec{p}}(t)$$
(B.66)

The energy eigenvalues are then clearly given by Eq. B.64. Such a relativistic Schrödinger equation is quite useful, but it has one serious limitation that we should recognize. Let us Fourier transform both sides of Eq. B.66 back to position of space. Then, Eq. B.66 becomes:

$$i\hbar \frac{\partial}{\partial t} \Psi(\vec{r},t) = \int d^3 \vec{r}'\, K(\vec{r} - \vec{r}')\Psi(\vec{r}',t)$$
(B.67)

where

$$K(\vec{r} - \vec{r}') = \int \frac{d^3 p}{(2\pi\hbar)^3} e^{i\vec{p}\cdot(\vec{r}-\vec{r}')/\hbar} c\sqrt{p^2 + m^2 c^2}$$
(B.68)

Equation B.67 is non-local; this means that the value of the right side at $\vec{r}$ depends on the value of Ψ at other points $\vec{r}'$. The kernel $K(\vec{r} - \vec{r}')$ is sizable if $\vec{r}'$ is within a distance $\sim \hbar/mc$ from $\vec{r}$. Note that $\hbar/mc$, the Compton wavelength of the particle, is the only unit of length in this problem. As a consequence of this non-locality, the rate of change in time of Ψ at spacetime point $\vec{r}, t$ depends on the values of Ψ at points $\vec{r}'$, t outside the light cone centered on $\vec{r}, t$. If we construct an initial wave packet localized well within a Compton wavelength of the origin, say, then the packet will be non-zero an arbitrarily short time later at points as distant as the Compton wavelength.

Thus, Eq. B.65 violates relativistic causality when used to describe particles localized to within more than a Compton wavelength and is useful only in describing non-highly localized states. We will not pursue this equation further here, but observe that this is not the last we may see of problems of localization in relativistic quantum mechanics.

We can form a local wave equation by first squaring Eq. B.64 to obtain the following equation:

$$E^2 = p^2 c^2 + m^2 c^4 \tag{B.69}$$

and then use Eq. B.62. This implies that wave function Ψ obeys this equation:

$$\left(\frac{i\hbar}{c}\frac{\partial}{\partial t}\right)^2 \Psi(\vec{r}, t) = \left(\frac{\hbar}{i}\nabla\right)^2 \Psi(\vec{r}, t) + m^2 c^2 \Psi(\vec{r}, t) \tag{B.70}$$

This equation is known as the *Klein-Gordon equation*. We can write Eq. B.70 in the following form as well:

$$\left[\frac{1}{c^2}\frac{\partial^2}{\partial t^2} - \nabla^2 + \left(\frac{mc}{\hbar}\right)^2\right]\Psi(\vec{r}, t) = 0 \tag{B.71}$$

In this form it looks like a classical wave equation with the extra term $(mc/\hbar)^2$.

The coupling of a charged particle to the EM field can be included in a relativistically covariant manner in the Klein-Gordon equation by making the substitutions of Eq. B.63, because ϕ and $\vec{A}$ transfer relativistically as a 4-vector. Thus, in an EM field Eq. B.70 becomes:

$$\frac{1}{c^2}\left[i\hbar\frac{\partial}{\partial t} - e\phi(r, t)\right]^2 \Psi(\vec{r}, t) = \left\{\left[\frac{\hbar}{i}\nabla - \frac{e}{c}\vec{A}(\vec{r}, t)\right]^2 + m^2 c^2\right\}\Psi(\vec{r}, t) \tag{B.72}$$

Before exploring the consequences of the Klein-Gordon equation, let us note its behavior under a Lorentz transformation. The operator,

$$\frac{1}{c^2}\frac{\partial^2}{\partial t^2} - \nabla^2 \tag{B.73}$$

Appendix B: The Schrödinger Wave Equation

takes the same form in any Lorentz frame and thus for a *spiniest* particle, $\Psi'(\vec{r}',t')$, the wave function in the new frame is given in terms of $\Psi(\vec{r},t)$ in the original frame by

$$\Psi'(\vec{r}',t') = \Psi(\vec{r},t) \tag{B.74}$$

where $\vec{r}'$, t' are the coordinates in the new frame of the spacetime point $\vec{r}$, t; in other words, $\Psi(\vec{r},t)$ is a scalar.

The Klein-Gordon equation has several unusual features. First, unlike the non-relativistic Schrödinger equation, it is *second order* in time. Therefore, to predict the future behavior of a particle we must know both $\Psi(\vec{r},t)$ and $\partial\Psi(\vec{r},t)/\partial t$ at any one time; because of these time-independent quantities we must basically know "twice" as much information about the particle to specify its state as non-relativistically. To put it another way, the particle has essentially an extra degree of freedom. We will see shortly that this extra degree of freedom corresponds to specifying the *charge* of the particle, and that the Klein-Gordon equation describes both a particle and its *anti-particle* in one fell swoop.

A related consequence of the equation being second order is the functions:

$$\Psi(\vec{r},t) = e^{i(\vec{p}\cdot\vec{r}-Et)/\hbar} \tag{B.75}$$

solve the free particle Eq. B.70 with *either* sign of E as:

$$E = \pm c\sqrt{p^2 + m^2 c^2} \tag{B.76}$$

In other words, the Klein-Gordon equation has negative energy solutions! These have the strange property that, as we *increase* the magnitude of $\vec{p}$, the energy of the particle decreases. We will see that the negative energy eigenstates of the Klein-Gordon equation describe anti-particles, while the positive energy eigenstates describe particles.

Yet another related peculiarity is the fact that one cannot construct a positive probability density, like $\Psi\Psi^*$ non-relativistically, that is conserved in time. For the Klein-Gordon equation $\int d^3\vec{r}\,\Psi\Psi^*$ generally changes in time; thus, we are at a loss to interpret $\Psi(\vec{r},t)\Psi^*(\vec{r},t)$ as being the probability for finding a particle at $\vec{r}$.

There is one conserved density that we can construct though. Because the operator acting on Ψ in Eq. B.71 is real, Ψ^* also satisfies the same equation. Thus, we can write the following relationship:

$$\Psi^*\left[\frac{1}{c^2}\frac{\partial^2}{\partial t^2} - \nabla^2 + \left(\frac{mc}{\hbar}\right)^2\right]\Psi - \Psi\left[\frac{1}{c^2}\frac{\partial^2}{\partial t^2} - \nabla^2 + \left(\frac{mc}{\hbar}\right)^2\right]\Psi^* = 0 \tag{B.77}$$

and by some slight rearrangement we find the continuity equation:

$$\frac{\partial \rho(\vec{r},t)}{\partial t} + \vec{\nabla} \cdot J(\vec{r},t) = 0 \tag{B.78}$$

where the current density $J(\vec{r},t)$, as non-relativistically, is given as:

$$J(\vec{r},t) = \frac{\hbar}{2im}[\Psi^* \nabla \Psi - \Psi \nabla \Psi^*] \tag{B.79}$$

and the density $\rho(\vec{r},t)$ is given as:

$$\rho(\vec{r},t) = \frac{i\hbar}{2mc^2}\left[\Psi^* \frac{\partial \Psi}{\partial t} - \Psi \frac{\partial \Psi^*}{\partial t}\right] \tag{B.80}$$

From Eq. B.79 the integral of this density over all space does not change with time. Nevertheless, Eq. B.80 need not be positive; for example, $\rho < 0$ for a negative energy free particle eigenstate. Thus, we cannot interpret $\rho(\vec{r})$, neither as being the particle (probability) density at $\vec{r}$ nor can we interpret $J(\vec{r})$ as the particle current. The interpretation that will emerge is that charged particles $e\rho(\vec{r})$ represent the *charge density* at $\vec{r}$, which can have either sign, and $eJ(\vec{r})$ represents the *electric current* at $\vec{r}$.

One can verify that Eq. B.78 is still satisfied in the presence of an EM field with the current and density given by:

$$J(\vec{r},t) = \frac{1}{2m}\left[\Psi^*\left(\frac{\hbar}{i}\nabla - \frac{e}{c}\vec{A}\right)\Psi + \Psi\left(-\frac{\hbar}{i}\nabla - \frac{e}{c}\vec{A}\right)\Psi^*\right] \tag{B.81}$$

as non-relativistically, and

$$\rho(\vec{r},t) = \frac{1}{2mc^2}\left[\Psi^*\left(i\hbar\frac{\partial}{\partial t} - e\phi\right)\Psi + \Psi\left(-i\hbar\frac{\partial}{\partial t} - e\phi\right)\Psi^*\right] \tag{B.82}$$

B.6.1 Anti-particles

In particle physics every type of particle has an associated anti-particle with the same mass but with opposing physical charges (e.g., electric charges). For example, the anti-particle of the electron is the anti-electron, often referred to as positron. Even though the electron has a negative electric charge, the positron has a positive electric charge and is produced naturally in certain types of radioactive decay. The opposite is also true: the anti-particle of the positron is the electron.

Some particles, such as the photon, are their own anti-particle. Otherwise, for each pair of anti-particle partners, one is designated as normal matter (the kind we are made of), and the other (usually given the prefix "anti-") as anti-matter. Particle–anti-particle pairs can annihilate each other, producing photons; because the charges

Fig. B.5 Illustration of particles (left) and anti-particles (right)

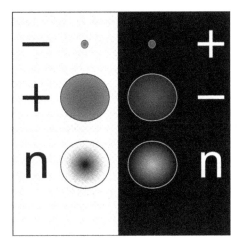

of the particle and anti-particle are opposite, the total charge is conserved. For example, the positrons produced in natural radioactive decay quickly annihilate themselves with electrons, producing pairs of gamma rays—a process used in positron emission tomography.

The laws of nature are very nearly symmetrical with respect to particles and anti-particles. For example, an anti-proton and a positron can form an anti-hydrogen atom, which is believed to have the same properties as a hydrogen atom. This leads to the question of why the formation of matter after the Big Bang resulted in a Universe consisting almost entirely of matter, rather than being a half-and-half mixture of matter and anti-matter. The discovery of charge parity violation helped to shed light on this problem by showing that this symmetry, originally thought to be perfect, was only approximate. Figure B.5 shows the electric charge of particles (*left*) and anti-particles (*right*). From top to bottom: electron/positron, proton/anti-proton, neutron/anti-neutron.

Because charge is conserved, it is not possible to create an anti-particle without either destroying another particle of the same charge (e.g., in the case when anti-particles are produced naturally via beta decay or the collision of cosmic rays with Earth's atmosphere), or by the simultaneous creation of both a particle and its anti-particle, which can occur in particle accelerators such as the Large Hadron Collider at CERN.

Although particles and their anti-particles have opposite charges, electrically neutral particles need not be identical to their anti-particles. The neutron, for example, is made of quarks, the anti-neutron from anti-quarks, and they are distinguishable from one another because neutrons and anti-neutrons annihilate each other on contact. Nonetheless, other neutral particles are their own anti-particles, such as photons, Z^0 bosons, π^0 mesons, and hypothetical gravitons—some hypothetical.

B.6.2 Negative Energy States and Anti-particles

If we now construct a consistent interpretation of the Klein-Gordon equation, we should start by considering a free particle at rest, where the momentum is set to zero as $\vec{p} = 0$. For a positive energy solution, the wave function of the particle is given as:

$$\Psi(\vec{r},t) = \exp\left(-\frac{imc^2 t}{\hbar}\right) \tag{B.83}$$

because the energy of a particle at rest is given by the Einstein equation, $E = mc^2$.

Now let us make a Lorentz transformation to a frame moving with velocity $-\vec{v}$ with respect to a particle at rest. In that frame the particle appears to have velocity $\vec{v}$, momentum $\vec{p} = m\vec{v}/\sqrt{1 - v^2/c^2}$, and energy $E_p = mc^2/\sqrt{1 - v^2/c^2}$. Thus, in the new frame, denoted by primes, for the wave function of the $\Psi'(\vec{r}',t)$, we can write the following equation:

$$\Psi'(\vec{r}',t) = \exp\left(-\frac{imc^2 t}{\hbar}\right) = \exp\left(\frac{i(\vec{p}\cdot\vec{r}' - E_p t')}{\hbar}\right) \tag{B.84}$$

Because the quantity $(\vec{p}\cdot\vec{r}' - E_p t')$ is a *Lorentz scalar*, it is equal to $-mc^2 t$ in the rest frame and therefore in all terms. Equation B.84 is the expected result that a particle with momentum $\vec{p}$ and energy E_p has the wave function such as:

$$\Psi'(\vec{r}',t) = \exp\left(\frac{i(\vec{p}\cdot\vec{r}' - E_p t')}{\hbar}\right) \tag{B.85}$$

From Eqs. B.78 and B.79, we can see that density ρ for this particle is written as:

$$\rho(\vec{r}',t') = \frac{E_p}{mc^2} \tag{B.86}$$

and the current density $J(\vec{r}',t')$ is given by:

$$J(\vec{r}',t') = \frac{p}{m} = \frac{pc^2}{E_p}\rho(\vec{r}',t') \tag{B.87}$$

Thus, the current is just the density times the relativistic velocity of the particle, yielding:

$$\vec{v} = \frac{\vec{p}c^2}{E_p} \tag{B.88}$$

Notice that $\rho(\vec{r},t)$ transforms as the time component of a 4-vector (see Appendix C as well).

Appendix B: The Schrödinger Wave Equation

This is perfectly reasonable because a unit volume in the rest frame appears smaller by a factor of $\sqrt{1 - v^2/c^2}$ when seen from a moving frame. Thus, a unit density in the rest frame will appear as a density of $\sqrt{1 - v^2/c^2} = E_p/mc^2$ in a frame in which the particle is moving. Now we can try to understand the concept of a *negative energy* solution. Again, we start with a particle at rest described by the wave function as:

$$\Psi(\vec{r}, t) = \exp\left(\frac{imc^2 t}{\hbar}\right) \tag{B.89}$$

This equation corresponds to a solution of the Klein-Gordon equation with energy $-mc^2$. The density $\rho(\vec{r}, t)$ in this state is from Eq. B.79, yielding:

$$\rho(\vec{r}, t) = -1 \tag{B.90}$$

One way to interpret a state with a negative particle density is to say that it is a state with a positive density of *anti-particles*. We use this interpretation and say that a particle at rest with a negative energy of $-mc^2$ actually is an anti-particle at rest with positive energy mc^2. This interpretation of the negative energy states will lead to a consistent theoretical picture, which has been amply confirmed experimentally.

In a Lorentz frame of reference traveling at velocity $-\vec{v}$ with respect to the anti-particle, the wave function in Eq. B.89 then becomes:

$$\Psi'(\vec{r}', t') = \exp\left(\frac{imc^2}{\hbar}\right) = \exp\left(\frac{i(\vec{p} \cdot \vec{r}' - E_p t')}{\hbar}\right) \tag{B.91}$$

where $\vec{p} = m\vec{v}/\sqrt{1 - v^2/c^2}$ is the momentum of the anti-particle and $E_p = mc^2 \sqrt{1 - v^2/c^2}$ is its energy. In this new frame, the anti-particle has velocity $\vec{v}$ and momentum $\vec{p}$ as well as energy E_p. Notice that the wave function in Eq. B.91 describes the *anti-particle* as a *particle* of energy $-E_p$ and momentum $\vec{p}$.

From Eqs. B.87 and B.88, the density $\rho(\vec{r}', t')$ in the Lorentz frame yields the following form:

$$\rho(\vec{r}', t') = -\frac{E_p}{mc^2} \tag{B.92}$$

Consequently, the current density $\vec{J}(\vec{r}', t')$ is:

$$\vec{J}(\vec{r}', t') = -\frac{\vec{p}}{m} = \frac{\vec{p} c^2}{E_p} \rho(\vec{r}', t') \tag{B.93}$$

An anti-particle moving with velocity $\vec{v}$ has associated with it a current moving in the opposite direction; a flow of anti-particles in one direction is equivalent to a flow of particles in the opposite direction.

For a charged particle $e\rho(\vec{r},t)$ is the charge density, which is positive for a free particle (i.e., $e > 0$) and negative for a free anti-particle because an anti-particle has the opposite charge as a particle. Similarly, $e\vec{J}(\vec{r},t)$ is the electrical current density of the state $\Psi(\vec{r},t)$; for a particle the electric current density is in the direction of the particle's velocity. For an anti-particle, $e < 0$, the electrical current density is opposite to the velocity.

Thus, the interpretation of negative energy solutions as corresponding to anti-particles is consistent with the interpretation of density ρ as a charge density and $\vec{J}$ as a current density of charge, which has the opposite sign for anti-particles. Now if this interpretation is really correct, we must show that EM fields in fact couple to anti-particles with the opposite sign of e. To show this let us take the *complex conjugate* of Eq. B.72 and fields ϕ and $\vec{A}$ that are real, and only the explicit i's change sign. Thus, we have the following form of the equation [4]:

$$\frac{1}{c^2}\left[i\hbar\frac{\partial}{\partial t} + e\phi(\vec{r},t)\right]^2 \Psi^*(\vec{r},t) = \left[\left(\frac{\hbar}{i}\nabla + e\vec{A}(\vec{r},t)\right)^2 + m^2c^2\right]\Psi^*(\vec{r},t) \quad (B.94)$$

This equation says that if $\Psi(\vec{r},t)$ is a solution to the Klein-Gordon equation with a certain sign of the charge, then $\Psi^*(\vec{r},t)$ is a solution to the Klein-Gordon equation with the *opposite* sign of the charge and the same mass. In this sense the relativistic theory of a spin zero particle predicts the existence of its anti-particle with an opposite charge and the same mass; at the same time, that the theory has solutions to describe a particle, it also has solutions to describe its anti-particle [4].

This is a general feature of relativistic quantum mechanics. Now, because complex conjugation reserves the signs of all frequencies and momenta in $\Psi(\vec{r},t)$, the complex conjugate of a negative energy solution to original Eq. B.72 is a positive energy solution to Eq. B.94 with an opposite sign of the charge. Thus, to interpret a negative energy solution, we simplify the complex conjugate of the wave function and interpret it as a positive energy solution of the opposite charge. One example of this is our interpretation of the wave functions, Eq. B.91, with a complex conjugate that is $\exp[i(\vec{p}\cdot\vec{r} - E_p)/\hbar]$, as describing an anti-particle of energy E_p and momentum $\vec{p}$. The wave function $\Psi^*(\vec{r},t)$ is known as the *charge conjugate* wave function:

$$\rho(\vec{r},t) = -\frac{1}{2mc^2}\left[\Psi(\vec{r},t)\left(i\hbar\frac{\partial}{\partial t} + e\phi\right)\Psi^*(\vec{r},t) + \Psi^*(\vec{r},t)\left(-i\hbar\frac{\partial}{\partial t} + e\phi\right)\Psi(\vec{r},t)\right]$$
(B.95)

where ρ_c is the density evaluated with the wave function $\Psi^*(\vec{r},t)$ and the opposite sign of e, as indicated in Eq. B.79. Thus, the quantity ρ_c is minus the density ρ. Similarity, $\vec{J}_c$ is the current evaluated with the wave function $\Psi^*(\vec{r},t)$ and the opposite sign of e in Eq. B.80 is given [4] by:

$$\vec{J}_c(\vec{r},t) = -\vec{J}(\vec{r},t) \quad (B.96)$$

Appendix B: The Schrödinger Wave Equation

We can normalize the solution to the Klein-Gordon equation by the condition that the state represents one unit of charge:

$$\int \rho(\vec{r}, t) d^3 \vec{r} = \pm 1 \tag{B.97}$$

That is, the upper sign is for the particle solution and the lower is the anti-particle solution. The normalization Eq. B.97 is conserved in time and is invariant under a Lorentz transformation. The complex conjugate of a normalized solution to the Klein-Gordon equation of Eq. B.72 has the opposite sign of the normalization because $\int \rho = -\int \rho_c$.

B.6.3 Neutral Particles

We always have interpreted $\rho(\vec{r}, t)$ as a charge density. How do neutral particles fit into the theory? There are two possible cases to consider. First, suppose that the particle is different from its anti-particle (e.g., a K^0 meson) with strangeness 1, with an anti-particle, $\bar{K}^0$, that has strangeness -1. The statement that the particle and anti-particle are different particles means that there exist interactions that distinguish them. Then we always can define a charge, though not EM ones; that is, $+1$ for the particles and -1 for the anti-particles, and $\rho(\vec{r}, t)$ can be interpreted as a density of this charge. For K^0 mesons this charge is simply their strangeness.

An example is the π^0 meson—the π^+ and π^- are anti-particles. Then there must be complete symmetry between positive and negative energy solutions of the *Klein-Gordon equation*, as in the case of the scalar coupling S. If $\Psi(\vec{r}, t)$ satisfies the Klein-Gordon equation so must $\Psi^*(\vec{r}, t)$—the "charge conjugate" wave function. Thus, the wave function can always be chosen to be real, and for a real wave function density $\rho(\vec{r}, t)$ is zero. This situation is analogous to the description of the photon, a neutral spin one particle, by the EM field, a real field. The physical interpretation of the solution of the Klein-Gordon equation in terms of particles is quite analogous to the interpretation of the solutions of Maxwell's equations in terms of photons.

Appendix C: Four Vectors and the Lorentz Transformation

The *kinematics* of special relativity is the part of special relativity that pertains purely to space, time, and motion without reference to matter and its interactions—although some reference to dynamics has in fact crept in. Together we replaced Isaac Newton's static concept of space (i.e., an infinite array of points, all motionless relative to each other, through which bodies move) with Albert Einstein's dynamic concept (i.e., an infinite array of inertial frames, each comprising an infinite number of points that are motionless relative to each other, occupied and marked by bodies that are shifted from one frame to another by forces). After this subtle change in worldview Einstein founded *relativity*—to which the concept of the inertial frame of reference is as necessary and as fundamental as the concept of the straight line is to plane geometry.

C.1 Introduction

The *Lorentz transformation*, for which this appendix is named, is the coordinate transformation that replaces the Galilean transformation, as we saw in Chap. 2. Now that we have seen the main consequences of the postulates of *special relativity* (i.e., the relativity of simultaneity, time dilation, and length contraction) it is clear that the Galilei transformation, with its absolute time, is incorrect.

These important physical phenomena can be seen as direct consequences of the correct transformation relating inertial frames—the Lorentz transformation. This transformation is the key for the formulation of special relativity in an enlightening four-dimensional formalism, which we will discuss in the next appendix.

Here we study the Lorentz transformation and its properties and derive length contraction and time dilation directly from it, in addition to the transformation property of velocities. We must emphasize that, although it was discovered by studying Maxwell's equations, its validity is more general. The Lorentz transformation relates inertial frames without reference to the kind of physics studied in them.

Lorentz invariance is a general requirement for *any* physical theory, not just for electromagnetism.

The Lorentz transformation comes into play in a situation where the Galilei transformation is not going to be valid for speeds that are not small enough in comparison to the speed of light. Thus, consider two frames of reference—namely, S and S'—allowing for a corresponding coordinate system designated by (x, y, z, t) and (x', y', z', t'); accordingly, then we can induce the Lorentz transformation in *relativistic mechanics*, where the Galilean transformation in *classical mechanics* fails for speed v, which is not negligible compared to the speed of light c. With such a tool at our disposal, we can take the 4-vector coordinates (x, y, z, ct) of an event into consideration while we are studying the field of electrodynamics within the principle of relativity theory and *Minkowski geometrical space* and time.

Therefore, the correct transformation relating space and time coordinates occurs in the two inertial frames S and S', moving with relative velocity v in a *standard configuration*, as was shown in Figs. 2.9 and 2.10. Thus, aligning their corresponding axes with the x and x' axes along the line of relative motion, as depicted in Fig. C.1, so that the origin of the two reference frames coincide at $t = t' = 0$, is called the standard configuration of a pair of reference frames; the planes $y = 0$ and $z = 0$ always coincide with the planes $y' = 0$ and $z' = 0$. In this configuration we can set the zero points on the clock so that $t = t' = 0$ is the instant when $x = x'$.

In such a standard configuration, if an event has coordinates (x, y, z, t) in S, then its coordinates in S' are given by the following relations:

$$t' = \gamma(t - vx/c^2) \quad \text{(C.1)}$$

$$x' = \gamma(x - vt) \quad \text{(C.2)}$$

$$y' = y \quad \text{(C.3)}$$

$$z' = z \quad \text{(C.4)}$$

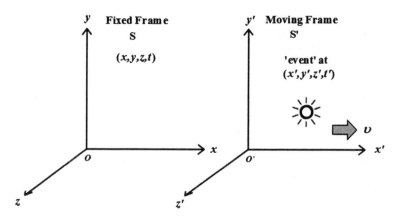

Fig. C.1 Two frames of reference, S and S'

Appendix C: Four Vectors and the Lorentz Transformation

where $\gamma = \gamma(v) = 1/(1 - v^2/c^2)^{1/2}$ is known as the *Lorentz factor* and in some of the literature or textbooks is sometimes also presented by the character L. See Sect. C.2 for proof and derivations of Lorentz factor γ.

By solving Eqs. C.1 and C.2 for (x, y, z, t) in terms of (x', y', z', t'), we can easily derive the inverse Lorentz transformation by exchanging $x \leftrightarrow x'$ and $v \leftrightarrow v'$ according to the *Principle of Relativity*:

$$t = \gamma\left(t' + vx'/c^2\right) \tag{C.5}$$

$$x = \gamma(vt' + x') \tag{C.6}$$

$$y = y' \tag{C.7}$$

$$z = z' \tag{C.8}$$

This also can be obtained by replacing v with $-v$ and swapping primed and unprimed symbols in the first set of equations. This is how it must turn out: if S' has velocity $\vec{v}$ in S, then S has velocity $-\vec{v}$ in S'; both are equally valid inertial frames. This again is the application of the Principle of Relativity, which sometimes is called the "principle of reciprocity" and is a consequence of the isotropy of space [5, 6].

Now, if we immediately extract from the Lorentz transformation the events of time dilation and Lorentz contraction, then the former simply will pick up two events at the same spatial location in fixed frame S, separated by time τ. We may as well pick the origin point, $x = y = z = 0$, and times $t = 0$ and $t = \tau$ in frame S. Now applying Eq. C.1 to the two events, we find that the first event occurs at time $t' = 0$, and the second at time $t' = \gamma\tau$, so the time interval between them in frame S' is $L\tau$ (i.e., longer than in the first frame by the *Lorentz factor* γ). This is time dilation.

For Lorentz contraction one must consider not two events but two world lines. These are the the two ends in the x-direction of some object fixed in S. Place the origin on one of these world lines, and then the other end lies at $x = L_0$ for all t, where L_0 is the rest length. Now consider these world lines in frame S_0 and pick time $t' = 0$. At this moment the world line passing through the origin of S is also at the origin of S' (i.e., at $x' = 0$). Using the Lorentz transformation, the other world line is found at

$$\begin{aligned} t' &= L(t - vL_0/c^2) \\ x' &= L(-vt + L_0) \end{aligned} \tag{C.9}$$

Because we have considered the situation at $t' = 0$, we deduce from the first equation set in Eq. C.9 that $t = vL_0/c^2$. Substituting this into the second equation we obtain $x' = \gamma L_0(1 - v^2/c^2) = L_0/\gamma$. Thus, in the primed frame at a given instant the two ends of the object are at $x' = 0$ and $x' = L_0/\gamma$. Therefore, the length of the object is reduced from L_0 by a factor γ—this is Lorentz contraction.

For relativistic addition of velocities, Eq. (21.8), consider a particle moving along the x'-axis with speed u in frame S_0. Its world line is given by $x' = ut'$. Substituting it into Eq. C.6, we obtain $x = \gamma(vt' + ut') = \gamma^2(v + u)(t - vx/c^2)$. Solving for x as a function of t, one obtains $x = wt$ with w as given [7] by:

$$w = \frac{u+v}{1+(uv/c^2)} \tag{C.10}$$

This equation is the formula for *relativistic addition of velocities*. The formula predicts that w is never greater than c as long as u and v are both less than or equal to c; this is in agreement with the *light speed postulate*. The result predicted by classical physics is $w = u + v$; the relativistic formula reproduces this in the limit $uv \ll c^2$.

Note that the velocities in the formula in Eq. C.10 are all what we call *relative velocities*, and they concern three different reference frames. There is another type of velocity that can be useful in calculations, which we will refer to as *closing velocity*. The concept of "closing velocity" applies in a *single* reference frame, and it refers to the *rate of change of distance* between two objects—all distances and times being measured in a single reference frame. When both objects are moving relative to the reference frame, a closing velocity is *not* necessarily the velocity of any physical object or signal, and it can exceed the speed of light [7].

In summary, the postulates of relativity, taken together, lead to a description of *spacetime*; see Chap. 2 in which the notions of simultaneity, time duration, and spatial distance are clearly defined in each inertial reference frame, but their values, for a given pair of events, can vary from one reference frame to another. In particular, objects evolve more slowly and are contracted along their direction of motion when observed in a reference frame relative to which they are in motion [7].

A good way to think of the Lorentz transformation is to regard it as a kind of "translation" from the x, y, z, t "language" to the x', y', z', t' language. The basic results given in the preceding serve as an introduction to increase our confidence with the transformation and its use. In the rest of this appendix we will use it to treat more general situations (e.g., addition of non-parallel velocities), the Doppler effect for light emitted at a general angle to the direction of motion, and other phenomena.

C.2 Lorentz Transformation Factor Derivation

The Lorentz transformation factor γ can be derived on the basis of the two postulates of special relativity. First, because of the isotropy of space contained in the second postulate, we can orient the spatial axes of an inertial frame S' with those of another inertial frame S and limit ourselves to considering the motion of the two frames in standard configuration. Using Fig. C.1 as a starting point for deducing the transformation relating two inertial frames S with events in (x, y, z, t) and S' with events at (x', y', z', t') in relative motion with velocity v in standard configuration, it is reasonable to assume that the transformation is linear and is written:

$$x' = G(x - vt) \tag{C.11}$$

where element G is a dimensionless constant that depends only on v/c. This assumption corresponds to the homogeneity of space and time because G does not

Appendix C: Four Vectors and the Lorentz Transformation

depend on (x, t). Because the spatial part of the Galilei transformation, $x' = x - vt$, must be recovered in limit $v/c \to 0$, G must tend to unity in this limit.

The laws of physics must have the same form in S and S', and then one must obtain the inverse Lorentz transformation by the exchange $(t', \vec{x}') \leftrightarrow (t, \vec{x})$ and $v \leftrightarrow -v$ (i.e., the Principle of Relativity again); thus, we can write:

$$x = G(x' + vt') \quad (\text{C.12})$$

There is no relative motion in the y- and z-directions, thus these coordinates must not be affected by y, the transformation, or:

$$\begin{cases} y' = y \\ z' = z \end{cases} \quad (\text{C.13})$$

Consider now a spherical pulse of EM radiation emitted at the origin of S at $t = 0$. It is received at a point on the x-axis and, during its propagation,

$$x = ct \quad (\text{C.14})$$

The same law must be true in S' because of the constant of the speed of light c; thus, we can write:

$$x' = ct' \quad (\text{C.15})$$

Or from Eq. C.11, we get $ct' = G(ct - vt)$, which leads to the following relation:

$$t' = G\left(t - \frac{v}{c}t\right) \quad (\text{C.16})$$

while

$$ct = G(ct' + vt') = G(c + v)t' \quad (\text{C.17})$$

By Substituting Eq. C.16 into Eq. C.12, we can obtain the following:

$$ct = G(c + v)G\left(t + \frac{v}{c}t\right) = \frac{G^2}{c}(c + v)(c + v)t \quad (\text{C.18})$$

and

$$G^2 = \frac{c^2}{(c + v)(c - v)} \quad (\text{C.19})$$

Therefore, we have:

$$G = \frac{1}{\sqrt{1 - \frac{v^2}{c^2}}} \equiv \gamma \quad (\text{C.20})$$

As we can see, Eq. C.20 is results in the *Lorentz factor* γ and

$$x' = \gamma(x - vt) \tag{C.21}$$

Implementing $x = \gamma(x' + vt')$ and $x' = \gamma(x - vt)$, then substituting x' into x to obtain:

$$x = \gamma\left[\underbrace{\gamma(x - vt)}_{x'} + vt'\right] \tag{C.22}$$

$$x = \gamma^2 x - \gamma^2 vt + \gamma vt'$$

from which, we conclude that

$$t' = \frac{x - \gamma^2 x + \gamma^2 vt}{\gamma v} \tag{C.23}$$

and

$$t' = \frac{x}{v\gamma}(1 - \gamma^2) + \gamma t \tag{C.24}$$

Meanwhile, using Eq. C.19, we get:

$$1 - \gamma^2 = 1 - \frac{1}{1 - \frac{v^2}{c^2}} = \frac{1 - \frac{v^2}{c^2} - 1}{1 - \frac{v^2}{c^2}} = \frac{-\left(\frac{v^2}{c^2}\right)}{1 - \frac{v^2}{c^2}} \tag{C.25}$$

Then, we have:

$$\frac{1 - \gamma^2}{\gamma} = \sqrt{1 - \frac{v^2}{c^2}}\left\{\frac{-\left(\frac{v^2}{c^2}\right)}{\left(1 - \frac{v^2}{c^2}\right)}\right\} = \frac{-\frac{v^2}{c^2}}{\sqrt{1 - \frac{v^2}{c^2}}} = -\gamma\left(\frac{v^2}{c^2}\right) \tag{C.26}$$

$$t' = \frac{1 - \gamma^2}{\gamma}\frac{x}{v} + \gamma t = -\gamma\frac{v^2}{c^2} + \gamma t$$

and, finally

$$t' = \gamma\left(t - \frac{v}{c^2}x\right) \tag{C.27}$$

which completes the derivation. Equations C.13, C.21, and C.27 constitute the *Lorentz transformation*.

The fact that it can be derived from the two postulates of special relativity is conceptually important; it means that these two postulates constitute the physical

Appendix C: Four Vectors and the Lorentz Transformation

explanation of the mathematical transformation and that this transformation should not be assumed in place of the two postulates, as Poincaré seems to have intended. Although *Poincaré*, Lorentz, and *FitzGerald* stopped at the transformation, which is an important ingredient of special relativity and unveils the mixing of space and time of the *four-dimensional world view*), they tried to explain it with an ether and with length contraction. It was Einstein's genius that reduced the physical explanation of the transformation into two simple and general postulates and that led to a reexamination of the concepts of space and time, thus developing the full theory, which was missed by other researchers [8].

C.3 Mathematical Properties of the Lorentz Transformation

We are the point where we can examine the properties of the Lorentz transformation as described in the following.

- Qualitatively, the Lorentz transformation mixes t and x, therefore three-dimensional lengths and time intervals cannot be left invariant. Quantitatively, length contraction and time dilation can be derived from the Lorentz transformation as a consequence, which is done in the Chap. 2, Sect. 2.9 (i.e., spacetime, interval, and diagram) of this book. The quantity is:

$$d\mathsf{a}^2 = dx^2 + dy^2 + dz^2 - c^2 dt^2 \tag{C.28}$$

Bear in mind that thr *Minkowski line element" is invariant* under Lorentz transformations. This invariance is easy to see, by using it we have:

$$\begin{aligned}(d\mathbf{a}')^2 &\equiv (dx')^2 + (dy')^2 + (dz')^2 - c^2(dt')^2 \\ &= \gamma^2(dx - vdt)^2 + dy^2 + dz^2 - c^2\gamma^2\left(dt - \tfrac{v}{c^2}dx\right)^2 \\ &= -c^2\gamma^2\left(1 - \tfrac{v^2}{c^2}\right)dt^2 + 2c^2\gamma^2\tfrac{v}{c^2}dtdx + \gamma^2 dx^2\left(1 - \tfrac{v^2}{c^2}\right) \\ &\quad - 2\gamma^2 vdtdx + dy^2 + dz^2 \\ &= -c^2 dt^2 + dx^2 + dz^2 \equiv (d\mathbf{a})^2\end{aligned} \tag{C.29}$$

For extensive discussion of special relativity, refer to the book by Faraoni [8].

- The Lorentz transformation is symmetric under the exchange of $x \leftrightarrow ct$ as:

$$\begin{cases} x' = \gamma(x - vt) \\ y' = y \\ z' = z \\ ct' = \gamma\left(ct - \dfrac{vx}{c}\right) \end{cases} \quad (C.30)$$

where Eq. C.30, becomes:

$$\begin{cases} ct' = \gamma\left(ct - \dfrac{vx}{c}\right) \\ y' = y \\ z' = z \\ x' = \gamma(x - vt) \end{cases} \quad (C.31)$$

In standard configuration, the Lorentz transformation is also symmetric under the exchange $y \leftrightarrow z$.

- The Galilei transformation can somehow be recovered from the Lorentz transformation in the limit of small velocities ($|v|/c \ll 1$), although the derivation is a bit finicky. First, expand the Lorentz factor γ to first order in v/c, using the *Taylor expansion*:

$$\gamma \equiv \frac{1}{\sqrt{1 - \frac{v^2}{c^2}}} = 1 + \frac{v^2}{2c^2} + \cdots \approx 1 \quad (C.32)$$

and

$$\begin{cases} x' \approx x - vt \\ y' = y \\ z' = z \end{cases} \quad (C.33)$$

Strictly speaking, the transformation of the time coordinate gives to first order in v/c as:

$$t' = t - \frac{v}{c^2} x \quad (C.34)$$

not $t' = t$ as in the *Galilei transformation*; the relativity of simultaneity persists to first order (it is a first order effect in v/c). If the Lorentz transformation reduced to the Galilei transformation to first order, then infinitesimal Lorentz transformations and infinitesimal Galilei transformations would coincide, which is not the case. This point is very clear in Furry [9] and Baierlein [10].

Still, *time dilation* is computed by considering time differences and recording two spatial events at the same location. Because $\Delta t' = \Delta t - \frac{v}{c^2}\Delta x$, by setting $\Delta x = 0$ time dilation is eliminated to first order. In practice when speeds are small, *time intervals* Δt are measured over spatial distances Δx such that $c\Delta t \gg \Delta x \gg (v/c)\Delta x$ and the Δx term can be dropped. Although the Lorentz transformation does not quite reduce to the Galilei transformation, which is recovered only in the limit $(v/c) \to 0$, Newtonian mechanics and the Galilei transformation turn out to be adequate in the limit $|v| \ll c$.

Note that the derivation of the Lorentz transformation from the postulate of special relativity requires only that the *spatial part* of $x_1 = G(v)(x - vt)$ reduces to the spatial part of the Galilei transformation $x' = x - vt$ in the limit $(|v|/c) \ll 1$, from which we can deduce that $G \to 1$. We did not assume recovery of $t' = t$ in this limit; thus, the proof is correct.

- Because the Lorentz transformation is linear and homogeneous, finite coordinate differences transform in the same way as infinitesimal coordinate differences:

$$\begin{aligned}\Delta x' &= \gamma(\Delta x - v\Delta t) & dx' &= \gamma(dx - vdt)\\ \Delta y' &= \Delta y & dy' &= dy\\ \Delta z' &= \Delta z & \text{and} \quad dz' &= dz\\ \Delta t' &= \gamma\left(\Delta t - \frac{v}{c^2}\Delta x\right) & dt' &= \gamma\left(dt - \frac{v}{c^2}dx\right)\end{aligned} \qquad (C.35)$$

In summary, we know by now that the Lorentz factor $\gamma \equiv 1/\left(\sqrt{1 - v^2/c^2}\right)$, the ratio between coordinate and proper times, diverges as $v \to c$. The inequality $v > c$ leads to a purely imaginary γ; therefore, the *relative velocity* of two inertial frames must be *strictly smaller than c*.

Because an inertial frame can be associated with any non-accelerated particle or object moving with subluminal (i.e., $|v| > c$) speed, this statement translates into the requirement that the speed of particles and of all physical signals be limited by c, which is the speed of light in vacuum; and the speed of particles traveling in a medium can be larger than the speed of light in that medium. If the particle traveling faster than the speed of light in that medium is charged (i.e., emitting radiation) this is known as *Cherenkov radiation*. For further explanation see the next section, Cherenkov Radiation.

Never mind the fact that the Lorentz factor becomes imaginary; we can agree to define γ as the modulus as follows:

$$\left(\sqrt{1 - v^2/c^2}\right) \quad \text{if} \quad |v| > c \qquad (C.36)$$

What is truly important in this situation is that the restriction $|v| \leq c$ preserves the notion of cause and effect. In fact, consider a process in which an event P causes, or affects, an event Q by sending a signal containing some information from P to Q.

If a signal were sent from P to Q at superluminal speed, $u > c$ in some inertial frame S, we could orient the axes of S so that both events P and Q occur on the x-axis and their time and spatial separations satisfy $\Delta t > 0$ and $\Delta x > 0$ in this frame. Then, in an inertial frame S' moving with respect to S with speed v in standard configuration, we would have:

$$\Delta t' = \gamma\left(\Delta t - v\frac{\Delta x}{c^2}\right) = \gamma \Delta t \left(1 - \frac{uv}{c^2}\right) \qquad (C.37)$$

where we have used $\Delta x = u \Delta t$. Now, because $u > c$, it also is obvious that $-u > -c$ which, together with $0 < v < c$, implies that $-uv < -c^2$ or $(-uv/-c^2) < 1$; thus, we can write:

$$\Delta t' = \gamma \Delta t \left(1 - \frac{uv}{c^2}\right) < 0 \qquad (C.38)$$

According to this result, in the frame S' the event Q precedes P; cause and effect are reversed or the signal goes backward in time. The signal reaches Q before being emitted by P, which creates a logical problem. The fact that there is an absolute speed limit c comes to the rescue and enforces causality; if both $|u|$ and $|v| < c$, then $\Delta t'$ in Eq. C.37 has the same sign as Δt. The possibility of reversing cause and effect and traveling in time would lead to logical paradoxes, which have been discussed at length in the literature (see Lockwood [9] and the references therein).

C.4 Cherenkov Radiation

Cherenkov radiation, also known as *Vavilov-Cherenkov radiation* (VCR), named after Sergey Vavilov and Pavel Cherenkov (i.e., alternative spelling forms: Cherenkov or Cerenkov, and Vavilo or Wawilow). It is EM radiation emitted when a charged particle (e.g., an electron) passes through a dielectric medium at a speed greater than the phase velocity of light in that medium. The characteristic blue glow of an underwater nuclear reactor is because of Cherenkov radiation. The Soviet scientist, Pavel Cherenkov, the 1958 Nobel Prize winner, was the first to detect it experimentally [10]. A theory of this effect was later developed within the framework of Einstein's special relativity theory by Igor Tamm and Ilya Frank, who also shared the Nobel Prize. Cherenkov radiation had been theoretically predicted by English polymath Oliver Heaviside in papers published in 1888–1889 [11].

Although electrodynamics holds that the speed of light in a vacuum is a universal constant (c), the speed at which light propagates in a material may be significantly less than c. For example, the speed of the propagation of light in water is only $0.75\,c$. Matter can be accelerated beyond this speed (although still to less than c) during nuclear reactions and in particle accelerators. Cherenkov radiation results when a charged particle, most commonly an electron, travels through a dielectric (i.e., electrically polarizable) medium with a speed greater than that at which light propagates in the same medium.

Moreover, the velocity must be exceeded is the phase velocity of light rather than the group velocity of light. The phase velocity can be altered dramatically by employing a periodic medium, and in that case one can even achieve Cherenkov radiation with no minimum particle velocity—a phenomenon known as the *Smith-Purcell effect*. In a more complex periodic medium, such as a photonic crystal, one also can obtain a variety of other anomalous Cherenkov effects, such as radiation in a backward direction, whereas ordinary Cherenkov radiation forms an acute angle with the particle velocity [12].

As a charged particle travels, it disrupts the local EM field in its medium. In particular, the medium becomes electrically polarized by the particle's electric field. If the particle travels slowly, then the disturbance elastically relaxes back to mechanical equilibrium as the particle passes. When the particle is traveling fast enough, however, the limited response speed of the medium means that a disturbance is left in the wake of the particle, and the energy contained in this disturbance radiates as a coherent shockwave.

A common analogy is the sonic boom of a supersonic aircraft or bullet. The sound waves generated by the supersonic body propagate at the speed of sound itself; as such, the waves travel slower than the speeding object and cannot propagate forward from the body, instead forming a shock front. In a similar way, a charged particle can generate a light shock wave as it travels through an insulator. Figure C.2 illustrates the geometry of Cherenkov radiation (shown for the ideal case of no dispersion).

In the figure the particle (*red arrow*) travels in a medium with speed v_p as phase velocity, such that $c/n < v_p < c$, where c is the speed of light in vacuum and n is the *refractive index* of the medium. (If the medium is water, the condition is $0.75 < v_p < c$ because $n = 1.33$ for water at 20 C.) We define the ratio between the speed of the particle and the speed of light as $\beta = v_p/c$. The emitted light waves (*blue arrows*) travel at speed $v_{em} = c/n$.

Fig. C.2 The geometry of the Cherenkov radiation

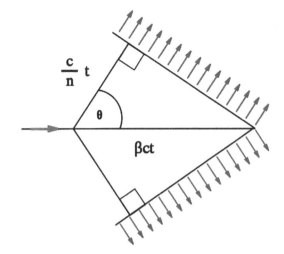

In Fig. C.2 the left corner of the triangle represents the location of the superluminal particle at some initial moment ($t = 0$). The right corner of the triangle is the location of the particle at some later time t. In the given time t, the particle travels the distance of x_p, as shown here:

$$x_p = v_p t = \beta c t \qquad (C.39)$$

whereas the emitted EM waves are constricted to travel the distance of x_{em}:

$$x_{em} = v_{em} t = \frac{c}{n} t \qquad (C.40)$$

So

$$\cos \theta = \frac{1}{n\beta} \qquad (C.41)$$

Note that because this ratio is independent of time, one can take arbitrary times and achieve similar triangles. The angle stays the same, meaning that subsequent waves generated between the initial time $t = 0$ and final time t will form similar triangles with coinciding right endpoints to the one shown.

C.4.1 Arbitrary Cherenkov Emission Angle

Cherenkov radiation also can radiate in an arbitrary direction using properly engineered one-dimensional meta-materials [13]. The latter is designed to introduce a gradient of phase retardation along the trajectory of the fast traveling particle, reversing or steering Cherenkov emission at ($d\phi/dx$) arbitrary angles given by the generalized relation as:

$$\cos \theta = \frac{1}{n\beta} + \frac{n}{k_0} \cdot \frac{d\phi}{dx} \qquad (C.42)$$

C.4.2 Reverse Cherenkov Effect

A reverse Cherenkov effect can be experienced using materials called negative-index meta-materials—that is, materials with a subwavelength microstructure that gives them an effective "average" property very different from their constituent materials—in this case having *negative permittivity* and *negative permeability*). This means that when a charged particle (usually electrons) passes through a medium at a speed greater than the phase velocity of light in that medium, that particle will emit trailing radiation from its progress through the medium rather than in front of it—as is the case in normal materials with both positive permittivity and permeability [13]. One also can obtain such reverse-cone Cherenkov radiation in meta-material that is not a periodic media, where the periodic structure is on the same scale as the wavelength, so it cannot be treated as an effectively homogeneous meta-material [12].

C.4.3 Cherenkov Radiation Characteristics

The frequency spectrum of Cherenkov radiation by particle q is given by the *Frank-Tamm formula*:

$$\frac{d^2\mathcal{E}}{dxd\omega} = \frac{q^2}{4\pi}\mu(\omega)\omega\left(1 - \frac{c^2}{v^2 n^2 \omega}\right) \tag{C.43}$$

The formula describes the amount of energy emitted from Cherenkov radiation per wavelength, per unit length traveled. $\mu(\omega)$ is the permeability and $n(\omega)$ is the *index of refraction* of the material the charge particle moves through.

Unlike fluorescence or emission spectra that have characteristic spectral peaks, Cherenkov radiation is continuous. Around the visible spectrum, the relative intensity per unit frequency is approximately proportional to the frequency. That is, higher frequencies (shorter wavelengths) are more intense in Cherenkov radiation. This is why visible Cherenkov radiation is observed to be brilliant blue. In fact, most Cherenkov radiation is in the *ultraviolet spectrum*—it is only with sufficiently accelerated charges that it even becomes visible; the sensitivity of the human eye peaks at green and is very low in the violet portion of the spectrum.

There is a *cut-off frequency* above which the equation $\cos\theta = 1/(n\beta)$ can no longer be satisfied. The refractive index n varies with frequency (thus with wavelength) in such a way that the intensity cannot continue to increase at ever shorter wavelengths, even for very relativistic particles, where v/c is close to 1. At X-ray frequencies the refractive index becomes less than unity (note that in media the phase velocity may exceed c without violating relativity); thus, no X-ray emission or shorter wavelength emissions (e.g., gamma rays) would be observed. Nevertheless, X-rays can be generated at special frequencies just below the frequencies corresponding to core electronic transitions in a material because the index of refraction is often greater than 1 just below a resonant frequency (see *Kramers-Kronig relation* [14] and *anomalous dispersion* [15]).

As in sonic booms and bow shocks, the angle of the shock cone is directly related to the velocity of the disruption. The Cherenkov angle is zero at the threshold velocity for the emission of Cherenkov radiation. The angle takes on a maximum as the particle speed approaches the speed of light. Therefore, observed angles of incidence can be used to compute the direction and speed of a Cherenkov radiation-producing charge.

Cherenkov radiation can be generated in the eye by charged particles hitting the *vitreous humour*, giving the impression of flashes [16], as in cosmic ray visual phenomena and possibly some observations of criticality accidents.

C.4.4 Cherenkov Radiation Applications

There are many industrial and medical applications that are using the physics of Cherenkov radiation and a few can be described here.

1. *Detection of labeled biomolecules*: Cherenkov radiation is widely used to facilitate the detection of small amounts and low concentrations of biomolecules [16]. Radioactive atoms, such as phosphorus-32, are readily introduced into biomolecules by enzymatic and synthetic means and subsequently may be easily detected in small quantities for the purpose of revealing biological pathways and in characterizing the interaction of biological molecules such as affinity constants and dissociation rates.
2. *Medical imaging of radioisotopes and external beam radiotherapy*: More recently, Cherenkov light has been used to image substances in the body [17–19]. These discoveries have led to intense interest around the idea of using this light signal to quantify and/or detect radiation in the body, either from internal sources, such as injected radiopharmaceuticals, or from external beam radiotherapy in oncology. Radioisotopes, such as the positron emitters ^{18}F and ^{13}N, or beta emitters ^{32}P or ^{90}Y have measurable Cherenkov emission [20], and isotopes ^{18}F and ^{131}I have been imaged in humans for diagnostic value demonstrations [21–22]. External beam radiation therapy has been shown to induce a substantial amount of Cherenkov light in the tissue being treated because of the photon beam energy levels used in the 6 MeV to 18 MeV ranges. Figure C.3 shows Cherenkov light emission imaged from the chest wall of a patient undergoing whole breast

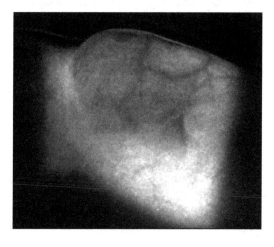

Fig. C.3 The Cherenkov light emission image of a human chest

Appendix C: Four Vectors and the Lorentz Transformation 545

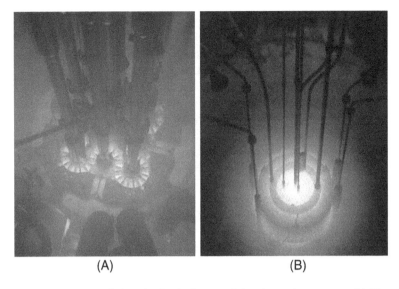

Fig. C.5 (**a**) Cherenkov radiation glowing in the core of the advanced test reactor, (**b**) Cherenkov radiation in the Reed Research Reactor

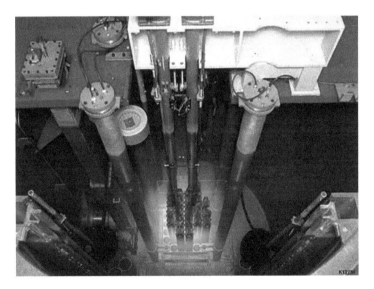

Fig. C.6 Cherenkov radiation in a TRIGA reactor pool

radioactivity of spent fuel rods. This phenomenon is used to verify the presence of nuclear fuel in spent fuel pools for nuclear safeguarding purposes [26] (Fig. C.6).

Figure C.5, also shows that application of Cherenkov radiation in *Training, Research, Isotopes, General Atomics* (TRIGA), which is a class of small nuclear reactors, designed and manufactured by *General Atomic* (GA); the project was led by the physicist Freeman Dyson in collaboration with Edward Teller [27].

C.5 Vacuum Cherenkov Radiation

The Cherenkov effect can occur in vacuum. In a slow-wave structure the phase velocity v_p decreases and the velocity of charged particles can exceed the phase velocity while remaining lower than the speed of light c in vacuum. In such a system this effect can be derived from conservation of the energy and momentum, where the momentum of a photon should be $p = \hbar\beta$ and β is phase constant [28] rather than the de Broglie relation $\vec{p} = \hbar\vec{k}$, and $\hbar = h/2\pi$ is called the *reduced Plank constant*. Here $\vec{k}$ is the *wave vector*. This type of radiation—that is, *vacuum Cherenkov radiation* (VCR)—is used to generate *high power microwaves* (HPM) [29].

Note that from college physics, the de Broglie relation is the relation where de Broglie wavelength λ is associated with a massive particle and is related to its momentum p through the Planck constant, h, and is written as $\lambda = h/p$, and energy $\mathcal{E}$ is given by the *Plank-Einstein relation*, as $\mathcal{E} = h\nu$, where ν is the light frequency; thus, the momentum is expressed by the following equation, knowing that $\nu = 1/\lambda$:

$$p = \frac{\mathcal{E}}{c} = \frac{h}{\lambda} \tag{C.44}$$

If we take the wave equation into consideration in the form of one-dimension (1D), yet complex wave function Ψ (i.e., *x*-direction) as $\Psi = Ae^{i(px - \omega t)}$, where ω is the angular frequency, we can draw the propagation of de Broglie waves in one-dimensional space (Fig. C.7). As demonstrated in the figure, the propagation of the de Broglie waves in 1D for the real part of complex wave function Ψ amplitude is blue and the imaginary part is green.

The probability (shown as the color opacity) of finding the particle at a given point *x* is spread out like a waveform; there is no definite position of the particle. As the amplitude increases above zero, the curvature decreases, so the amplitude decreases again, and vice versa. The result is an alternating amplitude, as shown in Fig. C.7. Note that the relationship $\lambda = h/p$ is now known to hold for all types of matter; all matter exhibits properties of both particles and waves.

C.6 Lorentz Invariance and 4-Vectors

We have seen the basic physical consequences of the two postulates of *special relativity*, and we know how to derive them from the mathematical transformation relating two inertial frames—the Lorentz transformation. The mathematics gives us an insight into how space and time are inextricably mixed and the most natural way to see this is in a representation of the world with four dimensions: three spatial and one temporal. This is not just mathematics; the physics just makes so much more sense when viewed in four dimensions than in three spatial dimensions with time as a parameter, just because that is all the time in Newtonian physics.

Fig. C.7 Propagation of a de Broglie wave in 1D: Top: plane wave. Bottom: wave packet

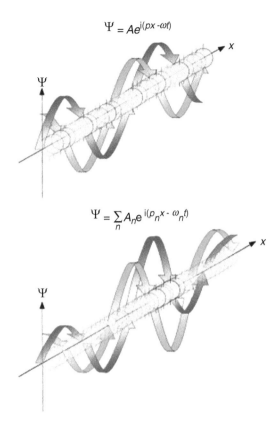

To better understand the meaning of these invariant quantities, let us introduce a new class of objects called *4-vectors*; they are four-component column vectors and are characterized by their linear transformation properties when going from one inertial frame to another, which for a Lorentz boost is given in a different format, as described in the following.

As part of Lorentz transformations components in four dimensions of *Cartesian coordinates*, and what we have learned so far—that is, from the main consequences that were presented in Eqs. C.1 through C.5 and expanded to Eqs. C.30, C.31, and C.33—is that most general coordinate transformations are between inertial systems, which is compatible with the relativity principle. Yet, there were many constraints that we imposed on the transformation, which are satisfied by Galilean transformations as well, before we introduced the invariance of the speed of light in Sect. C.3 in Appendix C. This was done by considering Fig. C.1 with the two inertial systems of O and O' for coordinates of (x, y, z, t) in the fixed frame of O and (x', y', z', t') in moving frame of O' with velocity v in the x-direction in respect to fixed frame O.

Nevertheless, if we set up the coordinates in both systems of O and O' in such a way that event $x = y = z = t = 0$ coincides with event $(x', y', z', t' = 0)$, which is a possible scenario by a translation of the origins, we are faced with a linear transformation between two vector spaces of dimension four. This is specified by a 4×4 matrix $\vec{\Lambda}$, thus the transformation under Lorentz conditions will be written as:

$$\begin{cases} x' = \Lambda_{xt}t + \Lambda_{xx}x + \Lambda_{xy}y + \Lambda_{xz}z \\ y' = \Lambda_{yt}t + \Lambda_{yx}x + \Lambda_{yy}y + \Lambda_{yz}z \\ z' = \Lambda_{zt}t + \Lambda_{zx}x + \Lambda_{zy}y + \Lambda_{zz}z \\ t' = \Lambda_{tt}t + \Lambda_{tx}x + \Lambda_{ty}y + \Lambda_{tz}z \end{cases} \quad (C.45)$$

Again, form the four-dimensional world, Eq. C.45 allows us to compare the Galilei and the Lorentz transformations:

$$\begin{array}{ll} x' = x - vt & x' = \gamma(x - vt) \\ y' = y & y' = y \\ z' = z & z' = z \\ t' = t & ct' = \gamma\left(ct - \frac{v}{c}x\right) \end{array} \quad (C.46)$$

Galilei Transformation
leaves Newtonian mechanics invariant
intervals of absolute time are invariant
3-D lengths are invariant

Lorentz Transformation
leaves Maxwell's theory invariant
time intervals are not invariant
3-D lengths are not invariant

Because the Lorentz transformation mixes the time and space coordinates, it implicitly suggests treating these quantities on the same footing and to contemplate a four-dimensional space (x, y, z, ct). The four-dimensional *worldview* was developed by Hermann Minkowski after the publication of Einstein's theory. In Minkowski's words: "Henceforth space by itself and time by itself are doomed to fade away into mere shadows, and only a kind of union of the two will preserve an independent reality." In the Newtonian picture of the world, space and time are separate entities with time playing the role of a parameter in the Newtonian equations of motion. An *event* is now simply a point in the four-dimensional Minkowski spacetime.

Now, back to Eq. C.45 and continuing on the subject of Lorentz invariance and 4-vectors, as a first constraint linked to this particular choice where the coordinate axes in O and O' are parallel to each other and $\vec{v}_{OO'}$ is directed along the x-axis (i.e., we set $\vec{v}_{OO'} = (v, 0, 0)$); we must neglect all terms mixing dissimilar spatial coordinates; otherwise, the coordinate axes would not be parallel to each other. So, we can write:

$$\Lambda_{xy} = \Lambda_{xz} = \Lambda_{yx} = \Lambda_{yz} = \Lambda_{zx} = \Lambda_{zy} = 0 \quad (C.47)$$

Next, we obtain a set of constraints based on the isotropy of space. Λ cannot depend on the particular orientation of the y, z-axes; thus, any combined rotation around the x and x' axes for both systems (i.e., leaving the y, y', z, z' still parallel to each other) must leave Λ unchanged. If we consider in particular a rotation by π,

Appendix C: Four Vectors and the Lorentz Transformation

which changes the signs of y, y', z, and z', while leaving x, x', t and t' unchanged, we deduce that no terms can exist that mixes the two different sets of coordinates—that is, we have:

$$\Lambda_{xy} = \Lambda_{xz} = \Lambda_{yx} = \Lambda_{zx} = \Lambda_{zx} = 0 \tag{C.48}$$

By going through this exercise, we already have reduced the free parameters of Λ from 16 to 6 as:

$$\begin{aligned} x' &= \Lambda_{xt}(v)t + \Lambda_{xx}(v)x \\ y' &= \Lambda_{yy}(v)y \\ z' &= \Lambda_{zz}(v)z \\ t' &= \Lambda_{tt}(v)t + \Lambda_{tx}(v)x \end{aligned} \tag{C.49}$$

This equation explicitly indicates the dependency of the coefficients on velocity v, the relative speed of moving frame O' with respect to fixed frame O.

Isotropy can be used again by considering the inversion of both the x and the x' axis; if at the same time we invert the sign of v, Λ must remain unchanged. From that we deduce that $\Lambda_{tt}(v)$, $\Lambda_{xx}(v)$, $\Lambda_{yy}(v)$, and $\Lambda_{zz}(v)$ are even functions of v, while $\Lambda_{tx}(v)$ and $\Lambda_{xt}(v)$ are odd. A general requirement that we do is that Λ be a continuous function of v and that it tends to the identity transformation for $v \to 0$ so that $\Lambda_{tt}(0) = \Lambda_{xx}(0) = \Lambda_{yy}(0) = \Lambda_{zz}(0) = 1$ and $\Lambda_{tx}(0) = \Lambda_{xt}(0) = 0$.

Now, we are at the point where we can consider the relative velocities of fixed frame O and moving frame O' and their relation to the coefficients. The origin of frame O', represented by $x' = y' = z' = 0$, moves in O according to $y = z = 0$ with $x = vt$ [28], so that:

$$\Lambda_{xt}(v) = -v\Lambda_{xx}(v) \tag{C.50}$$

In a similar fashion we can take into consideration the motion of the origin of O, $x = y = z = 0$, as seen from reference frame O', where it moves with velocity vector $\vec{v}_{OO'}$. Using the same argument of invariance as before, under combined rotations around the x, x' axes, one sees that velocity $\vec{v}_{OO'}$ cannot point to any direction but x' itself; thus, we can write $\vec{v}_{OO'} = (w, 0, 0)$, where w is some velocity equal to $-v$, while the world line of the origin of O in O' that is given as the result of $y' = z' = 0$ and $x' = wt'$, so that:

$$\Lambda_{xt}(v) = w\Lambda_{tt}(v) \tag{C.51}$$

Yet, such a conclusion is not so obvious and would be an incorrect assumption in a non-isotropic universe. Thus, let us follow the precise chain of deductions leading to such a conclusion by applying the following steps.

First, velocity w should not be just some velocity, and it must be well defined as a function of velocity v and as a universal function of v, $w = f(v)$, which according to the relativity principle must be independent of the particular starting reference fixed frame O. Given the task in hand we further state the following properties of $f(v)$:

1. If O' applies f to w, the result must be v (i.e., $f(f(v)) = v$, f coincides with its inverse).
2. From Eqs. C.50 and C.51 and from the even–odd properties of the transformation coefficients, we deduce $f(-v) = -f(v)$. This can be linked directly to space isotropy because the even–odd properties of the coefficients are a consequence of that.
3. Finally, if we ask that for $v \to 0$ the transformation goes to identity, then $f(0) = 0$.

We can easily prove that the only two continuous functions that satisfy the properties in the preceding are $f(v) = \pm v$. On the other hand, if we set $w = f(v) = v$, then we obtain, from Eqs. C.50 and C.51, that $\Lambda_{tt}(v) = -\Lambda_{xx}(v)$, which is not compatible with Λ going to the identity for $v \to 0$. Therefore, the only possibility that seems trivial would be $w = -v$, as we claimed before, so that:

$$\Lambda_{tt}(v) = \Lambda_{xx}(v) \tag{C.52}$$

Now let us consider the inverse transformation of Λ^{-1}, which brings us back from moving frame O' to fixed frame O, again using Fig. C.1; it must be of the same form as Λ that was described by Eq. C.49, and it will be obtained by substitution of $v \to w = -v$ (i.e., $(\Lambda(-v))^{-1} = \Lambda(-v)$). If we impose this condition, together with the known even–odd properties of the coefficients, we obtain:

$$\begin{aligned}
t &= \Lambda_{tt}(-v)t' + \Lambda_{tx}(-v)x' = \Lambda_{tt}(v)t' - \Lambda_{tx}(v)x' \\
x &= \Lambda_{xt}(-v)t' + \Lambda_{xx}(-v)x' = -\Lambda_{xt}(v)t' + \Lambda_{xx}(v)x' \\
y &= \Lambda_{yy}(-v)y' = \Lambda_{yy}(v)y' \\
z &= \Lambda_{zz}(-v)z' = \Lambda_{zz}(v)z'
\end{aligned} \tag{C.53}$$

That compared to Eqs. C.49, C.50, and C.51 will lead us to a relationship such as $\Lambda_{yy}(v) = \Lambda_{zz}(v) = 1$, and if we set the *Lorentz factor* $\gamma(v) \equiv \Lambda_{xx}(v)$ to the following form:

$$\Lambda_{tx} = \frac{1-\gamma^2}{\gamma v} \tag{C.54}$$

Then, we are left with only one unknown function, $\gamma(v)$, defining the most general pure boost transformation between inertial systems:

$$\begin{pmatrix} t' \\ x' \\ y' \\ z' \end{pmatrix} = \begin{pmatrix} \gamma & \frac{-\gamma^2-1}{\gamma v} & 0 & 0 \\ -\gamma v & \gamma & 0 & 0 \\ 0 & 0 & 1 & 0 \\ 0 & 0 & 0 & 1 \end{pmatrix} \begin{pmatrix} t \\ x \\ y \\ z \end{pmatrix} \tag{C.55}$$

Appendix C: Four Vectors and the Lorentz Transformation

where we have expressed spacetime event coordinates as column vectors and the notation stands for the standard row-by-column product. Notice that up to now the invariance of the speed of light has not been used anywhere; therefore, Galilean transformations must be of the same form as in Eq. C.55, which is indeed the case by setting Lorentz factor $\gamma = 1$.

We can impose the invariance of c by considering a particular light beam trajectory in O (i.e., $x = ct$ and $y = z = 0$) and by requiring that Eq. C.55 transforms it into a trajectory with the same speed (i.e., $x' = ct'$ and $y' = z' = 0$). After some elementary algebra we obtain $1/\gamma^2 = 1 - v^2/c^2$, which requires $\lim_{v \to 0} \gamma(v) = 1$ and has the unique solution that results in the Lorentz factor as:

$$\gamma = \frac{1}{\sqrt{1 - v^2/c^2}} \tag{C.56}$$

Now we can rewrite the transformation in a simple form by adopting ct in place of t, which permits us to deal with a homogeneous set of coordinates, and by defining $\beta \equiv v/c$ as:

$$\begin{pmatrix} ct' \\ x' \\ y' \\ z' \end{pmatrix} = \begin{pmatrix} \gamma & -\gamma\beta & 0 & 0 \\ -\gamma\beta & \gamma & 0 & 0 \\ 0 & 0 & 1 & 0 \\ 0 & 0 & 0 & 1 \end{pmatrix} \begin{pmatrix} ct \\ x \\ y \\ z \end{pmatrix} \tag{C.57}$$

This equation represents the Lorentz transformation for a pure boost along the x-axis when $c \to \infty$ for speed of light.

We thus can look at Galilean transformations as the correct realization of the relativity principle as long as we do not know about the existence of a universal, invariant velocity of nature, if $c = \infty$; it is electromagnetism that has made us aware of such a velocity for the first time. We notice that the transformation, as shown in Eq. C.67, becomes ill-defined or complex for $|\beta| \geq 1$; thus, it is suggesting that relative velocities cannot exceed c. Evidence is not compelling, however, and the question about c as a limiting velocity of nature requires a more careful discussion, as it is claimed that a *scalar wave* travels much faster than the speed of light.

As we illustrated in the preceding, it is possible to continue by finding equations describing the transformation of acceleration, and then introducing force and its transformation, which we show in the following sections, using Eq. C.67 as it was derived before. Still, we can have a better and completer understanding of these issues as a whole; the subject has gained prominence with the new approach presented as Eq. C.57, where time and space are handled together. See the section related to spacetime in Chap. 2 of this book.

As we mentioned at the beginning of this section and induction of Eq. C.67, via introduction of Fig. C.1 in fixed frame of reference O, the *4-vector technique* allowed us to arrange the coordinates t, x, y, z into a vector of four components and the conclusion of that resulted in the development of Eq. C.57, which has the same physical dimensions, and we can write it as:

$$A \equiv \begin{pmatrix} ct \\ x \\ y \\ z \end{pmatrix} \tag{C.58}$$

We always will use a capital letter and the plain font as in "A" for 4-vector quantities. For the familiar 3-vectors we use a vector font as in "$\vec{a}$" and mostly but not always a small letter. You should think of 4-vectors as column vectors not row vectors so that *Lorentz transformation* equations can be written:

$$A' = \mathcal{L}A \tag{C.59}$$

With symbol $\mathcal{L}$ the presentation of coefficients on the right side of Eq. C.57 is:

$$\mathcal{L} = \begin{pmatrix} \gamma & -\gamma\beta & 0 & 0 \\ -\gamma\beta & \gamma & 0 & 0 \\ 0 & 0 & 1 & 0 \\ 0 & 0 & 0 & 1 \end{pmatrix} \tag{C.60}$$

where again β is given as:

$$\beta \equiv \frac{v}{c} \tag{C.61}$$

The right side of Eq. C.58 does represent the product of a 4 × 4 matrix for Lorentz transformation $\mathcal{L}$ with a 4 × 1 vector A, using the standard rules of matrix multiplication. With such an operation, we easily can check the results presented by Eq. C.31. The inverse Lorentz transformation obviously can be written as:

$$A = \mathcal{L}^{-1}A' \tag{C.62}$$

All we need to do in order to show L^{-1} is to multiply both sides of Eq. C.59 by L^{-1} so that we can find the following result:

$$\mathcal{L}^{-1} = \begin{pmatrix} \gamma & \gamma\beta & 0 & 0 \\ \gamma\beta & \gamma & 0 & 0 \\ 0 & 0 & 1 & 0 \\ 0 & 0 & 0 & 1 \end{pmatrix} \tag{C.63}$$

It should not surprise us that this is simply $\mathcal{L}$ with a change of sign of β. You can confirm that $\mathcal{L}^{-1}\mathcal{L} = I$, where I is designated for the *identity matrix*.

Appendix C: Four Vectors and the Lorentz Transformation

When we want to refer to the components of a *4-vector technique*, we use the following notation, or any other designation than X^μ, depending on your flavor of symbols as:

$$A^i = A^0, A^1, A^2, A^3 \quad \text{or} \quad A^t, A^x, A^y, A^z \tag{C.64}$$

where the zeroth component is the "time" component, ct for the case of A as defined by Eq. C.58, and the other three components are the "spatial" components, x, y, z, for the case of Eq. C.58 as well.

As part of the 4-vectors infrastructure in a general form, we are developing the *four-dimensional vector* and expanding Eq. C.64 further by again considering the coordinates of an event in fixed frame O, as (ct, x, y, z), presenting the components of a four-dimensional radius vector or for short a four-radius vector in a four-dimensional space, using the *Cartesian schema*. We can annotate its components by x^i, where the *index i* takes on the values 0, 1, 2, 3, and 4 so that we can write:

$$x^0 = ct, \quad x^1 = x, \quad x^2 = y, \quad x^3 = z \tag{C.65}$$

The square of the "length" of the radius 4-vector is then given by:

$$(x^0)^2 - (x^1)^2 - (x^2)^2 - (x^3)^2 \tag{C.66}$$

This equation's format does not change under any rotation of a four-dimensional coordinate system, in particular, under *Lorentz transformation*.

Thus, in general a set of four quantities defined in Eq. C.64—namely, $A^i = A^0, A^1$, A^2, A^3—transform like the components of the radius 4-vector x^i under transformations of the four-dimensional coordinate system; this is called a *four-dimensional vector*, or for short *4-vector*, A^i; under Lorentz transformations, we are able to write the following result [29]:

$$\begin{cases} A^0 = \dfrac{A'^0 + \dfrac{v}{c}A'^1}{\sqrt{1 - \dfrac{v^2}{c^2}}} \\[2ex] A^1 = \dfrac{A'^1 + \dfrac{v}{c}A'^0}{\sqrt{1 - \dfrac{v^2}{c^2}}} \\[2ex] A^2 = A'^2 \\ A^3 = A'^3 \end{cases} \tag{C.67}$$

The square magnitude of any 4-vector is defined analogously to the square of the radius 4-vector that is presented in Eq. C.66 and it is:

$$(A^0)^2 - (A^1)^2 - (A^2)^2 - (A^3)^2 \tag{C.68}$$

For further convenience of notation, we can introduce two "types" of components of 4-vectors, denoting them by the symbols A^i and A_j with superscripts and subscripts i and j, respectively, and they are related by the following presentation:

$$A_0 = A^0 \quad A_1 = -A^1 \quad A_2 = -A^2 \quad A_3 = -A^3 \tag{C.69}$$

The quantity A^i is called the *contravariant* and A_j is called the *covariant*, which are components of the 4-vector. The square of the 4-vector then presents itself in the form:

$$\sum_{i=0}^{3}\sum_{j=0}^{3} A^i A_j = A^0 A_0 + A^1 A_1 + A^2 A_2 + A^3 A_3 \tag{C.70}$$

It is clear that Eq. C.70 can be written as either $A^i B_j$, $A^i B_i$, $A_i B^j$, or $A_i B^i$, where the result is the same under 4-vector analyses. In general, one can switch upper and lower indices in any pair or dummy as the individual wishes.

C.7 Transformation Laws for Velocities

As we stated at the beginning of this appendix, one of the main consequences of Lorentz transformations is a different additional law for velocities, which is expected from the invariance of the speed of light. Let us consider a particle that, as seen from reference frame O, is in (x, y, z) at time t and in $(x + \Delta x, y + \Delta y, z + \Delta z)$ at time $t + \Delta t$, thus moving with an average velocity ($V_x = \Delta x/\Delta t$, $V_y = \Delta y/\Delta t$, $V_z = \Delta z/\Delta t$). Because the coordinate transformation is linear, it applies to coordinate differences as well; therefore, in reference frame O' we have, using Lorentz transformation in a simpler form than given by Eq. C.57, $\Delta y' = \Delta y$, $\Delta z' = \Delta z$ and

$$\Delta x' = \gamma(\Delta x - v\Delta t) \quad \Delta t' = \gamma\left(\Delta t - \frac{v}{c^2}\Delta x\right) \tag{C.71}$$

From which we obtain the following relationship:

$$V'_x \equiv \frac{\Delta x'}{\Delta t'} = \frac{\Delta x - vt}{\Delta t - \frac{v}{c^2}\Delta x} = \frac{V_x - v}{1 - \frac{vV_x}{c^2}} \quad \text{and} \quad V'_{y/z} = \frac{1}{\gamma}\frac{V_{y/z}}{1 - \frac{vV_x}{c^2}} \tag{C.72}$$

instead of $V'_x = V_x - v$ and $V'_{y/z} = V_{y/z}$, as predicted by Galilean laws. It requires only some simple algebra to prove that, according to Eq. C.72, if $|\vec{V}| = c$ then also $|\vec{V'}| = c$ because it should be in agreement with the invariance of the speed of light.

C.8 Faster than the Speed of Light

The discussion about the geometric structure of spacetime leads us to some further considerations. First, we can now clarify why we abhor signals propagating faster than light. Time-ordering is basis of our way to classify natural phenomena and deduce the laws of nature from experimental observations; we need to establish causality relationships between various events (e.g., event A causes event B) and for that we need an absolute time-ordering—that is, A cannot be the cause of B if it happens after B in some reference frame. This usually is called the *causality principle*. If we had signals able to propagate a cause–effect relationship and traveling faster than light, they could connect events at space-like distances, thus destroying the causality principle. That means that things moving faster than light are allowed, in principle, but they cannot bring any information with them (i.e., anything capable of establishing a cause–effect relationship between events.

Think, for instance, of a very large circular room of radius R with a laser apparatus in the center of it that projects a beam on the internal circular wall. If the laser starts rotating with angular velocity ω, the light spot corresponding to the beam projection on the wall will move with velocity ωR. Nothing prevents ωR from being larger than c; however, this is not a problem at all. The light spot carries information from the laser apparatus, but it cannot carry any information from one point of the wall to the other; it is just a projection. Such types of fastly rotating beams exist in our Universe; they are called *pulsars*.

It is clear, however, that should it happen that any future experiment demonstrates the existence of signals carrying information (e.g., particles) and traveling faster than light, then we should seriously reconsider the causality principle itself. The second consideration is that we can see the end of the concept of time as an absolute quantity. Although this already is clear from the form of *Lorentz transformations*, we have learned that even time-ordering may be a relative concept.

Appendix D: Vector Derivatives

Some vector derivative entities are listed here for further use and application. In case you need other entities that are not listed here, refer to any Table of Mathematics textbook.

Cartesian

$$d\vec{l} = dx\hat{x} + dy\hat{y} + dz\hat{z} \qquad d\tau = dxdydz \tag{D.1}$$

$$\nabla f = \frac{\partial f}{\partial x}\hat{x} + \frac{\partial f}{\partial y}\hat{y} + \frac{\partial f}{\partial z}\hat{z} \tag{D.2}$$

$$\vec{\nabla} \cdot \vec{E} = \frac{\partial E_x}{\partial x} + \frac{\partial E_y}{\partial y} + \frac{\partial E_z}{\partial z} \tag{D.3}$$

$$\vec{\nabla} \times \vec{E} = \left(\frac{\partial E_z}{\partial y} - \frac{\partial E_y}{\partial z}\right)\hat{x} + \left(\frac{\partial E_x}{\partial z} - \frac{\partial E_z}{\partial x}\right)\hat{y} + \left(\frac{\partial E_y}{\partial x} - \frac{\partial E_x}{\partial y}\right)\hat{z} \tag{D.4}$$

$$\nabla^2 f = \frac{\partial^2 f}{\partial x^2} + \frac{\partial^2 f}{\partial y^2} + \frac{\partial^2 f}{\partial z^2} \tag{D.5}$$

Cylindrical

$$d\vec{l} = dr\hat{r} + rd\varphi\hat{\varphi} + dz\hat{z} \qquad d\tau = rdrd\varphi dz \tag{D.6}$$

$$\nabla f = \frac{\partial f}{\partial r}\hat{r} + \frac{1}{r}\frac{\partial f}{\partial \varphi}\hat{\varphi} + \frac{\partial f}{\partial z}\hat{z} \tag{D.7}$$

$$\vec{\nabla} \cdot \vec{E} = \frac{\partial E_r}{\partial r} + \frac{E_r}{r} + \frac{1}{r}\frac{\partial E_\varphi}{\partial \varphi} + \frac{\partial E_z}{\partial z} \tag{D.8}$$

$$\vec{\nabla} \times \vec{E} = \left(\frac{1}{r}\frac{\partial E_z}{\partial \varphi} - \frac{\partial E_\varphi}{\partial z}\right)\hat{r} + \left(\frac{1}{r}\frac{\partial E_r}{\partial z} - \frac{\partial E_z}{\partial r}\right)\hat{\varphi} + \left(\frac{\partial E_\varphi}{\partial r} + \frac{E_\varphi}{r} - \frac{1}{r}\frac{\partial E_r}{\partial \varphi}\right)\hat{z}$$
(D.9)

$$\nabla^2 f = \frac{\partial^2 f}{\partial r^2} + \frac{1}{r}\frac{\partial f}{\partial r} + \frac{1}{r^2}\frac{\partial^2 f}{\partial \varphi^2} + \frac{\partial^2 f}{\partial z^2}$$
(D.10)

Spherical

$$d\vec{l} = dr\hat{r} + rd\theta\hat{\theta} + r\sin\theta d\varphi\hat{\varphi} \qquad d\tau = r^2 \sin\theta dr d\theta d\varphi$$
(D.11)

$$\vec{\nabla} f = \frac{\partial f}{\partial r}\hat{r} + \frac{1}{r}\frac{\partial f}{\partial \theta}\hat{\theta} + \frac{1}{r\sin\theta}\frac{\partial f}{\partial \varphi}\hat{\varphi}$$
(D.12)

$$\vec{\nabla} \cdot \vec{E} = \frac{\partial E_r}{\partial r} + \frac{2E_r}{r} + \frac{1}{r}\frac{\partial E_\theta}{\partial \theta} + \frac{\cot\theta}{r}E_\theta + \frac{1}{r\sin\theta}\frac{\partial E_\varphi}{\partial \varphi}$$
(D.13)

$$\vec{\nabla} \times \vec{E} = \left(\frac{1}{r}\frac{\partial E_\varphi}{\partial \theta} + \frac{\cot\theta}{r}E_\varphi - \frac{1}{r\sin\theta}\frac{\partial E_\theta}{\partial \varphi}\right)\hat{r}$$
$$+ \left(\frac{1}{r\sin\theta}\frac{\partial E_r}{\partial \varphi} - \frac{\partial E_\varphi}{\partial r} - \frac{1}{r}E_\varphi\right)\hat{\theta}$$
$$+ \left(\frac{\partial E_\theta}{\partial r} - \frac{E_\theta}{r} - \frac{1}{r}\frac{E_r}{\partial \theta}\right)\hat{\varphi}$$
(D.14)

$$\nabla^2 f = \frac{\partial^2 f}{\partial r^2} + \frac{2}{r}\frac{\partial f}{\partial r} + \frac{1}{r^2}\frac{\partial^2 f}{\partial \theta^2} + \frac{\cot\theta}{r^2}\frac{\partial f}{\partial \theta} + \frac{1}{r^2 \sin^2\theta}\frac{\partial^2 f}{\partial \varphi^2}$$
(D.15)

Appendix E: Second-Order Vector Derivatives

Some vector derivative entities are listed here for further use and application. In case you need other entities that are not listed here, refer to any Table of Mathematics textbook.

Cartesian

$$\vec{\nabla}(\vec{\nabla} \cdot \vec{E}) = \left(\frac{\partial^2 E_x}{\partial x^2} + \frac{\partial^2 E_y}{\partial x \partial y} + \frac{\partial E_z}{\partial x \partial z}\right)\hat{x}$$

$$+ \left(\frac{\partial^2 E_x}{\partial y \partial x} + \frac{\partial^2 E_y}{\partial y^2} + \frac{\partial^2 E_z}{\partial y \partial z}\right)\hat{y} \quad \text{(E.1)}$$

$$+ \left(\frac{\partial^2 E_x}{\partial z \partial x} + \frac{\partial^2 E_y}{\partial z \partial y} + \frac{\partial^2 E_z}{\partial z^2}\right)\hat{z}$$

$$\vec{\nabla} \times \vec{\nabla} \times \vec{E} = \left(\frac{\partial^2 E_y}{\partial y \partial x} - \frac{\partial^2 E_x}{\partial y^2} - \frac{\partial^2 E_x}{\partial z^2} + \frac{\partial^2 E_z}{\partial z \partial x}\right)\hat{x}$$

$$+ \left(\frac{\partial^2 E_z}{\partial z \partial y} - \frac{\partial^2 E_y}{\partial z^2} - \frac{\partial^2 E_y}{\partial x^2} + \frac{\partial^2 E_x}{\partial x \partial y}\right)\hat{y} \quad \text{(E.2)}$$

$$+ \left(\frac{\partial^2 E_x}{\partial x \partial z} - \frac{\partial^2 E_z}{\partial x^2} - \frac{\partial^2 E_z}{\partial y^2} + \frac{\partial^2 E_y}{\partial y \partial z}\right)\hat{z}$$

$$\nabla^2 \vec{E} = \left(\frac{\partial^2 E_x}{\partial x^2} + \frac{\partial^2 E_x}{\partial y^2} + \frac{\partial E_x}{\partial z^2} \right) \hat{x}$$
$$+ \left(\frac{\partial^2 E_y}{\partial x^2} + \frac{\partial^2 E_y}{\partial y^2} + \frac{\partial E_y}{\partial z^2} \right) \hat{y} \quad \text{(E.3)}$$
$$+ \left(\frac{\partial^2 E_z}{\partial x^2} + \frac{\partial^2 E_z}{\partial y^2} + \frac{\partial E_z}{\partial z^2} \right) \hat{z}$$

Cylindrical

$$\vec{\nabla}(\vec{\nabla} \cdot \vec{E}) = \left(\frac{\partial^2 E_r}{\partial r^2} + \frac{1}{r} \frac{\partial E_r}{\partial r} - \frac{1}{r^2} E_r + \frac{1}{r} \frac{\partial^2 E_\varphi}{\partial r \partial \varphi} - \frac{1}{r^2} \frac{\partial E_\varphi}{\partial \varphi} + \frac{\partial^2 E_z}{\partial r \partial z} \right) \hat{r}$$
$$+ \left(\frac{1}{r} \frac{\partial^2 E_r}{\partial r \partial \varphi} + \frac{1}{r^2} \frac{\partial E_r}{\partial \varphi} + \frac{1}{r^2} \frac{\partial^2 E_\varphi}{\partial \varphi^2} + \frac{1}{r} \frac{\partial^2 E_z}{\partial \varphi \partial z} \right) \hat{\varphi} \quad \text{(E.4)}$$
$$+ \left(\frac{\partial^2 E_r}{\partial r \partial z} + \frac{1}{r} \frac{\partial E_r}{\partial z} + \frac{1}{r} \frac{\partial^2 E_\varphi}{\partial z \partial \varphi} + \frac{\partial^2 E_z}{\partial z^2} \right) \hat{z}$$

$$\vec{\nabla} \times \vec{\nabla} \times \vec{E} = \left(-\frac{1}{r^2} \frac{\partial^2 E_r}{\partial \varphi^2} - \frac{\partial^2 E_r}{\partial z^2} + \frac{1}{r} \frac{\partial^2 E_\varphi}{\partial r \partial \varphi} + \frac{1}{r^2} \frac{\partial E_\varphi}{\partial \varphi} + \frac{\partial^2 E_z}{\partial r \partial z} \right) \hat{r}$$
$$+ \left(\frac{1}{r} \frac{\partial^2 E_r}{\partial r \partial \varphi} - \frac{1}{r^2} \frac{\partial E_r}{\partial \varphi} - \frac{\partial^2 E_\varphi}{\partial r^2} - \frac{1}{r} \frac{\partial E_\varphi}{\partial r} + \frac{1}{r^2} E_\varphi - \frac{\partial^2 E_\varphi}{\partial z^2} + \frac{1}{r} \frac{\partial^2 E_z}{\partial z \partial \varphi} \right) \hat{\varphi}$$
$$+ \left(\frac{\partial^2 E_r}{\partial r \partial z} + \frac{1}{r} \frac{\partial E_r}{\partial z} + \frac{1}{r} \frac{\partial^2 E_\varphi}{\partial \varphi \partial z} - \frac{1}{r^2} \frac{\partial^2 E_z}{\partial \varphi^2} - \frac{\partial^2 E_z}{\partial r^2} - \frac{1}{r} \frac{\partial E_z}{\partial r} \right) \hat{z}$$
$$\text{(E.5)}$$

$$\nabla^2 \vec{E} = \left\{ \left(\frac{\partial^2 E_r}{\partial r^2} + \frac{1}{r} \frac{\partial E_r}{\partial r} + \frac{1}{r^2} \frac{\partial^2 E_r}{\partial \varphi^2} + \frac{\partial^2 E_r}{\partial z^2} \right) - \frac{1}{r^2} E_r - \frac{2}{r^2} \frac{\partial E_\varphi}{\partial \varphi} \right\} \hat{r}$$
$$+ \left\{ \frac{2}{r^2} \frac{\partial E_r}{\partial \varphi} + \left(\frac{\partial^2 E_\varphi}{\partial r^2} + \frac{1}{r} \frac{\partial E_\varphi}{\partial r} + \frac{1}{r^2} \frac{\partial^2 E_\varphi}{\partial \varphi^2} + \frac{\partial^2 E_\varphi}{\partial z^2} \right) - \frac{1}{r^2} E_\varphi \right\} \hat{\varphi} \quad \text{(E.6)}$$
$$+ \left\{ \frac{\partial^2 E_z}{\partial r^2} + \frac{1}{r} \frac{\partial E_z}{\partial r} + \frac{1}{r^2} \frac{\partial^2 E_z}{\partial \varphi^2} + \frac{\partial^2 E_z}{\partial z^2} \right\} \hat{z}$$

Appendix E: Second-Order Vector Derivatives

Spherical

$$\vec{\nabla}(\vec{\nabla}\cdot\vec{E}) = \left[\begin{array}{c}\dfrac{\partial^2 E_r}{\partial r^2}+\dfrac{2}{r}\dfrac{\partial E_r}{\partial r}-\dfrac{2}{r^2}E_r+\dfrac{1}{r}\dfrac{\partial^2 E_\theta}{\partial r\partial\theta}-\dfrac{1}{r^2}\dfrac{\partial E_\theta}{\partial\theta}\\[6pt] +\dfrac{\cot\theta}{r}\dfrac{\partial E_\theta}{\partial r}-\dfrac{\cot\theta}{r^2}E_\theta+\dfrac{1}{r\sin\theta}\dfrac{\partial^2 E_\varphi}{\partial r\partial\varphi}-\dfrac{1}{r^2\sin\theta}\dfrac{\partial E_\varphi}{\partial\varphi}\end{array}\right]\hat{r}$$
$$+\left[\begin{array}{c}\dfrac{1}{r\sin\theta}\dfrac{\partial^2 E_r}{\partial r\partial\varphi}+\dfrac{2}{r^2\sin\theta}\dfrac{\partial E_r}{\partial\varphi}+\dfrac{1}{r^2\sin\theta}\dfrac{\partial^2 E_\theta}{\partial\theta\partial\varphi}\\[6pt] +\dfrac{\cos\theta}{r^2\sin^2\theta}\dfrac{\partial E_\varphi}{\partial\varphi}+\dfrac{1}{r^2\sin^2\theta}\dfrac{\partial^2 E_\varphi}{\partial\varphi^2}\end{array}\right]\hat{\varphi}$$
$$+\left[\begin{array}{c}\dfrac{1}{r}\dfrac{\partial^2 E_r}{\partial r\partial\theta}+\dfrac{2}{r^2}\dfrac{\partial E_r}{\partial\theta}+\dfrac{1}{r^2}\dfrac{\partial^2 E_\theta}{\partial\theta^2}+\dfrac{\cot\theta}{r^2}\dfrac{\partial E_\theta}{\partial\theta}-\dfrac{1}{r^2\sin^2\theta}E_\theta\\[6pt] +\dfrac{1}{r^2\sin\theta}\dfrac{\partial^2 E_\varphi}{\partial\theta\partial\varphi}-\dfrac{\cos\theta}{\partial\theta\partial\varphi}\dfrac{\partial E_\varphi}{\partial\varphi}\end{array}\right]\hat{\theta}$$

(E.7)

$$\vec{\nabla}\times\vec{\nabla}\times\vec{E} = \left[\begin{array}{c}-\dfrac{1}{r^2\sin^2\theta}\dfrac{\partial^2 E_r}{\partial\varphi^2}-\dfrac{1}{r^2}\dfrac{\partial^2 E_r}{\partial\theta^2}-\dfrac{\cot\theta}{r^2}\dfrac{\partial E_r}{\partial\theta}+\dfrac{1}{r}\dfrac{\partial^2 E_\theta}{\partial r\partial\theta}\\[6pt] +\dfrac{1}{r^2}\dfrac{\partial E_\theta}{\partial\theta}+\dfrac{\cot\theta}{r}\dfrac{\partial E_\theta}{\partial r}+\dfrac{\cot\theta}{r^2}E_\theta+\dfrac{1}{r\sin\theta}\dfrac{\partial^2 E_\varphi}{\partial r\partial\varphi}+\dfrac{1}{r^2\sin\theta}\dfrac{\partial E_\varphi}{\partial\varphi}\end{array}\right]\hat{r}$$
$$+\left[\begin{array}{c}\dfrac{1}{r\sin\theta}\dfrac{\partial^2 E_r}{\partial r\partial\varphi}+\dfrac{1}{r^2\sin\theta}\dfrac{\partial^2 E_\theta}{\partial\varphi\partial\theta}-\dfrac{\cos\theta}{r^2\sin^2\theta}\dfrac{\partial E_\theta}{\partial\varphi}-\dfrac{\partial^2 E_\varphi}{\partial r^2}\\[6pt] -\dfrac{1}{r^2}\dfrac{\partial^2 E_\varphi}{\partial\theta^2}-\dfrac{2}{r}\dfrac{\partial E_\varphi}{\partial r}-\dfrac{\cot\theta}{r^2}\dfrac{\partial E_\varphi}{\partial\theta}+\dfrac{1}{r^2\sin^2\theta}E_\varphi\end{array}\right]\hat{\varphi}$$
$$+\left[\begin{array}{c}\dfrac{1}{r}\dfrac{\partial^2 E_r}{\partial r\partial\theta}-\dfrac{\partial^2 E_\theta}{\partial r^2}-\dfrac{1}{r^2\sin^2\theta}\dfrac{\partial^2 E_\theta}{\partial\varphi^2}-\dfrac{2}{r}\dfrac{\partial E_\theta}{\partial r}\\[6pt] +\dfrac{1}{r^2\sin\theta}\dfrac{\partial^2 E_\varphi}{\partial\theta\partial\varphi}+\dfrac{\cos\theta}{r^2\sin^2\theta}\dfrac{\partial E_\varphi}{\partial\varphi}\end{array}\right]\hat{\theta}$$

(E.8)

$$\nabla^2 \vec{E} = \left\{ \left[\frac{\partial^2 E_r}{\partial r^2} + \frac{2}{r}\frac{\partial E_r}{\partial r} + \frac{1}{r^2}\frac{\partial^2 E_r}{\partial \theta^2} + \frac{\cot\theta}{r^2}\frac{\partial E_r}{\partial \theta} + \frac{1}{r^2 \sin^2\theta}\frac{\partial^2 E_r}{\partial \varphi^2} \right] \right.$$
$$\left. - \frac{2}{r^2}E_r - \frac{2}{r^2}\frac{\partial E_\theta}{\partial \theta} - \frac{2\cot\theta}{r^2}E_\theta - \frac{2}{r^2 \sin\theta}\frac{\partial E_\varphi}{\partial \varphi} \right\}\widehat{r}$$
$$+ \left\{ \left[\frac{\partial^2 E_\varphi}{\partial r^2} + \frac{2}{r}\frac{\partial E_\varphi}{\partial r} + \frac{1}{r^2}\frac{\partial^2 E_\varphi}{\partial \theta^2} + \frac{\cot\theta}{r^2}\frac{\partial E_\varphi}{\partial \theta} + \frac{1}{r^2 \sin^2\theta}\frac{\partial^2 E_\varphi}{\partial \varphi^2} \right] \right.$$
$$\left. + \frac{2}{r^2 \sin\theta}\frac{\partial E_r}{\partial \varphi} + \frac{2\cos\theta}{r^2 \sin^2\theta}\frac{\partial E_\theta}{\partial \varphi} - \frac{1}{r^2 \sin^2\theta}E_\varphi \right\}\widehat{\varphi} \quad \text{(E.9)}$$
$$+ \left\{ \left[\frac{\partial^2 E_\theta}{\partial r^2} + \frac{2}{r}\frac{\partial E_\theta}{\partial r} + \frac{1}{r^2}\frac{\partial^2 E_\theta}{\partial \theta^2} + \frac{\cot\theta}{r^2}\frac{\partial E_\theta}{\partial \theta} + \frac{1}{r^2 \sin^2\theta}\frac{\partial^2 E_\theta}{\partial \phi^2} \right] \right.$$
$$\left. + \frac{2}{r^2}\frac{\partial E_r}{\partial \theta} - \frac{1}{r^2 \sin^2\theta}E_\theta - \frac{2\cos\theta}{r^2 \sin^2\theta}\frac{\partial E_\varphi}{\partial \varphi} \right\}\widehat{\theta}$$

References

1. W. Rindler, *Introduction to Special Relativity*, 2nd edn. (Oxford Science Publications, Oxford, 1991)
2. D. Griffiths, *Introduction to Electrodynamics*, 3rd edn. (Prentice-Hall, Inc, New York, 1989)
3. D.A. Woodside, Classical four-vector fields in the relativistic longitudinal gauge. J. Math. Phys. **41**(7), 4622 (2000)
4. *Mechanics* (The Benjamin/Cummings Publishing Company, Inc. 1969)
5. G.F.R. Ellis, J.-P. Uzan, Fundamental constants, gravitation and cosmology. Am. J. Phys. **73**, 240 (2005)
6. V. Berzi, V. Gorini, J. Math. Phys. **13**, 665 (1969)
7. A. Steane, *Relativity Made Relatively Easy*, 1st edn. (Oxford University Press, Oxford, 2012)
8. V. Faraoni, *Special Relativity, Undergraduate Lecture Notes in Physics* (Springer, Switzerland, 2013)
9. M. Lockwood, *The Labyrinth of Time* (Oxford University Press, Oxford, 2005)
10. P.A. Cherenkov, Visible emission of clean liquids by action of γ radiation. Dok. Akad. Nauk SSSR **2**, 451 (1934). Reprinted in Selected Papers of Soviet Physicists, Usp. Fiz. Nauk 93 (1967) 385. V sbornike: Pavel Alekseyevich Cerenkov: Chelovek, Otkrytie pod redaktsiej A. N. Gorbunova, E. P. Cerenkovoj, M., Nauka, 1999, s. 149–153. (ref Archived October 22, 2007, at the Wayback Machine)
11. P.J. Nahin, *Oliver Heaviside: The life, work, and times of an electrical genius of the victorian age* (John Hopkins University Press, Maryland, 1988), pp. 125–126. ISBN 9780801869099
12. C. Luo, M. Ibanescu, S.G. Johnson, J.D. Joannopoulos, Cerenkov radiation in photonic crystal. Science **299**(5605), 368–371 (2003). Bibcode: 2003 Sci...299...368L
13. P.F. Schewe, B. Stein, Topsy turvy: The first true "left handed" material. Am. Inst. Phys (2004). retrieved 1 December 2008
14. https://en.wikipedia.org/wiki/Kramers%E2%80%93Kronig_relations
15. https://en.wikipedia.org/wiki/Dispersion_(optics)#Material_dispersion_in_optics
16. H. Liu, X. Zhang, B. Xing, P. Han, S.S. Gambhir, Z. Cheng, R, adiation-luminescence-excited quantum dots for in vivo multiplexed optical imaging. Small **6**(10), 1087–1091 (2010). https://doi.org/10.1002/smll.200902408 PMID 20473988
17. H. Liu, G. Ren, S. Liu, X. Zhang, L. Chen, P. Han, Z. Cheng, Optical imaging of reporter gene expression using a positron-emission-tomography probe. J. Biomed. Optics **15**(6), 060505 (2010). https://doi.org/10.1117/1.3514659 Bibcode:2010JBO.15f0505L. Freely Accessible. PMC 3003718 Freely Accessible. PMID 21198146

18. J. Zhong, C. Qin, X. Yang, S. Zhu, X. Zhang, J. Tian, Cerenkov luminescence tomography for in vivo radiopharmaceutical imaging. Int. J. Biomed. Imaging. **2011**(641618), 1–6 (2011). https://doi.org/10.1155/2011/641618
19. C.L. Sinoff, Radical irradiation for carcinoma of the prostate. South African Med J **79**(8), 514 (1991). PMID 2020899
20. G.S. Mitchell, R.K. Gill, D.L. Boucher, C. Li, S.R. Cherry, In vivo Cerenkov luminescence imaging: A new tool for molecular imaging. Phil. Trans. R. Soc. A **369**(1955), 4605–4619 (2011)
21. S. Das, D.L.J. Thorek, J. Grimm, Cerenkov imaging. Emerging applications of molecular imaging to oncology. Adv. Cancer Res. **124**, 213 (2014). https://doi.org/10.1016/B978-0-12-411638-2.00006-9 ISBN 9780124116382
22. A.E. Spinelli, M. Ferdeghini, C. Cavedon, E. Zivelonghi, R. Calandrino, A. Fenzi, A. Sbarbati, F. Boschi, First human Cerenkography. J Biomed Optics **18**(2), 020502 (2013). Bibcode:2013JBO....18b0502S
23. L.A. Jarvis, R. Zhang, D.J. Gladstone, S. Jiang, W. Hitchcock, O.D. Friedman, A.K. Glaser, M. Jermyn, B.W. Pogue, Cherenkov video imaging allows for the first visualization of radiation therapy in real time. Int. J. Radia. Oncol*Biol*Phys **89**(3), 615–622 (2014). https://doi.org/10.1016/j.ijrobp.2014.01.046
24. https://en.wikipedia.org/wiki/Super-Kamiokande
25. https://en.wikipedia.org/wiki/STACEE
26. E. Branger, On Cherenkov light production by irradiated nuclear fuel rods. J. Instrum **12**, T06001. Bibcode:2017JInst.12.6001B. https://doi.org/10.1088/1748-0221/12/06/T06001
27. https://en.wikipedia.org/wiki/TRIGA
28. C.M. Becchi, M. D'Elia, *Introduction to the Basic Concepts of Modern Physics, Special Relativity, Quantum and Statistical Physics*, 3rd edn. (Springer Publishing Company, Dordrecht, 2010)
29. Landau, L. D., Lifshitz, E. M., *The Classical Theory of Fields*, 3rd revised english edn, (Pergamon Press, Addison-Wesley Publication, 1971)

Index

A
Aharonov-Bohm effect, 269, 273, 289, 320, 471
All-electronic (AE), 459
Alternating current (AC), 473
Ampère's circuital law, 80
Ampère's Law, 91, 124
Anomalous dispersion, 541
Anti-submarine warfare (ASW), 461
Artificial ionospheric mirror (AIM), 318
Axicon angle, 341

B
Backward light cone, 169
Bessel beam, 333
$B^{(3)}$ field, 470
Biot-Savart Law, 77, 82, 118, 123, 200, 260, 261, 264
Birkeland currents, 206

C
Canonical momentum, 268
Cartesian coordinates, 33, 547
Causality principle, 555
Charge conservation, 503
Charge continuity equation, 262
Charge couple device (CCD), 311, 358
Charge density, 124, 131
Charge density function, 20
Charge field, 253
Cherenkov radiation, 537
Classical electricity and magnetism, 429
Classical electrodynamics, 431
Classical electromagnetism, 266
Classical Hamilton-Jacobi equation, 158
Classical mechanics, 155, 158, 266, 413, 530
Classical wave equation, 180
Closing velocity, 532
Common Sense' law of velocity addition, 500
Conjugate momentum, 149
Conservative electrostatic field, 135
Constrained propagation, 253
Contiguous circuits loop, 142
Continuity equation, 262
Contravariant, 554
Contravariant tensor, 165
Coulomb barrier, 459
Coulomb gauge, 75, 80, 283, 286
Coulomb's Law, 19, 200
Coulomb radiation, 203
Covariant, 554
Covariant tensor, 165
Cross product, 4
Curl-free vector potential (CFVP), 286, 287, 289
Current density, 124, 261
Cut-off frequency, 541

D
D'Alembert, 339
D'Alembertian, 294
D'Alembert operator, 466
Damping term, 406
de Broglie relation, 546
Degeneracy parameter, 516
Delta-function, 49
Dirac delta function, 125
Dirac equation, 431
Displacement current, 37

D

Distance theory, 42
Divergence theorem, 13
Dot product, 4
Durnin's beams, 335

E

Earnshaw's theorem, 53
Einstein-Cartan-Evans theory, 474
Einstein's Law of light propagation, 500
Electrical primitivity, 445
Electric displacement, 36, 143, 221
Electric field, 21, 23, 43, 138, 143
Electric potential, 138
Electric susceptibility, 36
Electric-wave, 231
Electromagnetic (EM), 487
Electromagnetic compatibility (EMC), 223
Electromagnetic field, 193
Electromagnetic pulse (EMP), 302
Electromagnetic waves (EMWs), 207, 409, 461, 472
Electromotive force (emf), 134, 139
Electromechanical (EM), 459
Electroscalar wave, 444
Electrostatic scalar Potential (ESP), 321
Energy continuity equation, 250
Energy density, 147
Energy equations, 215
Energy flux, 147
Epilimnion layer, 192
Euclidean space, 159, 214, 498
Euler equation, 251
Euler's formula, 185
Euler-Lagrange, 266
Euler-Lagrange equations, 148, 266, 267, 429, 431
Extensive air shower experiment (HAWC), 544
Extremely low frequency (ELF), 238

F

Faraday cages, 209, 268, 479, 481, 488, 490
Faraday's law, 118, 466
Feynman, R.P., 70, 114
Field equations, 214, 215
Finite aperture approximations (FAA), 330
First-order partial differential equations, 199
FitzGerald, 535
Forward light cone, 169
Four-current density, 503
Four-dimensional vector, 553
Four-dimensional world view, 535

Four-force, 500
Four-momentum, 157
Four-vector, 152, 497
Four-vector geometry, 441
Four-vector momentum, 157
Four-vector technique, 551, 553
Four-vector tensor rank, 164
Four-velocity, 152
Fourier transform spectroscopy, 309
Fourier transformation, 102
Four-irrotational, 464
Four-solenoidal, 464
Frank–Tamm formula, 541
Friction, 494
Fundamental tensor, 164

G

Galilean framework, 494
Galilean transformation, 158, 163, 518
Galilei transformation, 536
Gauge invariance, 202
Gauge transformation, 202
Gaussian pulses, 331, 341
Gauss's law, 80, 417
Gauss's theorem, 85, 91
General atomic (GA), 545
Generalized force, 268
General Lorentz transformation, 163, 164
Gradient-driven, 458
Green's function, 82, 114

H

Hamilton's Canonical equation, 268
Hamiltonian, 266
Hamiltonian density, 149, 430
Hamiltonian dynamic, 147
Hamiltonian function, 155
Heinrich Hertz, 207, 237
Helmholtz's theorem, 69, 73
Hertzian dipole, 487
Hertzian part, 488
High frequency (HF), 317
High Frequency Active Auroral Research Program (HAARP), 317
High Momentum Particle Identification Detector (HMPID), 543
High power microwaves (HPM), 546
High temperature superconductivity (HTS), 468
Homogeneous scalar wave equation, 221
Horizonal axis, 168
Hypolimnion layer, 192

I

Identity matrix, 552
Imaging Atmospheric Cherenkov Technique (IACT), 543
Index of refraction, 196, 541
Index of refractive, 539
Inhomogeneous scalar wave equation, 221
Inner product, 4
Instantaneous Coulomb potential, 203
Institute of Electrical and Electronics Engineers (IEEE), 460
Invariant condition, 493
Invariant interval, 173
Inverse Faraday effects, 470
Inverse transformation, 500
Inversion of Biot-Savart law, 124
Ionospheric Research Instrument (IRI), 317
Irritational current, 203

J

Joe Blogg's law, 93

K

Klein-Gordon equation, 281, 520, 527
Kramers-Kronig relation, 541

L

Lagrange function, 152
Lagrangian, 149, 266
Lagrangian density, 431, 433, 465
Lagrangian density equation, 440, 462
Lagrangian of a discrete system, 147
Lagrangian dynamics, 147
Lagrangian function, 152
Langmuir waves, 221, 484
Laplace's equation, 26
Laplacian operator, 167
Large hadron collider (LHC), 444, 464, 543
LED's signals, 488
Lenz's law, 84
Lienard-Wiechert potential, 419
Light-like, 170, 173
Light quanta, 156
Linearly conductive media, 467
Localized waves (LW), 317, 331
Local vector potential, 142
Long frequency (LW), 238
Longitudinal current, 203
Longitudinal oscillations, 484
Longitudinal pressure waves (LPWs), 445
Longitudinal propagation, 218
Longitudinal scalar wave (LSW), 205, 306
Longitudinal scalar wave fields, 487
Longitudinal waves (LWs), 206, 219, 446, 471
Longmuir oscillation, 221
Lorentz condition, 202
Lorentz contracted, 417
Lorentz contraction, 495
Lorentz factor, 498, 531, 534, 536, 537, 550, 551
Lorentz force, 31, 134, 269, 493, 503
Lorentz force law, 58, 424, 493, 503
Lorentz formula, 100
Lorentz gauge, 89, 441
Lorentz gauge condition, 502
Lorentz invariants, 424, 439, 494, 500
Lorentz scalar, 524
Lorentz transformation matrix, 424
Lorentz transformation factor, 173
Lorentz transformations, 150, 158, 281, 419, 426, 498, 500, 529, 530, 534–536, 552, 553, 555
Lorentz vector, 426
Low-energy nuclear reactions (LENRs), 444, 458, 459

M

Magnetic field, 141, 143, 147
Magnetic flux density, 470
Magnetic intensity, 131, 141–143, 147, 201
Magnetic permeability, 445
Magnetic susceptibility, 36
Magnetic vector potential, 86
Magnetic-wave, 231
Magneto hydrodynamics, 1
Magnifying transmitter, 482
Mass density, 148
Maxwell's equations, 1, 33, 199, 214, 424, 427, 504
Maxwell's theory, 498
Medium wave (MW), 238
Merian's formula, 189
Millikan, R., 237
Minkowski diagram, 164, 168, 413
Minkowski field, 441
Minkowski force, 497
Minkowski geometrical space, 150, 530
Minkowski line element, 535
Minkowski space, 214, 439
Mitochondria, 211
Mixed tensor, 165
Möbius coil, 211

Monochromatic waves, 195
More complete electromagnetic (MCE), 440, 441, 458, 462, 465–468

N
NASA Jet Propulsion Laboratory (NASA-JPL), 313, 314
National Solar Thermal Test Facility (NSTTF), 544
Near-field, 230
Negative energy, 525
Negative permeability, 540
Negative permittivity, 540
Neutrinos, 156
Newton's gravitational, 498
Newtonian gravitational force, 498
Newton's second law, 509
Newton's third law of motion, 42
Nikola Tesla, 463
Non-diffracting wave (NDW), 336
Non-linear X-wave (NLX), 335
Nuclear magnetic resonance (NMR), 470

O
Ohmic materials, 402
Outer product, 5

P
Particle equation, 255
Pauli exclusion principle, 290
Permeability, 541
Permittivity of free space, 416
Perspex, 113
Plank's constant, 431
Plasma wave, 484
Poincaré, 535
Point charge, 19
Poisson's equation, 26, 55, 76, 88
Potential energy, 415
Potential vortex, 487
Potential waves, 276
Poynting's theorem, 31, 461
Poynting vector, 250, 462
Principle of Least Action, 148, 150, 156, 158, 266, 267
Principle of Relativity, 531

Q
Quantum-electro-dynamics (QED), 320
Quantum electromagnetics (QE), 320
Quantum mechanics (QM), 266, 320

Quantum wave, 325
Quantum wave theory (QWT), 325

R
Reduced Plank constant, 546
Relative velocities, 532
Relativistic addition of velocities, 532
Relativistic Hamilton-Jacobi equation, 158, 471
Relativistic mechanics, 155–158, 530
Relativistic quantum mechanics, 413
Relativistic velocity additional, 500
Rest energy, 154
Rest mass energy, 415
Restricted gauge transformation, 202
RICH detector, 543
Right-hand-side (RHS), 125
Rotationally invariant, 79, 81
Rydberg's Law, 508, 509

S
Scalar, 2
Scalar factor, 466
Scalar field, 149, 214, 215, 218, 262, 266, 484
Scalar invariant, 501
Scalar Lagrangian, 428
Scalar longitudinal electrodynamic, 465
Scalar longitudinal waves (SLWs), 440, 444–446, 457, 458, 460, 466, 467
Scalar magnetic field, 214
Scalar potential, 200, 203, 270, 329
Scalar product, 4
Scalar waves (SWs), 206, 219, 239, 242, 274, 306, 461, 471, 474, 489, 551
Schrödinger equation, 505, 509–511, 514
Scientific Consciousness Interface Operation (SCIO), 213
Shalimar Treaty, 330
Short wave (SW), 238
Skin depth, 409
Solar Tower Atmospheric Cherenkov Effect Experiment (STACEE), 544
Solenoidal current, 203
Soliton wave, 444
Source charges, 137
Space-like, 170, 173
Spacetime, 532
Spacetime diagram, 164
Special relativistic EM field, 431
Special relativity, 424, 546
Special unitary (SU), 280
Spherically symmetric wave solutions, 467
Standard configuration, 163, 530, 532, 536
Standard Galilean transformation, 499

Standard Lorentz transformation, 163
Standing wave, 184, 211, 230, 239
Stationary field theory, 215
Stationary waves, 186
Stock's theorem, 14
Strategic Defense Initiative (SDI), 205
Stueckelberg-Lagrangian density, 440
Sudbury Neutrino Observatory (SNO), 544
Superluminal phenomenon, 330
Surface charge density, 416
Symmetric wave solution, 466

T
Taylor expansion, 536
Taylor expansion method, 152
Taylor series, 117
Telluric currents, 206, 472
Tension of space, 478
Tensor object, 426
Three-current density, 503
Three-dimensional vector, 157
Three-dimensional velocity vector, 152
Three-force, 498, 500
Three-scalar potential, 504
Three-vector potential, 504
Threshold of hearing, 449
Test charge, 137, 172, 175
Time-dependent Maxwell's equations configuration, 200
Time-dependent Schrödinger equation, 512
Time dilation, 537
Time intervals, 172, 537
Time-like, 170, 173
Total work, 27
Training, Research, Isotopes, General Atomics (TRIGA), 545
Transformation matrix, 161
Transverse current, 203
Transverse electric (TE), 317
Transverse electromagnetic (TEM), 458
Transverse gauge, 203

Transverse waves (TW), 208, 472, 473
Tunneling, 489

U
Ultra high frequency (UHF), 238
Ultra-relativistic, 156
Ultraviolet spectrum, 541
Undistorted progressive waves (UPWs), 330
Unified field, 443
Unified field theory, 489
Uniqueness theorem, 81

V
Vacuum Cherenkov Radiation (VCR), 538, 546
Vacuum state, 475
Vector, 2
Vector analysis, 2
Vector gradient, 8
Vector potential, 200, 203, 270, 329, 476
Vector product, 4
Vertical axis, 168
Very high frequency (VHF), 317
Very large array (VLA), 314
Very low frequency (VLF), 238
Vitreous humour, 542

W
Wave equations, 218
Wave function, 509
Wavelength, 481
Wave-packet, 184
Wave vector, 546
Werner Heisenberg, 235
Wheeler, J.A., 114
Worldline, 168

Y
Young's modules, 148